Sexual Selection

Perspectives and Models from the Neotropics

Sexual Selection

Perspectives and Models from the Neotropics

Edited by

Regina H. Macedo
Departamento de Zoologia, Instituto de Biologia,
Universidade de Brasilia, Brasilia, Brazil

Glauco Machado
Departamento de Ecologia, Instituto de Biociências,
Universidade de São Paulo, São Paulo, Brazil

AMSTERDAM • BOSTON • HEIDELBERG • LONDON
NEW YORK • OXFORD • PARIS • SAN DIEGO
SAN FRANCISCO • SINGAPORE • SYDNEY • TOKYO

Academic Press is an imprint of Elsevier

Academic Press is an imprint of Elsevier
32 Jamestown Road, London NW1 7BY, UK
225 Wyman Street, Waltham, MA 02451, USA
525 B Street, Suite 1800, San Diego, CA 92101-4495, USA

British Library Cataloguing-in-Publication Data
A catalogue record for this book is available from the British Library

Library of Congress Cataloging-in-Publication Data
A catalog record for this book is available from the Library of Congress

ISBN: 978-0-12-416028-6

For information on all Academic Press publications
visit our website at elsevierdirect.com

Typeset by TNQ Books and Journals Pvt Ltd.
www.tnq.co.in

Printed and bound in United States of America

14 15 16 17   10 9 8 7 6 5 4 3 2 1

Contents

3. **Secondary Sexual Traits, Immune Response, Parasites, and Pathogens: The Importance of Studying Neotropical Insects**

Jorge Contreras-Garduño and Jorge Canales Lazcano

4. **Territorial Mating Systems in Butterflies: What We Know and What Neotropical Species Can Show**

Paulo Enrique Cardoso Peixoto and Luis Mendoza-Cuenca

5. **Macroecology of Harvestman Mating Systems**

Bruno A. Buzatto, Rogelio Macías-Ordóñez and Glauco Machado

6.   Adventurous Females and Demanding Males:
    Sex Role Reversal in a Neotropical Spider
    Anita Aisenberg

7.   Sexual Selection, Ecology, and Evolution of Nuptial
    Gifts in Spiders
    Maria J. Albo, Søren Toft and Trine Bilde

8. Paternal Care and Sexual Selection in Arthropods

*Gustavo S. Requena, Roberto Munguía-Steyer and
Glauco Machado*

9. Underestimating the Role of Female Preference and
    Sexual Conflict in the Evolution of ARTs in Fishes:
    Insights from a Clade of Neotropical Fishes

Molly R. Morris and Oscar Ríos-Cardenas

10. Mode of Reproduction, Mate Choice, and Species
     Richness in Goodeid Fish

Constantino Macías Garcia

11. Parental Care, Sexual Selection, and Mating Systems in Neotropical Poison Frogs

Kyle Summers and James Tumulty

12. Testosterone, Territoriality, and Social Interactions in Neotropical Birds

John C. Wingfield, Rodrigo A. Vasquez and Ignacio T. Moore

During the Brazilian Ethology Meeting in 2010 in Alfenas, a small city in the state of Minas Gerais, Brazil, we sat over coffee and discussed the various talks presented during the meeting. We were simultaneously discouraged and hopeful: discouraged because we could not fathom why some of the terrific talks and exciting information on sexual selection and mating systems presented during the meeting remained virtually unknown to the scientific world outside Latin America, and then hopeful because we could imagine producing a book based on the studies of sexual selection being conducted by researchers throughout the Neotropics. Thus was hatched the idea of a book about sexual selection of neotropical animals. It has since become apparent that putting together such a book is a somewhat harder endeavor than we had originally envisioned. It has been a lengthy process, and we have had to nag colleagues about deadlines, and also question ourselves about preconceived concepts that we thought were well-established. But now that we are on the verge of publication, we definitely think this has been a thoroughly worthwhile mission.

Initially, we scrutinized the literature to see what was available relative to tropical biology, sexual selection, and mating systems. Although there are many recent academic books that cover diverse aspects of tropical biology, and many more that concentrate on sexual selection and mating systems, there are none that integrate these two areas. One that approaches this merging of theoretical framework with geographical scope is *Behavioral Ecology of Tropical Birds* (Stutchbury and Morton, 2001), but the book is restricted to birds and broad-based, covering behavioral ecology as a whole. We wondered why no book had ever been published examining sexual selection of neotropical organisms, given the explosive growth in publications about sexual selection in general, and the enormous diversity of tropical organisms and biological interactions available for study in tropical America. Furthermore, we pondered why even those books that concentrate on sexual selection rarely cite studies of tropical animals. Below, we attempt to answer these questions by addressing two issues. First, we rationalize why this book will contribute to the burgeoning theoretical and empirical foundations of sexual selection. Second, we focus on why the study of the abundant animals and systems in the Neotropics has been appallingly neglected. We follow up by introducing the concepts and themes explored in each chapter.

Which explanations validate examination of the interface between sexual selection and the Neotropics? Is there a reason to assume major differences in the pressures that shape sexual adaptations in temperate and tropical organisms? The neotropical region, broadly defined for the purposes of this book, encompasses vast tropical areas as well as large altitudinal variations. Climatic and geographic factors promote elevated faunal and vegetational diversity, which in turn results in complex selective pressures upon organisms, generating assorted and unique reproductive behaviors and morphological traits. The vast diversity harbored by the Neotropics is both a difficulty and a wonderful phenomenon, resulting in many organisms lacking even basic descriptions and taxonomic classification. The enormous diversity of animal life yields another type of diversity, that of biological interactions, which peaks in neotropical regions but is much less varied in temperate regions. The study of this type of diversity is markedly absent from the literature.

Despite the diversity of species, interactions, and behaviors harbored by the Neotropics, most biological research has focused upon temperate region species and systems. Dobzhansky mentioned this problem over 60 years ago: "Temperate faunas and floras, and species domesticated by or associated with man, have supplied, up to now, practically all the material for studies on population genetics and genetical ecology" (Dobzhansky, 1950). In the field of sexual selection, in particular, most theoretical concepts are indeed based on studies of the less diverse North American and European faunas. Even these studies in temperate regions of the northern hemisphere focus on a relatively small number of model species that, because of logistical factors or abundance, became the object of numerous studies.

The main goal of this book, then, is to present relatively new sexual selection study models based upon neotropical species that have been intensively studied during the past 20 years. These studies have shown both convergences as well as unexpected divergences relative to the traditional models that have been studied in North America and Europe. We hope that by providing comparisons with well-known species and models, this book will help to integrate various concepts in sexual selection. Another objective of the book is to combine different perspectives and levels of analysis using a broad taxonomic basis. These approaches distinguish this book from current publications in sexual selection mainly because of the geographic and taxonomic focus, so that readers will be introduced to systems and information largely unknown outside the tropics, some of which diverge from well-established patterns for temperate regions.

Given the potential for important contributions to biological research, why then has this wealth of knowledge been overlooked? Although this book is not about history and politics, it is a fact that such issues can shape (or misshape) the value placed on as well as the progress of science. Thus, acknowledging the role of politics and history in science is an important step toward understanding, and perhaps changing, the scientific scenario. Historically, most of Latin America has been both politically and economically turbulent, usually with unfortunate consequences for the establishment of educational systems and modern research. Using Brazil, the country we are most familiar with, we can illustrate how politics and economic turbulence have dominated cultural and educational initiatives, and generated obstacles to education, research, and scientific progress.

The Portuguese arrived in Brazil in 1500, only 8 years after Christopher Columbus set foot in the Caribbean islands. Portugal did not allow higher education institutions to be established in its colonies, in contrast with England, which was mostly responsible for colonizing North America. The three oldest universities in the United States, for instance, are Harvard, established in 1636, the University of Pennsylvania, established in 1740, and the College of William and Mary, which opened its doors in 1779. In comparison, the oldest higher education establishment in Brazil was inaugurated over 300 years later, in 1934 – only 80 years ago! In this simple historical fact, it is easy to perceive how politics and knowledge are intrinsically linked.

Political instability has posed overwhelming challenges to the establishment of educational institutions and economic prosperity all across Latin America. In comparison to the United States, where citizens over the past 200 years have grown used to the idea of democracy and casting votes to choose their leaders, Brazilian citizens have had only intermittent access to voting rights for approximately 77 years, the past 30 of which have had an enormous and positive impact upon the country. In general, throughout Latin America, dictatorships have been a pervasive feature of the political landscape since the colonial period. With the exception of Costa Rica, Mexico, and French Guiana, all other former colonies of France, Portugal, and Spain experienced periods of military dictatorship during the second half of the 20th century (Calleros, 2009). Dictatorships have lasted from a few years to decades, with the record going to Cuba (54 years), followed by Paraguay (35 years) and the Dominican Republic (31 years). But even short dictatorships are disruptive, when one considers the 9000–30,000 disappeared and dead individuals in Argentina between 1976 and 1983, or the approximately 60,000 people (2% of the population) detained during the military regime in Uruguay from 1973 to 1983. Considering political instability, one has only to realize that about 200,000 people died during the Guatemalan civil war, which lasted 36 years (1960–1996), to appreciate that research and science can very much become low priorities for a society.

During the 1980s and 1990s, elected democratic governments coming into power after periods of dictatorship in Latin America were faced with high economic inequality, high inflation, and macroeconomic instability (Bittencourt, 2012). We contend that economic instability has also been a hindrance to the establishment of educational systems and modern research. In the second half of the 1980s, monthly inflation in Brazil ranged from 6% to 83%. Thus, a researcher who submitted a research proposal to a funding agency in that period would have to have been granted approximately four times the amount requested to compensate for the loss due to inflation during the 3-month proposal evaluation period of the funding agency. Obviously, such an adjustment was never incorporated into the funding mechanism, resulting in serious negative consequences to successful implementation of most projects. Inflation not only engulfed a large part of research funds, but also influenced governmental investment policies. In Brazil, this meant fewer resources devoted to education, particularly for research funding.

Thankfully, this disheartening scenario is rapidly changing throughout Latin America in the 21st century. The Democracy Index (a measure of the state of democracy worldwide; compiled by the Economist Intelligence Unit [EIU, 2013]) suggests that most of Latin America and the Caribbean currently boast what is termed a "flawed democracy", which is second only to the "full democracies" of North America and Western Europe. This state of democratic transition ranks above "hybrid regimes" (Asia, Australasia, Central and Eastern Europe, sub-Saharan Africa) and "authoritarian regimes" (Middle, East, and North Africa). And while a "flawed democracy" may still be far from ideal, its favorable impact can already be felt in the surging educational institutions, universities, and research outputs.

In Brazil, the past decade has seen a phenomenal increase in the number of universities and a significant impact upon research. Moreover, the economic stability attained nearly 15 years ago promoted fast and profound changes in academia, with great benefits to research funding. For the first time in Brazilian history, students and researchers are being encouraged to publish their studies in high-impact journals in their fields of expertise. As an initial result of this investment-and-reward policy, the number of peer-reviewed publications by Brazilian researchers jumped from 6000 per year in 1995 to 32,000 in 2009 (Regalado, 2010). Increasingly, research is being conducted by native researchers who have now accumulated many years of data on neotropical model organisms. As more long-term data emerge, so do surprising and interesting details that either enrich known ecological models or pose challenges to existing paradigms.

Despite the enormous advances observed, it is naïve to assume that a centuries-old gap in scientific knowledge can be eradicated in a few years. Scientific activity in many countries in Latin America continues to require strenuous effort to overcome commercial barriers or high taxes that complicate the acquisition of scientific equipment and supplies crucial to research. Additionally, in some historically neglected areas of research, such as behavioral ecology, the scientific community is small and sparsely distributed within and among countries, making it difficult for researchers to exchange information and to organize themselves into societies with political leverage, both of which are essential for the advancement of knowledge. Environmental laws that are extremely restrictive to researchers, particularly foreigners, also have slowed research in areas such as ecology and systematics by making data collection and international collaboration more difficult. Finally, some cultural barriers remain in various Latin American countries – for example, where members of the scientific community resist publishing their studies in languages other than their own.

We believe that the strong and weak points in this book, along with the taxonomic gaps and oversights in relation to some topics, reflect the current state of knowledge about sexual selection in the Neotropics. We attempted to include the largest possible taxonomic diversity and number of topics in the book, but in many cases we met with lack of information or specialists to contribute to our proposal. Our final list of authors includes researchers that live in the Neotropics as well as researchers from other regions. All of these, however, have been engaged in research on the neotropical fauna for a considerable number of years, and publish in excellent international journals. Authors were asked to highlight how their studies of neotropical fauna can contribute toward the theoretical and empirical advancement of sexual selection theory. In what follows, we briefly summarize the content of the chapters. We decided to open the book with two general chapters, which encompass broad theoretical issues, and then to structure the remaining contents following a taxonomic order, given the great array of topics presented by authors.

In Chapter 1, Macías-Ordóñez and collaborators provide a theoretical framework to approach the study of sexual selection on a broad geographic scale. They call this approach *macroecology of sexual selection*, formally defined as large-scale influence of climatic conditions on sexually selected traits. Using the Neotropics as a reference, the authors postulate broad predictions concerning the effect of biotic and abiotic factors on the life history, reproductive behavior, and sexually selected traits of arthropods, and ectothermic and endothermic vertebrates. The main goal of this chapter is to show that variation in climatic conditions, in terms of either precipitation or temperature, should be a powerful predictor of sexually selective forces.

Chapter 2 addresses an important conceptual issue in science: our ability to recognize differences between established biological patterns and exceptions. Peretti begins by examining what are rules and exceptions in biology and, particularly, in reproduction. Then he explores the two most important changes that may occur in criteria and concepts within subject areas: when a rule comes to be viewed as an exception, and when something that was considered an exception becomes a rule. Studies of scorpions are used to exemplify these two major changes, and to highlight how new data from neotropical species may provide information to confirm, limit, or refute general rules in sexual selection.

Chapter 3 outlines the contribution of neotropical insects to the study of ecoimmunology – a relatively new field that investigates immunity from an ecological and evolutionary perspective. Contreras-Garduño and Canales Lazcano focus on the importance of pathogen/parasite pressure on the evolution of secondary sexual characteristics, reviewing the available data and discussing how future studies should take into account natural parasite or pathogen infections. The authors also stimulate interpopulation comparisons using neotropical insect species to investigate the existence of possible latitudinal gradients in the intensity of both immune responses and secondary sexual characteristics.

The theoretical ideas proposed in Chapter 1 are revisited in two other chapters, which provide empirical tests for some macroecological predictions related to mating systems. In Chapter 4, Peixoto and Mendoza-Cuenca use butterflies to analyze how the frequency of territorial mating systems and the intensity of male–male competition for access to females may vary according to climate. In Chapter 5, Buzatto and collaborators show that the occurrence of maternal care and the length of the mating season in harvestmen may be influenced by an interaction between environmental temperature and precipitation, and discuss how these influences may extend to the types of mating systems found in species of the order Opiliones.

Chapters 6, 7, and 8 focus on arthropod species in which males exhibit high reproductive investment, in the form of either exclusive post-oviposition paternal care or expensive nuptial gifts. In Chapter 6, Aisenberg presents the reproductive ecology of the sand-dwelling wolf spider *Allocosa brasiliensis*, which shows reversal in both the sex roles and the sexual size dimorphism found in most spiders. She discusses hypotheses about the evolution of reversed sexual patterns, and shows how this species provides a unique opportunity to explore the pressures driving sex roles in spiders. In Chapter 7, Albo and collaborators compare the function and evolution of nuptial gifts in two spider species, the neotropical *Paratrechalea ornata* and the palearctic *Pisaura mirabilis*. Although these two species belong to different families and show several differences in their ecology, their reproductive behavior is remarkably similar, suggesting convergent evolution of several sexually selected traits. In Chapter 8, Requena and collaborators review the theoretical background for the evolution of parental investment and sex roles, contrasting classical views with the most recent criticisms and advances proposed by mathematical models. They then introduce the cases in which male arthropods care exclusively for offspring, stressing the contribution of neotropical species to our understanding of the role of paternal care in male attractiveness and in sex role reversal.

Chapters 9 and 10 focus on fish, but their approaches are entirely different. In Chapter 9, Morris and Ríos-Cardenas stress the importance of incorporating selection due to female mate preference and sexual conflict when considering the evolution of alternative reproductive tactics. They use neotropical live-bearing fishes belonging to the genus *Xiphophorus* to investigate some fundamental yet poorly explored questions, such as to what extent one of the alternative tactics relies on the presence of the other to gain reproductive success. In Chapter 10, Macías Garcia explores how the interaction of complex geological history and mode of reproduction may have promoted the diversification of a Mexican fish clade, the Goodeinae. He suggests that viviparity is linked to a consensual method of sperm transfer that allows the evolution of costly male ornaments via female mate choice. Moreover, he argues that both male mating selectivity and female ornamentation provide a mechanism of morphological diversification promoting speciation.

In Chapter 11, Summers and Tumulty discuss the ecology and evolution of parental care in neotropical poison frogs, highlighting how sexual selection may affect and be affected by parental care. Although this chapter focuses on frogs, the theoretical background is analogous to that presented in Chapter 8 for arthropods. In fact, both chapters share a very similar structure, wherein the authors first review relevant theory and then present empirical research on the forms of parental care, stressing their consequences for mating systems and sexual selection. Contrary to the situation for arthropods, however, knowledge concerning poison frogs has benefited from a large number of field studies, a great amount of naturalistic information available in the literature, and well-resolved phylogenies. The group, therefore, offers unique opportunities to test hypotheses on the evolution of parental care, and the chapter fully explores this subject.

The next four chapters (Chapters 12–15) focus on birds, but emphasize entirely different theoretical and empirical topics. In Chapter 12, Wingfield and collaborators examine hormonal patterns of birds in the Neotropics. Questions surrounding differences in hormonal cycles in birds of the southern versus northern hemispheres have scarcely been explored. Given the differences in the endocrine regulatory systems of birds adapted to northern and southern photoperiods, it is expected that many reproductive and social patterns should be profoundly different. This issue is explored by examining the limited evidence currently available, especially with regard to testosterone production, and in a few better-studied species, such as the rufous-collared sparrow (*Zonotrichia capensis*).

Chapter 13 explores the evolution of vocal signaling in neotropical birds. Podos evaluates whether, and in what ways, the neotropical environment may have shaped patterns of vocal evolution that differ from those known for temperate birds. He examines this question by contrasting vocal behavior and acoustic structure of song of neotropical birds with their temperate counterparts. He ends the chapter by arguing that sexual selection has not operated differently in neotropical versus temperate zone birds. Rather, he suggests that in the Neotropics sexual selection achieves a greater range of outcomes, given the more extensive scale of diversity and interactions among organisms.

Chapters 14 and 15 differ from the two previous ones in that they focus on long-term studies of single species that exhibit remarkable sexual ornamentation and behavior: the long-wattled umbrellabird (*Cephalopterus penduliger*) and the blue-black grassquit (*Volatinia jacarina*). In Chapter 14, Karubian and Durães examine sexual traits from a fresh angle and ask how these may impact interspecific ecological interactions that include predation, parasitism, and mutualistic interactions. They use the long-wattled umbrellabird, a most unusual looking lek-breeding bird, to examine how its mating behavior may be important in seed dispersal in a neotropical forest. They first present a theoretical overview of lekking birds and point out an important difference between lekking species of tropical versus temperate regions: nearly all tropical species are frugivorous, whereas temperate species are primarily granivorous and insectivorous. The rest of the chapter focuses on the demographic and genetic consequences of animal-mediated seed dispersal, and how sexual selection and mating systems, particularly lek-breeding, can influence seed dispersal outcomes. They end the chapter by illustrating these concepts using empirical results from their longstanding study of the long-wattled umbrellabird.

In the last of the four bird chapters, Chapter 15, Manica and collaborators review the evolution of ornaments through sexual selection and examine the mechanisms that may drive polygamy in the socially monogamous blue-black grassquit. The authors begin by appraising concepts relative to sexual selection in general, and then zoom in on sexual selection and mating systems in birds before finally focusing the discussion on extra-pair paternity in tropical birds. Long-term studies of the blue-black grassquit mating system provide the basis for the various considerations that follow, covering the evolution of sexual ornamentation, ranging from iridescent, UV-based structural plumage to acoustic signals and display behavior.

Finally, in Chapter 16, Voigt reviews case studies of neotropical bats to identify morphological traits and behaviors that have been shaped by sexual selection, and also discusses the possible underlying mechanisms of sexual selection in these specific cases. Each one of the six bat species treated in the chapter sheds light on a specific aspect of sexual selection, such as female mate choice, including the precise sensory modalities used by a species for mate-choice decisions, or mechanisms of male–male competition.

REFERENCES

Bittencourt, M., 2012. Democracy, populism and hyperinflation: some evidence from Latin America. Econ. Gov. 13, 311–332.

Calleros, J.C., 2009. The Unfinished Transition to Democracy in Latin America. Routledge, New York.

Dobzhansky, T., 1950. Evolution in the tropics. Am. Scientist 38, 209–221.

Economist Intelligence Unit, 2013. Democracy Index 2012: Democracy at a Standstill. Retrieved from the Economist Intelligence Unit database.

Regalado, A., 2010. Brazilian science: riding a gusher. Science 330, 1306–1312.

Stutchbury, B.J.M., Morton, E.S., 2001. Behavioral Ecology of Tropical Birds. Academic Press, London.

Acknowledgments

We would like to thank numerous colleagues who contributed with their expertise in reviewing drafts of various chapters and parts of the book: Shelly Adamo, John Alcock, Maydianne Andrade, Elizabeth Derryberry, Hugh Drummond, James Gilbert, Sarah Goodwin, Rogelio Macías-Ordóñez, Paulo Inácio Prado, Robert Ricklefs, Ronald Rutowski, Julia Schad, Luiz Ernesto Costa Schmidt, Silke Voigt-Heucke, Leonardo Wedekin, and Marlene Zuk. Our editor, Pat Gonzalez, and the staff at Elsevier were patient and encouraging even when we were impatient and discouraged. During our work, we were funded by the Conselho Nacional de Desenvolvimento Científico e Tecnológico (CNPq) and Fundação de Amparo à Pesquisa do Estado de São Paulo.

Regina H. Macedo
Glauco Machado

Contributors

Anita Aisenberg   Laboratorio de Etología, Ecología y Evolución, Instituto de Investigaciones Biológicas Clemente Estable, Montevideo, Uruguay

Maria J. Albo   Laboratorio de Etología, Ecología y Evolución, Instituto de Investigaciones Biológicas Clemente Estable, Montevideo, Uruguay; Department of Bioscience, Aarhus University, Aarhus, Denmark

Trine Bilde   Department of Bioscience, Aarhus University, Aarhus, Denmark

Bruno A. Buzatto   Centre for Evolutionary Biology, School of Animal Biology, The University of Western Australia, Crawley, WA, Australia

Jorge Contreras-Garduño   Departamento de Biología, División de Ciencias Naturales y Exactas, Universidad de Guanajuato, Noria Alta, Guanajuato, Mexico

Renata Durães   Department of Ecology and Evolutionary Biology, Tulane University, New Orleans, Louisiana, USA

Constantino Macías Garcia   Laboratorio de Conducta Animal, Instituto de Ecología, Universidad Nacional Autónoma de México, Mexico

Jefferson Graves   School of Biology, University of St Andrews, St Andrews, Fife, UK

Jordan Karubian   Department of Ecology and Evolutionary Biology, Tulane University, New Orleans, Louisiana, USA

Jorge Canales Lazcano   Departamento de Biología, División de Ciencias Naturales y Exactas, Universidad de Guanajuato, Noria Alta, Guanajuato, Mexico

Regina H. Macedo   Departamento de Zoologia, Universidade de Brasília, Brasília, Brazil

Glauco Machado   LAGE, Departamento de Ecologia, Instituto de Biociências, Universidade de São Paulo, São Paulo, Brazil

Rogelio Macías-Ordóñez   Red de Biología Evolutiva, Instituto de Ecología, A.C., Xalapa, Veracruz, Mexico

Lilian T. Manica   Departamento de Zoologia, Universidade de Brasília, Brasília, Brazil

Luis Mendoza-Cuenca   Laboratorio de Ecología de la Conducta, Facultad de Biología, Universidad Michoacana de San Nicolás de Hidalgo, Morelia, Michoacán, Mexico

Ignacio T. Moore   Department of Biological Sciences, Virginia Tech. University, Blacksburg, Virginia, USA

Molly R. Morris   Department of Biological Sciences, Ohio University, Athens, Ohio, USA

Roberto Munguía-Steyer   Departamento de Ecología Evolutiva, Instituto de Ecología, Universidad Nacional Autónoma de México, México, DF, Mexico

Paulo Enrique Cardoso Peixoto   Laboratório de Entomologia, Departamento de Ciências Biológicas, Universidade Estadual de Feira de Santana, Feira de Santana, Brazil

Alfredo V. Peretti   Instituto de Diversidad y Ecología Animal, Laboratorio de Biología Reproductiva y Evolución, Universidad Nacional de Córdoba, Argentina

Jeffrey Podos   Department of Biology, University of Massachusetts, Amherst, Massachussetts, USA

Gustavo S. Requena   Departamento de Ecologia, Instituto de Biociências, Universidade de São Paulo, São Paulo, Brazil

Oscar Ríos-Cardenas   Departamento de Biología Evolutiva, Instituto de Ecología, A.C., Xalapa, Veracruz, Mexico

Kyle Summers   Department of Biology, East Carolina University, Greenville, North Carolina, USA

Søren Toft   Department of Bioscience, Aarhus University, Aarhus, Denmark

James Tumulty   Department of Ecology, Evolution and Behavior, University of Minnesota, Saint Paul, Minnesota, USA

Rodrigo A. Vasquez   Institute of Ecology and Biodiversity, Departamento de Ciencias Ecologicas, Facultad de Ciencias, Universidad de Chile, Santiago, Chile

Christian C. Voigt   Leibniz Institute for Zoo and Wildlife Research, Berlin, Germany

John C. Wingfield   Department of Neurobiology, Physiology and Behavior, University of California, Davis, California, USA

Macroecology of Sexual Selection: Large-Scale Influence of Climate on Sexually Selected Traits

Rogelio Macías-Ordóñez,[1] Glauco Machado[2] and Regina H. Macedo[3]

[1]Red de Biología Evolutiva, Instituto de Ecología, A.C., México, [2]Departamento de Ecologia, Instituto de Biociências, Universidade de São Paulo, Brazil, [3]Departamento de Zoologia, Instituto de Biologia, Universidade de Brasília, Brasília, Brazil

INTRODUCTION

The exuberant variety of life forms in the Neotropics, with their many shapes, sounds, smells, and colors, have lured naturalists for centuries. Darwin himself was amazed by this complexity (Darwin, 1862), and was among the first to suggest that many of these traits were not the result of natural (viability) selection, but rather of an additional and sometimes opposite selective force: sexual selection (Darwin, 1871). Does this imply that sexual selection is stronger in tropical environments? Or at least that sexual selection acts differently in the tropics when compared to more temperate or colder regions? Although sexual selection is arguably the most studied evolutionary mechanism nowadays, a broad geographic perspective is seldom applied to answer these kinds of questions. The aim of this chapter is to provide a theoretical framework to approach the study of sexual selection in a broad geographic scale using the Neotropics as a reference to compare what we know of sexual selection and reproductive strategies in this and other environments of the planet. In this chapter, we will postulate broad predictions concerning the effect of environmental factors on reproductive behavior and sexually selected traits of several animal groups, with the objective of stimulating research along these lines.

DEFINING THE NEOTROPICS

A strictly geographic definition of the Neotropics would exclude any area of the American continent north of the Tropic of Cancer and south of the Tropic

Sexual Selection. http://dx.doi.org/10.1016/B978-0-12-416028-6.00001-3

1

of Capricorn. Yet the common perception for this region clearly extends beyond the 23° latitudes. The limits, though, vary widely depending on the criteria used when attempting a formal definition. On the one hand, the *Neotropical Floristic Kingdom* excludes southern South America and northern Mexico (Good, 1964; Takhtajan, 1969). On the other hand, Udvardy's (1975) *Neotropical Realm* includes all of South America and the southern tip of Florida in the southeastern United States, but excludes not only northern Mexico but also the highlands of central and southern Mexico, Guatemala, and eastern Honduras. *Biogeographical Realms* were later adopted by the World Wildlife Fund (WWF) with modifications as *ecoregions* (Bailey, 1998). A strictly climatic approach would include only those regions in the Americas with a *tropical* climate, defined as those areas with an average temperature of the coldest month above 18°C (Peel *et al.*, 2007). This approach would lead to the inclusion of areas from the southern tip of Florida, in the southeastern United States, to Brazil, Paraguay, Bolivia, and Peru, but would exclude Chile, Argentina, and Uruguay altogether. A common concept in the literature is that the Neotropics equates to Latin America, encompassing the areas from Mexico southwards. This interpretation of what constitutes the Neotropics corresponds roughly to one of the eight biogeographic realms defined by Olson *et al.* (2001), but it excludes southern Florida while including temperate and cold regions of the extreme south of South America. This view was explored under a historical perspective in the Preface of the book.

As we hope will become clear, for the scope of this chapter and the rest of this book we really do not need to choose (and thus restrict ourselves to) one definition for the Neotropics; neither do we need to coin a new one. In fact, we will avoid discussing the limits of the Neotropics and focus on the patterns expected along environmental gradients. When addressing macroevolution at large geographical scales, Darlington (1958) suggested that "we do not need a full or precise definition (of *the tropics*), but one that will emphasize the significant differences between the tropics and the (north) temperate zone". We are interested in the study of selective forces shaping reproductive behavioral, morphological, and physiological traits, and in order to do so we need to contrast environmental conditions shaping those forces. We have decided to use the Neotropics as a landmark or reference for this comparison for several reasons. This is the area we are most knowledgeable about, both from personal as well as professional perspectives, since it is where we carry out our research, and this is also true for almost all authors of this book. The Neotropics also encompass a region relatively well studied compared to other tropical regions of the world, although still relatively understudied when compared to most parts of the northern hemisphere (see below). Whatever limits we may adopt for the Neotropics, it is clear the region is far from homogeneous in terms of climatic conditions, landscapes, and vegetation types, and thus offers a great opportunity to explore how the action of sexual selection in different environments results in distinctive arrays of traits. Furthermore, we are interested in comparing what we know from the Neotropics with what we know from other regions.

This yields the possibility of exploring a yet wider range of environmental variables, encompassing the greater wealth of research on sexual selection available from temperate and colder regions.

Tropical environments, as will be more formally defined later, present very different challenges and opportunities from those in more extreme latitudes (see review in Schemske *et al.*, 2009). However, there is an oversimplified view of the Neotropics (and the tropics overall) that, in our opinion, severely constrains the potential to address macro-evolutionary patterns in broad environmental gradients. Although Darlington (1958) defined *the tropics* as "the zone within which, when other conditions are suitable, warmth and stability of temperature permit development of what we call tropical rain forest with its associated complex fauna", the same author acknowledged that "many definitions are possible and no simple one is entirely satisfactory, for the tropics vary in climate (wet to dry), vegetation (rain forest to desert), and animal life". We need an environmental framework with enough variability to accommodate predictions along environmental gradients, yet broad enough to be able to reflect nearly global trends. We have chosen to define this environmental framework using the current climate classification (Peel *et al.*, 2007). Climate types provide an independent set of environmental variables that should cover major selective forces shaping sexual selection at a broad geographical scale.

CLIMATIC REGIONS

A good place to start is the Köppen–Geiger climate classification (Table 1.1; Fig. 1.1). With few modifications, this climate classification has endured the test of time among climatologists and is now widely accepted as the standard in climatology (Peel *et al.*, 2007). This section introduces this classification to the reader, as it is an important foundation point for the chapter. Climate has long been suggested as a powerful axis to address macro-evolutionary patterns (Darlington, 1958). Although "latitude" has more recently been used as a simpler proxy of climatic gradients (see, for example, Blanckenhorn *et al.*, 2006; Schemske *et al.*, 2009; Moya-Loraño, 2010), it will become evident in this chapter why, for our purposes, latitude is an oversimplification that does not allow predictions in specific combinations of environmental conditions, especially in such a rich mosaic of climate types as can be found within the Neotropics.

Although not entirely independent from biological factors, climate probably offers the most independent array of environmental variables defining plant (and thus biomes; Audesirk and Audesirk, 1996) and animal distributions, and thus key environmental factors of natural (viability) and sexual selection. The Köppen–Geiger climate classification is based on a nested set of climatic regimes defining 29 climate types identified by two- or three-letter combinations (Table 1.1). The first level is the broader climate classification of five climate types: *tropical* (A), *arid* (B), *temperate* (C), *cold* (D), and *polar* (E)

TABLE 1.1 Description of Köppen Climate Symbols and Defining Criteria (modified from Peel *et al.*, 2007)

1st	2nd	3rd	Description	Criteria[a]
A			**Tropical**	$T_{cold} \geq 8$
	f		Rainforest	$P_{dry} \geq 60$
	m		Monsoon	Not (Af) & $P_{dry} \geq 100 - MAP/25$
	w		Savannah	Not (Af) & $P_{dry} < 100 - MAP/25$
B			**Arid**	$MAP < 10 \times P_{threshold}$
	W		Desert	$MAP < 5 \times P_{threshold}$
	S		Steppe	$MAP \geq 5 \times P_{threshold}$
		h	• Hot	$MAT \geq 18$
		k	• Cold	$MAT < 18$
C			**Temperate**	$T_{hot} > 10$ & $0 < T_{cold} < 18$
	s		Dry summer	$P_{sdry} < 40$ & $P_{sdry} < P_{wwet}/3$
	w		Dry winter	$P_{wdry} < P_{swet}/10$
	f		Without dry season	Not (Cs) or (Cw)
		a	• Hot summer	$T_{hot} \geq 22$
		b	• Warm summer	Not (a) & $T_{mon10} \geq 4$
		c	• Cold summer	Not (a or b) & $1 \leq T_{mon10} < 4$
D			**Cold**	$T_{hot} > 10$ & $T_{cold} \leq 0$
	s		Dry summer	$P_{sdry} < 40$ & $P_{sdry} < P_{wwet}/3$
	w		Dry winter	$P_{wdry} < P_{swet}/10$
	f		Without dry season	Not (Ds) or (Dw)
		a	• Hot summer	$T_{hot} \geq 22$
		b	• Warm summer	Not (a) & $T_{mon10} \geq 4$
		c	• Cold summer	Not (a, b or d)
		d	• Very cold winter	Not (a or b) & $T_{cold} < -38$

TABLE 1.1 Description of Köppen Climate Symbols and Defining Criteria (modified from Peel *et al.*, 2007)—cont'd

1st	2nd	3rd	Description	Criteria[a]
E			Polar	$T_{hot} < 10$
	T		Tundra	$T_{hot} > 0$
	F		Frost	$T_{hot} \leq 0$

[a]*MAP, mean annual precipitation; MAT, mean annual temperature; T_{hot}, temperature of the hottest month; T_{cold}, temperature of the coldest month; T_{mon10}, number of months in which the temperature is above 10°C; P_{dry}, precipitation of the driest month; P_{sdry}, precipitation of the driest month in summer; P_{wdry}, precipitation of the driest month in winter; P_{swet}, precipitation of the wettest month in summer; P_{wwet}, precipitation of the wettest month in winter. $P_{threshold}$ varies according to the following rules: if 70% of MAP occurs in winter then $P_{threshold} = 2 \times MAT$, if 70% of MAP occurs in summer then $P_{threshold} = 2 \times MAT +$, otherwise $P_{threshold} = 2 \times MAT + 14$. Summer (winter) is defined as the warmer (cooler) 6-month period of Oct–Nov–Dec–Jan–Feb–Mar and Apr–May–Jun–Jul–Aug–Sep.*

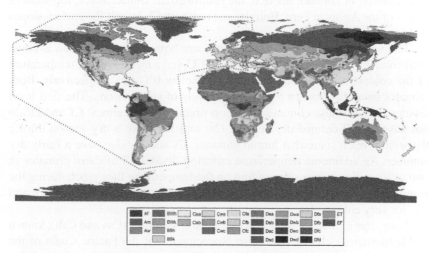

Climate	Tropical	Arid	Temperate	Cold	Polar
Area	11.9	6.0	8.9	14.4	6.0
% Area	25	13	19	31	13
Sites	35	26	92	109	10
% sites	13	10	34	40	4
% bias	-95	-34	45	24	-246

FIGURE 1.1   The Köppen–Geiger climate classification map (modified from Peel *et al.*, 2007) and frequency of field studies on sexual selection in America and Europe (dotted line). The area where each climate type occurs is expressed in millions of km². For descriptions and criteria of each climate type, see Table 1.1. See color plate at the back of the book.

(Peel *et al.*, 2007; Fig. 1.1). With the exception of B (arid), all major climate types are defined using only temperature criteria.

As stated above, a tropical (A) climate includes those regions with monthly average temperature of the coldest month above 18°C. A second nested classification of tropical climate is defined by precipitation regime. The tropical rainforest climate (Af) is defined by precipitation in the driest month above 60 mm. The western Amazon basin is emblematic of this climate (Fig. 1.1). The tropical monsoon climate (Am) is defined by a somewhat more seasonal but still relatively humid driest month precipitation. The eastern Amazon basin, for instance, is characterized by this climate (Fig. 1.1). The tropical savannah climate (Aw) includes a severe dry season contrasting with a heavy rainy season. Most of southern Mexico and central Brazil have this climate type (Fig. 1.1).

The arid (B) climate is defined using a precipitation criterion and comprises regions with very low annual precipitation. A first division of this climate defines two different environments associated with two rain regimes: steppes (BS) and deserts (BW). Within those environments, a temperature criterion defines cold steppes (BSk) or deserts (BWk), and hot steppes (BSh) or deserts (BWh), depending on whether the mean annual temperature is below or above 18°C, respectively. A mosaic of the four different types of arid climate is characteristic of northern Mexico, the southwestern United States, the western slopes of the Andes in Peru, Bolivia, and northern Chile, and the eastern slopes in Argentina (Fig. 1.1).

Temperate (C) and cold (D) climates are both defined by mean average temperature of the hottest month above 10°C, but a mean average temperature of the coldest month either above (C) or below 0°C (D), respectively. Both climates have the same two additional levels of subdivisions. The first level divides each of these climates based on precipitation regimes. Cf and Df do not have a well-defined dry season. Cw and Dw have a dry season during the winter, but a somewhat humid summer. Cs and Ds also have a fairly dry summer. An additional temperature criterion divides all of these climates in two to four climate types depending on the temperature they reach during the summer: "a" for hot summer, "b" for warm summer, "c" for cold summer, and "d" for very cold winter.

Temperate climates with dry, hot and warm summers (Csa and Csb), known as Mediterranean climates, are also characteristic of the Pacific Coast of the United States and central Chile (Fig. 1.1). Temperate humid climates with hot summers (Cfa) are characteristic of the southeastern United States, southern Brazil, and northern Argentina (Fig. 1.1). Temperate humid climates with warm summers (Cfb) are characteristic of western Europe (Fig. 1.1). Temperate climates with dry winters and hot or warm summers (Cwa and Cwb) are restricted to the central Mexican plateau, the eastern slope of the central Andes, and southeastern Brazil (Fig. 1.1). Temperate climates with cold summers are extremely rare in the Neotropics, and are restricted to very small regions around mountain peaks. Cold humid climates with hot, warm, or cold summers (Dfa, Dfb, and

Dfc) are dominant among other cold climates in northern North America and Europe. They cover immense areas in the northern United States, Canada, and northern and eastern Europe. As might be expected, the summer temperature defining each of these climate types shows a clear latitudinal pattern (Fig. 1.1).

The polar (E) climate characterizes those regions where the average monthly temperature of the hottest month is below 10°C. Another temperature criterion divides this climate into polar tundra (ET) if the average monthly temperature is above 0°C, and polar frost (EF) if it is not. Within Europe and North America, polar frost is restricted to Greenland (Fig. 1.1). The greatest extensions of polar tundra can be found in the northern and southern extremes of the American continent, and also in isolated spots in northern Europe and around mountain peaks in the Andes, central Mexico, and central Europe (Fig. 1.1).

In the Neotropics, an intricate mosaic of climate types is the result of complex orographic systems: the two Sierras in North America that converge in central Mexico, a volcanic belt in Central America, the Andean ridge along the Pacific coast, and the long mountain chains along the Brazilian Atlantic coast in South America. The strong but very different influences of the cold Pacific and the warmer Atlantic oceans result in a fragmented pattern of climates, very distinct from the same latitudes in Africa or Asia (Fig. 1.1). Tropical Africa is mostly covered by a continuous area of arid climate and another of tropical climate, with some temperate regions in the southern end of the continent. A similar pattern describes tropical Asia, Australia, and the Pacific Islands (Fig. 1.1). Thus, the Neotropics forms clearly the most diverse and complex region in terms of climates in the world.

HOW MUCH DO WE KNOW ABOUT THE INFLUENCE OF SEXUAL SELECTION IN EACH ENVIRONMENT?

One of the most frequent claims about the Neotropics is that our knowledge about its diversity is very limited. Indeed, the number of new plant and animal species to be discovered in this region is probably very high when compared with temperate and cold regions (see, for example, Adis, 1990; Fouquet *et al.*, 2007). Richness, however, is only part of this diversity, closely related to taxonomy. Ecologists may also ask how much we know about the behavior and evolutionary mechanisms of neotropical species. The answer is probably "very little", but we would like to know how little, particularly if we restrict this fundamental question to a specific field of animal behavior: sexual selection. A recent extensive review of latitudinal patterns of biological interactions (Schemske *et al.*, 2009) describes this topic as one of the less addressed under a broad geographic–environmental perspective (see also Twiss *et al.*, 2007). For this chapter, we looked for an answer by reviewing the past 14 years of the four journals of animal behavior with the highest impact factors: *Animal Behaviour*, *Behavioral Ecology*, *Behavioral Ecology and Sociobiology*, and *Ethology*. These journals publish high-quality international research in many different fields of animal behavior, and there is no *a priori* reason to suspect that there is any bias in the acceptance of papers according to taxonomic group or author nationality.

Using the *Web of Science* database, the first step of our search was to select a set of key words that should appear in the title of the papers: "breed*" or "mate" or "mating" or "reproduct*" or "sexual". Because we needed the precise location where the studies were conducted, we restricted the search to the period from 1998 to October 2011, for which all four journals have pdf files available. We then filtered the results, selecting papers whose abstracts contained the words "field" or "nature" or "natural" or "population" or "wild", but did not contain the words "cage*" or "captiv*" or "laboratory". Our aim was to focus our search only on studies conducted under natural conditions, where study animals were subjected to the effects of environmental variables. We read the abstracts of all the selected papers and removed those that were not related to sexual selection. The remaining papers were downloaded and their content was searched for two basic pieces of information: geographic coordinates of the study site (when this was not available in the text, we obtained them using Google Maps®); and studied taxon, which we lumped into major categories or "functional groups" – arthropods, other invertebrates, ectothermic and endothermic vertebrates (see below). We only selected papers for studies conducted in the Americas and Europe (delineated by the dotted line in Fig. 1.1). Finally, we plotted the coordinates for study sites on the most recent version of the Köppen–Geiger climate map (Peel *et al.*, 2007) to obtain the number of studies conducted in each climate type. Directly from the authors of this map, we obtained a detailed database containing the total area covered by each climate type in the Americas and Europe, and compared the proportion of studies undertaken in each climate type to the relative area covered by the respective climate to provide a %bias index that represented any research bias relative to specific climate types. This index represents the percentage by which the number of studies should increase (negative values) or decrease (positive values) in order to be proportional to the area covered by each climate type.

We came up with 254 studies conducted in 272 sites in 36 countries. Endothermic vertebrates (birds and mammals) account for nearly 57% of the studied taxa, while ectothermic vertebrates and arthropods account for 19.5% and 22.4%, respectively. This pattern still holds when we individually consider the five major types of climate (A to E). Even though the tropical climate (A) covers around 25% of the combined area of the Americas and Europe, only 13% ($n=35$) of the studies were carried out in this climate. The arid climate (B) covers 13% of the area, while 10% ($n=26$) of the studies were performed in arid environments. Temperate climates cover 19% of the area, but 34% ($n=92$) of the studies were carried out in this environment. Cold climates cover 31% of the area, with 40% ($n=109$) of the studies. Finally, polar climates cover 13% of the area, but only 4% ($n=10$) of the studies were conducted in this environment (Fig. 1.1). There are many ways to read these data, but overall we can say that studies in tropical and arid environments are somewhat under-represented according to their areas in the Americas and Europe. Studies in temperate and cold environments are somewhat over-represented, and polar environments are greatly under-represented. Despite the fact that one out of four studies on sexual selection in the wild has been carried out in the (climatically defined) Neotropics, these results still provide support for the widespread notion that our knowledge on sexual selection is biased towards species living in temperate regions. This is especially true if one considers the ratio of the number of studies to the potentially very high number of species available in the Neotropics, instead of area. The consequences of this bias relative to our assumptions about general patterns are discussed in detail in Chapter 2, which is devoted to rules and exceptions in sexual selection.

MACROECOLOGY OF SEXUAL TRAITS

Studies at large geographic scales, covering areas such as whole continents, are relatively recent in the ecological literature. The great majority of such studies investigate large-scale variation in species richness, abundances, distributions, and body sizes through space (Ricklefs and Schluter, 1993; Gaston and Blackburn, 2000). Macroecological studies make several assumptions, some of which are closely connected to large-scale variation in physiological responses (Chown and Nicolson, 2004). Two widely known examples are the Rapoport effect, i.e., the increase in the latitudinal range sizes of species towards higher latitudes as a consequence of wider climatic tolerances (Stevens, 1989), and Allen's rule, which states that endotherms from colder climates have shorter limbs than their relatives from warmer climates as a mechanism of heat conservation (Alho et al., 2011, but see also Nudds and Oswald, 2007). We argue here that combining some of these large-scale ecological patterns may result in emergent trends directly related to reproductive ecology. Blanckenhorn et al. (2006), for instance, explored the topic of latitudinal changes in sexual size dimorphism both in vertebrates and in invertebrates by combining the effect of Bergmann's rule (organisms are larger at higher altitudes or in colder climates) and Rensch's rule (male body size varies, or evolutionarily diverges, more than female body size among species within a lineage).

In contrast to large-scale studies concerning the variation of morphological, physiological, or diversity attributes exemplified above, and despite the fact that sexual selection is one of the most intensively studied evolutionary forces, there is no solid theoretical framework directly relating the behavioral ecology of sexual traits to their variation in space, i.e., a "macroecology of sexual traits". Large-scale rules are virtually absent, and only very recent empirical studies have drawn attention to how spatial environmental heterogeneity can produce great fluctuations in both the strength and the direction of sexual selection (reviewed in Cornwallis and Uller, 2009). Environmental conditions, including both biotic and abiotic factors, may exert a marked influence on many behavioral, morphological, physiological, and life-history traits (Bradshaw, 2003; Chown and Nicolson, 2004; Fig. 1.2). At least a set of these traits are directly or indirectly related to reproduction and are under sexual selection (Fig. 1.2). For instance, it has recently been suggested that a latitudinal gradient of temperature and humidity could be related to movement rate, resulting in more frequent and more diverse biotic interactions, such as parasitism and predation (Schemske, 2009; Moya-Loraño, 2010; Fig. 1.2). In fact, there is strong empirical evidence that parasitism is more prevalent in tropical climates (reviewed in Schemske *et al.*, 2009), and therefore tropical hosts should invest more heavily in parasite defense than their non-tropical counterparts (Møller, 1998). Given that life-history theory predicts a trade-off between investment in defense against parasites versus investment in other fitness components, intense parasitism may

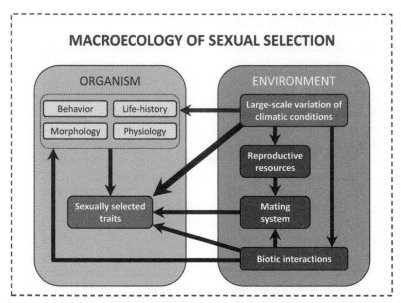

FIGURE 1.2   Scheme showing the influence of large-scale variation of environmental conditions (biotic and abiotic factors) on sexually selected traits. It is possible to suppose a direct influence (wider arrow) of climate on the costs of expression of sexually selected traits, i.e., in harsh climate conditions the energetic trade-off between self-maintenance and elaborate ornaments and weaponry may be shifted. Climate may also become an indirect (arrows) agent of sexual selection on physiological, morphological, behavioral, or life-history traits. Climatic conditions also affect the importance of biotic interactions, such as predation and parasitism, which in turn have a strong influence on the expression of conspicuous sexually selected traits. Finally, climatic conditions shape the spatial and temporal distribution of key reproductive resources, which have a central role in determining the type of mating system. Given that the intensity of sexual selection may differ among different mating systems, climatic conditions may have an indirect influence on the expression of sexually selected traits. The large-scale influence of climatic conditions on sexually selected traits, either by direct influence of the environment on the phenotype or through the action of sexual selection, is what we call "macroecology of sexual selection."

affect both the evolution and the expression of condition-dependent sexually selected traits (Møller, 1990; Lochmiller and Deerenberg, 2000).

Life-history theory undeniably offers a good starting point to build a general framework for the macroecology of sexual traits, more specifically for how large-scale variations in climatic conditions may influence reproductive effort (Ricklefs and Wikelski, 2002). Traditionally, reproductive effort has been partitioned into three main components: (1) *parental effort*, i.e., actions that increase offspring fitness; (2) *mating effort*, i.e., actions to acquire additional matings; and (3) *somatic effort*, i.e., actions that increase parental survival (Clutton-Brock, 1991). Given that each individual has only a fixed amount of energy to expend in reproduction, an increase in effort in one component must imply a decrease in the allocation of effort to one or both of the other components (Stearns, 1992). Individuals, therefore, should be selected to maximize their lifetime reproductive success by optimizing energy allocation to these three components in each breeding attempt (Trivers, 1972). Any environmental factor that changes the fitness profit for one component will have a direct effect on one or both of the other components. For example, if climatic seasonality constrains the reproductive season to only a few months, when fertile females are abundant in the population, the net benefit of mating effort compared with parental effort is expected to be high, and males should allocate more time to searching for additional matings (Magrath and Komdeur, 2003). By contrast, in places where the reproductive season lasts several months and fertile females appear asynchronously in the population, the low availability of potential mates decreases the rewards of mating effort compared with those from parental and/or somatic effort. In this situation, males should not invest in searching for additional mates and could enhance their reproductive success by providing some parental care (Kokko and Jennions, 2008) or by investing in somatic structures that increase their resource-holding power, such as elaborate weaponry (Emlen, 2008).

Mating system theory also offers the opportunity to explore how large-scale variations in environmental conditions may exert direct or indirect influence on a wide range of sexually selected traits. The distribution and limitation of key resources for reproduction in time and space predict the optimal set of mating strategies in a population (Emlen and Oring, 1977; Shuster and Wade, 2003). In other words, these resources drive the evolution of mating strategies and any trait related to reproductive ecology (Fig. 1.2). Under a gradient of environmental conditions we may also expect a gradient of mating strategies as a result of a gradual change of selective pressures on reproductive traits including not only behavior, but also morphology and physiology (Fig. 1.2). Given that the tropics are, in general, strikingly different from extreme latitudes in climatic conditions, and thus in the relative amount and diversity of available resources, as well as their temporal and spatial distribution, we expect markedly different selective pressures on reproductive strategies, and thus on mating systems and a whole array of aspects associated with reproductive ecology. These include, among others, length of breeding season, amount and distribution of time allocated to mate-searching, foraging or parental care, and patterns of sperm competition. Furthermore, since areas of homogeneous environmental conditions seem to be more fragmented in the Neotropics compared to other tropical regions (Fig. 1.1), single species may experience a wider set of environmental variables and we could expect higher intraspecific variation in mating strategies compared to resident species living in the Paleotropics. In turn, this may result in higher rates of speciation due to sexual selection (Ritchie, 2007; Seddon et al., 2008).

Much lower winter temperatures in more extreme latitudes usually constrain many species' life cycles to yearly, synchronous generations that either die or enter dormancy or diapause in the winter. In the tropics, however, many invertebrate and vertebrate life cycles are driven by water availability more than temperature. For invertebrates, wet seasons frequently trigger reproduction in dry places where host plant species depend on seasonal rain, and individuals either die or diapause during the dry season in some extreme climates. Air humidity is usually a limiting factor for insect flight, thus restricting dispersal and mate- or resource-searching (Zachariassen, 1991; Chown and Nicolson, 2004). For many vertebrates, especially insectivorous birds and mammals, the dependency upon invertebrates as a source of food for offspring also restricts breeding to the rainy season (see, for example, Poulin et al., 1992; Martins et al., 2006; Visser et al., 2006). In the humid tropics near the equator, however, temperature and humidity fluctuations are much smaller, thus generations are frequently less synchronous in the absence of climatic constraints. For invertebrates, the availability of some abundant host plants throughout the year may imply that monopolizing them may not be a profitable mating strategy since both male competitors and females may easily find undefended patches. In such cases, mating strategies such as leks or direct mate search and courtship may be more profitable for males in terms of reproductive opportunities (Thornhill and Alcock, 1983).

We suggest that searching for macroecological patterns of variation in sexual traits and selective forces along broad environmental variables is a fruitful field of investigation with important repercussions upon our understanding of the classic sexual selection models (see Twiss et al., 2007). A similar effort has been carried out recently in a review by Schemske et al. (2009) approaching biological interactions in general. Sexual selection itself is a byproduct of several types of interactions starting with intraspecific and somewhat mutualistic interactions between sexual morphs, i.e., males and females, and between parents and offspring. Studying the interrelationship at broad environmental scales between these and other types of intra- and interspecific interactions such as parasitism, competition, or predation, to name a few, could fill several volumes. Nevertheless, we only found a handful of studies approaching sexual selection at broad environmental scales.

Based on current theoretical and empirical knowledge on sexual selection and mating systems theory, we have developed general predictions relative to the variation of reproductive traits over a wide variety of climatic conditions.

In order to approach this task, we developed these predictions within an over-simplified 2×2 environmental matrix composed of two interacting variables, temperature and precipitation, which generates the following four conditions: hot and humid, hot and dry, cold and humid, cold and dry (Fig. 1.3). We argue that these extreme values represent the most useful data points to test our predictions within this bidimensional plane. The literature review we described in the previous section, however, shows that the extremes (the coldest, the driest, and the hottest) happen to attract relatively less effort to the study of sexual selection in the field.

Some broad environmental gradients may be suggested in a bidimensional plane covering most climatic conditions (see Fig. 1.3). Energetic demands, for instance, should be higher in cold and dry climates, or during cold seasons, compared to regions in which air or water temperatures are closer to those required for most metabolic processes (Willmer *et al.*, 2000). This may have diverse consequences not only on self-maintenance but also on offspring demands and mortality (Schemske *et al.*, 2009). Water loss in drier areas also results in physiological constraints that may very well shape maintenance and reproduction, especially in water-sensitive groups such as fish, amphibians, or aquatic arthropods, but to some degree in all animal groups (Willmer *et al.*, 2000). Even latitudinal trends in seasonal differences in light cycle may be closely associated with the onset of reproductive or hibernating seasons. An overall abundance

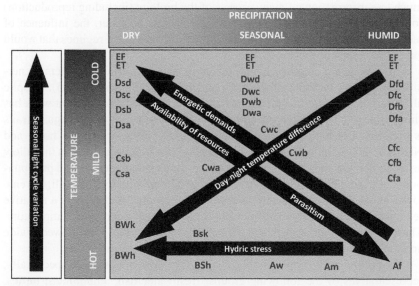

FIGURE 1.3 **Environmental plane of climatic conditions in terms of temperature and precipitation regimes, and selective pressures derived from such combination of climatic regimes.** Climate types are distributed depending on their defining temperature and precipitation regime. Gradients of selective pressures (arrows) are used to predict patterns of reproductive strategies in different functional groups.

of resources may be expected in hot and humid (i.e., tropical) regions due to higher nutrient and water availability (Hawkins *et al.*, 2003; Moya-Loraño, 2010), which most likely results in less resource-restricted mating systems as compared to more extreme environments. Finally, biotic interactions are clearly more important in tropical regions, where vector-borne parasites and pathogenic diseases are more frequent (Schemske *et al.*, 2009).

Based on this organizational diagram, we developed predictions relative to sexually selected traits, which we present below as broadly as possible, and avoiding details on the reproductive biology of specific taxa (Table 1.2). Some of the taxon-oriented chapters of the book will explore some of these predictions in more detail and even provide empirical tests of some of them.

ANIMAL FUNCTIONAL GROUPS TO TEST PREDICTIONS ON SEXUAL SELECTION

For the sake of simplicity and generality, instead of adopting a strictly phylogenetic classification, we classified animals into three functional groups that share basic physiological features: (1) arthropods, (2) ectothermic vertebrates, and (3) endothermic vertebrates (Table 1.2). This classification refers mainly to terrestrial animals and freshwater fish because our predictions rely on a bidimensional plane composed of temperature and precipitation. Our classification ignores non-arthropod invertebrates and exclusively marine fauna. There is much less information on many aspects of the biology (including reproduction) for these two major groups; also, particularly for the latter, the influence of most environmental variables follows different geographical regimes that would exceed the scope of this book.

Arthropods are a conspicuous element of any terrestrial ecosystem, generally comprising the bulk of diversity and biomass (Sanways, 2005). Despite all variation associated with feeding habit, age, and body size, recent work has shown that a considerable proportion of the variation in physiological traits is partitioned at high taxonomic levels, such as family or order (Chown and Nicolson, 2004). Indeed, species share general physiological features that allow us to predict how they should respond to variations in temperature and precipitation. For example, water loss rates in arthropods vary with total rainfall, extreme lower lethal limits decline with proximity to the poles, and there is a negative relationship between standard metabolic rate and environmental temperature. These patterns of variation would not be detectable if arthropods showed a wide array of responses to the environment (Chown and Nicolson, 2004).

The second group comprises the ectothermic vertebrates, more specifically, representatives of the following groups: freshwater fish, amphibians, turtles, alligators, lizards, and snakes. In general, ectotherms are dependent on environmental heat sources and control body temperature by means of both behavioral and physiological processes. Moreover, they have relatively low metabolic rates and are resistant to prolonged starvation, so that individuals are less vulnerable

TABLE 1.2 Predictions about the Effect of Environmental Factors (Temperature and Precipitation) on the Reproductive Behavior and Sexually Selected Traits of Three Functional Groups of Animals that Share Basic Physiological Features: Arthropods, Ectothermic Vertebrates, and Endothermic Vertebrates

Hot Environments				Humid Environments			
Dry Environments							
Traits	*Arthropods*	*Ectotherms*	*Endotherms*	*Traits*	*Arthropods*	*Ectotherms*	*Endotherms*
Abundance	Very scarce	Nearly absent	Moderate	Abundance	Scarce	Scarce	Moderate
Body size	Small	Small	Small to large	Body size	Small	Small	Small to large
Time to sexual maturity	Usually in 1 year	Several years	From 1 to several years	Time to reach sexual maturity	Usually in 1 year	Several years	From 1 to several years
Longevity	Usually 1 year	Usually 1 year	Usually >1 year	Longevity	Usually 1 year	1 to few years	1 to a several years
Gamete replenishing (time-out)	Long	Long	Long	Gamete replenishing	Long to moderate	Long	Long to moderate
Offspring development	Fast	Fast	Fast	Offspring development	Fast	Fast	Fast
Parental care	Unusual	Unusual	Usual	Parental care	Unusual	Unusual	Usual
Reproductive events	Usually 1	Usually 1	Usually 1	Reproductive events	Usually 1	Usually 1	Usually 1
Hibernation/diapause	Frequent	Frequent	Rare	Hibernation/diapause	Frequent	Frequent	Frequent

Continued

TABLE 1.2 Predictions about the Effect of Environmental Factors (Temperature and Precipitation) on the Reproductive Behavior and Sexually Selected Traits of Three Functional Groups of Animals that Share Basic Physiological Features: Arthropods, Ectothermic Vertebrates, and Endothermic Vertebrates—cont'd

<table>
<thead>
<tr><th></th><th colspan="6">Hot Environments</th></tr>
<tr><th></th><th colspan="3">Dry Environments</th><th colspan="3">Humid Environments</th></tr>
<tr><th>Traits</th><th>Arthropods</th><th>Ectotherms</th><th>Endotherms</th><th>Arthropods</th><th>Ectotherms</th><th>Endotherms</th></tr>
</thead>
<tbody>
<tr><td>Abundance</td><td>Abundant</td><td>Abundant</td><td>Abundant</td><td>Abundant</td><td>Abundant</td><td>Abundant</td></tr>
<tr><td>Body size</td><td>Small to mid-size</td><td>Small to mid-size</td><td>Small to large</td><td>Mid- to large size</td><td>Mid- to large size</td><td>Small to large</td></tr>
<tr><td>Time to reach sexual maturity</td><td>From 1 to several years</td><td>1 to few years</td><td>From 1 to several years</td><td>1 to several years</td><td>1 to few years</td><td>1 to several years</td></tr>
<tr><td>Longevity</td><td>Usually >1 year</td><td>Very long</td><td>Usually >1 year</td><td>Usually 1 year</td><td>Long</td><td>Usually long</td></tr>
<tr><td>Gamete replenishing</td><td>Long to moderate</td><td>Moderate</td><td>Moderate</td><td>Short</td><td>Short</td><td>Short</td></tr>
<tr><td>Offspring development</td><td>Fast</td><td>Fast</td><td>Slow to fast</td><td>Slow</td><td>Fast</td><td>Slow</td></tr>
<tr><td>Parental care</td><td>Frequent, but short</td><td>Unusual or short</td><td>Usual</td><td>Frequent</td><td>Frequent</td><td>Usual</td></tr>
<tr><td>Reproductive events</td><td>1 to several</td><td>From 1 to several</td><td>From 1 to several</td><td>Usually several</td><td>Usually several</td><td>Usually several</td></tr>
<tr><td>Hibernation/diapause</td><td>Absent or short</td><td>Absent or short</td><td>Absent</td><td>Absent</td><td>Absent</td><td>Absent</td></tr>
</tbody>
</table>

to fluctuations in food supply (Randall *et al.*, 1997; Willmer *et al.*, 2000). Within the ectothermic vertebrates, we can highlight at least three groups that differ in relation to their resistance to hydric stress. Freshwater fish are totally vulnerable to dehydration since, with rare exceptions in the tropics, they spend their whole life cycle associated with water. Amphibians usually spend part of their life cycle strictly associated with water. Even as adults, they have a permeable tegument, which renders them particularly vulnerable to desiccation. At the other extreme, the polyphyletic "reptiles" (turtles, lizards, alligators, and snakes) have a waterproof tegument that allows them to occupy the driest places on Earth (Randall *et al.*, 1997; Willmer *et al.*, 2000). Whenever necessary, we will consider this fundamental difference to derive appropriate predictions (see below).

Finally, our third group refers to endothermic vertebrates, in which heat is primarily generated as a result of internal metabolic processes. Although endothermy is known to occur in some scattered species of "reptiles", fish, and insects, it is the main physiological strategy of birds and mammals (Koteja, 2004). For the sake of generality, therefore, we will use the expression "endothermic vertebrates" to refer exclusively to typically warm-blooded animals – i.e., birds and mammals. Endothermy has evolved independently in these two vertebrate groups, leading to striking convergences such as: (1) high aerobic capacity allowing intense physical activity; (2) high costs of maintaining the increased capacity of the visceral organs necessary to support high rates of total daily energy expenditures; and (3) capacity of the parent to control incubation temperature, which is probably related to the evolution of parental care (Yu-shan *et al.*, 2003). Endothermic vertebrates are less vulnerable to fluctuations in external temperature, but a high metabolic rate generates an equally high demand for food, so individuals cannot resist prolonged starvation (Randall *et al.*, 1997; Willmer *et al.*, 2000).

Arthropods

Terrestrial arthropods comprise representatives of four major groups: insects, arachnids, myriapods (centipedes and millipedes), and crustaceans of the order Isopoda (woodlice). There is no doubt that diversity of morphologies, behaviors, and perhaps physiological strategies can thwart generalizations. Nonetheless, we will explore some general common features of these groups to develop large-scale predictions. Like many other groups of organisms, abiotic factors, such as severe temperature, limit the occurrence of arthropod species at extreme latitudes, whereas limits of species towards the equator are set by biotic interactions, such as predation, diseases, and competition (Schemske, 2002, 2009; Chown and Nicolson, 2004), although this latter factor, though potentially significant, has not been substantiated by any studies to date. In fact, arthropods are almost absent from the most severe climates that combine cold temperatures and low precipitation levels (Sanways, 2005). Although some arthropod species with specialized physiological mechanisms endure freezing temperatures in temperate and cold zones, most species go through their entire life cycles in the warm months, when they grow fast, reproduce, and eventually die (Chown and Nicolson, 2004). On the other hand, hot and humid environments, such as the Amazon forest, hold a huge diversity of long-lived arthropods, some of them reaching 10 years or more (Adis, 2002).

In temperate and cold regions, a short period of favorable climatic conditions probably constrains development time, leading to fast sexual maturity and small body size (especially among predators). Moreover, time-consuming activities, such as post-zygotic parental care, can be expected to be rare. Females are expected to lay a large number of eggs and hide them in protected places, where they probably will overwinter (see, for example, Tallamy and Schaeffer, 1997; Machado and Macías-Ordóñez, 2007). Semelparity, therefore, is expected to be the rule, and for the few iteroparous species the time for egg replenishing is likely to be long since resources are scarce and concentrated in time (Tallamy and Brown, 2001). A final consequence of the short mating season, when a large number of individuals are reproductively active at the same time, is that the most frequent mating system should be a scramble competition polygyny, in which temperature may play an important role influencing the relative mating success of different phenotypes (see, for example, Moya-Loraño et al., 2007). As predicted by theory, a large number of males within the searching area makes territoriality unprofitable (Thornhill and Alcock, 1983). In some cases, however, protandrous males of low-density species may be able to monopolize some key reproductive resource so that the observed mating system is a resource-defense polygyny.

At the other extreme, arthropod species living in hot and humid environments, with resources available throughout the year, are predicted to have a slow development time when compared to their relatives from temperate and cold regions. Intense predation and fungal infection on eggs select for post-zygotic parental care, which indeed appears to be more common among tropical species (Costa, 2006). High temperatures, however, should accelerate embryonic development, so that the period of parental care is not expected to be long. For predatory arthropods, which are generally food deprived while caring for the offspring (see, for example, Thomas and Manica, 2003; Machado and Macías-Ordóñez, 2007), parental care is expected to be followed by a fast period of egg replenishing, which is a direct consequence of abundant food sources. Even for species that do not exhibit any form of post-zygotic parental care, favorable and stable climatic conditions should favor iteroparity, so that individuals are expected to have several reproductive events throughout the year. Since the mating season is long, there would be no selective pressure on reproductive synchrony and, at any given moment, populations should be composed of reproductive females and non-reproductive females. A higher degree of female reproductive asynchrony would result in a more male-biased operational sex-ratio and thus in potentially stronger sexual selection (Shuster and Wade, 2003). A common mating system under this scenario is expected to be some kind of resource- or female-defense polygyny (Thornhill and Alcock, 1983). In such circumstances, male–male competition is expected to be fierce.

As a result of intense male–male competition, we also predict that weaponry, such as antlers, horns, forceps, and claws employed in agonistic interactions between rivals for female access, will be more common in arthropods in hot and humid environments than in any other types of environments. Moreover, since the expression of weapons in males is costly, and there is a trade-off between the investment in weapons and somatic structures (Nijhout and Emlen, 1998), only rich food environments may offer enough resources to allow males to invest in elaborate weaponry. Parasites can also be expected to have better environmental conditions in the tropics (Schemske *et al.*, 2009), generating high immunological costs in their host species. Thus, costly secondary sexual traits in host species, which may or may not be used as weapons, would become honest signals in these environments.

Whenever a subset of males bearing well-developed weaponry or large body size is able to monopolize most of the copulations, there is opportunity for alternative male tactics to evolve (Shuster and Wade, 2003; Taborsky *et al.*, 2008). Individuals of the alternative male morph (sneakers) generally do not invest in weapons or large body size, but rather allocate their scarce resources to optimizing their mating opportunities by maximizing the fertilization of gametes (Simmons *et al.*, 2001). A final consequence is that alternative male morphs are expected to be more common and the intensity of sperm competition is expected to be higher among species living in hot and humid environments than in any other environment.

Ectothermic Vertebrates

Freshwater fishes, amphibians, and "reptiles" (snakes, lizards, alligators, and turtles) share their dependency on heat for their metabolism, as well as their vulnerability to overheating. For that reason we have grouped them in a single functional group since our first climatic axis is temperature. However, given their strikingly different dependence on water, we may be forced to make some distinctions among them. With the rare exception of lungfishes in the seasonal rainforest (Lomholt *et al.*, 1975), freshwater fish spend their whole life cycle in water and are highly dependent on precipitation regimes. Their reproductive behavior is closely linked to stream flow and the rate of evaporation of temporal ponds, and the population dynamics is governed by the temporal water connections among these ponds in highly seasonal environments (Chapman and Kramer, 1991; Faulks *et al.*, 2010). Most amphibians are still highly dependent on precipitation at least for reproduction, to the point that they are usually more conspicuous during their characteristic explosive mating seasons coupled with precipitation peaks (Grundy and Storey, 1998; Saenz *et al.*, 2006). Although most are still somewhat dependent on water due to their highly permeable skin, several species manage to occupy fairly dry areas, using diapauses and other physiological strategies to overcome the dry periods (Harvey *et al.*, 1997; Grundy and Storey, 1998). Indeed, a few amphibians have managed to colonize some very dry regions, but they still depend on water for reproduction, sometimes waiting years before breeding (Withers, 1995). "Reptiles", on the other hand, are relatively independent of precipitation, and can subsist in the driest regions in the planet (Bentley and Schmidt-Nielsen, 1966).

All ectothermic vertebrates, however, depend on environmental heat to fuel their metabolism, which results in a tight relationship between developmental time and temperature (Willmer *et al.*, 2000). In cold regions terrestrial species must rely on long hibernations, which still has some metabolic costs. Thus, time allocated to foraging during the short warmer season is yet another constraint. A first set of nearly obvious predictions is that ectothermic vertebrate abundance, size, and diversity should be clearly lower in cold environments, and, conversely, they should be more abundant, larger, and more diverse in hot environments – as indeed they are (Duellman, 1999). Temperature conditions throughout the year are so severe in polar regions that ectothermic vertebrates are almost absent in these zones (Duellman, 1999; Sindaco and Jeremčenko, 2008). Fish and amphibians should do slightly better in cold but humid regions, whereas "reptiles" should thrive in hot and dry areas, and indeed they do (Vitt *et al.*, 2003; Sindaco and Jeremčenko, 2008).

Despite the above ecological patterns for many ectothermic vertebrates, the consequences of all these patterns on sexually selected traits and overall reproductive strategies are less obvious and have seldom been addressed (for an exception, see Olsson *et al.*, 2010). Fish and amphibian explosive mating seasons in hot but seasonally dry regions are usually time-constrained. Water becomes a key reproductive resource for egg-laying, but it is frequently hard to monopolize. As a result, scramble competition and lek polygyny mating systems are expected to be very frequent. In more humid environments water is usually not restricted, so other more defendable environmental resources, such as favorable oviposition or thermoregulation sites, may become disputed territories under resource-defense polygynous mating systems. For "reptiles", precipitation is usually not an issue in hot environments, and their mating system may be more associated with their metabolic needs. Top predators, such as snakes, have moderately low densities and probably relatively low encounter rates (Uller and Olsson, 2008), thus they may exhibit scramble mating competition mating systems. Insectivorous groups such as lizards, on the other hand, may be more abundant, and appropriate display sites may become reproductive territories. In temperate and cold climates temperature seasonality becomes an issue, and short to extremely short mating seasons may result in reproductive trends for snakes and lizards similar to those experienced by fish and amphibians under seasonal precipitation regimes.

The length of reproductive seasons for all ectothermic vertebrates, either water- or heat-dependent, results in interactions between traits, such as offspring development time, risk of parasitism and predation, and parental care. A higher risk of offspring predation in the hot and humid tropics (Schemske *et al.*, 2009) may select for longer periods of parental care (or gestational time in viviparous species), favored by short offspring development time without compromising adult size. In cold regions, although temperature limited, short mating seasons may also select for short offspring development time and time to sexual maturity, resulting in smaller adults. This would be especially relevant in species in which size is either intra- or intersexually selected, so that size would be an honest signal for mates and competitors in cold environments. Similarly, conspicuous and costly ornaments may be honest signals for mates and competitors in hot and humid areas where resources must be allocated among these traits and immunological effort against intense parasitism (Møller, 1990; Lochmiller and Deerenberg, 2000).

Resources are more abundant for ectotherms in the tropics, leading to a lower search time overall. However, ectotherm abundance relative to the number of competitors is hard to predict, as the latter depend on many other factors. In contrast, although resources may generally be scarcer in drier and colder places, when temperature or rain defines a short and explosive mating season, reproductive resources per capita may actually be more abundant. This would lead to lower levels of competition, thus decreasing the strength of sexual selection. The rate of energetic expenditure to look for resources is highly temperature-dependent, and must be allocated between that and other physiological demands such as recovery or preparation for the next reproductive event (e.g., gamete production, territory acquisition, mate search). Regions where water or heat constrain the duration of the mating season for ectothermic vertebrates probably select for a semelparous life history associated with explosive mating strategies. Iteroparity and overlapping generations may be selected in warmer and more humid regions, in which case sexual selection may favor delaying sexual maturation until a larger size or endurance in courtship or territorial defense has been reached. This may in fact set the stage for the evolution of alternative life-history mating strategies, just as predicted for arthropods in the previous section.

Endothermic Vertebrates

Despite the common attribute of endothermy, birds and mammals differ in fundamental ways that have shaped morphological and behavioral attributes associated with their reproductive biology. First and foremost, birds and mammals differ in the incubation process of their zygotes, which affects parental investment strategies and, therefore, mating systems (Emlen and Oring, 1977). Female mammals incubate their fertilized eggs internally, usually with little or no assistance from the male, whereas birds incubate their fertilized eggs externally, a process that frequently requires both sexes (Clutton-Brock, 1991). Secondly, birds and mammals usually differ in the need for male versus female parental investment in the feeding of offspring. Lactating female mammals generally are able to supply all of their brood's early nutritional needs, which may explain the prevalence of polygyny in the group (Clutton-Brock, 1989). Offspring of birds, on the other hand, typically are fed by both parents, and this need for biparental care is commonly associated with social monogamy (Emlen and Oring, 1977). These two fundamental biological peculiarities dictate many of the differences in reproductive biology and sexually selected traits in these two groups of vertebrates, which will be addressed later. From a strictly physiological perspective, however, birds and mammals have been grouped together because they are "warm-blooded", i.e., they can regulate body temperature through internal metabolic processes. This common trait allows birds and mammals a relative independence from ambient temperatures, so that they are able to occupy extremely cold environments, such as polar regions. The costs associated with heat production, however, result in an exceptional dependence upon a constant and high supply of food (Randall et al., 1997; Willmer et al., 2000), which has profound implications for the breeding biology of both birds and mammals.

At least for birds, differences in life histories of tropical and temperate species have been repeatedly emphasized in the literature, with considerable contrasts in areas associated with breeding, predation, territoriality, mortality, and singing, to name a few (see, for example, Skutch, 1985; Hau, 2001; Ricklefs and Wikelski, 2002; Slater and Mann, 2004; Fedy and Stutchbury, 2005; Schemske et al., 2009). However, one common misconception is that tropical species breed year-round (reviewed in Wikelski et al., 2000). As pointed out before, the Neotropics encompasses an elevated heterogeneity of landscapes, temperatures, and precipitation regimes (Fig. 1.1), which sets the stage for very different selective pressures. In those areas where there is little variation in photoperiod, temperature, and precipitation, such as the western Amazonian basin (Af climate), food resources for insectivorous and frugivorous species remain fairly constant, resulting in optimal breeding conditions for females throughout the year. However, even in the evergreen tropical rainforest, dramatic seasonal fluctuations in water flow due to ice melting from the Andes result in annual cycles that change environmental conditions and dictate fluctuations of different resources, which in turn may influence the reproductive activity of many birds and mammals (Junk, 1997). A similar or even more pronounced pattern is also expected for the vast regions of the world with marked seasonal variation in precipitation, such as the eastern and southern Amazonian Basin, with Am and Aw climates (Fig. 1.1). A well-defined rainy season leads to seasonal availability of food resources, which restricts the breeding periods of both birds and mammals (see, for example, Poulin et al., 1992; Martins et al., 2006; Visser et al., 2006).

In areas with little or no climatic seasonality, where resource stability may prevail continuously, animals can establish year-long territories, population densities are expected to remain fairly stable, and female asynchrony in breeding can be expected as the rule. This should be true for both birds and mammals that depend upon resources that can be monopolized through territorial defense – for example, insects and other small prey items, fruits, seeds, or pasture. Temporal asynchrony of receptive females associated with little climatic seasonality should increase the opportunity for sexual selection (see, for example, Twiss *et al.*, 2007). Among birds, in which social monogamy prevails and weaponry is mostly absent, male–male competition will include elaborate behaviors and social systems that optimize access to females, such as extra-pair copulations (EPC), sperm competition, and mate guarding (Stutchbury and Neudorf, 1997; Dunn *et al.*, 2001). These types of mating strategies, therefore, should be much more common among species living in tropical than in temperate or cold climates. Moreover, some resources, such as fruits and flowering plants, are also much more abundant overall in tropical climates, despite being temporally ephemeral. Males of species that depend upon such fleeting resources in regions with little seasonality may be unable to establish territorial systems, however. When resources are not defendable, male dominance mating systems are expected to emerge (Oring, 1982). Indeed, lekking species are prevalent in the tropical climates, and are almost always dependent upon ephemeral resources such as fruits (e.g., birds of paradise, manakins, cotingas) or nectar (e.g., hummingbirds) (Höglund and Alatalo, 1995). One explanation that has been suggested is that small clutch sizes of tropical bird species (Cody, 1966; Ricklefs, 1969; Skutch, 1985), which may be due to high nest predation pressure (Skutch, 1949) as well as several other environmental and life-history traits (see Martin, 1996; Martin *et al.*, 2001, 2006), allow male emancipation from paternal care and the evolution of leks (Snow, 1962; Lill, 1974, 1976; Beehler, 1987). The evolution of small clutch sizes in tropical species is a highly debated subject that remains inconclusive despite the many proposed hypotheses (Martin, 1996). Like many other obvious life-history differences between tropical and temperate organisms, more data are needed to provide the tests of the different paradigms that exist in the literature.

Mammals, on the other hand, have predominantly polygynous mating systems, and male weaponry is fairly common (Clutton-Brock, 1989). This conventional situation may be more often true for temperate or cold region mammals than for their tropical counterparts, despite the basic biological framework that favors polygynous mating systems in mammals. In cold environments, temporally clumped resources defining short mating seasons may generate a gregarious distribution of females, and thus the opportunity for males to defend female aggregations or resources (grazing or hunting areas), which may increase the advantages of large but expensive weapons such as antlers, horns, and long teeth (Clutton-Brock, 1989). In tropical regions of little seasonality, resources are more readily available and distributed more uniformly. The distribution of food in such circumstances may allow females to live non-gregariously, thus minimizing competition over food (Alexander, 1974; van Schaik, 1983). Male reproduction is less dependent upon food, but depends upon access to fertile females (Clutton-Brock, 1989). In such a scenario, with females spatially separated, males may be forced into social monogamy, especially if such a mating system also leads to other benefits, such as more effective care of the offspring. Social monogamy is especially common among primates, and this group of mammals presents an excellent opportunity to examine the contrasts in the evolution of mating systems and secondary sexual attributes. Although primates are distributed primarily in tropical and subtropical regions of the world, the contrasts provided by seasonal versus non-seasonal tropical regions would allow the testing of specific hypotheses relative to sexual selection. Generally, we can expect a trend toward more ornamentation in the tropics than weaponry in several mammalian taxa, as males may compete more with other stimuli, including courtship from other males. Given that the prevalence of parasites among tropical birds and mammals is very high (see below), this ornamentation may also be an honest signal of immunologically expensive parasite resistance (Møller, 1990; Lochmiller and Deerenberg, 2000).

Birds and mammals in hot and humid areas should have relatively low mortality rates associated with less severe abiotic pressures, compared to their counterparts from temperate or cold climates. Therefore, we can expect abiotic factors to be less relevant than biotic selective pressures such as predation and parasitism (Schemske *et al.*, 2009). Indeed, nest predation rates in birds and prevalence of blood parasites and vector-borne parasites in both birds and mammals are known to be higher in the tropics (Schemske *et al.*, 2009). Despite such assumed high rates of parasites and other possible sources of mortality for neotropical birds, mortality rates of adults tend to be low in the tropics compared to higher latitudes (Yom-Tov *et al.*, 1992; but see also Karr *et al.*, 1990), although the causes of such lower mortality remain nebulous. High brood predation among tropical birds may select for high rates of paternal investment and, consequently, lower rates of EPC (reviewed in Griffith *et al.*, 2002). An alternative behavioral strategy that could result from high nest predation, leading however to high rates of EPC, is that pairs may nest in groups and breed synchronously, which may result in less predation through two mechanisms. First, the higher density and synchronous breeding would dilute the per capita rate of nest predation (if a higher number of attracted predators did not cancel out such an effect). Second, this could lead to higher rates of EPC, whereby females and males may copulate randomly (or not), spreading their offspring over many nests, thus also diluting the chances of complete failure of nesting attempts due to predation. In addition to increased EPC, we might also expect to see increased intraspecific brood parasitism in such species, as this too would provide a mechanism to dilute predation risk.

The influence of high prevalence of parasites on the mating system of tropical endotherms has been well studied. In mammals, for instance, body size (Morand and Poulin, 1998), mating system, and social organization (Altizer *et al.*, 2003) are known to be related to parasite load. As all three traits may be related to environmental climatic conditions, we expect parasite loads also to fluctuate with climate regime. Furthermore, continuous reproduction and other forms of social interactions in mammals with high densities, relatively high temperatures year round, and high water availability result in ideal conditions for parasite contagion, and has been suggested as one of the factors leading to high mortality in the tropics (Schemske *et al.*, 2009). Sex bias in parasite loads has been shown in mammals and birds, since estrogens stimulate immunity whereas androgens depress it (Schalk and Forbes, 1997). As suggested by Møller (1990), this would further enhance sexual dimorphism of secondary sexual traits in species more vulnerable to parasite infection (e.g., tropical species) as "health-certifying traits". The stimuli competition for female attention plus the higher exposure to parasites may help to explain the array of elaborate acoustic and visual male sexual traits in the tropics. A higher exuberance of sexually selected traits of tropical animals, compared to their temperate counterparts, is a common view, though this has not been closely examined through comparative studies using worldwide distributed species. This may be especially important when considering bird coloration that has evolved due to different selective pressures. For example, a study using barn owl species and subspecies (*Tyto* genus) as a model organism found that those in the tropics exhibited larger eumelanic spots than those in temperate regions, and the authors speculate that this indicates a link between eumelanic coloration and greater resistance to parasites (Roulin *et al.*, 2009).

Severe winters of temperate and cold climates may result in a higher mortality for birds and mammals when compared to their counterparts living in tropical climates (Schemske *et al.*, 2009), even when considering the biotic pressures of higher predation and parasitism in the tropics. Thus, adult longevity should be higher for tropical species, resulting in more densely packed populations with fewer breeding opportunities for juveniles trying to establish territories. One of the possible consequences of decreased mortality with extended longevity and high population density in warm and stable environments is for the offspring of many species to delay dispersal, which may lead to overlapping generations living within the same territory or neighborhood areas (Emlen, 1982; Brown, 1987). At least among birds, such demography in many areas with tropical climate may have contributed to the evolution or maintenance of a social system known as cooperative breeding, characterized by collaboration among adult group members in rearing offspring produced either by one breeding pair or by several breeding adults (Arnold and Owens, 1998, 1999). Although most of the focus relative to cooperative breeding has been on birds, there are many mammalian taxa that exhibit cooperative breeding, including primates, canids, viverrids, and rodents (reviewed in Solomon and French, 1997). Despite the intrinsic biological differences between birds and mammals, we can still expect that the ecological characteristics of certain geographic areas will constitute selective pressures leading to cooperative breeding, including traits shared by many tropical birds and mammals: year-round residency, high adult survivorship, and overlapping generations (Fry, 1977; Brown, 1987). At least in cooperative breeding birds, sexual selection can be strong and associated with the many observed mating systems, ranging from monogamous species to highly promiscuous ones, and generating both extreme sexual ornamentation as well as behavioral strategies such as EPC (reviewed in Pruett-Jones, 2004).

Hibernation is a well-known strategy to deal with extreme temperature seasonality that allows mammals to colonize cold environments while minimizing metabolic costs during the winter, for which both males and females allocate long periods to store reserves (Willmer *et al.*, 2000). Hibernation is usually followed by a well-marked mating season, followed in turn by an intense season of maternal nurture of very altricial offspring that, when coincident with larger body sizes, results in maternal care that may last several months or even years. Long periods of maternal care should result in high risks of infanticide by males of predatory species as a strategy to end the nurturing phase of females and induce sexual receptivity. In tropical regions, shorter maternal care due to more precocious offspring and higher and more continuous resource availability may result in lower offspring mortality due to infanticide. However, in hot and dry environments, such as steppes and deserts, water seasonality may result in long periods of resource limitations that would also select for long periods of maternal or social care, and thus high infanticide risk as well.

Because birds can cover long distances through flight, the winter in high altitudes frequently leads to migration to lower latitudes. In fact, within the savannas and regions with temperate climate in the Neotropics, many bird species migrate during the dry/cold season to other areas with high food supply, or breed only seasonally, during the most favorable months (Stutchbury and Morton, 2001). An additional pressure for diurnal birds to migrate to lower latitudes during the winter is that the number of light hours to forage is greatly reduced in the winter, but even more so at higher altitudes. If this is in fact a selective pressure, we may expect a relatively lower frequency or distance of migration in bird species that forage at night, such as owls. Migration has important implications for the reproductive biology of birds, affecting energy dynamics and metabolism, reproductive success, immune condition, sperm competition, and, ultimately, fitness, among other possibilities (reviewed in Rappole, 1995 and Able, 1999). The settlement hypothesis, for instance, suggests that if migratory birds establish reproductive territories within a short time period, then chance rather than male quality may determine territory ownership (Weatherhead and Yezerinac, 1998). In these circumstances, females socially mated to low quality males may be expected to engage in EPC, especially given the high concentration of males at the start of the breeding season. This hypothesis leads to the conjecture that EPC should be more common in temperate and cold than in tropical climates, when migratory birds return to their breeding habitats and quickly establish territories. However, female choice mechanisms should not necessarily depend upon large male aggregations and synchronous breeding, and should be pertinent to species in the tropics. For example, females of resident tropical species that flock (and mate) during the winter may have more time to evaluate and compare potential mates during the non-breeding season, when many species aggregate in flocks. Flocking may provide females with a greater opportunity for evaluating males in two ways: (1) more males are exposed to the females in flocks; and (2) females can assess males in food-lean times and during molt, both of which may be more honest indicators of male quality.

CONCLUDING REMARKS

Although marked seasonality in temperate regions offers an opportunity to monitor complete mating seasons in relatively safe and somewhat comfortable working conditions, frequently close to research institutions, working conditions in harsher environments may be quite different. Researchers themselves are subject to the energetic demands of the polar tundra or intense parasitism in the tropical rain forest, and this affects the viability of research in these areas. Moreover, human density is commonly low in harsher environments, which means fewer research institutions. In such conditions, field study sites are frequently far from the closest research institutions, and research is more expensive due to transportation and survival costs. Despite the limitations and obstacles to collecting field data in such difficult and challenging regions, we hope that this chapter, stimulates research and collaboration among scientists at broad geographic scales. Our purpose is to draw attention to a still mostly unexplored and promising field, which we term "macroecology of sexual selection", generating controversies and the testing of new ideas. We hope to have shown convincingly that variation in climatic conditions, in terms of either precipitation or temperature, should be a powerful predictor of sexually selective forces. There are regions with year-round cold, hot, humid, or dry conditions, but the great majority of the world has some form of seasonality. Very cold places near the poles have a strong seasonality in photoperiod. Most arid areas have short but critical precipitation pulses. The great majority of cold and temperate areas have hot to mild summers. Even the most humid and hot areas in the Neotropics near the equator have a strong seasonality of water flow from melting ice and snow from the Andes. All these climatic pulses define the temporal and spatial patterns of food, water, mates, parasites, predators, competitors, and solar heat, to name a few. Understanding how seasonality affects sexual selection at large scales depends on understanding how time and resource constraints shape reproductive strategies.

ACKNOWLEDGMENT

This chapter was greatly improved by the suggestions made by Dr Robert Ricklefs, whose ongoing interest in and enthusiasm for the tropics is a source of inspiration to us all.

REFERENCES

Able, K.P., 1999. Gatherings of Angels: Migrating Birds and their Ecology. Comstock Books, Ithaca.

Adis, J., 1990. Thirty million arthropod species: too many or too few? J. Trop. Ecol. 6, 115–118.

Adis, J., 2002. Amazonian Arachnida and Myriapoda. Pensoft Publishers, Sofia-Moscow.

Alexander, R.D., 1974. The evolution of social behavior. Annu. Rev. Ecol. Syst. 5, 325–383.

Alho, J.S., Herczeg, G., Laugen, A.T., Räsänen, K., Laurila, A., Merilä, J., 2011. Allen's rule revisited: quantitative genetics of extremity length in the common frog along a latitudinal gradient. J. Evol. Biol. 24, 59–70.

Altizer, S., Nunn, C.L., Thrall, P.H., Gittleman, J.L., Antonovics, J., Cunningham, A.A., Dobson, A.P., Ezenwa, V., Jones, K.E., Pedersen, A.B., Poss, M., Pullian, J.R.C., 2003. Social organization and parasite risk in mammals: integrating theory and empirical studies. Annu. Rev. Ecol. Syst. 34, 517–547.

Arnold, K.E., Owens, I.P.F., 1998. Cooperative breeding in birds: a comparative test of the life history hypothesis. Proc. R. Soc. London B. 265, 739–745.

Arnold, K.E., Owens, I.P.F., 1999. Cooperative breeding in birds: the role of ecology. Behav. Ecol. 10, 465–471.

Audesirk, T., Audesirk, G., 1996. Biology: Life on Earth. Prentice Hall, Upper Saddle River.

Bailey, R.G., 1998. Ecoregions: the Ecosystem Geography of Oceans and Continents. Springer-Verlag, New York.

Beehler, B.M., 1987. Birds of paradise and mating system theory – predictions and observations. Emu 87, 78–89.

Bentley, P.J., Schmidt-Nielsen, K., 1966. Cutaneous water loss in reptiles. Science 151, 1547–1549.

Blanckenhorn, W.U., Stillwell, R.C., Young, K.A., Fox, C.W., Ashton, K.G., 2006. When Rensch meets Bergmann: does sexual size dimorphism change systematically with latitude? Evolution 60, 2004–2011.

Bradshaw, S.D., 2003. Vertebrate Ecophysiology: an Introduction to its Principles and Applications. Cambridge University Press, New York.

Brown, J.L., 1987. Helping and Communal Breeding in Birds. Princeton University Press, Princeton.

Chapman, L.J., Kramer, D.L., 1991. The consequences of flooding for the dispersal and fate of poeciliid fish in an intermittent tropical stream. Oecologia 87, 299–306.

Chown, S., Nicolson, S.W., 2004. Insect Physiological Ecology: Mechanisms and Patterns. Oxford University Press, Oxford.

Clutton-Brock, T.H., 1989. Mammalian mating systems. Proc. R. Soc. London B. 236, 339–372.

Clutton-Brock, T.H., 1991. The Evolution of Parental Care. Princeton University Press, Princeton.

Cody, M.L., 1966. A general theory of clutch size. Evolution 20, 174–184.

Cornwallis, C.K., Uller, T., 2009. Towards an evolutionary ecology of sexual traits. Trends Ecol. Evol. 25, 145–152.

Costa, J.T., 2006. The Other Insect Societies. The Belknap Press of Harvard University Press, Cambridge.

Darlington, P.J., 1958. Area, climate and evolution. Evolution 13, 488–510.

Darwin, C., 1862. The Voyage of the Beagle. Doubleday, Garden City.

Darwin, C., 1871. The Descent of Man, and Selection in Relation to Sex. Murray, London.

Duellman, W.E., 1999. Patterns of Distribution of Amphibians: a Global Perspective. Johns Hopkins University Press, Baltimore.

Dunn, P.O., Whittingham, L.A., Pitcher, T.E., 2001. Mating systems, sperm competition, and the evolution of sexual dimorphism in birds. Evolution 55, 161–175.

Emlen, D.J., 2008. The evolution of animal weapons. Annu. Rev. Ecol. Syst. 39, 387–413.

Emlen, S.T., 1982. The evolution of helping. I. An ecological constraints model. Amer. Nat. 119, 29–39.

Emlen, S.T., Oring, L.W., 1977. Ecology, sexual selection, and the evolution of mating systems. Science 197, 215–223.

Faulks, L.K., Gilligan, D.M., Beheregaray, L.B., 2010. Islands of water in a sea of dry land: hydrological regime predicts genetic diversity and dispersal in a widespread fish from Australia's arid zone, the golden perch (*Macquaria ambigua*). Mol. Ecol. 19, 4723–4737.

Fedy, B.C., Stutchbury, B.J.M., 2005. Territory defence in tropical birds: are females as aggressive as males? Behav. Ecol. Sociobiol. 58, 414–422.

Fouquet, A., Gilles, A., Vences, M., Marty, C., Blanc, M., Gemmell, N.J., 2007. Underestimation of species richness in neotropical frogs revealed by mtDNA analyses. PLoS ONE 2, e1109.

Fry, C.H., 1977. The evolutionary significance of cooperative breeding in birds. In: Stonehouse, B., Perrins, C.M. (Eds.), Evolutionary Ecology, University Park Press, Baltimore, pp. 127–136.

Gaston, K., Blackburn, T., 2000. Pattern and Process in Macroecology. Blackwell Science, Oxford.

Good, R., 1964. The Geography of Flowering Plants, third ed. Longman, London.

Griffith, S.C., Owens, I.P.F., Thuman, K.A., 2002. Extra-pair paternity in birds: a review of interspecific variation and adaptive function. Mol. Ecol. 11, 2195–2212.

Grundy, J.E., Storey, K.B., 1998. Antioxidant defenses and lipid peroxidation damage in aestivating toads, *Scaphiopus couchii*. J. Comp. Physiol. B. 168, 132–142.

Harvey, L.A., Propper, C.R., Woodley, S.K., Moore, M.C., 1997. Reproductive endocrinology of the explosively breeding desert spadefoot toad, *Scaphiopus couchii*. Gen. Comp. Endocrinol. 105, 102–113.

Hau, M., 2001. Timing of breeding in variable environments: tropical birds as model systems. Horm. Behav. 40, 281–290.

Hawkins, B.A., Field, R., Cornell, H.V., Currie, D.J., Guegan, J.F., Kaufman, D.M., Kerr, J.T., Mittelbach, G.G., Oberdorff, T., O'Brien, E.M., Porter, E.E., Turner, J.R.G., 2003. Energy, water, and broad-scale geographic patterns of species richness. Ecology 84, 3105–3117.

Höglund, J., Alatalo, R.V., 1995. Leks. Princeton University Press, Princeton.

Junk, W.J., 1997. The Central Amazon Foodplain: Ecology of a Pulsing System. Springer, Berlin.

Karr, J.R., Nichols, J.D., Klimkiewicz, M.K., Brawn, J.D., 1990. Survival rates of birds of tropical and temperate forests: will the dogma survive? Am. Nat. 136, 277–291.

Kokko, H., Jennions, M.D., 2008. Parental investment, sexual selection and sex ratios. J. Evol. Biol. 21, 919–948.

Koteja, P., 2004. The evolution of concepts on the evolution of endothermy in birds and mammals. Physiol. Biochem. Zool. 77, 1043–1050.

Lill, A., 1974. Sexual behavior of the lek-forming white-bearded manakin (*Manacus manacus trinitatis* Hartert). Zeitschrift Tierpsychologie 36, 1–36.

Lill, A., 1976. Lek behavior in the golden-headed manakin, *Pipra erythrocephala* in Trinidad (West Indies). Zeitschrift Tierpsychologie 18, 1–84.

Lochmiller, R.L., Deerenberg, C., 2000. Trade-offs in evolutionary immunology: just what is the cost of immunity? Oikos 88, 87–98.

Lomholt, J.P., Johansen, K., Maloiy, G.M.O., 1975. Is the aestivating lungfish the first vertebrate with suctional breathing? Nature 257, 787–788.

Machado, G., Macías-Ordóñez, R., 2007. Reproduction. In: Pinto da Rocha, R., Machado, G., Giribet, G. (Eds.), Harvestmen: the Biology of Opiliones, Harvard University Press, Cambridge, pp. 414–454.

Magrath, M.J.L., Komdeur, J., 2003. Is male care compromised by additional mating opportunity? Trends Ecol. Evol. 18, 424–430.

Martin, T.E., 1996. Life history evolution in tropical and south temperate birds: what do we really know? J. Avian Biol. 27, 4.

Martin, T.E., Møller, A.P., Merino, S., Clobert, J., 2001. Does clutch size evolve in response to parasites and immunocompetence? Proc. Natl. Acad. Sci. U. S. A. 98, 2071–2076.

Martin, T.E., Bassar, R.D., Bassar, S.K., Fontaine, J.J., Lloyd, P., Mathewson, H.A., Niklison, A.M., Chalfoun, A., 2006. Life-history and ecological correlates of geographic variation in egg and clutch mass among passerine species. Evolution 60, 390–398.

Martins, E.G., Bonato, V., Silva, C.Q., dos Reis, S.F., 2006. Seasonality in reproduction, age structure and density of the gracile mouse opossum *Gracilinanus microtarsus* (Marsupialia: Didelphidae) in a Brazilian cerrado. J. Trop. Ecol. 22, 461–468.

Møller, A.P., 1990. Parasites and sexual selection: current status of the Hamilton and Zuk hypothesis. J. Evol. Biol. 3, 319–328.

Møller, A.P., 1998. Evidence of larger impact of parasites on hosts in the tropics: investment in immune function within and outside the tropics. Oikos 82, 265–270.

Morand, S., Poulin, R., 1998. Density, body mass and parasite species richness of terrestrial mammals. Evol. Ecol. 12, 717–727.

Moya-Loraño, J., El-Sayyid, M.E.T., Fox, C.W., 2007. Smaller beetles are better scramble competitors at cooler temperatures. Biol. Lett. 3, 475–478.

Moya-Loraño, J., 2010. Can temperature and water availability contribute to the maintenance of latitudinal diversity by increasing the rate of biotic interactions? Open Ecol. J. 3, 1–13.

Nijhout, H.F., Emlen, D.J., 1998. Competition among body parts in the development and evolution of insect morphology. Proc. Natl. Acad. Sci. U. S. A. 95, 3685–3689.

Nudds, R.L., Oswald, S.A., 2007. An interspecific test of Allen's rule: evolutionary implications for endothermic species. Evolution 61, 2839–2848.

Olson, D.M., Dinerstein, E., Wikramanayake, E.D., Burgess, N.D., Powell, G.V.N., Underwood, E.C., D'Amico, J.A., Itoua, I., Strand, H.E., Morrison, J.C., Loucks, C.J., Allnutt, F., T., Ricketts, T.H., Kura, Y., Lamoreux, J.F., Wettengel, W.W., Hedao, P., Kassem, K.R., 2001. Terrestrial ecoregions of the world: a new map of life on Earth. BioScience 51, 933–938.

Olsson, M., Wapstra, E., Schwartz, T., Madsen, T., Ujvari, B., Uller, T., 2010. In hot pursuit: fluctuating mating system and sexual selection in sand lizards. Evolution 65, 574–583.

Oring, L.W., 1982. Avian mating systems. In: Farner, D.S., King, J.R., Parkes, K.C. (Eds.), Avian Biology, Academic Press, New York, pp. 1–92.

Peel, M.C., Finlayson, B.L., McMahon, T.A., 2007. Updated world map of the Köppen–Geiger climate classification. Hydrol. Earth Syst. Sci. 4, 439–473.

Poulin, B., Lefebvre, G., McNeil, R., 1992. Tropical avian phenology in relation to abundance and exploitation of food resources. Ecology 73, 2295–2309.

Pruett-Jones, S., 2004. Summary. In: Koenig, W.D., Dickinson, J.L. (Eds.), Ecology and Evolution of Cooperative Breeding in Birds. Cambridge University Press, Cambridge, pp. 228–238.

Randall, D., Burggren, W., French, K., 1997. Eckert – Animal Physiology. Mechanisms and Adaptations. W. H. Freeman and Company, New York.

Rappole, J.H., 1995. The Ecology of Migrant Birds: a Neotropical Perspective. Smithsonian Institution Press, Washington.

Ricklefs, R.E., 1969. An Analysis of Nesting Mortality in Birds. Smithsonian Contributions to Zoology 9, 1–48.

Ricklefs, R.E., Schluter, D., 1993. Species Diversity in Ecological Communities – Historical and Geographical Perspectives. University of Chicago Press, Chicago.

Ricklefs, R.E., Wikelski, M., 2002. The physiology/life-history nexus. Trends Ecol. Evol. 17, 462–468.

Ritchie, M.G., 2007. Sexual selection and speciation. Annu. Rev. Ecol. Syst. 38, 79–102.

Roulin, A., Wink, M., Salamin, N., 2009. Selection on a eumelanic ornament is stronger in the tropics than in temperate zones in the worldwide-distributed barn owl. J. Evol. Biol. 22, 345–354.

Saenz, D., Fitzgerald, L.A., Baum, K.A., Conner, R.N., 2006. Abiotic correlates of anuran calling phenology: the importance of rain, temperature, and season. Herpetological Monographs 20, 64–82.

Sanways, M.J., 2005. Insect Diversity Conservation. Cambridge University Press, Cambridge.

Schalk, G., Forbes, M.R., 1997. Male biases in parasitism of mammals: effects of study type, host age, and parasite taxon. Oikos 78, 67–74.

Schemske, D.W., 2002. Ecological and evolutionary perspectives on the origins of tropical diversity. In: Chazdon, R.L., Whitmore, T.C. (Eds.), Foundations of Tropical Forest Biology, University of Chicago Press, Chicago, pp. 163–173.

Schemske, D.W., 2009. Biotic interactions and speciation in the tropics. In: Butlin, R.K., Bridle, J.R., Schluter, D. (Eds.), Speciation and Patterns of Diversity, Cambridge University Press, Cambridge, pp. 219–239.

Schemske, D.W., Mittelbach, G.G., Cornell, H.V., Sobel, J.M., Roy, K., 2009. Is there a latitudinal gradient in the importance of biotic interactions? Annu. Rev. Ecol. Syst. 40, 245–269.

Seddon, N., Merrill, R.M., Tobias, J.A., 2008. Sexually selected traits predict patterns of species richness in a diverse clade of suboscine birds. Am. Nat. 171, 620–631.

Shuster, S.M., Wade, M.J., 2003. Mating systems and strategies. Princeton University Press, Princeton.

Simmons, L.W., 2001. Sperm Competition and Its Evolutionary Consequences in the Insects. Princeton University Press, Princeton.

Sindaco, R., Jeremčenko, V.K., 2008. The Reptiles of the Western Palearctic: Annotated Checklist and Distributional Atlas of the Turtles, Crocodiles, Amphisbaenians and Lizards of Europe, North Africa, Middle East and Central Asia. Edizioni, Belvedere.

Skutch, A.F., 1949. Do tropical birds rear as many young as they can nourish? Ibis 91, 430–455.

Skutch, A.F., 1985. Clutch size, nesting success, and predation on nests of neotropical birds, reviewed. Ornithological Monographs 36, 575–594.

Slater, P.J.B., Mann, N.I., 2004. Why do the females of many bird species sing in the tropics? J. Avian Biol. 35, 289–294.

Snow, D.W., 1962. A field study of the black and white manakin, *Manacus manacus*, in Trinidad. Zoologica 47, 65–104.

Solomon, N.G., French, J.A., 1997. Cooperative Breeding in Mammals. Cambridge University Press, Cambridge.

Stearns, S.C., 1992. The Evolution of Life Histories. Oxford University Press, Oxford.

Stevens, G.C., 1989. The latitudinal gradient in geographical range: how so many species coexist in the tropics. Am. Nat. 133, 240–256.

Stutchbury, B.J.M., Morton, E.S., 2001. Behavioral Ecology of Tropical Birds. Academic Press, London.

Stutchbury, B.J.M., Neudorf, D.L., 1997. Female control breeding synchrony, and the evolution of extra-pair mating systems. Ornithological Monographs 49, 103–121.

Taborsky, M., Oliveira, R.F., Brockmann, H.J., 2008. The evolution of alternative reproductive tactics: concepts and questions. In: Oliveira, R.F., Taborsky, M., Brockmann, H.J. (Eds.), Alternative Reproductive Tactics: an Integrative Approach. Cambridge University Press, Cambridge, pp. 1–21.

Takhtajan, A., 1969. Flowering Plants: Origin and Dispersal. Oliver & Boyd, Edinburgh.

Tallamy, D.W., Brown, W.P., 1999. Semelparity and the evolution of maternal care in insects. Anim. Behav. 57, 727–730.

Tallamy, D.W., Schaefer, C., 1997. Maternal care in the Hemiptera: ancestry, alternatives, and current adaptive value. In: Choe, J.C., Crespi, B.J. (Eds.), The Evolution of Social Behavior in Insects and Arachnids, Cambridge University Press, Cambridge, pp. 94–115.

Thomas, L.K., Manica, A., 2003. Filial cannibalism in an assassin bug. Anim. Behav. 66, 205–210.

Thornhill, R., Alcock, J., 1983. The Evolution of Insect Mating Systems. Harvard University Press, Cambridge.

Trivers, R.L., 1972. Parental investment and sexual selection. In: Campbell, B. (Ed.), Sexual Selection and the Descent of Man, Aldine, Chicago, pp. 136–179.

Twiss, S.D., Thomas, C., Poland, V., Graves, J.A., Pomeroy, P., 2007. The impact of climatic variation on the opportunity for sexual selection. Biol. Lett. 3, 12–15.

Udvardy, M.D., 1975. A Classification of the Biogeographical Provinces of the World. International Union for Conservation of Nature and Natural Resources, Morges.

Uller, T., Olsson, M., 2008. Multiple paternity in reptiles: patterns and processes. Mol. Ecol. 17, 2566–2580.

Van Schaik, C.P., 1983. Why are diurnal primates living in groups? Behaviour 87, 120–144.

Visser, M.E., Holleman, L.J.M., Gienapp, P., 2006. Shifts in caterpillar biomass phenology due to climate change and its impact on the breeding biology of an insectivorous bird. Oecologia 147, 164–172.

Vitt, L.J., Pianka, E.R., Cooper, W.E., Schwenk, K., 2003. History and the global ecology of squamate reptiles. Am. Nat. 162, 44–60.

Weatherhead, P.J., Yezerinac, S.M., 1998. Breeding synchrony and extra-pair matings in birds. Behav. Ecol. Sociobiol. 40, 151–158.

Wikelski, M., Hau, M., Wingfield, J.C., 2000. Seasonality of reproduction in a neotropical rain forest bird. Ecology 81, 2458–2472.

Willmer, P., Stone, G., Johnston, I., 2000. Environmental Physiology of Animals. Blackwell Publishing, Boston.

Withers, P., 1995. Cocoon formation and structure in the aestivating Australian desert frogs, *Neobatrachus* and *Cyclorana*. Aust. J. Zool. 43, 429–441.

Yom-Tov, Y., McCleery, R., Purchase, D., 1992. The survival rate of Australian passerines. Ibis 134, 374–379.

Yu-shan, W., Zu-wang, W., De-hua, W., Zhi-bin, Z., 2003. Evolution of endothermy in animals: a review. Zoolog. Res. 24, 480–487.

Zachariassen, K.E., 1991. The water relations of overwintering insects. In: Lee, R.E., Denlinger, D.L. (Eds.), Insects at Low Temperatures. Chapman and Hall, New York, pp. 47–63.

Chapter 2

Sexual Selection in Neotropical Species: Rules and Exceptions

Alfredo V. Peretti

Instituto de Diversidad y Ecología Animal, Laboratorio de Biología Reproductiva y Evolución, Universidad Nacional de Córdoba, Argentina

INTRODUCTION

Science is a changing system in which ideas, hypotheses, theories, and paradigms are in constant change. Science is inexorably opposed to static systems, such as dogmas, that remain unchangeable in their aspects. Regardless of the epistemological schools, the scientific community usually agrees that changes occur, but at different rates depending upon the area of knowledge. The analysis of the details in the construction of a paradigm, its acceptance during a certain period of time, and the crisis resulting from its eventual fall, constitute a fluid cycle that has already been exemplified in various research fields, and from different perspectives (Kuhn, 1996; Bussard, 2005). Indeed, the history of science is full of examples of paradigms that have been widely accepted in the past but are no longer valid nowadays. Similarly, ideas currently accepted were rejected or even unimagined in the past. In evolutionary biology, for instance, sexual selection was not completely accepted until a few decades ago, remaining almost unknown for a long period after Darwin formulated its theoretical framework (Gould and Gould, 1989; Andersson, 1994).

The shift from "not acceptable" to "acceptable" (and vice versa) for a scientific concept may happen for multiple reasons, some of which we will examine in this chapter. In this context, it is interesting to contemplate two opposing situations regarding what is thought about a particular fact or concept at a certain time within a scientific discipline. In other words, it may be insightful to consider the existence of rules and exceptions. But what is a rule? There is no single definition, but it can be summarized as "a generalized statement that describes what is true in most or all cases" (*The American Heritage® Dictionary of the English Language*, 2003). A term associated with a rule is "scientific law", which is a scientific statement that claims a constant relationship between two or more variables or factors, each of which represents (at least partially and indirectly) a property of concrete systems (Honderich, 1995). Typically, the

Sexual Selection. http://dx.doi.org/10.1016/B978-0-12-416028-6.00002-5

terms rules, norms, and laws are used as mere synonyms. Indeed, in natural sciences a scientific law is a rule that links events that have a joint, usually causal, occurrence. Rules are very valuable because they are key parts in the construction of scientific theories. The enunciation of rules in biology (e.g., Mendelian laws in genetics), however, is not frequent when compared to other areas of science, such as physics and chemistry (see discussion in Dhar and Giuliani, 2010).

But within the context of scientific rules, what is an exception? Useful definitions are: (1) "a case that does not conform to a rule or generalization" (*The American Heritage Dictionary of the English Language*, 2003), and (2) "anything excluded from or not in conformance with a general rule, principle, class, etc." (*Collins English Dictionary – Complete and Unabridged*, 2003). Therefore, *rule* and *exception* are closely related concepts that scientists have in mind, formally or intuitively, during their research routine. Typically, however, there is not a unique type of exception, but several – in other words, gradients or levels to which the word "exception" applies. In biology, this diversity of meanings is clearly evident. For example, cases of aberrant morphologies illustrate the extreme concept of the term. The study of such extreme morphologies led to teratology, a scientific discipline that, within zoology, studies abnormal physiological development in diverse creatures, "such as those individuals in a species which do not respond to the common pattern" (Dicke, 1989). Thus, the last part of Dicke's explanation is almost synonymous with the definition of exception. Teratological observations have proved to be of great value in understanding particular issues in developmental biology, as shown in an extraordinary manner by Mark Blumberg in his book *Freaks of Nature* (Blumberg, 2009). Nevertheless, the extreme expression of a trait usually does not imply teratology, but rather part of the natural variability within the normal range of a particular phenotype. Given their usefulness for a better understanding of adaptive and developmental processes, greater consideration should be given to both teratological cases and those at the extremes in the distribution of a trait. Identifying phenomena that constitute general or specific rules has been the pattern of study in most areas, including sexual selection, while identifying those cases that represent exceptions has been mostly neglected.

INTERACTIVE LINK BETWEEN RULES AND EXCEPTIONS

From a historical perspective, the fluid connection between rules and exceptions has been more or less evident depending on the field of knowledge. At least four situations can occur in this changing system: (1) rules of the past that persist to the present; (2) exceptions of the past that persist to the present; (3) rules of the past that are now considered exceptions; and (4) exceptions of the past that are now considered rules. Although there are variations to these options, they serve as guidelines for examining a number of issues. Many fields in evolutionary biology, such as animal behavior and sexual selection, offer good examples of such fluid connection between rules and exceptions.

Below I discuss the conceptual framework for these four basic scenarios, which are summarized and exemplified in Table 2.1. In the next section, the

TABLE 2.1 Schematic Diagram of the Time Dynamics between Rules and Exceptions

Situation	Historical Moment		Main Characteristics	Examples	
	Past	Present		General	Scorpion Sexual Biology
1	RULE	RULE	Generalized or limited. Flexible, not fixed	Eco-geographic rules	Spermatophore-mediated sperm transfer. Sexual stimulation by cheliceral massage
2	EXCEPTION	EXCEPTION	Subestimated? Important for adaptation. Could it become a rule in some contexts?	Teratologies. Conditional alternative strategies (e.g., sneaker males)	Teratological males that court
3	RULE	EXCEPTION	Based on few data. Insufficient technology. Influence of prevailing paradigm	Monogamy *sensu stricto*	Post-mating sexual cannibalism
4	EXCEPTION	RULE	Based on few data. Insufficient technology. Influence of prevailing paradigm	Learning capacity of arthropods	Copulatory courtship. Sexual sting

interactive link between rules and exceptions will be illustrated with specific examples on reproductive biology of scorpions, a group in which I conducted much of my research. Finally, to broaden the scope of this chapter, some examples of neotropical organisms discussed by other authors in this book will be added to the conceptual framework.

Situation "Rule–Rule"

This is a conservative situation, and the most likely to evolve into dogmatism, but also provides the relatively "stable" principles over time allowing biological facts to be understood and interpreted. It is important to mention that the status "past" can refer to different time scales, ranging from decades or more in some cases to just a couple of years or months (or less) in other cases. The situation "present", however, invariably refers to what is currently considered relative to a specific issue.

A rule may be general or restricted. The problem with any rule, especially in biology, is that it inevitably should be taken as a principle or norm with some degree of flexibility. Otherwise, the first exception found would justify the rejection of the rule as inadequate, when realistically it should be restricted relative to its influence to include additional cases. Although biological rules (from ecology, genetics, paleontology, etc.) provide general explanations for a diversity of biological phenomena, many "exceptions to the rule" have continuously been found over time. For example, Allen's classic rule, which states that endotherms from colder climates usually have shorter limbs compared with endotherms from warmer climates, was criticized for some taxa and/or regions (Serrat *et al.*, 2008; Alho *et al.*, 2011). Similarly, Cope's rule, which states that population lineages tend to increase in body size over evolutionary time, does not hold true for all taxonomic groups (Hone and Benton, 2005). However, the important point to consider is that new information does not necessarily mean that such rules are wrong, but some of the original statements should be revisited and discussed more carefully according to specific organisms and/or regions (Hone *et al.*, 2008; Alho *et al.*, 2011).

Situation "Exception–Exception"

From the fine work of William Bateson in developmental genetics to the creative review of Blumberg (2009), the utility of teratology for biological sciences has been rescued from the bag of "not useful". Indeed, teratology provides a better understanding of what is considered "normal" in biology. No words illustrate this better than William Bateson's famous quotation, cited in Cock and Forsdyke's (2008) biography on Bateson, which summarizes this central point:

Nevertheless, if I may throw out a word of council to beginners, it is: Treasure your exceptions! When there are none, the work gets so dull that no one cares to carry it further. Keep them always uncovered and in sight. Exceptions are like the rough brickwork of a growing building which tells that there is more to come and shows where the next construction is to be.

(Bateson, 1908)

The discovery of a teratological feature may shed some light or obscure facts, depending on its interpretation. A classical example in sexual selection was the study of killer sperm (kamikaze sperm hypothesis; Baker and Bellis, 1988), in which the rare event of random sperm agglutination in the female's genital tract was incorrectly interpreted as a phenomenon with adaptive value, i.e., a male strategy for sperm competition (Moore *et al.*, 1999).

Certainly, an exception is not synonymous with teratology. Perhaps much more common for those working in biology is that the majority of rare observations are merely interpreted as outliers in their statistical analyses. Common definitions are: "an outlying observation, or outlier, is one that appears to deviate markedly from other members of the sample in which it occurs" (Grubbs, 1969); alternatively, "an outlier is an observation that lies outside the overall pattern of a distribution" (Moore and McCabe, 1999). Interestingly, the existence of an outlier may also imply some sort of problem – for example, a case that does not fit the model under study, or perhaps a measurement error. A formal statistical definition of an outlier is a point that falls more than 1.5 times beyond the interquartile range, either above the third or below the first quartile (Renze, 2011). If the outlier results from using a scatterplot to compare the relationships between two sets of data, something quite common in biology (e.g., male weight vs paternity), we tend to remove these outliers when running a "least squares fitting to data" technique because they may strongly influence the outcome during calculation of the best fitting line (Renze, 2011).

Although these definitions and tips are useful from a statistical perspective, it is clear that if the outliers are not measurement errors, we should not view them as a problem or just remove them when calculating best fitting models. An evolutionary biologist should rescue these points from graphs and try to examine them in light of other associated characteristics of the individual, encompassing interpretations somehow within the scope of the study. Otherwise, valuable information is lost when this could be useful to other researchers who encounter similar situations in their own studies. In any case, extreme values of a given trait within the natural distribution of a population, whether or not considered as outliers *sensu stricto*, are potentially valuable from adaptive perspectives. This was perceived early in the history of evolutionary biology by Charles Darwin in his classic studies on variation and selection forces (Darwin, 1859, 1868, 1871), as well as by William Bateson in his analyses of non-teratological variability during development (Cock and Forsdyke, 2008; Peterson, 2008). Traditionally, evolutionary biology has shown that some of the "rare" expressions of a trait or set of traits may be affected by directional selection in certain contexts (Westneat and Fox, 2010). Therefore, what yesterday was the tail-end of the distribution may well be the "average" value of a certain trait today, and vice versa. Indeed, this dynamism is the essence of both natural and sexual selection.

Another scenario is the existence of minorities within a certain species that represent clearly defined alternative tactics. In sexual selection the example of sneaker males is very instructive, since until a few decades ago this phenomenon was completely ignored. This changed with the strengthening of the sexual conflict hypotheses in which sneaking tactics were more clearly recognized and interpreted within an evolutionary framework (Gross, 1982, 1991). In general, these tactics are employed by a minor part of the male population, as is the case for some fish (e.g., salmons, guppies) and invertebrates (e.g., water-striders, dragonflies), among others (Oliveira *et al.*, 2008). Typically, alternative tactics are generally discovered after the "main" pattern of the species has already been described. However, a serious mistake is to overemphasize a minor alternative tactic, which could erroneously lead to it being interpreted as the most relevant for a particular mating system (see discussions in Peretti and Córdoba-Aguilar, 2007; Peretti and Aisenberg, 2011).

Teratological observations, outliers, and alternative mating tactics primarily refer to an intraspecific level of analysis. However, the term is also used for interspecific comparisons within a clade, in which the emphasis relies on particular differences of a species relative to related taxa. In sexual selection, a typical example is the presence of sex-role reversal in species of a particular genus or family, such as has been observed among some spiders, fish, and birds (Berglund and Rosenqvist, 2003; Aisenberg *et al.*, 2007; Maurer *et al.*, 2011).

Situation "Rule–Exception"

Three aspects may be associated with the change in the historical consideration of a rule: (1) insufficient technological resources at the time the rule was proposed; (2) rules originally based on partial data or unrepresentative information for most of the group; and (3) strong theoretical bias during experimental design and/or interpretation of results. The lack of advanced technology for addressing specific issues is probably easy to envision as a common factor in situations where rules have become exceptions. The second aspect, i.e., rules established from limited data, is far from being rare. A central question here is: how much information should be gathered before explicitly or implicitly establishing a rule? This is a crucial detail, and it is where studies including a great diversity of organisms, rather than focusing on single "model" species, can make a big difference. The partial and/or unrepresentative data problem would also decrease if biological data from many world regions were to be included. The fact is that many biological rules were developed using model species of the Nearctic region – probably for the obvious reasons associated with the proximity of traditional scientific centers in this region. Understandably, this implies an inherent and involuntary bias, but it has resulted in the application of ideas and explanations that emerged based on studies of organisms (populations or species) from regions outside the Neotropics. As this book intends to show, this has proved to be a mistake in some cases.

In this context, the transition rule–exception could be related to two different and non-mutually exclusive processes: geographic and phylogenetic inadequacies. Geographic (intraspecific) inadequacy occurs when we extend the study area to unexplored regions for a certain species and we find that a traditional rule is an exception in new places – for example, if we find that populations inhabiting distinct elevational and/or climatic zones exhibit eco-behavioral patterns that are very different from those previously described for "the species". Phylogenetic (interspecific) inadequacy is probably the most common issue determining the transition rule–exception. This occurs when we find that by increasing the lineage diversity in our studies a rule originally applied to some groups is an exception for most of the groups. Of course, both limitations may occur simultaneously in many cases, a fact that may obscure their individual influences. Nevertheless, this discrimination could be useful when analyzing and discussing emergent patterns. Some taxonomic-oriented chapters of this book will focus on the importance of these effects in certain groups.

A third aspect leading to historical changes of rules may include biases imposed by a prevailing paradigm. It is always difficult to interpret data outside the theoretical framework in which we are directly or indirectly immersed. Scientific "inertia" may occur if one only pursues established ideas without exercising critical thinking. In contrast, diverging either wholly or at least partially from traditional explanations is a challenge that involves a great deal of courage and creativity. Emblematic examples in sexual selection include a step-by-step analysis of traditional hypotheses on the evolution of animal genitalia by Eberhard (1985), and the recent critical discussion on traditional thinking concerning sex roles, parental investment, and sex ratios by Kokko and Jennions (2008).

Many studies on sexual selection in the Neotropics are faced with the challenge of breaking with preconceived ideas. It is uncertain, however, whether the theory will change much as a result of new information from the Neotropics. Possibly, the most general theories work well anywhere in the world. What does change is perhaps the knowledge of new patterns. For example, neotropical species may exhibit patterns or strategies that differ from temperate species, usually studied as "model organisms". In other words, different climate zones involve changes in patterns, but not necessarily a change in theory (see Chapter 1).

Inadequate technology, limited information, and geographical, taxonomical, or theoretical bias are all variables that may act alone or in combination to erroneously transform an observation into a widespread rule or norm when it should be more restricted than originally supposed. In sexual selection, the study of monogamy in birds illustrates this point quite well (Petrie and Kempenaers, 1998; Bennett and Owens, 2002). Until a few decades ago, monogamy was a type of mating system widely accepted and unquestioned for more than 90% of all bird species (Lack, 1968). Progressively, genetic monogamy was restricted to certain groups, among which seagulls were always presented as the best example. With the advance of molecular techniques in behavioral ecology, however, it was confirmed that this rule did not work in the strictest sense (i.e., 100% of paternity for a single male) for either seagulls or other seabirds (e.g., Birkhead, 2000; Ismar *et al.*, 2010). Indeed, studies suggest that fewer than 25% of birds appear to be genetically monogamous (reviewed in Griffith *et al.*, 2002). Therefore, in this example, a rule of the past is now considered an exception (monogamy) and, simultaneously, an exception of the past is now considered a rule (extra-pair copulations). In the next topic, I will focus on this latter situation.

Situation "Exception–Rule"

The factors mentioned above for the change "from rule to exception" also apply to the inverse situation. Few data, insufficient technology, and/or strong influence of a paradigm can lead to the omission of elements that could indicate that observed patterns do not represent rare phenomena. An example is the ability of arthropods to learn – a capacity historically underestimated by most behavioral biologists, with the exception of a few cases (e.g., Thorpe, 1943; Monteith, 1963). However, several studies have shown that crustaceans, spiders, and insects can modify their behavior according to previous experiences (e.g., Reaka, 1980; Jackson and Pollard, 1996; Dukas, 1998, 2008). The influence of experience is not restricted to simple tasks, but also involves complex behaviors, such as predatory and sexual behavioral patterns (e.g., Cronin et al., 2006; Dukas et al., 2006; Jakob et al., 2011). Another example is the discovery of "personalities" in many animal species, a fact that appears to be a common feature (Sih et al., 2004; but see also Schuett et al., 2010; Morgan and Cézilly, 2011). Of course, as previously mentioned, the changing of an exception into a rule does not imply that the latter has to be generalized to all organisms or regions. However, as exemplified by the case of the demonstrated capacity of learning among arthropods, this does mean that researchers should consider this novel information in new experimental designs.

AN EXAMPLE FROM THE NEOTROPICS: SCORPION SEXUAL BIOLOGY

In general, studies on the reproductive biology of scorpions have been historically very diverse and curious, in part because many of them resulted from miscellaneous observations conducted by researchers mainly focused on the systematics of the group. Most of the data on reproductive behavior lacked accuracy and details. For this reason, researchers from other disciplines have often misinterpreted certain aspects of scorpion reproduction when using them as examples for certain arguments. Below, I briefly summarize these interesting situations, following the order shown in Table 2.1.

"Rule–Rule": On Spermatophores and Cheliceral Massages

Perhaps the most classic example of the rule–rule situation is that all scorpions possess an indirect sperm transfer mechanism by means of a sclerotized spermatophore that the male deposits on the substrate (Fig. 2.1). The male then induces the female to pick up the spermatophore, and the partners separate. This is the typical rule relative to the reproductive biology of scorpions (Polis and Sissom, 1990; Peretti, 2010). Indeed, there is always a spermatophore, formed by two halves,

FIGURE 2.1   Example of universal rule in scorpions: indirect sperm transfer by spermatophore with temporary pair formation. A male (right) of chaquean bothriurid scorpion, *Timogenes elegans*, is depositing a sclerotized spermatophore (arrow) on the soil. *Photograph courtesy of D. Vrech.*

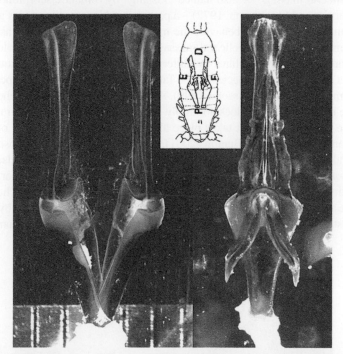

FIGURE 2.2   Scorpion spermatophore (right) is formed by two almost identical parts, named hemispermatophores (left), that occupy a great portion of the male's body. Example: the pampean scorpion, *Bothriurus bonariensis*. Abbreviations: posterior (P), dorsal (D), and external (E) portions of the hemispermatophores in the male. *Photograph courtesy of D. Vrech.*

inside the male's reproductive tract (Fig. 2.2). Of course, there are variants of spermatophores, from relatively simple (e.g., flagelliform in the family Buthidae) to more complex (e.g., lamelliform in the families Vaejovidae and Bothriuridae).

Surprisingly, it was not until the middle of the 20th century that this rule was established. One late and curious idea suggested that the male might use the two hemispermatophores (the component parts of a spermatophore) as "temporal penises" during copulation, removing them immediately after sperm transfer ended (Bücherl, 1956). However, this and other speculative explanations were rejected by studies carried out independently in different regions of the world (Uruguay: Zolessi, 1956; South Africa: Alexander, 1957; Germany: Angerman, 1957; Israel: Shulov and Amitai, 1958) which described the basic structure of the scorpion spermatophore as well as its functional variants. This pattern has continued to be confirmed in further studies on other species and/or regions (Maury, 1968; Francke, 1979; Peretti, 2010). Currently, it is certain that, independent of spermatophore type, the general events of production and functioning are quite similar for all scorpions. If in the future we were to find that some scorpions do not use spermatophores, we would be facing an exception that could overturn the rule.

Another example of the rule–rule situation, in this case of sexual behavior, is the "cheliceral massage" – also named "kissing" by romantic scorpionologists (e.g., Garnier and Stockmann, 1972) – a pattern used by courting males to stimulate females. During this behavior the male rubs his chelicerae against those of the female (Fig. 2.3). Cheliceral massage, together with pedipalp holding, is one of the most common behavioral patterns among scorpions during courtship. It occurs in all the basal species of the family Buthidae and several species of other families (Polis and Sissom, 1990). Although the occurrence of cheliceral massage is invariable, this does not imply that its performance is the same overall, since in some species it takes place almost continuously while in others it occurs only sporadically (Benton, 2001). Exceptionally, complete absence of cheliceral massage has been observed in some species of Bothriuridae (e.g., *Bothriurus chacoensis*) in which males use other behavioral patterns for sexual stimulation (Peretti, 1993). This exception does not mean rejecting cheliceral massage as a "common" behavioral pattern, as it remains widely used by males of almost all species. This is an example of a flexible rule, with variants and

FIGURE 2.3 **Male (right) of the chaquean bothriurid scorpion *Timogenes elegans*, performing cheliceral massages to the female.** *Photograph courtesy of D. Vrech.*

even occasional exceptions for certain species. The point here, however, is to identify the exceptions to limit the scope of the rule, and perhaps to contextualize the evolutionary and ecological basis for such exceptions.

"Exception–Exception": On Males that Want but Can't

Among scorpions, teratological cases have occupied a small but almost permanent place in the studies of the group. Developmental anomalies are well known, the most common being the duplication of various posterior body segments, including specimens with two telsons or deformities of the pedipalps (Vachon, 1952; Polis and Sissom, 1990). In the bothriurid *Brachistosternus pentheri*, a hermaphroditic individual was found with embryos as well as hemispermatophores (Maury, 1983). There is no knowledge, however, about how these teratologies are associated with behavior – a typical lack of information not only for scorpions but also for most animal groups. Some years ago, however, I conducted behavioral observations directly for two teratological cases (Peretti, 2000). Two males were observed normally courting females in the field despite them having just one of the two hemispermatophores. However, functional mating stopped at the moment of spermatophore deposition, although the males persisted behaviorally with intense body movements (e.g., shaking telson) in order to expel the spermatophore from the genital aperture, an act that never occurred. When examining the gonopore of these specimens, I found that a segment of the basal part (foot) of the spermatophore was emerging from the genital opening. These males were then completely dissected and were found to lack not only one paraxial organ, but also the seminal vesicle, accessory glands, and testis branch on the same side (a normal reproductive system is shown in Fig. 2.4). This

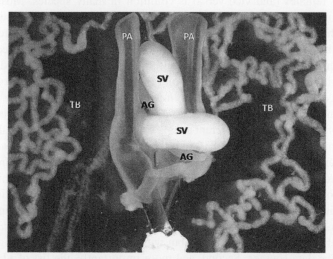

FIGURE 2.4   Male reproductive system of the pampean scorpion *Bothriurus bonariensis*. The system is composed of two identical sides, each of which presents a paraxial organ (PA) (with a hemispermatophore), a seminal vesicle (SV), accessory glands (AG), and a testis branch (TB). *Photograph courtesy of D. Vrech.*

anomaly was useful to show that in this scorpion there is physiological independence between predisposition to court and complete development of the reproductive system. Anecdotally, it is interesting to note that publication of these observations (even as a research note) was a very arduous task because referees insisted on arguing that such data had no value because it was "just a teratology".

"Rule–Exception": Facts and Fiction on Sexual Cannibalism

Perhaps the most well-known situation in scorpions regarding the change from rule to exception is the finding that post-copulatory sexual cannibalism does not exist in the group. Sexual cannibalism is a very attractive feature for behavioral biologists because of its ecological and evolutionary implications. Among arachnids, both pre- and post-copulatory sexual cannibalism have been analyzed in detail and demonstrated in some spiders. For example, pre-copulatory sexual cannibalism has been described in the sand-dwelling wolf-spider *Allocosa brasiliensis*, a species with sex role reversal exhibiting choosy males and competitive, courting females (Aisenberg *et al.*, 2007; see also Chapter 6). In this spider, males frequently attack and eat non-virgin courting females. Post-copulatory cannibalism has its classic example in redback spiders (Andrade, 1996). In this group, occurrence of this type of cannibalism is generally based upon low remating opportunities for males (Andrade and Banta, 2002). Above all, males increase paternity by offering their body as food to inseminated females.

Cannibalism is not a rare phenomenon in scorpions (Polis and Sissom,1990). Many observations in the field and in captivity have shown that cannibalism of males by females (and vice versa) is common in many species. Most of the descriptions of mating behavior in the group were made in captivity, mainly by placing the male and female together in small containers. Post-copulatory cannibalism has been observed during the same day of mating or, more frequently, in subsequent days, and in the same captivity situation, with male and female crowded in the same container. The point is that "real" post-copulatory sexual cannibalism, by definition, must occur immediately after sperm transfer. Attacks at other moments are merely cases of ordinary cannibalism (e.g., between individuals of the same sex or different ages). Therefore, it is pointless to talk about "post-copulatory" sexual cannibalism if the individuals cannot move away from each other when mating is finished. Otherwise, the intersexual encounters are being forced and this artificial situation obviously favors attacks, which also occur in other arachnids such as solpugids and many spiders. This misinterpretation gained importance on two occasions. First, the term sexual cannibalism was included in a chapter on life history in the classic book *The Biology of Scorpions* (Polis and Sissom, 1990), but without highlighting the strong limitations of the data sources. Second, and more importantly, the list of 11 species from four families provided in the book was later used as a notable example to illustrate post-copulatory sexual cannibalism in animals (Elgar, 1992).

The mistake in such descriptions was detected by reviewing the original papers and adding fine-scaled observations on sexual behavior of the species in that famous list (Peretti *et al.*, 1999). It was confirmed that sexual cannibalism, if present in particular species, is of the "pre-copulatory" type, whereas "post-copulatory" cannibalism could be completely discarded, at least in those species in which it had been reported. In conclusion, a very attractive rule from some evolutionary perspectives (e.g., sexual selection) was dropped due to new evidence and critical analyses of the available information.

"Exception–Rule": Subtle vs Conspicuous Displays

Two examples illustrate this situation in scorpions: the existence of copulatory courtship, and of a sexual sting. The former involves subtle male signals occurring during mating that can pass unnoticed by the observer. In contrast, the latter refers to a very striking behavioral pattern that may tempt us to speculate about its adaptive implications. Let us examine them in detail.

Copulatory courtship in scorpions exemplifies a very special situation. This particular behavioral pattern was first identified in a single species and subsequently, after looking into previous data from other species, was found to be more wide-ranging and not restricted to a particular species or group. In other words, although it is an exception that cannot yet be regarded as a general rule, copulatory courtship may well become a much broader explanation as more groups are added to the list, based upon detailed observations of their sexual behavior. The fact that courtship does not end when copulation begins, but can continue throughout and even after copulation ends, is a pattern that was first detected by Eberhard (1991) for several insects. Occurrence of copulatory courtship has been underestimated mostly because observations were originally focused on description of the more obvious movements associated with mating without including data on other more subtle movements of males (e.g., tapping, kicks, rubbing, and subtle cheliceral massage). Several studies over the past decade or so have shown that copulatory courtship increases the male's fertilization success within a context of cryptic female choice (e.g., Eberhard, 1996, 2009; Edvarsson and Arnqvist, 2000; Aisenberg and Eberhard, 2009; Peretti and Eberhard, 2010). To date, copulatory courtship has been clearly confirmed for five scorpion families (Peretti, 1997; Carrera, 2008) as a result of fine details provided by previous studies on their sexual behavior (see, for example, Polis and Farley, 1979; Benton, 1992, 1993). It is very likely that copulatory courtship occurs in many other species as well (Peretti, 1997). Further studies should confirm whether the male strategy of copulatory courtship represents a general rule for the order Scorpiones.

The sexual sting has been described as a typical male behavior among scorpion families, except for Buthidae (Polis and Sissom, 1990). The description of the sexual sting is that the male apparently punctures the female's body with the aculeous of his telson. Depending on the species, the single or multiple sexual stings occur at different moments of mating (Polis and Sissom, 1990; Benton, 2001), and the body region affected in the female varies for different species (Benton, 1993; Peretti, 1993; Carrera, 2008; Toscano-Gadea, 2010). Recently, this behavior was linked to the discovery of a surprising fact: the existence of a "prevenom" in Buthidae (Inceoglu *et al.*, 2003). Indeed, in *Parabuthus transvaalicus* the venom is composed of two fractions: a prevenom that immobilizes the prey, and another venom containing toxins that kill the prey. Although that study was focused on the chemical properties of the venom and, mainly, concerned a family in which males never perform the sexual sting (see Polis and Sissom, 1990; Peretti and Carrera, 2005), Inceoglu *et al.* (2003) speculated on the relationship between the prevenom and the sexual sting behavior. They suggested that during the sexual sting males could "drug" the female by inoculating the prevenom fraction. Given the aggressiveness shown by many scorpions, this hypothesis is logical, especially from the perspective of conflict of interest between males and females. Indeed, that idea was included and discussed even more extensively in a book on sexual conflict (Arnqvist and Rowe, 2005). Unfortunately, nobody investigated whether males actually pierced the female body – an obvious and indispensable condition to inoculate the prevenom.

These ideas have been questioned by recent observations, particularly concerning neotropical species. Detailed studies of bothriurid scorpions carried out by two researchers from the Neotropics (Carrera, 2008; Toscano-Gadea, 2010), using video analyses and detailed examination of the female's body, showed that the male just places the sting on the female body without piercing the cuticle. In addition, the same pattern occurs in a species from Europe, *Euscorpius flavicaudis* (P. Carrera and C. A. Toscano-Gadea, personal communication). Therefore, this stinging "ritual" is not restricted to neotropical groups such as the Bothriuridae. Curiously, these recent studies with neotropical scorpions initially intended to analyze how the prevenom inoculation might affect female behavior. At the outset these researchers were puzzled (P. Carrera and C. A. Toscano-Gadea, personal communication) given the fact that their observations did not conform to the previous widespread "belief". What they found is that in some species the tip of the male sting is gently positioned between the pleurae or articulated joints of the female body, which gives the impression that the female is being punctured (Fig. 2.5). Whether the ritual of the sexual sting is a widespread rule within the order Scorpiones is a key issue to be investigated in further studies.

CONCLUDING REMARKS

In this chapter, I propose a basic framework for thinking about the changes in ideas and facts supporting our interpretation concerning patterns and rarities. Of course there are feasible alternatives, since this framework may be perceived as too simple – i.e., I considered just two components (rules and exceptions), and only straightforward associations between them. Thus, additional scenarios

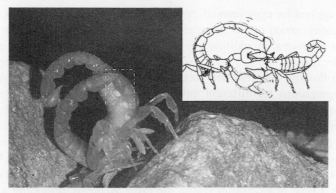

FIGURE 2.5   Examples of the sexual sting in scorpions. Photo: *Bothriurus bonariensis* male (left) inserts the aculeous in the space between two tergites without puncturing the cuticle. In *Timogenes elegans* (upper right), the sting is placed on a pedipalp. *Photograph courtesy of P. Olivero.*

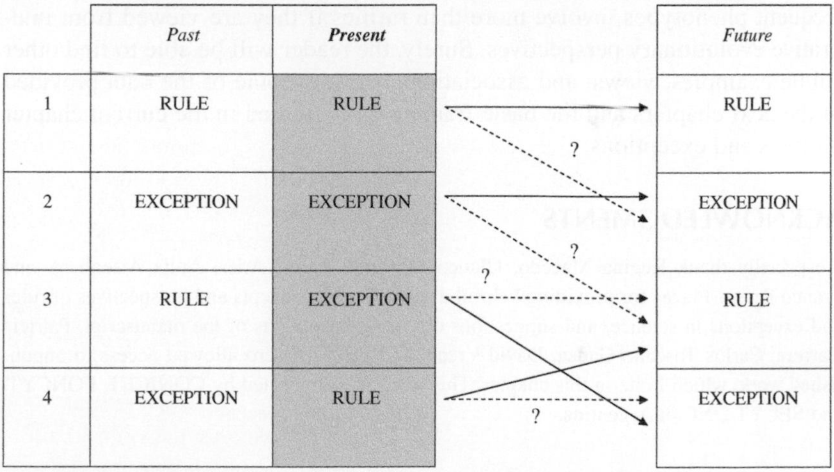

	Past	Present		Future
1	RULE	RULE		RULE
2	EXCEPTION	EXCEPTION		EXCEPTION
3	RULE	EXCEPTION		RULE
4	EXCEPTION	RULE		EXCEPTION

FIGURE 2.6   Potential further changes in the dynamism between rules and exceptions within a scientific discipline.

would be very welcome to improve this starting point. However, one fact is clear: current situations may change in the future (Fig. 2.6). In this context, a constructive habit to reinforce the dynamism between rules and exceptions would be to include more data on exceptions in our publications. For example, we could add an online appendix as supplementary material, which is potentially useful to other researchers. Currently, with all the digital and online editorial apparatus, it would be easy to find and use this kind of information, such as is currently available with databases of images in taxonomic collections or morphological surveys in different groups. Therefore, a researcher could compare any "rare" data with other similar information from those appendices. This could be applied for experimental, descriptive, and comparative studies. Surely, investigations carried out on neotropical organisms can provide keys to resolve important questions, in addition to confirming or refuting paradigms extrapolated to the region.

The following chapters will focus on this challenge, offering novel data and interpretations for different model organisms. Probably the most common situation prevailing among the chapters is the confirmation of general rules when exploring mating systems and associated contexts, such as sex role and parental care, among neotropical species. The duality between the situations rule–exception is evident in some chapters of this book. In Chapter 6, for instance, the case of a neotropical wolf-spider with sex role reversal is shown as a type of exception within the order Araneae, providing a unique opportunity for discussing pressures driving sex roles in spiders. In addition, some similarities between alternative reproductive tactics of swordtail fishes (e.g., "bourgeois" vs "parasitic" males – Chapter 9) and harvestmen ("major" vs "minor" males – Chapter 5) basically show that inconspicuous and/or less frequent phenotypes involve more than rarities if they are viewed from integrative evolutionary perspectives. Surely, the reader will be able to find other subtle examples, views, and associations between some of the data provided in the next chapters and the basic framework presented in the current chapter on rules and exceptions.

ACKNOWLEDGMENTS

I especially thank Regina Macedo, Glauco Machado, Cesar Ades, Anita Aisenberg, and Franco Peretti-Plazas for providing helpful discussions on concepts and perspectives of rules and exceptions in science, and suggestions on previous versions of the manuscript. Patricia Carrera, Carlos Toscano-Gadea, David Vrech, and Paola Olivero allowed access to unpublished work, which I cite in this chapter. This work was supported by CONICET, FONCYT, and SECYT-UNC of Argentina.

REFERENCES

Aisenberg, A., Eberhard, W.G., 2009. Possible cryptic female choice in a spider: female cooperation in making a copulatory plug depends on male copulatory courthip. Behav. Ecol. 20, 1236–1241.

Aisenberg, A., Viera, C., Costa, F.G., 2007. Daring females, devoted males and reversed sexual size dimorphism in the sand-dwelling spider *Allocosa brasiliensis* (Araneae, Lycosidae). Behav. Ecol. Sociobiol. 62, 29–35.

Alexander, A.J., 1957. The courtship and mating of the scorpion, *Opisthophthalmus latimanus*. Proc. Zool. Soc. Lond. 128, 529–544.

Alho, J.S., Herczeg, G., Laugen, A.T., Räsänen, K., Laurila, A., Merilä, J., 2011. Allen's rule revisited: quantitative genetics of extremity length in the common frog along a latitudinal gradient. J. Evol. Biol. 24, 59–70.

Andersson, M., 1994. Sexual Selection. Princeton University Press, Princeton.

Andrade, M.C.B., 1996. Sexual selection for male sacrifice in the Australian redback spider. Science 240, 70–72.

Andrade, M.C.B., Banta, E.M., 2002. Value of male remating and functional sterility. Anim. Behav. 63, 857–870.

Angermann, H., 1957. Über Verhalten, Spermatophorenbildung und Sinnesphysiologie von *Euscorpius italicus* Herbst und vertwandten Arten (Scorpiones, Chactidae). Z. Tierpsychol. 14, 276–302.

Arnqvist, G., Rowe, L., 2005. Sexual Conflict. Princeton University Press, Princeton.

Baker, R.R., Bellis, M.A., 1988. Kamikaze sperm in mammals? Anim. Behav. 36, 936–939.

Bateson, W., 1908. The Methods and Scope of Genetics: An Inaugural Lecture Delivered on 23 October 1908. Cambridge University Press, Cambridge.

Bennett, P.M., Owens, I.P.F., 2002. Evolutionary Ecology of Birds: Life Histories, Mating Systems, and Extinction. Oxford University Press, Oxford.

Benton, T.G., 1992. Courtship and mating in *Leiurus quinquestriatus*. In: Cooper, J.E., Pearce-Kelly, P., Williams, P. (Eds.), Arachnida, Scorpiones; Buthidae. Chiron, London, pp. 83–98.

Benton, T.G., 1993. The courtship behavior of the scorpion, *Euscorpius flavicaudis*. Bull. Br. Arachnol. Soc. 9, 137–141.

Benton, T.G., 2001. Reproductive biology. In: Brownell, P., Polis, G.A. (Eds.), Scorpion Biology and Research, Oxford University Press, Oxford, pp. 278–301.

Berglund, A., Rosenqvist, G., 2003. Sex role reversal in pipefish. Adv. Stud. Behav. 32, 131–167.

Birkhead, T.G., 2000. Promiscuity: An Evolutionary History of Sperm Competition. Harvard University Press, Cambridge, MA.

Blumberg, M., 2009. Freaks of Nature. What Anomalies Tell Us about Development and Evolution. Oxford University Press, New York.

Bücherl, W., 1956. Escorpiões e escorpionismo no Brasil: observações sobre o aparelho reprodutor masculino e o acasalamento de *Tityus bahiensis*. Mem. Inst. Butantan. 27, 121–155.

Bussard, A.E., 2005. A scientific revolution? EMBO Rep. 6, 691–694.

Carrera, P., 2008. Estudio comparado de los patrones de selección sexual en escorpiones y ácaros acuáticos: evaluación de hipótesis de conflicto sexual y cooperación intersexual. Doctoral Thesis, Facultad de Ciencias Exactas, Físicas y Naturales, Universidad Nacional de Córdoba, Argentina.

Cock, A.G., Forsdyke, D.R., 2008. Treasure Your Exceptions. The Science and Life of William Bateson. Springer, New York.

Collins English Dictionary – Complete and Unabridged (1991, 1994, 1998, 2000, 2003). Retrieved October 25 2011 from http://www.thefreedictionary.com/exception.

Cronin, T.W., Caldwell, L.R., Marshall, H., 2006. Learning in stomatopod crustaceans. Int. J. Comp. Psych. 19, 297–317.

Darwin, C., 1859. On the Origin of Species by Means of Natural Selection, or the Preservation of Favoured Races in the Struggle for Life. John Murray, London.

Darwin, C., 1868. The Variation of Animals and Plants under Domestication. John Murray, London.

Darwin, C., 1871. The Descent of Man, and Selection in Relation to Sex. John Murray, London.

Dhar, P.K., Giuliani, A.L., 2010. Laws of biology: why so few? Syst. Synth. Biol. 4, 7–13.

Dicke, J.M., 1989. Teratology: principles and practice. Med. Clin. North Am. 73, 567–582.

Dukas, R., 1998. Evolutionary ecology of learning. In: Dukas, R. (Ed.), Cognitive Ecology, University of Chicago Press, Chicago, pp. 129–174.

Dukas, R., 2008. Evolutionary biology of insect learning. Ann. Rev. Entomol. 53, 145–160.

Dukas, R., Clark, C.W., Abbott, K., 2006. Courtship strategies of male insects: when is learning advantageous? Anim. Behav. 72, 1395–1404.

Eberhard, W.G., 1985. Sexual Selection and Animal Genitalia. Harvard University Press, Cambridge.

Eberhard, W.G., 1991. Copulatory courtship and cryptic female choice in insects. Biol. Rev. 66, 1–31.

Eberhard, W.G., 1996. Female Control: Sexual Selection by Cryptic Female Choice. Princeton University Press, Princeton.

Eberhard, W.G., 2009. Postcopulatory sexual selection: Darwin's omission and its consequences. Proc. Nat. Acad. Sci. U. S. A. 106, 10025–10032.

Edvardsson, M., Arnqvist, G., 2000. Copulatory courtship and cryptic female choice in red flour beetles *Tribolium castaneum*. Proc. R. Soc. Lond. B. 267, 559–563.

Elgar, M.A., 1992. Sexual cannibalism in spiders and other invertebrates. In: Elgar, M.A., Crespi, B.J. (Eds.), Cannibalism. Ecology and Evolution among Diverse Taxa, Oxford University Press, Oxford, pp. 128–155.

Francke, O.F., 1979. Spermatophores of some North American scorpions (Arachnida, Scorpiones). J. Arachnol. 7, 19–32.

Garnier, G.R., Stockmann, R., 1972. Étude comparative de la pariade chez differentes espèces de scorpions et chez *Pandinus imperator*. Annals de l'Université d'Abidján, Séries E 5(1), 475–497.

Gould, J.L., Gould, C.G., 1989. Sexual Selection. Scientific American Library, New York.

Griffith, S.C., Owens, I.P.F., Thuman, K.A., 2002. Extra pair paternity in birds: a review of interspecific variation and adaptive function. Mol. Ecol. 11, 2195–2212.

Gross, M.R., 1982. Sneakers, satellites and parentals: polymorphic mating strategies in North American sunfishes. Z. Tierpsychol. 60, 1–26.

Gross, M.R., 1991. Evolution of alternative reproductive strategy: frequency-dependent sexual selection in male bluegill sunfish. Phil. Trans. R. Soc. London B. 332, 59–66.

Grubbs, F.E., 1969. Procedures for detecting outlying observations in samples. Technometrics 11, 1–21.

Honderich, B., 1995. Laws, natural or scientific. Oxford Companion to Philosophy, Oxford University Press, Oxford, pp. 474–476.

Hone, D.W., Benton, M.J., 2005. The evolution of large size: how does Cope's Rule work? Trends Ecol. Evol. 20, 4–6.

Hone, W., Dyke, J., Haden, M., Benton, M.J., 2008. Body size evolution in Mesozoic birds: little evidence for Cope's rule. J. Evol. Biol. 21, 1673–1682.

Inceoglu, B., Lango, J., Jing, J., Chen, L., Doymaz, F., Pessah, I., Bruce, D., 2003. One scorpion, two venoms: prevenom of *Parabuthus transvaalicus* acts as an alternative type of venom with distinct mechanism of action. Proc. Natl. Acad. Sci. U. S. A. 100, 922–927.

Ismar, S.M.H., Daniel, C., Stephenson, B.M., Hauber, M.E., 2010. Mate replacement entails a fitness cost for a socially monogamous seabird. Naturwissenschaften 97, 109–113.

Jackson, R.R., Pollard, S.D., 1996. Predatory behavior of jumping spiders. Ann. Rev. Entomol. 41, 287–308.

Jakob, E.M., Christa, D., Skow, C.D., Long, S., 2011. Plasticity, learning and cognition. In: Herberstein, M.E. (Ed.), Spider Behaviour: Flexibility and Versatility, Cambridge University Press, Cambridge, pp. 307–347.

Kokko, H., Jennions, M.D., 2008. Parental investment, sexual selection and sex ratios. J. Evol. Biol. 21, 919–948.

Kuhn, T.S., 1996. The Structure of Scientific Revolutions, third ed. University of Chicago Press, Chicago.

Lack, D., 1968. Ecological Adaptations for Breeding in Birds. Chapman and Hall, London.

Maurer, G., Double, M.C., Milenkaya, O., Süsser, M., Magrath, R.D., 2011. Breaking the rules: sex roles and genetic mating system of the pheasant coucal. Oecologia 167, 413–425.

Maury, E.A., 1968. Aportes al conocimiento de los escorpiones de la República Argentina I. Observaciones biológicas sobre *Urophonius brachycentrus* (Thorell, 1877) (Bothriuridae). Physis 27, 407–418.

Maury, E.A., 1983. Singular anomalía sexual en un ejemplar de *Brachistosternus pentheri* Mello-Leitão 1931 (Scorpiones, Bothriuridae). Rev. Soc. Entomol. Arg. 42, 155–156.

Moore, D.S., McCabe, G.P., 1999. Introduction to the Practice of Statistics, 3rd edn. W. H. Freeman, New York.

Moore, H.D., Martin, M.M., Birkhead, T.R., 1999. No evidence for killer sperm or other selective interactions between human spermatozoa in ejaculates of different males in vitro. Proc. R. Soc. Lond. B. 266, 2343–2350.

Monteith, L.G., 1963. Habituation and associative learning in *Drino bohemica* Mesn. (Diptera: Tachinidae). Can. Entomol. 95, 418–426.

Morgan, D., Cézilly, F., 2011. Personality may confound common measures of mate-choice. PLoS One 6, 1–5.

Oliveira, R.F., Taborsky, M., Brockmann, H.J., 2008. Alternative Reproductive Tactics: An Integrative Approach. Cambridge University Press, Cambridge.

Peretti, A.V., 1993. Estudio de la biología reproductiva en escorpiones Argentinos (Arachnida, Scorpiones): un enfoque etológico. Doctoral Thesis, Facultad de Ciencias Exactas, Físicas y Naturales, Universidad Nacional de Córdoba, Argentina.

Peretti, A.V., 1997. Evidencia de cortejo copulatorio en el orden Scorpiones (Arachnida), con un análisis en *Zabius fuscus* (Buthidae). Rev. Soc. Entomol. Arg. 56, 21–30.

Peretti, A.V., 2000. Existencia de cortejo en el campo de machos de *Bothriurus bonariensis* (Scorpiones, Bothriuridae) que carecen de un órgano paraxial. Rev. Soc. Entomol. Arg. 59, 96–98.

Peretti, A.V., 2010. An ancient indirect sex model: single and mixed patterns in the evolution of scorpion genitalia. In: Leonard, J.L., Córdoba-Aguilar, A. (Eds.), The Evolution of Primary Sexual Characters in Animals, Oxford University Press, New York, pp. 218–248.

Peretti, A.V., Aisenberg, A., 2011. Communication under sexual selection hypotheses: challenging prospects for future studies under extreme sexual conflict. Acta Ethol. 14, 109–116.

Peretti, A.V., Carrera, P., 2005. Female control of mating sequences in the mountain scorpion *Zabius fuscus*: males do not use coercion as a response to unreceptive females. Anim. Behav. 69, 453–462.

Peretti, A.V., Córdoba-Aguilar, A., 2007. On the value of fine-scaled behavioural observations for studies of sexual coercion. Ethol. Ecol. Evol. 19, 77–86.

Peretti, A.V., Eberhard, W.G., 2010. Cryptic female choice via sperm dumping favours male copulatory courtship in a spider. J. Evol. Biol. 23, 271–281.

Peretti, A.V., Acosta, L.E., Benton, T.G., 1999. Sexual cannibalism in scorpions: fact or fiction? Biol. J. Linn. Soc. 68, 485–496.

Peterson, E.L., 2008. William Bateson: from *Balanoglossus* to materials for the study of variation: the transatlantic roots of discontinuity and the (un)naturalness of selection. J. Hist. Biol. 41, 267–305.

Petrie, M., Kempenaers, B., 1998. Extra-pair paternity in birds: explaining variation between species and populations. Trends Ecol. Evol. 13, 52–58.

Polis, G.A., Farley, R.D., 1979. Behaviour and ecology of mating in cannibalistic scorpion *Paruroctonus mesaensis* Stahnke. J. Arachnol. 7, 33–46.

Polis, G.A., Sissom, W.D., 1990. Life history. In: Polis, G.A. (Ed.), The Biology of Scorpions, Stanford University Press, Palo Alto, pp. 161–223.

Reaka, M.L., 1980. On learning and living in holes by mantis shrimps. Anim. Behav. 28, 111–115.

Renze, J., 2011. Outlier. From MathWorld – A Wolfram Web Resource, created by Eric W. Weisstein. (http://mathworld.wolfram.com/Outlier.html).

Schuett, W., Tregenza, T., Dall, S.R.X., 2010. Sexual selection and animal personality. Biol. Rev. 85, 217–246.

Serrat, M.A., King, D., Lovejoy, C.O., 2008. Temperature regulates limb length in homeotherms by directly modulating cartilage growth. Proc. Natl. Acad. Sci. U. S. A. 105, 19348–19353.

Shulov, A., Amitai, P., 1958. On mating habits of three scorpions: *Leiurus quinquestriatus* (H. & E.), *Buthotus judaicus* E. Sim, and *Nebo hierochonticus* E. Sim. Arch. Inst. Pasteur d'Algérie 38, 117–129.

Sih, A., Bell, A., Johnson, J.C., 2004. Behavioral syndromes: an ecological and evolutionary overview. Trends Ecol. Evol. 19, 372–378.

The American Heritage® Dictionary of the English Language, 2003, fourth ed. Retrieved October 25 2011 from http://www.thefreedictionary.com/rule.

Thorpe, W.H., 1943. Types of learning in insects and other arthropods. Br. J. Psychol. 33, 220–234.

Toscano-Gadea, C.A., 2010. Sexual behavior of *Bothriurus buecherli* (Scorpiones, Bothriuridae) and comparison with the *B. prospicuus* group. J. Arachnol. 38, 360–363.

Vachon, M., 1952. Études sur les scorpions. Publications de l'Institut Pasteur d'Algérie, Algiers.

Westneat, D.F., Fox, C.W., 2010. Evolutionary Behavioral Ecology. Oxford University Press, New York.

Zolessi, L.C.de, 1956. Observaciones sobre el comportamiento sexual de *Bothriurus bonariensis* (Koch) (Scorpiones, Bothriuridae). Nota preliminar. Bol. Fac. Agr., Montevideo 35, 1–10.

Chapter 3

Secondary Sexual Traits, Immune Response, Parasites, and Pathogens: The Importance of Studying Neotropical Insects

Jorge Contreras-Garduño and Jorge Canales Lazcano

Departamento de Biología, División de Ciencias Naturales y Exactas, Universidad de Guanajuato, Noria Alta, Guanajuato, Mexico

INTRODUCTION

Secondary sexual traits show an enormous variation in nature. Such variation could arise because these traits signal their bearers' health and disease resistance, which depend on biotic (i.e., predation, parasitism, population density, and sex-ratio) and abiotic (i.e., temperature, humidity, and access to food) factors. Furthermore, health and disease resistance are strongly affected by immune response performance, which in turn is affected by density and diversity of parasitism (biotic), temperature, and/or amount of resources (abiotic), for example. The great biological variation in tropical ecosystems provides optimal conditions to study the relative importance of distinct biotic and abiotic factors in the relation between secondary sexual traits and immune response. In this chapter, we outline the contribution that tropical insects can make to our understanding of parasite/pathogen selective pressures on secondary sexual traits and the immune response. We have focused on insects because over half of all described species belong to this group, they have short life cycles and are usually clustered in dense populations that are widely distributed, and show an enormous species richness according to a latitudinal gradient (Mayhew, 2007). From the sexual selection point of view, neotropical insects have been poorly studied. For example, a recent review (Cueva del Castillo, 2007) states that only a few species of Coleoptera (*Estaminodeus vectoris*), Lepidoptera (*Caligo illioneus, Callophryx xami, Hypolimnas bolina*), Odonata (*Hetaerina americana, Perithemis mooma,* and *Orthemis discolor*), and Orthoptera (*Sphenarium purpurascens*) have been studied. Finally, insects

Sexual Selection. http://dx.doi.org/10.1016/B978-0-12-416028-6.00003-7

provide an excellent model to study the immune response from the molecular, physiological, ecological, and evolutionary points of view (Siva-Jothy *et al.*, 2004; Beckage, 2008; Rolff and Reynolds, 2009; Söderhäll, 2010; Adamo, 2011; McKean and Lazzaro, 2011).

Widely distributed species occupy diverse environments with distinct biotic and abiotic factors, and consequently confront diverse selection pressures. By studying tropical ecosystems, it should be possible to reveal general patterns at the macroecological level (as proposed in Chapter 1), and perhaps to discover rules and exceptions in the relation between secondary sexual traits and the immune system (as proposed in Chapter 2). To explain the importance of tropical insects for this endeavor, we first describe the hypothesis that relates secondary sexual traits with disease resistance. Second, we review the molecular and physiological mechanisms through which insects fight infections: the immune response. Third, the ecoimmunological fundaments are outlined. Fourth, the relationship between secondary sexual traits and insect immune responses is reviewed, and fifth, the underlying physiological mechanisms that mediate these relationships are presented. Finally, we discuss how future studies should examine the relationship between secondary sexual traits and the immune response, but considering natural parasite or pathogen infections and comparing these variables among populations. We suggest that using neotropical insects as a model system will reveal whether patterns reflect incremental changes along a latitudinal gradient and/or whether they are typical of local adaptations.

SEXUAL SELECTION: HAMILTON AND ZUK'S HYPOTHESIS

Charles Darwin proposed the sexual selection theory to explain why and how secondary sexual traits evolved, and suggested two mechanisms: (1) competition for access to the opposite sex (mainly by males), and (2) mate choice (mainly by females) (Darwin, 1871). The former is responsible for the evolution of elaborate weapons (e.g., spines, antlers, and horns) while the latter is responsible for the evolution of ornaments (songs, dances, and brightly colored and large structures). The evolution of weapons is clear: the more effective the weapon, the greater access males have to females, resulting in greater offspring production. However, why and how are ornaments maintained? And what benefits do females obtain through their choice? To answer these questions, Zahavi (1975, 1977) proposed the handicap principle. According to his model, these ornaments represent honest signals about the quality of the bearer. For example, colorful secondary sexual traits make males more visible not only to females but also to predators (Godin and Dugatkin, 1996). According to the handicap principle, only males of high quality can afford to spend energy on the development of elaborate attributes of this type while putting their survival at risk. Thus, as in the example of the guppy *Poecilia reticulata*, females desiring to pass good genes on to their offspring prefer males with the most elaborate secondary sexual traits, which also allow them to escape from predators (Godin and Dugatkin, 1996). The problem is that if all females have a strong bias regarding the same traits, their consistent choice can erode the genetic variation of their populations. This phenomenon, known as the lek paradox (Andersson, 1994; Rowe and Houle, 1996), raises the question of why all males do not end up with the same genetic makeup (i.e., one that successfully attracts females). The answer could be that secondary sexual traits are reliable in revealing the bearer's health.

Hamilton and Zuk (1982) proposed that one attribute of condition (i.e., quality) could be genes that confer resistance against invasion by parasites and pathogens ("good genes"), and that coadaptation cycles that characterize parasite–host interaction could drive the generation of variation in the adaptation of genotypes. For instance, there might be a natural population of cockroaches that included some males (**S**) that are resistant to attacks by *Serratia marcescens* (a Gram-negative bacterium) but susceptible to *Bacillus thuringiensis* (a Gram-positive bacterium). Other males (**B**) could be susceptible to *S. marcescens* but resistant to *B. thuringiensis*. If at a certain point in time the most common pathogen is *S. marcescens*, then females should prefer the **S** males to favor their offspring resistance. If at a later point in time environmental conditions favored the presence of *B. thuringiensis*, females should prefer the **B** males, once again passing on resistance to their offspring. In this manner, pressure exerted by parasites would generate new genotypes in the hosts and males should use their secondary sexual traits to advertise their resistance against particular parasites or pathogens. Assuming that only the healthiest males are able to expend energy on developing secondary sexual traits, there would be a negative correlation between the degree of expression of these traits and the parasite or pathogen load (Hamilton and Zuk, 1982). More resistant males would require less energy in fighting infections, allowing them to invest more energy in sexual traits. The males would then signal or show their resistance to females by the degree of ornamentation of these characteristics.

According to Hamilton and Zuk's (1982) hypothesis, there will be little variation in secondary sexual traits among males in the presence of a lethal parasite from which only resistant males survive (Jacobs and Zuk, 2011). Similarly, if a parasite has little impact on fitness (i.e., if all males are resistant), there should also be little variation in secondary sexual traits across males (Jacobs and Zuk, 2011). The hypothesis will be most applicable to populations with parasites that generate chronic, debilitating diseases (Jacobs and Zuk, 2011). Finding such parasites or pathogens to adequately test this hypothesis requires multidisciplinary studies in the field and lab that analyze the type and frequency of parasitism that males confront, the level of expression of secondary sexual traits, and the relation between these traits and the immunological response. However, the immune system responds to a variety of biotic and abiotic factors, in addition to the type of parasites/pathogens that are encountered, resulting in a multifactorial effect of the relation between secondary sexual traits and immune response.

THE IMMUNE SYSTEM OF INSECTS

In terms of fighting pathogens, invertebrates possess a very efficient innate immune system (Iwanaga and Lee, 2005) that resembles that of vertebrates (Litman *et al.*, 2005; Söderhäll, 2010; Loker, 2012; Criscitiello and de Figueiredo, 2013). As an example, recent evidence suggests that the immune systems of invertebrates have some ability for immunological specificity and memory (Kurtz, 2005; Little *et al.*, 2005). In addition, it has been proposed an epigenetic effect on the insects resistance (Gómez-Díaz *et al.*, 2012; Mukherjee *et al.*, 2012; Durdevic *et al.*, 2013). The following section briefly reviews mechanisms of the insect immune response, but this is covered in greater detail in Beckage (2008), Rolff and Reynolds (2009), and Söderhäll (2010).

The immune system of invertebrates is composed of humoral and cellular responses. Both responses are triggered when the receptor molecules of the insect's immune system bind to the ligands of parasites, pathogens, or diseased cells (pathogen-associated molecular patterns, or PAMPs) (Lavine and Strand, 2002; Iwanaga and Lee, 2005). PAMPs are well-preserved molecular patterns that are found in microorganisms, including the lipopolysaccharides of Gram-negative bacteria, peptidoglycans of Gram-positive and negative bacteria, and beta-glucans of fungi (Beckage, 2008; Marmaras and Lampropoulou, 2009; Söderhäll, 2010). PAMPs activate the host pattern recognition receptors (PRRs), and this leads to either a humoral or cellular immune response, or both (Ferrandon *et al.*, 2007; Beckage, 2008; Söderhäll, 2010).

Humoral Response

The humoral response refers to the production of blood-borne molecules that cause the death of parasites or pathogens. Among these compounds, those that have been most studied are antimicrobial peptides, which are produced by the IMD and Toll intracellular signaling pathways. Gram-negative bacteria mainly activate the expression of peptides by the IMD and JAK/STAT pathways, Gram-positive bacteria activate the Toll and IMD pathways, while fungi activate the Toll pathway (Royet *et al.*, 2005; Tsakas and Marmaras, 2010). There are four peptide families that attack bacteria: cecropin, defensins, and two types of polypeptides, one rich in glycine and the other rich in proline (Hetru *et al.*, 1998). Cecropin and defensin fight Gram-positive and Gram-negative bacteria. Some polypeptides that are rich in glycine also have a wide spectrum and attack both Gram-positive and Gram-negative bacteria (Hetru *et al.*, 1998), whereas the polypeptides that are rich in prolin only attack Gram-negative bacteria. A fifth peptide family, drosomycin, is specific to fungi (Hetru *et al.*, 1998; Steiner, 2004). Insects generally synthesize peptides that are pathogen-specific (e.g., to Gram-positive bacteria, Gram-negative bacteria, or fungi). However, there are also some peptides that have a broader spectrum, such as those that attack both Gram-positive and Gram-negative bacteria (Lemaitre and Hoffmann, 2007).

On the other hand, Toll, IMD, and JAK/STAT pathways appear to be responsible for the elimination of viruses (Costa *et al.*, 2009; Sabin *et al.*, 2010), but another important antiviral response is carried out by small molecules of RNA (Fullaondo and Lee, 2012). In *Drosophila*, three types have been identified: micro RNA (miRNA), small interfering RNA (siRNA), and piwi associated RNA (piRNA) (Fullaondo and Lee, 2012). Recently it has been observed that RNAi genes have one of the greatest rates of change in the *Drosophila* genome, which leads to speculation that this system is experiencing strong pressure for selection by virus attack (Obbard *et al.*, 2006). Even more, it seems that epigenetic mechanism play a key role to combat viruses (Durdevic *et al.*, 2013).

The pro-phenoloxidase (PPO) cascade is a relatively well-studied phenomenon in invertebrates. It involves the formation of melanin from tyrosine and phenylalanine (Riley, 1997). This process is initiated after host recognition of bacterial lipopolysaccharides, peptidoglycans and β1–3 glucans (Cerenius and Söderhäll, 2004). Recognition triggers the serine–protease cascade, which activates the PPO enzyme, which in turn induces production of the phenoloxidase enzyme (PO; Cerenius and Söderhäll, 2004). PO in the hemolymph adheres to the body of pathogens and parasites (Rizki *et al.*, 1985; Lavine and Strand, 2002), and then converts phenols into quinones, which are transformed into melanin (Tzou *et al.*, 2002; Cerenius and Söderhäll, 2004; Nappi and Christensen, 2005). Melanin adheres to and coats invaders. It has been suggested that the main reason for the death of pathogens during an attack by PO is the toxicity of quinones and other reactive molecules derived from oxygen and nitrogen generated during this process (Christensen *et al.*, 2005; Nappi and Christensen, 2005). It is important to point out that although the PO system has been widely studied, recent reviews suggest that it may not be as effective as other immune response mechanisms against some pathogens (Kanost and Gorman, 2008). Moreover, its level of activity may not correlate with disease resistance (Adamo, 2004a).

Another humoral response includes the formation of reactive nitrogen species (RNS), such as nitrous anhydride, peroxynitrite anion, nitric oxide, and reactive oxygen species (ROS) including the hydroxyl radical and superoxide anion. All of these reactive species cause damage to genetic material, deamination of basic compounds, and alteration of enzymes that repair DNA (Nappi and Ottaviani, 2000). Nitric oxide (NO) is an important RNS in the immune response of both invertebrates (Luckhart *et al.*, 1998) and vertebrates (Fang, 1997; Bogdan, 2001). In insects, NO levels increase with a pathogen or parasite attack (Foley and O'Farrell, 2003; Herrera-Ortiz *et al.*, 2004). In *Drosophila*, this increase triggers the hemocyte response and stimulates fat bodies to synthesize antimicrobial peptides (Leclerc and Reichhart, 2004). Nitric oxide is not only a powerful antimicrobial agent; its derivatives, such as peroxynitrite and nitrogen dioxide, also favor the expression of other potent free radicals (Fang, 1997). NO and these related molecules all cause oxidative damage to DNA, affecting enzymes for DNA repair and modifying deoxyribonucleotides

(Fang, 1997). In addition, superoxide anion is a ROS produced during melanotic encapsulation and this molecule could severely damage body invaders (Nappi et al., 1995). Some insect cells could possibly generate ROS and RNS by means of the dual oxidase (DUOX) system, and this system is triggered in Drosophila with a bacterial attack (i.e., converts H_2O_2 into the highly microbicidal HOCl; Lemaitre and Hoffmann, 2007), and is responsible for maintaining homeostasis in the epithelial mucous membrane of this host (Bae et al., 2010). Finally, Owusu-Ansah and Banerjee (2009) showed that an increase of ROS in the lymph gland of the last instar larvae of Drosophila melanogaster favored hemocyte differentiation. Hence, ROS and RNS seem to be an important mechanism against invaders, although in studies of sexual selection in insects they have rarely been studied.

Cellular Response

Cellular response refers to the ability of hemocytes, cells that are found in the hemolymph (i.e., blood), to attack pathogens. There are three basic types of cellular immune responses in insects: phagocytosis, nodulation, and encapsulation.

During phagocytosis, pathogens are introduced into cells by the invagination of their cytoplasmic membrane around the pathogen with the formation of a structure called a phagosome (Lavine and Strand, 2002). ROS (e.g., superoxide and hydrogen peroxide) and RNS (e.g., nitric oxide) are produced inside this structure, and at the same time hydrolytic enzymes are secreted (Lavine and Strand, 2002; Tzou et al., 2002; Hetru et al., 2003). ROS and RNS damage bacterial DNA and inhibit DNA repair enzymes (Nappi and Ottaviani, 2000), while the hydrolytic enzymes kill and degrade cells of microorganisms (Cheng, 1992; Tzou et al., 2002; Hetru et al., 2003). All of these antimicrobial compounds can also be released from the phagosome to attack parasites (Lavine and Strand, 2002). Hydrolytic enzymes modify the molecular conformation of pathogen surfaces, making them easily recognizable by phagocytes (Cheng, 1992; Cajaraville et al., 1995).

Nodulation is the aggregation of hemocytes around invading enemies such as bacteria, fungi, and viral-infected cells. This process apparently begins when lectins bind to the foreign agent (Lavine and Strand, 2002). The first step results in entrapment of the insect's enemies, and the second step consists of melanin production around microrganisms or virus-infected cells (Lavine and Strand, 2002).

Encapsulation is another process that forms melanin by means of the PO cascade. During this process, hemocytes bind to pathogens and form a melanin capsule, thus isolating them from the animal's body at the same time that toxic molecules are produced (Nappi et al., 1995; Lavine and Strand, 2002; Christensen et al., 2005; Nappi and Christensen, 2005). This response is very important in fighting macropathogens such as nematodes and parasitoids (Lemaitre and Hoffmann, 2007; Carton et al., 2008; Castillo et al., 2011).

Finally, a cellular mechanism known as autophagy is responsible for degrading cytoplasmic content and recycling it when the energy of the organism is diminished (Sabin *et al.*, 2010). But after a virus infection the JAK/STAT pathway signal induces autophagy of virus-infected cells in order to help the organism overcome the infection (Sabin *et al.*, 2010).

Thus, there are various mechanisms in the immune response of insects that eliminate micro- and macropathogens, parasites, and parasitoids. Examination of Table 3.1 shows that the immune system is not a monolithic entity, but is composed of multiple subsystems and interactions. Natural and sexual selection pressures may not be similar on each subsystem. For example, if a particular parasite/pathogen is a major selective force in a population, all the immune responses that are critical for destroying it will be under greater natural and sexual selection.

ECOIMMUNOLOGY: COSTS OF THE IMMUNE RESPONSE AND BIOTIC AND ABIOTIC FACTORS THAT FACILITATE ITS VARIATION

Ecoimmunology is the study of immunity from an ecological and evolutionary perspective (Rolff and Siva-Jothy, 2003; Schmid-Hempel, 2003, 2005, 2011; Demas and Nelson, 2011), whereas classical immunology describes the mechanisms that result in elimination of invading microbes without taking into account the variations of this response (Siva-Jothy *et al.*, 2004). Ecoimmunologists have shown that the immune response is subject to variations due to the costs of the immune response and the interaction of the organism with biotic and abiotic factors. Given such variation of immune response, ecoimmunological studies should consider the selection of the appropriate methodologies to characterize the immune system in vertebrates and invertebrates (see detailed information in Boughton *et al.*, 2011; Jacobs and Zuk, 2011), and should properly associate each mechanism with the specific parasite or pathogen that elicits a particular immune response (Adamo, 2004b).

Costs of the Immune Response

The assumption that the immune response is costly is the foundation of ecoimmunology (Rolff and Siva-Jothy, 2003; Schmid-Hempel, 2003, 2005) because costs limit the performance of the defense system. There are costs related to the maintenance and use of the immune response, and also evolutionary costs (Rolff and Siva-Jothy, 2003; Schmid-Hempel, 2003, 2005).

The costs of maintaining the immune response have been documented in *D. melanogaster*. In experimental conditions of starvation, mutant immune-deficient flies save energy by not investing in the immune response and survive longer than wild flies that invest in this response (Valtonen *et al.*, 2010). The cost of using or maintaining the immune response has also been demonstrated in damselflies. In the damselfly *Calopteryx virgo*, challenged males that had to invest in their immune response were less successful in mating and had shorter survival rates than control males, which were not required to make this energy

TABLE 3.1 X Denotes Mechanisms of the Immune Response that Have Been Implicated with Specific Immune Challenges (for Reference, see Text)

Note that the Effectiveness of these Mechanisms Could Change According to Host Species, Parasite or Pathogen Strains, and Possibly, According to Neotropical Populations and Environments

Mechanism of the Immune Response	Fungi	Bacteria (Gram-Positive)	Bacteria (Gram-Negative)	Protozoa	Nematodes	Parasitoids	Viruses	Ectoparasites (i.e., mites)
Toll	X	X		X	X		X	
IMD			X		X		X	
Jak-STAT							X	
PO (melanization)	X	X	X	X	X	X		X
DUOX	X	X	X	X	X	X		
Nodulation	X	X	X					
PO (encapsulation)				X	X	X		X
Autophagy							X	
Phagocytosis	X	X	X	X				
RNA of interference								X

investment (Rantala *et al.*, 2010). In *Hetaerina americana*, multiple immune challenges in teneral males led to male adults expressing less elaborate wing spots than in unchallenged males (Contreras-Garduño *et al.*, 2008). In this species the wing spot size has been considered a secondary sexual characteristic that males use during ritualized male–male contests to favor their mating outcome (Grether, 1996; Serrano-Meneses *et al.*, 2007; Raihani *et al.*, 2008).

Finally, evolutionary costs refer to the negative effect on fitness, which can be measured across generations. For instance, one line of *D. melanogaster* is resistant to the attack of the parasitoid wasp *Asobara tabida*. Interestingly, its greater success in eliminating the parasite, compared with the susceptible line, makes it less competitive in obtaining food (Kraaijeveld and Godfray, 1997). In the same way, if hosts are specialized to fight a particular parasite species with which they co-evolved, they could be susceptible to the attack of another parasite or pathogen species. This is termed the multiple-front cost hypothesis, and has been poorly studied to date (McKean and Lazzaro, 2011).

These findings can be summarized as follows: the costs of the immune response are a limiting factor in two important realms – the performance of this response (Demas and Nelson, 2011; Ardia *et al.*, 2012), and, if required, the development of secondary sexual traits (Sheldon and Verhulst, 1996; Jacobs and Zuk, 2011).

Biotic and Abiotic Factors Facilitating Variation in the Immune Response

Besides the costs of the immune response, biotic and abiotic factors are important in understanding immune response variability. These include factors such as temperature, food availability, and attacks by parasites and pathogens.

Temperature influences the fitness of organisms, particularly of ectotherms (Amarasekare and Savage, 2012). In ectotherms, temperature affects oviposition, body size, reproduction, nutrient assimilation, metabolism, and hormonal regulation. Drastic changes in temperature can even lead to the extinction of some species (Amarasekare and Savage, 2012). Recent studies show that temperature and humidity also affect the immune response of insects. For example, in the Mediterranean flour moth *Ephestia kuehniella* an increase in temperature favored cellular response, demonstrated by the fact that larvae fed with *B. thuringiensis* showed an increase in nodulation at 32°C but not at 15°C nor at 23°C. This nodulation also increased at a relative humidity of 85% but not at 43% (Mostafa *et al.*, 2005).

Food limitations can also reduce immune function. For example, if bumblebees *Bombus terrestris* are challenged with lipopolysaccharides (found on the surface of Gram-negative bacteria) or with a novel artificial challenge (microlatex beads), the survival of these insects is only reduced in conditions of nutritional stress (Moret and Schmid-Hempel, 2000). In mealworms *Tenebrio molitor*, the deprivation of food diminished the activity of phenoloxidase and melanization (Siva-Jothy and Thompson, 2002). Food deprivation not only affected the resistance of field crickets *Gryllus campestris* to the lipopolysaccharides of

S. marcescens, but also altered the singing of males used to attract females (Jacot *et al.*, 2004).

Finally, different parasites/pathogens induce different types of immune responses. As previously mentioned, Gram-negative bacteria elicit a response through the IMD and JAK/STAT pathways, while Gram-positive bacteria activate IMD and Toll pathways, and fungi prompt a response through the Toll pathway. In addition, Riddell *et al.* (2009) demonstrated a differential expression of antimicrobial peptides among distinct lines of *Bombus terrestris* colonies when they were challenged by different isolates of *Crithidia bombi*. Isolate I induced a greater expression of hymenoptaecin than isolate III when in contact with line 3 of the host. However, both of these isolates spurred a high expression of hymenoptaecin in line 4 of the host. This suggests that the immune response varies in different host colonies of the same species when faced with different pathogen strains. Under this perspective, it is possible that the immune response and the degree of elaboration of secondary sexual traits could vary among populations if pathogens also varied.

SIGNALING OF THE IMMUNE RESPONSE THROUGH SECONDARY SEXUAL TRAITS IN INSECTS

The notion of cuts within the context of sexual selection has been the foundation for establishing the relation between the immune response and secondary sexual traits, and, given that males are the sex more likely to produce such characters, they are the sicker sex (Zuk and Stoehr, 2002; Zuk, 2009). It has been proposed that females should prefer males that express their capacity for disease resistance through the exuberance/quality of secondary sexual traits because if these traits can be inherited, then offspring would exhibit attractive sexual traits and have a highly developed immune response (resistance) (Folstad and Karter, 1992). In insects, some aspects of the immune response have been shown to be heritable (Armitage and Siva-Jothy, 2005). Moreover, a study reported that the offspring inherited their immune response capacity from their fathers (Kurtz and Sauer, 1999). However, as far as we know, this is the only study that supports this idea in insects, making it critical to conduct similar studies, with other species, in order to determine whether this is a general principle by which females obtain indirect benefits of the male signaling immune response.

A recent paper also suggests that offspring could be specifically protected against the immune challenges confronted not only by their mothers (as maternal effects) but also by their fathers (Roth *et al.*, 2009). This phenomenon, termed cryptic paternal care, could be very important in the context of sexual selection and sexual conflict (Jokela, 2010). However, there is no information regarding the generality of this offspring immune care provided by males, and whether males could signal their immunological cryptic parental care to females through their secondary sexual traits. Studies related to male signaling and its relationship with their offspring immunity will be very important because mathematical models suggest that females could obtain more benefits by choosing for general male condition rather than solely for immune response (for discussion, see Adamo and Spiteri, 2005, 2009).

As mentioned above, mate choice is a key issue in the Hamilton and Zuk (1982) hypothesis, but mate competition for female access is also related to the relationship between secondary sexual traits and immune response (Lawniczak *et al.*, 2007), as revealed in damselflies (Contreras-Garduño *et al.*, 2006) and beetles (Pomfret and Knell, 2006). Hence, Table 3.2 lists studies encompassing secondary sexual traits and immune responses in insects with female choice and/or male–male competition. The table reveals that 9 out of 28 studies were carried out with neotropical species, and that all of them are odonates: *Erythemis vesiculosa, H. americana, H. titia, H. cruentata, H. vulnerata, H. occisa, Erythrodiplax funerea,* and *Paraphlebia zoe*. Of these, *H. americana* has been the most studied. In this species, the wing spot size (a sexual characteristic) reflects the male immune response (PO, NO, and melanization; see Table 3.2) and the relationship is dynamic. For example, population differences in immune response and secondary sexual trait elaboration have been reported: the population from the northern climate had smaller wing spot sizes and less elaborate immune responses than the southern population (Table 3.2). However, we are still far from understanding the biotic and abiotic factors related to the relationship between secondary sexual traits and immune response, not only because few studies have been carried out up until now but also because studies in the Neotropics are concentrated in odonates. Clearly, additional studies that consider different neotropical insect groups are required (i.e., Coleoptera, Lepidoptera, and Orthoptera).

The majority (23 out of 28) of the studies summarized in Table 3.2 show that secondary sexual traits are indeed indicators of at least one parameter of the immune response. Nevertheless, two problems arise: (1) in-depth analyses of immunocompetence have not been conducted, and (2) tests with the parasites and pathogens that actually co-exist under natural conditions with the focal study species have been rarely tested.

Regarding the first point, 13 of the 28 studies tested only one parameter of the immune response. Two parameters were evaluated in 12 studies, and more than two parameters were assessed in just 3 studies. The problem of considering only one parameter is that it provides an incomplete picture of the immune response, and hence results cannot support the idea of immunocompetence (Drury, 2010) or discriminate whether the immune response is correlated with a general condition. Among the 15 studies that considered more than one parameter, only two cases showed the development of secondary sexual traits related to more than one marker. This suggests that secondary sexual traits are not indicators of general immunocompetence of males. This could be explained because the immune response is traded off with other attributes of life history or secondary sexual traits, and a trade-off between different components of the immune response is also expected (Norris and Evans, 2000; Adamo, 2004b).

TABLE 3.2 Studies with Insects in which Secondary Sexual Traits (SST) and Immune Responses Were Analyzed

Species	SST/Sexual Selection Mechanism	Immune Marker/ Markers	Challenge	More Ornamented Males More Immunocompent?	Population or Phonological Variation	Physiological Mechanism	References
Acheta domesticus	Calling song/ female choice	Hemocyte count and encapsulation	Nylon monofilament	Yes	Not tested	Not tested	Ryder and Siva-Jothy, 2000
Calopteryx splendens	Wing spot size/ female choice and male–male competition	Phenol oxidase	Nylon monofilament	No	Not tested	Not tested but the amount of melanin was proposed	Siva-Jothy, 2000
Calopteryx splendens	Wing spot size/ female choice and male–male competition	Encapsulation	Nylon monofilament	Yes	Not tested	Not tested but the amount of melanin was proposed	Rantala et al., 2000
Tenebrio molitor	Pheromones/ female choice	Encapsulation and phenol oxidase	Nylon monofilament	Yes	Not tested	Juvenile hormone	Rantala et al., 2002, 2003a
Tenebrio molitor	Pheromones/ female choice	Encapsulation and phenol oxidase	Nylon monofilament	Yes, phenol oxidase	Not tested	Energy reserves (diet)	Rantala et al., 2003b
Gryllus bimaculatus	Calling song/ female choice	Encapsulation and lytic activity	Nylon monofilament	Yes, encapsulation	Not tested	Not tested	Rantala and Kortet, 2003

Calopteryx virgo	Wing spot size and territoriality/ female choice and male–male competition	Encapsulation and hemocyte count	Nylon monofilament	Yes, encapsulation	Not tested	Not tested	Koskimäki et al., 2004
Gryllus bimaculatus	Territoriality/ male–male competition	Encapsulation and lytic activity	Nylon monofilament	Yes, both measurements	Not tested	Energy reserves (fat)	Rantala and Kortet, 2004
Calopteryx xanthosoma	Wing spot size/ female choice and male–male competition	Phenol oxidase	Nylon monofilament	No	Not tested	Not tested but parasite load could be responsible	Rolff and Siva-Jothy, 2004
Teleogryllus commodus	Calling song/ female choice	Encapsulation	Nylon monofilament	Yes	Not tested	Not tested	Simmons et al., 2005
Teleogryllus oceanicus	Calling song/ female choice	Encapsulation	Nylon monofilament	Yes	Not tested	Not tested	Tregenza et al., 2006
Hetaerina americana	Wing pigmentation and territorial status/ male–male competition	Encapsulation	Nylon monofilament	Yes	Not tested	Energy reserves (fat)	Contreras-Garduño et al., 2006
Erythemis vesiculosa	Territorial status/ male–male competition	Encapsulation	Nylon monofilament	Yes	Not tested	Not tested	Córdoba-Aguilar and Méndez, 2006

Continued

TABLE 3.2 Studies with Insects in which Secondary Sexual Traits (SST) and Immune Responses Were Analyzed—cont'd

Species	SST/Sexual Selection Mechanism	Immune Marker/ Markers	Challenge	More Ornamented Males More Immunocompent?	Population or Phonological Variation	Physiological Mechanism	References
Euoniiticellus intermedius	Horn length/ male–male competition	Encapsulation and phenol oxidase	Nylon monofilament	Yes, phenol oxidase	Not tested	Not tested	Pomfret and Knell, 2006
Forficula auricularia	Forceps size/ male–male competition	Encapsulation and lytic activity	Nylon monofilament	Yes, encapsulation	Not tested	Not tested	Rantala et al., 2007
Tenebrio molitor	Pheromone/ female choice	Phenol oxidase	Nylon monofilament	Yes	Not tested	Not tested	Sadd et al., 2006
Hetaerina americana	Wing pigmentation and territorial status/ male–male competition	Encapsulation, phenol oxidase and hydrolytic enzymes	Nylon monofilament and *Serratia marcescens*	Yes	Not tested	Energy reserves (fat)	Contreras-Garduño et al., 2007
Hetaerina americana	Wing pigmentation and territorial status/ male–male competition	Encapsulation	Nylon monofilament	Yes	Not tested	Energy reserves (fat)	Contreras-Garduño et al., 2008

Calopteryx virgo	Territorial status/female choice and male–male competition	Phenol oxidase	Serratia marcescens	Yes	Not tested	Juvenile hormone	Contreras-Garduño et al., 2009a
Hetaerina americana	Wing pigmentation and territorial status/ male–male competition	Phenol oxidase and nitric oxide	Nylon monofilament	Yes	Yes	Energy reserves (fat)	Contreras-Garduño et al., 2009b
Hetaerina americana	Wing pigmentation and territorial status/ male–male competition	Phenol oxidase, encapsulation and nitric oxide	Nylon monofilament	Yes	Yes	Energy reserves (fat)	Córdoba-Aguilar et al., 2009
Hetaerina americana, H. titia, H. cruentata, H. vulnerata, and H. occisa	Wing pigmentation/ male–male competition	Phenol oxidase and encapsulation	Nylon monofilament and Serratia marcescens	Not always and variation within the same season was reported	Yes	Not tested	González-Santoyo et al., 2010
Teleogryllus oceanicus	Calling song/ female choice	Encapsulation, lysozyme-like activity and hemocyte count	Nylon monofilament	Yes, encapsulation and hemocyte count	Not tested	Not tested	Simmons et al., 2010

Continued

TABLE 3.2 Studies with Insects in which Secondary Sexual Traits (SST) and Immune Responses Were Analyzed—cont'd

Species	SST/Sexual Selection Mechanism	Immune Marker/ Markers	Challenge	More Ornamented Males More Immunocompent?	Population or Phonological Variation	Physiological Mechanism	References
Erythrodiplax funerea	Wing pigmentation/ male-male competition	Encapsulation	Nylon monofilament	No, a negative relationship	Not tested	Energy reserves (fat)	Contreras-Garduño et al., 2011b
Chorthippus biguttulus	Calling song/ female choice	Encapsulation	Nylon monofilament	Yes	Not tested	Not tested	Stange and Ronacher, 2012
Gnatocerus cornutus	Horn length	Pro Phenol oxidase and phenol oxidase	*Hymenolepis diminuta*	There was no relationship against challenge but without a challenge	Not tested	Not tested but Juvenile Hormone was proposed	Demuth et al., 2012
Paraphlebia zoe	Wing pigmentation/ female choice and male-male competition	Phenol oxidase and nitric oxide	No-challenge, basal response was recorded	Yes, phenol oxidase	Not tested	Not tested but the amount of melanin was proposed	Ruiz-Guzmán et al., 2012

However, such trends will only emerge if different immune response markers are considered.

The second problem is the lack of knowledge regarding which parasites or pathogens attack males in natural conditions. Only in one study was the natural parasite used. In *H. titia*, the gregarine burden was negatively correlated with the black wing spot size and positively correlated with melanization against a nylon implant (Córdoba-Aguilar *et al.*, 2007). Three other studies were carried out with the bacteria *S. marcescens*, and another one with the nematode *Hymenolepis diminuta*. Nonetheless, it is important to consider that these pathogens are not natural enemies of the host, but general enemies of insects. In the rest of the studies, the challenge was an artificial foreign agent (a nylon implant). The advantage of using a nylon implant is that it allows the determination of how the immune response would behave when facing a novel agent, without the confounding factor of virulence (Siva-Jothy *et al.*, 2004). However, the problem is that artificial challenges do not trigger the same immune reaction as natural infectious agents (Adamo, 2004b). In addition, without the use of natural enemies and the specific immune response it is impossible to determine whether females are selecting the immune response that signals the current parasites or pathogens of the males of a species, which is a cornerstone of the hypothesis proposed by Hamilton and Zuk (1982), or whether the immune response is an indicator of immunocompetence or general male condition, as suggested by Adamo and Spiteri (2005, 2009). Consequently, it is imperative to analyze the exuberance of secondary sexual traits and their relation with more than one parameter of the immune system's defense against one or more real parasites or pathogens in the males' natural environment.

MECHANISMS BEHIND THE IMMUNE RESPONSE AND SEXUAL SIGNALS RELATIONSHIP

Life-history theory assumes that a beneficial change in one trait comes with a negative effect on another trait (i.e., there is a trade-off between them; Stearns, 1992). Given that both the immune response and sexual signals are costly to produce and maintain, a trade-off between them is expected (Sheldon and Verhulst, 1996). Hence, only the fittest males can steadily maintain the expression of both mechanisms simultaneously. Consider, for example, the damselfly *Calopteryx splendens*, in which it was found that females prefer males with the most pigmented wings, which also are less prone to infection with gregarines (Siva-Jothy, 1999). Given that melanin (more properly phenylalanine and tyrosine) is the mediating mechanism related to both infection resistance (i.e., favoring melanization) and expression of elaborate wing spots, only males with more resources (amino acids) will favor both traits at the same time (Siva-Jothy, 2000). In this scenario, melanin production is the limited resource that should be incorporated in pigmentation or immune response. However, melanin is not always implicated in the elaboration of secondary sexual traits (i.e., pheromones or courtship), and not all components of the immune response derive from melanin production (i.e., antimicrobial peptides and ROS). A question therefore arises: what other mechanisms are involved in the trade-off between sexual traits' elaboration and immune response?

In general, there are three mechanisms that could promote immune response signaling associated with the exuberance of secondary sexual traits: (1) effect of hormones, (2) oxidative stress damage, and (3) availability of resources. In insects, only 10 of 28 studies have analyzed a potential physiological link between secondary sexual traits and immune response, and in 5 cases authors have suggested possible mechanisms (Table 3.2). However, all these studies focused only on the effect of hormones and availability of resources. The conceptual framework for each of the three mechanisms is briefly reviewed below.

Hormones

Hamilton and Zuk's (1982) publication led to a wave of experimental studies motivated by the question of whether males with more elaborate secondary sexual traits had fewer parasites and pathogens when compared to those with less elaborate traits (Milinski, 2001). This same publication spurred many researchers to carry out studies on sexual selection and the physiology that underlies resistance. Folstad and Karter (1992) put forward the immunocompetence handicap hypothesis (IHH), in which they propose that the endocrine and immune systems could provide the physiological basis for understanding the reliability of signs of health and viability, and suggest that testosterone could be a reliable link between the immune response and secondary sexual traits because it favors the expression of the latter at the expense of the former. For example, for courtship displays the development of attributes such as long feathers, bright colors, longer dances, and an increase in aggression are helpful in the competition for access to females, but at the same time diminish the efficiency of the immune response (Andersson, 1994; Demas and Nelson, 2011; but see also Roberts et al., 2004). A recent study in humans suggests that female preference for certain facial features in males is positively related to the amount of testosterone, and that this hormone is positively related to the production of antibodies against the hepatitis B virus (Rantala et al., 2012). These studies therefore suggest that the fittest males should be the ones to develop secondary sexual traits most intensively, despite the disadvantage imposed by the expression of testosterone upon the immune response (but see Roberts et al., 2004).

Insects, which lack testosterone (Chapman, 1998), also show a positive relation between the degree of expression of secondary sexual traits and the effectiveness of the immune response (Table 3.2). Thus, it has been proposed that a mechanism different from testosterone must exist to account for the IHH (Rantala et al., 2003a; Contreras-Garduño et al., 2009a). As the Juvenile Hormone (JH) favors reproduction (Rolff and Siva-Jothy, 2002) and is functionally analogous to testosterone (reviewed in Crook et al., 2008), it was suggested that JH mediates a trade-off between the expression of secondary sexual traits and the effectiveness of the immune response (Rantala *et al.*, 2003a; Contreras-Garduño *et al.*, 2009a). In *T. molitor*, Rantala *et al.* (2002, 2003b) found that males that produce more pheromones are preferred by females and have a better immune response. Subsequently, in another study it was shown that an increase in JH favors the production of pheromones but diminishes the immune response (Rantala *et al.*, 2003a). Additionally, in the damselfly *C. virgo*, Koskimäki *et al.* (2004) found that males with more pigmented wings won the competition for territories over males with less pigmented wings, and that a positive relation exists between melanization and the size of the spot on the wing. In a later study, it was shown that in this same species the presence of JH favors competitiveness for territories and the degree of wing pigmentation (Contreras-Garduño *et al.*, 2009a; 2011a), but diminishes the immune response (Contreras-Garduño *et al.*, 2009a). In addition, in the neotropical damselfly *H. americana*, phenoloxidase production was reduced 3 hours after treatment with a JH (Contreras-Garduño *et al.*, 2012), and in natural conditions non-JH treated males survived better than males treated with JH after both groups were challenged with bacteria (González-Tokman *et al.*, 2012).

The above evidence suggests that JH is a mechanistic link between secondary sexual traits, immune response, and resistance to immune challenge. However, the role of JH in the evolution of secondary sexual traits and immune response is a topic that still requires attention from molecular, biochemical, physiological, and evolutionary perspectives. First, so far, the effect of JH has only been evaluated in a handful of species. For instance, this hormone favors the development of horns (Emlen and Nijhout, 1999) and jaws (Gotoh *et al.*, 2011) in beetles, the stalks in stalk-eyed flies (Fry, 2006), wing pigmentation and aggressiveness in damselflies (Contreras-Garduño *et al.*, 2009a, 2011a), and pheromone production in mealworms (Rantala *et al.*, 2003a). Thus, it should be determined whether JH favors or decreases the expression of a wide variety of secondary sexual traits (and its potential role in primary sexual traits; see Fry, 2006). Second, studies in different species have determined that there are at least eight types of juvenile hormones, and although JHIII is the most common (Wheeler and Nijhout, 2003), studies of sexual selection should consider the specific JH of each species. Third, it is very difficult to determine physiological levels of JH. Various studies have used high doses of JH, the effect of which may be more pharmacological than physiological on the variables under study, since natural levels of this hormone are not known in many insects (Zera, 2007). Fourth, JH does not always have a negative effect on the immune response; it could even favor some immune markers (Tian *et al.*, 2010; Villanueva *et al.*, 2013) or might not affect other markers (Rantala *et al.*, 2003a; Villanueva *et al.*, 2013). To date, the causes of what triggers the positive, negative, or null effects remain unknown (Tian *et al.*, 2010; Adamo, 2011; Villanueva *et al.*, 2013). Fifth, only one study has tested the sexual dimorphism of JH upon immune response (Villanueva *et al.*, 2013), and more studies are needed to establish how general this observation is because if JH is the mechanisms behind the signal of males immunocompetence, then this hormone should only affect males and not the females (Jacobs and Zuk, 2011). Sixth, and most importantly from the evolutionary point of view, the consequence of the relationship between JH and secondary sexual traits on direct or indirect female benefits has not been studied.

Independently of the problems above that need to be addressed, a question emerges: Why would it be useful to study JH using neotropical insect populations? As reviewed in Chapter 1, temperature, pluviosity, and photoperiod vary throughout the Neotropics, and JH titers could vary accordingly, therefore possibly favoring the development of secondary sexual traits. Temperature affects JH titers: the colder the temperature gets, the lower the titer (Huang and Robinson, 1995). In the cotton bollworm *Helicoverpa armigera*, for example, reduced temperature influenced JH and pheromone production, affecting behavior and egg development (Zhou *et al.*, 2000; but see also Geister *et al.*, 2008). In addition, as suggested in Chapter 1, the availability of resources also changes according to the climate, and as JH could be favored by the amount of food consumed (see, for example, Hernández-Martínez *et al.*, 2007), the greater the food availability in a population, the higher the JH titers is expected. Clearly, these areas need further research.

The Oxidative Stress Damage Hypothesis

It has been postulated that accumulation of free radicals is linked to the inability of the least fit males to develop secondary sexual traits (Von Schantz *et al.*, 1999). For example, the high demand for oxygen during the production of secondary sexual traits, courtship, and competition for a mate could lead to an accumulation of reactive oxygen species (Von Schantz *et al.*, 1999, but see also Garratt and Brooks, 2012). The least fit males would have a lower production of antioxidants and a lower tolerance to the accumulation of free radicals, leading to a greater cost due to greater damage to proteins, lipids, and genetic material compared to the fittest males (Costantini and Verhulst, 2009; Dowling and Simmons, 2009; Monaghan *et al.*, 2009; Garratt and Brooks, 2012). According to this hypothesis, the accumulation of free radicals during metabolism would compromise the elaboration of secondary sexual traits (Von Schantz *et al.*, 1999; Dowling and Simmons, 2009; Mougeot *et al.*, 2010) or the adequate performance of the immune response because free radicals are also used during the immune response (Von Schantz *et al.*, 1999; see also the second section of this chapter), or oxidative stress could hinder the development of sexual traits (Von Schantz *et al.*, 1999).

It has been found that testosterone promotes oxidative stress in vertebrates (Von Schantz *et al.*, 1999). Although the oxidative damage hypothesis has not been tested in invertebrates, it could be pertinent to test this idea regarding JH, because this hormone favors oxidative stress resistance. For example, in *Drosophila melanogaster*, fertile females treated with methoprene produced more eggs but were less resistant to oxidative stress than solvent control (Salmon *et al.*, 2001). In males, methoprene also favors the expression of secondary sexual traits and reduces the immune response (Rantala *et al.*, 2003a; Contreras-Garduño *et al.*, 2009a). However, as far as we know, there are no studies in insects about the relationship between sexual traits, immune response, and stress resistance. Further studies with insects are required to explore the impact of stress damage on the relationship between immune response and the development of secondary sexual traits, and to determine whether JH is involved as a mechanistic link.

It has been suggested that oxidative stress could vary within species, environments, populations, age, and sexes (Buttemer, *et al.*, 2010; Constantini *et al.*, 2010; Metcalfe and Alonso-Alvarez, 2010; Archer *et al.*, 2013). Hence, neotropical insects offer an ideal opportunity to test these ideas. However, as with immune response and juvenile hormone, some methodological problems still require attention from molecular, biochemical, physiological, and evolutionary perspectives (for review, see Constantini *et al.*, 2010; Hõrak and Cohen, 2010).

Availability of Resources

Sheldon and Verhulst (1996) proposed a simple but elegant idea. They suggested that, given that immune response and sexual traits are costly and resources are finite, animals should allocate their resources to maximize their fitness. Hence, males could advertise honestly their amount of resources. Table 3.2 shows that 7 of 10 studies tested the energy reserves in the form of fat storage. According to this reasoning, the fittest males should have the greatest amount of reserves (i.e., fat stores). Fat bodies are storage organs for lipids, as well as the insect's main immune organ (Arrese and Soulages, 2010). Unfortunately, total body fat was analyzed with obsolete and unreliable methods in all the studies presented in Table 3.2, which may explain the contradictory results relative to the role of fat load as driver of the immune response. An interesting recent study suggested that it is not total body fat itself but instead its transporter, apolipophorin III, that serves as a mediator between immune response and lipid transport after flight-or-fight behavior (Adamo *et al.*, 2008). Further studies are required to determine whether this compound does in fact mediate the relationship between development of secondary sexual traits and immune response, and, if so, its generality in different insect species. In addition, as pointed out in Chapter 1, the amount of resources in the Neotropics is highly variable. Under this scenario, the population variation within and among neotropical insects could be helpful to test not only secondary sexual traits and immune response, but also whether apolipophorin III availability changes according to the availability of resources.

Testing the mechanisms reviewed here will not only provide evidence for the mechanisms behind the relationship between immune response and sexual traits, and how they vary in the Neotropics, but also discern whether such a relationship is better explained by the general condition of the males (i.e., energy stores) or their disease resistance (Jacobs and Zuk, 2011).

INTERACTION BETWEEN SEXUAL TRAITS, IMMUNITY, AND THE EFFECT OF PARASITES AND PATHOGENS

So far, we have discussed the following issues: (1) the mechanisms and relationship between immune response in insects according to the parasites and pathogens confronted; (2) the costs of the immune response; (3) the influence of different biotic or abiotic factors upon the expression of the immune response; (4) the relationship between secondary sexual traits and at least one immune response parameter; and (5) the potential physiological link between secondary sexual traits and the immune response. The theory that males in good condition produce both an immune response and secondary sexual traits at higher levels than males in poor condition has been discussed. However, at least in vertebrates (the most studied group), results are contradictory: some studies report that males with more elaborate secondary sexual traits have stronger immune response than males in poor condition, but in other studies the opposite trend is reported (see Jacobs and Zuk, 2011). A question inevitably follows: Can neotropical insects help resolve these discrepancies?

In order to develop a clear understanding about the relationship between secondary sexual traits and the male immune response, the spatial–temporal host–parasite interaction needs to be considered, determining how the expression of secondary sexual traits and the immune system are influenced by biotic (parasites and pathogens) and abiotic (temperature and water availability) factors, resulting, for example, in local adaptations. The biological effect of the relationship between immune response and secondary sexual traits may become clearer if natural parasites and pathogens are considered (Jacobs and Zuk, 2011), and this relationship could vary between populations in neotropical insects. Secondary sexual traits could vary according to temperature, humidity, and food availability (see Chapter 1) and at least temperature and food availability favor plasticity of the immune response (for review, see Lazzaro and Little, 2009; Demas and Nelson, 2011). Studies with birds have revealed that the immune response is more intense in the tropics when compared with temperate zones (Møller, 1998). Likewise, the prevalence and intensity of illnesses and parasitism increases with a decrease in latitude (Guernier et al., 2004; Douglas et al., 2009), and the greatest parasite richness and derived illnesses probably occur in the equatorial region (Guernier et al., 2004; Douglas et al., 2009). However, this pattern may not be a general rule, as there are contradictory reports in the literature. When considering the immune response, for example, studies with birds have found that in temperate zones the immune response is more intense than in the Neotropics (Martin et al., 2004), or that some but not all immune parameters are strongest near the equator (for example, the humoral response increased but the cellular response did not; Hassequist, 2007), contrary to some studies such as that of Møller (1998). Additionally, when considering infections, parasites, or pathogens, diversity is not associated with a decrease in latitude (Poulin and Mouritsen, 2003).

Studies with insects have also shown contradictory results. For example, a study with house cricket *Acheta domesticus* showed that infections caused by *Rickettsiella grylli* are not influenced by an increase in temperature, presumably because the host moved to a warmer climate and pathogens were more sensitive to higher than lower temperatures (Adamo, 1998). The same study reported that increased temperatures favored host resistance in other types of infections, such as those caused by *S. marcescens* (Adamo, 1998). In the cricket *Grillus texensis*, an increase in temperature resulted in greater phenoloxidase activity and a faster reaction of lysozyme-like enzymatic activity, which favored resistance against an attack by the *S. marcescens* bacteria, but not against *Bacillus cereus* (Adamo and Lovett, 2011).

The exceptions presented above could be explained by biological interactions, where the impact of virulence (from parasites and pathogens) and resistance (from the host) are not predictably expressed according to pluviosity and/or temperature gradients. This interaction has similarities to the geographic mosaic pattern, in which local conditions shape local adaptations (Thompson, 2005; Laine, 2009; Schmid-Hempel, 2011). For example, the virulence of the parasitoid wasp *Leptopilina boulardi* is negatively associated with the number of their host species (*Drosophila*) – in other words, if there is more than one host species in a specific area, the virulence is lower than when there is only one species (Dupas and Boscaro, 1999). Therefore, the parasitoid is locally adapted to the host spectrum (Dupas and Boscaro, 1999). In the case of the wasp *Asobara tabida* against *D. melanogaster*, the parasitoid showed lower virulence in northern, western, and central Europe than in the Mediterranean (Kraaijeveld and Van Der Wel, 1994). However, hosts are not passive agents, and in this case resistance drives the parasite–host interaction. For example, on the one hand the encapsulation ability of *D. melanogaster* from the Congo is better against different *L. boulardi* strains than is the encapsulation ability of flies from the south of France. On the other hand, the *D. melanogaster* encapsulation against *A. tabida* is stronger in central-southern Europe than in the north and southeast of the Iberian Peninsula (Kraaijeveld and Godfray, 1999). Further research revealed that *Drosophila* infected with *Lactococcus lactis* but not with *Pseudomonas aeruginosa* survived better in those populations where more bacteria species were present, but did not follow a latitudinal gradient, suggesting an important influence of biological interactions (Corby-Harris and Promislow, 2008).

It has been proposed that local adaptation generates population differences (Thompson, 2005). For instance, one macroecological study revealed that the species richness of trematodes is more related to local factors of infection and the parasite/host (i.e., snail) relationship than to latitude (Poulin and Mouritsen, 2003). Additionally, a study in the Neotropics revealed that of the 3408 isolates of *B. thuringiensis* from different populations in Brazil, only 62% killed between 80–100% of the moth *Spodoptera frugiperda*, and that there was no association between this parasite's virulence and latitude (Valicente and Barreto, 2003). Studying how neotropical insect populations vary in their immune investment and determining why this variability exists would reveal much about the selection pressures on immune system evolution. Given that the immune response is costly, males could not only signal their resistance but also their tolerance to infection (see Jacobs and Zuk, 2011), and this could vary among populations. Furthermore, the immune response is not solely affected by parasites and pathogens, and animals could be able to adapt rapidly to changes according to parasite species' diversity, virulence, regional abiotic conditions, and phenomenological changes (Møller *et al.*, 2003; Lazzaro and Little, 2009). This dynamism of the host–parasite (pathogen) relationship could explain why sexual traits or sexual selection might vary both temporally and spatially (see, for example, Baena and Macías-Ordóñez, 2012), and according to environmental pressures (either abiotic or biotic; see Cornwallis and Uller, 2010).

It can be concluded that integrated studies that consider numerous factors are required. These could include factors such as the number and virulence of parasites and pathogens afflicting a species, the types of immune responses, the biotic and abiotic factors that favor or diminish such responses, the degree of elaboration of secondary sexual traits, whether these traits are related to mate choice and mate competition and more recently, the contribution of epigenetics to the insects immune response, the elaboration of sexual signals, and the host–parasite relationship. Given the diversity of biotic and abiotic factors in the Neotropics (Chapter 1), descriptive studies of species are needed as well as studies regarding the richness, distribution, and abundance of insect parasites and pathogens. Such research would provide the cornerstone for studies about signaling of the immune response through secondary sexual traits, which would reveal whether in the Neotropics such relationships are related to the local (mosaic) adaptation hypothesis. Furthermore, such studies would reveal whether females are really choosing immune response as a signal of particular resistance, or whether they are choosing the male's general condition, which in turn correlates with resistance and immunity (Adamo and Spiteri, 2005, 2009). Considering the nature of pending research, the Neotropics represents an opportunity to consider multidisciplinary studies that encompass molecular biology, genetics and epigenetics, biochemistry, physiology, ecology, and evolution.

ACKNOWLEDGMENTS

We thank Regina Macedo and Glauco Machado for their invitation to write this chapter. Shelly Adamo and Marlene Zuk provided substantial comments. A grant from CONACyT (152666) was provided to JCG.

REFERENCES

Adamo, S.A., 1998. The specificity of behavioral fever in the cricket *Acheta domesticus*. J. Parasitol. 84, 529–533.

Adamo, S.A., 2004a. Estimating disease resistance in insects: phenoloxidase and lysozyme-like activity and disease resistance in the cricket *Gryllus texensis*. J. Insect Physiol. 50, 209–216.

Adamo, S.A., 2004b. How should behavioural ecologists interpret measurements of immunity? Anim. Behav. 68, 1443–1449.

Adamo, S.A., 2011. The importance of physiology for ecoimmunology: Lessons from the insects. In: Demas, G.E., Nelson, R.J. (Eds.), Ecoimmunology, Oxford University Press, Oxford, pp. 413–449.

Adamo, S.A., Lovett, M., 2011. Some like it hot: The effects of climate change on reproduction, immune function and disease resistance in the cricket *Gryllus texensis*. J. Exp. Biol. 214, 1997–2004.

Adamo, S.A., Spiteri, R., 2005. Female choice for male immunocompetence: when is it worth it? Behav. Ecol. 16, 871–879.

Adamo, S.A., Spiteri, R.J., 2009. He's healthy, but will he survive the plague? Possible constraints on female choice for disease resistance in addition to current health. Anim. Behav. 77, 67–78.

Adamo, S.A., Roberts, J.L., Easy, R., Ross, N.W., 2008. Competition between immune function and lipid transport for the protein apolipophorin III leads to stress-induced immunosuppression in crickets. J. Exp. Biol. 211, 531–538.

Amarasekare, P., Savage, V., 2012. A framework for elucidating the temperature dependence of fitness. Am. Nat. 179, 178–191.

Andersson, M., 1994. Sexual Selection. Princeton University Press, Princeton.

Archer, C.R., Sakaluk, S.K., Selman, C., Royle, N., Hunt, J., 2013. Oxidative stress and the evolution of sex differences in life span and ageing in the decorated cricket, *Gryllodes sigillatus*. Evolution 67, 620–634.

Ardia, D.R., Gantz, J.E., Schneider, B.C., Strebel, S., 2012. Costs of immunity in insects: an induced immune response increases metabolic rate and decreases antimicrobial activity. Funct. Ecol. 26, 732–739.

Armitage, S.A.O., Siva-Jothy, M.T., 2005. Immune function responds to selection for cuticular colour in *Tenebrio molitor*. Heredity 94, 650–656.

Arrese, E.L., Soulages, J.L., 2010. Insect fat body: Energy, metabolism, and regulation. Annu. Rev. Entomol. 55, 207–225.

Bae, Y.S., Choi, M.K., Lee, W.J., 2010. Dual oxidase in mucosal immunity and host-microbe homeostasis. Trends Immunol. 31, 278–287.

Baena, M.L., Macías-Ordóñez, R., 2012. Phenology of scramble polygyny in a wild population of chrysolemid beetles: the opportunity for and the strength of sexual selection. PLoS ONE 7, e38315. http://dx.doi.org/10.1371/journal.pone.0038315.

Beckage, N.E., 2008. Insect Immunology. Elsevier, San Diego.

Bogdan, C., 2001. Nitric oxide and the immune response. Nat. Immunol. 2, 907–916.

Boughton, R.K., Joop, G., Armitage, S.A.O., 2011. Outdoor immunology: methodological considerations for ecologists. Funct. Ecol. 25, 81–100.

Buttemer, W.A., Abele, D., Costantini, D., 2010. From bivalves to birds: oxidative stress and longevity. Funct. Ecol. 24, 971–983.

Cajaraville, M.P., Pal, S.G., Robledo, Y., 1995. Light and electron microscopical localization of lysosomal acid hydrolases in bivalve haemocytes by enzyme cytochemistry. Acta Histochem. Citochem. 28, 409–416.

Carton, Y., Poirié, M., Nappi, A., 2008. Insect immune response to parasitoids. Insect Sci. 15, 67–87.

Castillo, J.C., Reynolds, S.E., Eleftherianos, I., 2011. Insect immune responses to nematode parasites. Trends Parasitol. 27, 537–547.

Cerenius, L., Söderhäll, K., 2004. The prophenoloxidase-activating system in invertebrates. Immunol. Rev. 198, 116–126.

Chapman, R.F., 1998. The Insects. Structure and Function. Cambridge University Press, Cambridge.

Cheng, T.C., 1992. Selective induction of release of hydrolases from *Crassostrea virginica* hemocytes by certain bacteria. J. Invertebr. Pathol. 59, 197–200.

Christensen, B.M., Jianyong, L., Cheng, C.C., Nappi, A.J., 2005. Melanization immune responses in mosquito vectors. Trends Parasitol. 21, 192–199.

Contreras-Garduño, J., Canales-Lazcano, J., Córdoba-Aguilar, A., 2006. Wing pigmentation, immune ability and fat reserves in males of the rubyspot damselfly, *Hetaerina americana*. J. Ethol. 24, 165–173.

Contreras-Garduño, J., Lanz-Mendoza, H., Córdoba-Aguilar, A., 2007. The expression of a sexually selected trait correlates with different immune defense components and survival in males of the American rubyspot. J. Insect Physiol. 53, 612–621.

Contreras-Garduño, J., Buzatto, B., Serrano-Meneses, M., Nájera-Cordero, K., Córdoba-Aguilar, A., 2008. The size of the wing red spot as a heightened condition dependent trait in the American rubyspot. Behav. Ecol. 19, 724–732.

Contreras-Garduño, J., Córdoba-Aguilar, A., Lanz-Mendoza, H., Cordero Rivera, A., 2009a. Territorial behaviour and immunity are mediated by juvenile hormone: the physiological basis of honest signaling? Funct. Ecol. 23, 159–163.

Contreras-Garduño, J., Canales-Lazcano, J., Jiménez-Cortés, J.G., Juárez-Valdez, N., Lanz-Mendoza, H., Córdoba-Aguilar, A., 2009b. Spatial and temporal population differences in male density and condition in the American rubyspot, *Hetaerina americana* (Insecta: Calopterygidae). Ecol. Res. 24, 21–29.

Contreras-Garduño, J., Córdoba-Aguilar, A., Azpilicueta-Amorín, M., Cordero-Rivera, A., 2011a. Juvenile hormone favors sexually-selected traits in males and females but impairs fat reserves and abdomen mass. Evol. Ecol. 25, 845–856.

Contreras-Garduño, J., Córdoba-Aguilar, A., Martínez-Becerril, R.I., 2011b. The relationship between male wing pigmentation and condition in *Erythrodiplax funerea* (Hagen) (Anisoptera: Libellulidae). Odonatologica 40, 89–94.

Contreras-Garduño, J., Alonso-Salgado, A., Villanueva, G., 2012. Phenoloxidase production: the importance of time after juvenile hormone analogue administration in *Hetaerina americana* Fabricius. Odonatologica 41, 1–6.

Corby-Harris, V., Promislow, D.E.L., 2008. Host ecology shapes geographic variation for resistance to bacterial infection in *Drosophila melanogaster*. J. Anim. Ecol. 77, 768–776.

Córdoba-Aguilar, A., Méndez, V., 2006. Immune melanization ability and male territorial status in *Erythemis vesiculosa* (Fabricius) (Anisoptera: Libellulidae). Odonatologica 35, 193–197.

Córdoba-Aguilar, A., Lesher-Treviño, A.C., Anderson, H.N., 2007. Sexual selection in *Hetaerina titia* males: a possible key species to understand the evolution of pigmentation in calopterygid damselflies (Odonata: Zygoptera). Behaviour 144, 931–952.

Córdoba-Aguilar, A., Jiménez-Valdés, J.G., Lanz-Mendoza, H., 2009. Seasonal variation in ornament expression, body size, energetic reserves, immune response and survival in males of a territorial insect. Ecol. Entomol. 34, 228–239.

Cornwallis, C.K., Uller, T., 2010. Towards an evolutionary ecology of sexual traits. Trends Ecol. Evol. 25, 145–152.

Costa, A., Jan, E., Sarnow, P., Schneider, D., 2009. The IMD pathway is involved in antiviral immune responses in *Drosophila*. PLoS ONE 4, e7436. http://dx.doi.org/10.1371/journal.pone.0007436.

Costantini, D., Verhulst, S., 2009. Does high antioxidant capacity indicate low oxidative stress? Funct. Ecol. 23, 506–509.

Costantini, D., Rowe, M., Butler, M.W., McGraw, K.J., 2010. From molecules to living systems: historical and contemporary issues in oxidative stress and antioxidant ecology. Funct. Ecol. 24, 950–959.

Criscitiello, M.F., de Figueiredo, P., 2013. Fifty shades of immune defense. PLoS Pathogens 9, e1003110. http://dx.doi.org/10.1371/journal.ppat.1003110.

Crook, T.C., Flatt, T., Smiseth, P.T., 2008. Hormonal modulation of larval begging and growth in the burying beetle *Nicrophorus vespilloides*. Anim. Behav. 75, 71–77.

Cueva del Castillo, R., 2007. Sexual selection in tropical insects. In: Del Claro, K. (Ed.), International Commission on Tropical Biology and Natural Resources, Encyclopedia of Life Support Systems (EOLSS), Developed under the Auspices of the UNESCO, EOLSS Publishers, Oxford. http://www.colss.net.

Darwin, C., 1871. The Descent of Man, and Selection in Relation to Sex. Murray, London.

Demas, G., Nelson, R., 2011. Ecoimmunology. Oxford University Press, Oxford.

Demuth, J.P., Naidu, A., Mydlarz, L.D., 2012. Sex, war, and disease: the role of parasite infection on weapon development and mating success in a horned beetle (*Gnatocerus cornutus*). PLoS ONE 7, e28690. http://dx.doi.org/10.1371/journal.pone.0028690.

Douglas, W.S., Mittelbach, G.G., Cornell, H.V., Sobel, J.M., Roy, K., 2009. Is there a latitudinal gradient in the importance of biotic interactions? Annu. Rev. Ecol. Evol. Syst. 40, 245–269.

Dowling, D.K., Simmons, L.W., 2009. Reactive oxygen species as universal constraints in life-history evolution. Proc. R. Soc. B. 276, 737–1745.

Drury, J.P., 2010. Immunity and mate choice: a new outlook. Anim. Behav. 79, 539–545.

Dupas, S., Boscaro, M., 1999. Geographic variation and evolution of immunosuppressive genes in a *Drosophila* parasitoid. Ecography 22, 284–291.

Durdevic, A., Hanna, K., Gold, B., Pollex, T., Cherry, S., Lyko, F., Schaefer, M., 2013. Efficient RNA virus control in *Drosophila* requires the RNA methyltransferase Dnmt2. EMBO 14, 269–275.

Emlen, D.J., Nijhout, H.F., 1999. Hormonal control of male horn length dimorphism in the dung beetle *Onthophagus taurus* (Coleoptera: Scarabaeidae). J. Insect Physiol. 45, 45–53.

Fang, F.C., 1997. Mechanisms of nitric oxide-related antimicrobial activity. J. Clin. Invest. 100, S43–S50.

Ferrandon, D., Imler, J.L., Hetru, C., Hoffmann, J.A., 2007. The *Drosophila* systemic immune response: sensing and signaling during bacterial and fungal infections. Nat. Rev. Immunol. 7, 862–874.

Foley, E., O'Farrell, P.H., 2003. Nitric oxide contributes to induction of innate immune response to Gram-negative bacteria in *Drosophila*. Genes Dev. 17, 115–125.

Folstad, I., Karter, A.J., 1992. Parasites, bright males and the immunocompetence handicap. Am. Nat. 139, 604–622.

Fry, C.L., 2006. Juvenile hormone mediates a trade-off between primary and secondary sexual traits in stalk-eyed flies. Evol. Dev. 8, 191–201.

Fullaondo, A., Lee, S.Y., 2012. Regulation of *Drosophila*–virus interaction. Dev. Comp. Immunol. 36, 262–226.

Garratt, M., Brooks, R.C., 2012. Oxidative stress and condition-dependent sexual signals: more than just seeing red. Proc. R. Soc. 22, 3121–3130.

Geister, T.L., Lorenz, M.W., Meyering-Vos, M., Hoffmann, K.H., Fischer, K., 2008. Effects of temperature on reproductive output, egg provisioning, juvenile hormone and vitellogenin titres in the butterfly *Bicyclus anynana*. J. Insect Physiol. 54, 1253–1260.

Gómez-Díaz, E., Jordà, M., Peinado, M.A., Rivero, A., 2012. Epigenetics of host–pathogen interactions: the road ahead and the road behind. PLoS Pathog. 8, e100300.

Godin, J.G., Dugatkin, L.A., 1996. Female mating preference for bold males in the guppy, *Poecilia reticulata*. Proc. Natl. Acad. Sci. U. S. A. 93, 10262–10267.

González-Santoyo, I., Córdoba-Aguilar, A., González-Tokman, D.M., Lanz-Mendoza, H., 2010. Phenoloxidase activity and melanization do not always covary with sexual trait expression in *Hetaerina* damselflies (Insecta: Calopterygidae). Behaviour 147, 1285–1307.

González-Tokman, D.M., Munguía-Steyer, R.E., González-Santoyo, I., Baena-Díaz, F., Córdoba-Aguilar, A., 2012. Support for the immunocompetence handicap hypothesis in the wild: hormonal manipulation decreases survival in sick damselflies. Evolution 66, 3294–3301.

Gotoh, H., Cornette, R., Koshikawa, S., Okada, Y., Lavine, L.C., Emlen, D.J., Muran, T., 2011. Juvenile hormone regulates extreme mandible growth in male stag beetles. PLoS ONE 6, e21139. http://dx.doi.org/10.1371/journal.pone.0021139.

Grether, G.F., 1996. Intrasexual competition alone favors a sexually dimorphic ornament in the rubyspot damselfly Hetaerina americana. Evolution 50, 1949–1957.

Guernier, V., Hochberg, M.E., Guégan, J.-F., 2004. Ecology drives the worldwide distribution of human diseases. PLoS Biol. 2, e141. http://dx.doi.org/10.1371/journal.pbio.0020141.

Hamilton, W.D., Zuk, M., 1982. Heritable true fitness and bright birds: a role for parasites? Science 218, 384–386.

Hasselquist, D., 2007. Comparative immunoecology in birds: Hypotheses and tests. J. Ornithol. 148, 271–582.

Hernández-Martínez, S., Mayorala, J.G., Yiping, L., Noriega, F.G., 2007. Role of juvenile hormone and allatotropin on nutrient allocation, ovarian development and survivorship in mosquitoes. J. Insect Physiol. 53, 230–234.

Herrera-Ortíz, A., Lanz-Mendoza, H., Martínez-Bernetche, J., Hernández-Martínez, S., Villareal-Treviño, C., Aguilar-Marcelino, L., Rodríguez, M.H., 2004. Plasmodium berghei ookinetes induce nitric oxide production in Anopheles pseudopunctipennis midguts cultured in vitro. Insect Biochem. Mol. Biol. 34, 893–901.

Hetru, C., Hoffmann, D., Bulet, P., 1998. Antimicrobial peptides from insects. In: Brey, P.T., Hultmark, D. (Eds.), Molecular Mechanism of Immune Response in Insects, Chapman and Hall, London, pp. 40–46.

Hetru, C., Troxler, L., Hoffmann, J.A., 2003. Drosophila melanogaster antimicrobial defense. J. Infect. Dis. 187, S327–S334.

Hõrak, P., Cohen, A., 2010. How to measure oxidative stress in an ecological context: methodological and statistical issues. Funct. Ecol. 24, 960–970.

Huang, Z.Y., Robinson, G.E., 1995. Seasonal changes in juvenile hormone titers and rates of biosynthesis in honey bees. J. Comp. Physiol. B. 165, 18–28.

Iwanaga, S., Lee, B.L., 2005. Recent advances in the innate immunity of invertebrate animals. J. Biochem. Mol. Biol. 38, 128–150.

Jacobs, A.C., Zuk, M., 2011. Sexual selection and parasites: Do mechanisms matter? In: Demas, G., Nelson, R. (Eds.), Ecoimmunology, Oxford University Press, Oxford, pp. 468–496.

Jacot, A., Scheuber, J., Brinkhof, M.W.G., 2004. Costs of an induced immune response on sexual display and longevity in field crickets. Evolution 58, 2280–2286.

Jokela, J., 2010. Transgenerational immune priming as cryptic parental care. J. Anim. Ecol. 79, 305–307.

Kanost, M.R., Gorman, M.J., 2008. Phenoloxidases in insect immunity. In: Beckage, N.E. (Ed.), Insect Immunology, Academic Press/Elsevier, Oxford, pp. 69–98.

Koskimäki, J., Rantala, M.J., Suhonen, J., Taskinen, J., Tynkkynen, K., 2004. Immunocompetence and resource holding potential in the damselfly Calopteryx virgo L. Behav. Ecol. 15, 169–173.

Kraaijeveld, A., Godfray, H.C.J., 1997. Trade-off between parasitoid resistance and larval competitive ability in Drosophila melanogaster. Nature 389, 278–280.

Kraaijeveld, A., Godfray, H.C.J., 1999. Geographic patterns in the evolution of resistance and virulence in Drosophila and its parasitoids. Am. Nat. 153, S61–S74.

Kraaijeveld, A., Van Der Wel, N.N., 1994. Geographic variation in reproductive success of the parasitoid Asobara tabida in larvae of several Drosophila species. Ecol. Entomol. 19, 221–229.

Kurtz, J., 2005. Specific memory within innate immune systems. Trends Immunol. 26, 186–192.

Kurtz, J., Sauer, K.P., 1999. The immunocompetence handicap hypothesis: testing the genetic predictions. J. R. Soc. Med. B. 266, 2515–2522.

Laine, A.L., 2009. Role of coevolution in generating biological diversity: spatially divergent selection trajectories. J. Exp. Bot. 60, 2957–2970.

Lavine, M.D., Strand, M.R., 2002. Insect hemocytes and their role in immunity. Insect Biochem. Mol. Biol. 32, 1295–1309.

Lawniczak, M.K.N., Barnes, A.I., Linklater, J.R., Boone, J.M., Wigby, S., Chapman, T., 2007. Mating and immunity in invertebrates. Trends Ecol. Evol. 22, 48–55.

Lazzaro, B.P., Little, T.J., 2009. Immunity in a variable world. Philos. Trans. R. Soc. B. 364, 15–26.

Leclerc, V., Reichhart, J.M., 2004. The immune response of *Drosophila melanogaster*. Immunol. Rev. 198, 59–71.

Lemaitre, B., Hoffmann, J., 2007. The host defense of *Drosophila melanogaster*. Annu. Rev. Immunol. 25, 697–743.

Litman, G.W., Cannon, J.P., Dishaw, L.J., 2005. Reconstructing immune phylogeny: New perspectives. Nature 8, 866–879.

Little, T.J., Hultmark, D., Read, A.F., 2005. Invertebrate immunity and the limits of mechanistic immunology. Nat. Immunol. 6, 651–654.

Loker, E.S., 2012. Macroevolutionary immunology: A role for immunity in the diversification of animal life. Front. Immunol. 3, 25. http://dx.doi.org/10.3389/fimmu.2012.00025.

Luckhart, S., Vodovotz, Y., Cui, L., Rosenberg, R., 1998. The mosquito *Anopheles stephensi* limits malaria parasite development with inducible synthesis of nitric oxide. Proc. Natl. Acad. Sci. U. S. A. 95, 5700–5705.

Marmaras, V.J., Lampropoulou, M., 2009. Regulators and signaling in insect haemocyte immunity. Cell. Signal. 21, 186–195.

Martin, L.B., Pless, M., Svoboda, J., Wikelski, M., 2004. Immune activity in temperate and tropical house sparrows: A common-garden experiment. Ecology 85, 2323–2331.

Mayhew, P.J., 2007. Why are there so many insect species? Perspectives from fossils and phylogenies. Biol. Rev. 82, 425–454.

McKean, K.A., Lazzaro, B.P., 2011. The costs of immunity and the evolution of immunological defense mechanisms. In: Heyland, A., Flatt, T. (Eds.), Molecular Mechanisms of Life History Evolution, Oxford University Press, Oxford, pp. 299–310.

Metcalfe, N.B., Alonso-Alvarez, C., 2010. Oxidative stress as a life-history constraint: the role of reactive oxygen species in shaping phenotypes from conception to death. Funct. Ecol. 24, 984–996.

Milinski, M., 2001. Bill Hamilton, sexual selection, and parasites. Behav. Ecol. 12, 264–266.

Møller, A.P., 1998. Evidence of larger impact of parasites on hosts in the tropics: Investment in immune function within and outside the tropics. Oikos 82, 265–270.

Møller, A.P., Erritzøe, J., Saino, N., 2003. Seasonal changes in immune response and parasite impact on host. Am. Nat. 161, 657–671.

Monaghan, P., Metcalfe, N.B., Torres, R., 2009. Oxidative stress as a mediator of life history trade-offs: mechanisms, measurements and interpretation. Ecol. Lett. 12, 75–92.

Moret, Y., Schmid-Hempel, P., 2000. Survival for immunity: Activation of the immune system has a price for bumblebee workers. Science 290, 1166–1168.

Mostafa, A.M., Fields, P.G., Holliday, N.J., 2005. Effect of temperature and relative humidity on the cellular defense response of *Ephestia kuehniella* larvae fed *Bacillus thuringiensis*. J. Invertebr. Pathol. 90, 79–84.

Mougeot, F., Martínez-Padilla, J., Blount, J.D., Pérez-Rodríguez, L., Webster, L.M., Piertney, S.B., 2010. Oxidative stress and the effect of parasites on a carotenoid-based ornament. J. Exp. Biol. 213, 400–407.

Mukherjee, K., Fischer, R., Vilcinskas, A., 2012. Histone acetylation mediates epigenetic regulation of transcriptional reprogramming in insects during metamorphosis, wounding and infection. Front. Zool. 9, 25. http://dx.doi.org/10.1186/1742-9994-9-25.

Nappi, A.J., Christensen, B.M., 2005. Melanogenesis and associated cytotoxic reactions: applications to insect innate immunity. Insect Biochem. Mol. Biol. 35, 443–459.

Nappi, A.J., Ottaviani, E., 2000. Cytotoxicity and cytotoxic molecules in invertebrates. BioEssays 22, 469–480.

Nappi, A.J., Vass, E., Frey, F., Carton, Y., 1995. Superoxide anion generation in *Drosophila* during melanotic encapsulation of parasites. Eur. J. Cell Biol. 68, 450–456.

Norris, K., Evans, M.R., 2000. Ecological immunology: life history trade-offs and immune defense in birds. Behav. Ecol. 11, 19–26.

Obbard, D.J., Jiggins, F.M., Little, T.J., 2006. Rapid evolution of antiviral RNAi genes. Curr. Biol. 16, 580–585.

Owusu-Ansah, E., Banerjee, U., 2009. Reactive oxygen species prime *Drosophila* haematopoietic progenitors for differentiation. Nature 461, 537–541.

Pomfret, J.C., Knell, R.J., 2006. Immunity and the expression of a secondary sexual trait in a horned beetle. Behav. Ecol. 17, 466–472.

Poulin, R., Mouritsen, K.N., 2003. Large-scale determinants of trematode infections in intertidal gastropods. Mar. Ecol. Prog. Ser. 254, 187–198.

Raihani, G., Serrano-Meneses, M.A., Córdoba-Aguilar, A., 2008. Male mating tactics in the American rubyspot damselfly: territoriality, nonterritoriality and switching behaviour. Anim. Behav. 75, 1851–1860.

Rantala, M.J., Kortet, R., 2003. Courtship song and immune function in the field cricket *Gryllus bimaculatus*? Biol. J. Linnean Soc. 79, 503–510.

Rantala, M.J., Kortet, R., 2004. Male dominance and immunocompetence in a field cricket. Behav. Ecol. 15, 187–191.

Rantala, M.J., Koskimäki, J., Suhonen, J., Taskinen, J., Tynkkynen, K., 2000. Immunocompetence, developmental stability and wing spot size in *Calopteryx splendens*. Proc. R. Soc. B. 267, 2453–2457.

Rantala, M.J., Jokinen, I., Kortet, R., Vainikka, A., Suhonen, J., 2002. Do pheromones reveal immunocompetence? Proc. R. Soc. B. 269, 1681–1685.

Rantala, M.J., Kortet, R., Vainikka, A., 2003a. The role of juvenile hormone in immune function and pheromone production trade-offs: a test of the immunocompetence handicap principle. Proc. R. Soc. B. 270, 2257–2261.

Rantala, M.J., Kortet, R., Kotiaho, J.S., Vainikka, A., Suhonen, J., 2003b. Condition dependence of pheromones and immune function in the grain beetle *Tenebrio molitor*. Funct. Ecol. 17, 534–540.

Rantala, M.J., Roff, D.A., Rantala, L.M., 2007. Forceps size and immune function in the European earwig *Forficula auricularia*. Biol. J. Linnean Soc. 90, 509–516.

Rantala, M.J., Honkavaara, J., Suhonen, J., 2010. Immune system activation interacts with territory-holding potential and increases predation of the damselfly *Calopteryx splendens* by birds. Oecologia 163, 825–832.

Rantala, M.J., Moore, F.R., Skrinda, I., Krama, T., Kivleniece, I., Kecko, S., Krams, I., 2012. Evidence for the stress-linked immunocompetence handicap hypothesis in humans. Nat. Commun. 3, 694. http://dx.doi.org/10.1038/ncomms1696.

Riddell, C., Adams, S., Schmid-Hempel, P., Mallon, E.B., 2009. Differential expression of immune defenses is associated with specific host–parasite interactions in insects. PLoS ONE 4, e7621. http://dx.doi.org/10.1371/journal.pone.0007621.

Riley, P.A., 1997. Molecules in focus. Int. J. Biochem. Cell Biol. 29, 1235–1239.

Rizki, T.M.R., Rizki, M., Bellotti, R.A., 1985. Genetics of a *Drosophila* phenoloxidase. Mol. Genet. Genomics 201, 7–13.

Roberts, M.L., Buchanan, K.L., Evans, M.R., 2004. Testing the immunocompetence handicap hypothesis: a review of the evidence. Anim. Behav. 68, 227–239.

Rolff, J., Reynolds, S.E. (Eds.), 2009. Insect Infection and Immunity: Evolution, Ecology and Mechanisms, Oxford University Press, Oxford.

Rolff, J., Siva-Jothy, M.T., 2002. Copulation corrupts immunity: a mechanism for a cost of mating. Proc. Nat. Acad. of Sci. U. S. A. 99, 9916–9918.

Rolff, J., Siva-Jothy, M.T., 2003. Invertebrate ecological immunology. Science 301, 472–475.

Rolff, J., Siva-Jothy, M.T., 2004. Selection on insect immunity in the wild. Proc. R. Soc. B. 271, 2157–2160.

Roth, O., Joop, G., Eggert, H., Hilbert, J., Daniel, J., Schmid-Hempel, P., Kurtz, J., 2009. Strain-specific priming of resistance in the red flour beetle, Tribolium castaneum. Proc. R. Soc. B. 276, 145–151.

Rowe, L., Houle, D., 1996. The lek paradox and the capture of genetic variance by condition dependent traits. Proc. R. Soc. B. 263, 1415–1421.

Royet, J., Reichhart, J.M., Hoffmann, J.A., 2005. Sensing and signalling during infection in *Drosophila*. Curr. Opin. Immunol. 17, 11–17.

Ruiz-Guzmán, G., Canales-Lazcano, J., Jiménez-Cortés, J.G., Contreras-Garduño, J., 2012. Sexual dimorphism in immune response: testing the hypothesis in an insect species with two male morphs. Insect Sci.http://dx.doi.org/10.1111/j.1744-7917.2012.01551, Nov. 26.

Ryder, J., Siva-Jothy, M.T., 2000. Male calling song provides reliable information about immune function in a cricket. J. R. Soc. Med. B. 267, 1171–1175.

Sabin, L.R., Hanna, S.L., Cerry, S., 2010. Innate antiviral immunity in *Drosophila*. Curr. Opin. Immunol. 22, 4–9.

Sadd, B.M., Holman, L., Armitage, H., Lock, F., Marland, R., Siva-Jothy, M.T., 2006. Modulation of sexual signaling by immune challenged male mealworm beetles (*Tenebrio molitor*, L.): evidence for terminal investment and dishonesty. J. Evol. Biol. 19, 321–325.

Salmon, A.B., Marx, D.B., Harshman, L.G., 2001. A cost of reproduction in *Drosophila melanogaster*: stress susceptibility. Evolution 55, 1600–1608.

Schmid-Hempel, P., 2003. Variation in immune defence as a question in evolutionary ecology. J. R. Soc. B. 270, 357–366.

Schmid-Hempel, P., 2005. Evolutionary ecology of insect immune defenses. Annu. Rev. Entomol. 50, 529–551.

Schmid-Hempel, P., 2011. Evolutionary Parasitology: the Integrated Study of Infections, Immunology, Ecology, and Genetics. Oxford University Press, Oxford.

Serrano-Meneses, M.A., Córdoba-Aguilar, A., Méndez, V., Layen, S.J., Székely, T., 2007. Sexual size dimorphism in the American rubyspot: male body size predicts male competition and mating success. Anim. Behav. 73, 987–997.

Sheldon, B.C., Verhulst, S., 1996. Ecological immunology: costly parasite defenses and trade-offs. Trends Ecol. Evol. 11, 317–321.

Simmons, L.W., Zuk, M., Rotenberry, J.T., 2005. Immune function reflected in calling song characteristics in a natural population of the cricket *Teleogryllus commodus*. Anim. Behav. 69, 1235–1241.

Simmons, L.W., Tinghitella, R.M., Zuk, M., 2010. Quantitative genetic variation in courtship song and the covariation with immune function and sperm quality in the field cricket *Teleogryllus oceanicus*. Behav. Ecol. 21, 1330–1336.

Siva-Jothy, M.T., 1999. Male wing pigmentation may affect reproductive success via female choice in a calopterygid damselfly (Zygoptera). Behaviour 36, 1365–1377.

Siva-Jothy, M.T., 2000. A mechanistic link between parasite resistance and expression of a sexually selected trait in a damselfly. Proc. R. Soc. B. 267, 2523–2527.

Siva-Jothy, M.T., Thompson, J.W., 2002. Short-term nutrient deprivation affects adult immune function in the mealworm beetle *Tenebrio molitor* L. Physiol. Entomol. 27, 206–212.

Siva-Jothy, M.T., Moret, Y., Rolff, J., 2004. Evolutionary ecology of insect immunity. Adv. Insect Physiol. 32, 1–48.

Söderhäll, K., 2010. Invertebrate Immunity. Advances in Experimental Medicine and Biology Springer, Austin.

Stange, N., Ronacher, B., 2012. Grasshopper calling songs convey information about condition and health of males. J. Comp. Physiol. A. 198, 309–318.

Stearns, S.C., 1992. The Evolution of Life Histories. Oxford University Press, Oxford.

Steiner, H., 2004. Peptidoglycan recognition proteins: on and off switches for innate immunity. Immunol. Rev. 198, 83–86.

Tian, I., Guo, E., Diao, Y., Zhou, S., Peng, Q., Cao, Y., Ling, E., Li, S., 2010. Genome-wide regulation of innate immunity by juvenile hormone and 20-hydroxyecdysone in the *Bombyx* fat body. BMC Genomics 11, 549. http://dx.doi.org/10.1186/1471-2164-11-549.

Thompson, J.N., 2005. The Geographic Mosaic of Coevolution. University of Chicago Press, Chicago.

Tregenza, T., Simmons, L.W., Wedell, N., Zuk, M., 2006. Female preference for male courtship song and its role as a signal of immune function and condition. Anim. Behav. 72, 809–818.

Tsakas, T.S., Marmaras, V.J., 2010. Insect immunity and its signaling: an overview. Invert. Surviv. J. 7, 228–238.

Tzou, P., De Gregorio, E., Lemaitre, B., 2002. How *Drosophila* combats microbial infection: a model to study innate immunity and host-pathogen interactions. Curr. Opin. Microbiol. 5, 102–110.

Valicente, F.H., Barreto, M.R., 2003. *Bacillus thuringiensis* survey in Brazil: geographical distribution and insecticidal activity against *Spodoptera frugiperda* (J. E. Smith) (Lepidoptera: Noctuidae). Neotropical Entomol. 32, 639–644.

Valtonen, T.M., Kleino, A., Rämet, M., Rantala, M.J., 2010. Starvation reveals maintenence cost of humoral immunity. Evol. Biol. 37, 49–57.

Villanueva, G., Lanz-Mendoza, H., Hernández-Martínez, S., Sánchez Zavaleta, M., Manjarrez, J., Contreras-Garduño, J.M., Contreras-Garduño, J., 2013. In the monarch butterfly the juvenile hormone effect upon immune response depends on the immune marker and is sex dependent. Open J. Ecol. 3, 53–58. http://dx.doi.org/10.4236/oje.2013.31007.

Von Schantz, T., Bensch, S., Grahn, M., Hasselquist, D., Wittzell, H., 1999. Good genes, oxidative stress and condition-dependent sexual signals. Proc. R. Soc. B. 266, 1–12.

Wheeler, D.E., Nijhout, H.F., 2003. A perspective for understanding the modes of juvenile hormone action as a lipid signaling system. BioEssays 25, 994–1001.

Zahavi, A., 1975. Mate selection: a selection for a handicap. J. Theor. Biol. 53, 205–214.

Zahavi, A., 1977. The cost of honesty (further remarks on the handicap principle). J. Theor. Biol. 67, 603–605.

Zera, A., 2007. Endocrine analysis in evolutionary-developmental studies of insect polymorphism: hormone manipulation versus direct measurement of hormonal regulators. Evol. Dev. 9, 499–513.

Zhou, X., Coll, M., Applebaum, S.W., 2000. Effect of temperature and photoperiod on juvenile hormone biosynthesis and sexual maturation in the cotton bollworm, *Helicoverpa armigera*: implications for life history traits. Insect Biochem. Mol. Biol. 30, 863–868.

Zuk, M., 2009. The sicker sex. PLoS Pathogens 5, e1000267. http://dx.doi.org/10.1371/journal.ppat.1000267.

Zuk, M., Stoehr, A.M., 2002. Immune defense and life history. Am. Nat. 160, s9–s22.

Chapter 4

Territorial Mating Systems in Butterflies: What We Know and What Neotropical Species Can Show

Paulo Enrique Cardoso Peixoto[1] and Luis Mendoza-Cuenca[2]

[1]*Laboratório de Entomologia, Departamento de Ciências Biológicas, Universidade Estadual de Feira de Santana, Feira de Santana, Brazil,* [2]*Laboratorio de Ecología de la Conducta, Facultad de Biología, Universidad Michoacana de San Nicolás de Hidalgo, Morelia, Michoacán, Mexico*

WHAT ARE THE ORIGINAL CONCEPTS FOR TERRITORIAL MATING SYSTEMS?

Males and females may use a great diversity of behaviors to find mates (Andersson, 1994). Males, in particular, often have a relative lower reproductive investment per reproductive cell or parental effort per offspring compared with females (Williams, 1966; Trivers, 1972). Consequently, receptive females are often less available and choosier than males when selecting a mate – a pattern that increases the intensity of sexual selection on males (Bateman, 1948). Given this disparity, many of the most impressive sexual adaptations related to mate locating and mate acquisition are described for males (Andersson, 1994; Höglund and Alatalo, 1995; Oliveira *et al.*, 2008).

One of the most common male mate-locating strategies found in butterflies is the establishment of areas of exclusive access that presumably increase their chances to attract sexually receptive females (Thornhill and Alcock, 1983; Höglund and Alatalo, 1995; Oliveira *et al.*, 2008). This strategy defines what we generally call a *territorial mating system* (Emlen and Oring, 1977). For a territorial mating system to occur it is necessary for males and females to be attracted to the same places, to have the capacity to identify key environmental cues to establish territories, and also to have a sensory system capable of memorizing or locating the area daily. Due to this high degree of specialization, the selective forces that shaped the evolution of territorial mating systems have intrigued biologists for decades (Thornhill and Alcock, 1983; Höglund and Alatalo, 1995; Shuster and Wade, 2003).

Sexual Selection. http://dx.doi.org/10.1016/B978-0-12-416028-6.00004-9

The early theoretical basis that formally structured the conditions favoring the occurrence of territoriality for mating purposes began with the seminal work of Emlen and Oring (1977). Although they were mainly interested in mating system organization in birds, their work generated many implications for a great variety of organisms, including butterflies (Rutowski, 1991a; Wiklund, 2003). According to Emlen and Oring (1977), the evolution of territorial mating systems should depend on the interaction among four parameters: the ratio between sexually receptive males to sexually receptive females (called the operational sex ratio, OSR), the spatial and temporal distribution of the limiting sex (generally females), population density, and the species' capacity for showing territorial defense.

For species without parental care, the more biased the OSR is toward one sex, the higher will be the proportion of individuals of the other sex that will be unable to secure a mate, increasing the variance in mating success among them (Fisher, 1930; Shuster and Wade, 2003). Since males are often the more abundant sex, they often experience greater selective pressures associated with intra- and intersexual selection. Consequently, in the majority of species, any adaptive trait that increases male mating success should be strongly favored.

Although OSR should determine the intensity of sexual selection in species without parental care, it is not a sufficient condition for the evolution of territorial mating systems (Emlen and Oring 1977, Shuster and Wade 2003). Considering that males are often the more abundant sex, the spatial and temporal distribution of females may affect when, where, and how males are distributed. If receptive females are predictably found in space, males that aggregate near them should increase their mating rate (see, for example, Buzatto and Machado, 2008). This may be especially important if females become sexually receptive asynchronously, since the establishment of males in these areas should increase the chances of monopolizing females (e.g., Blanckenhorn *et al.*, 2003; Mendoza-Cuenca and Macías-Ordóñez, 2010). It is also possible that females are not spatially clumped, but important resources for feeding or oviposition are aggregated in particular areas or periods. In the same way, males that establish themselves in these sites should also increase their chances to intercept or attract mates and, consequently, this will increase their reproductive success (Wickman, 1985a; Meek and Herman, 1991; Fischer and Fiedler, 2001). In these conditions, if males have the strength and ability to expel rival males from such sites, territorial mating systems should evolve (Baker, 1983; Fitzpatrick and Wellington, 1983). Such mating systems may be disrupted if population density becomes very high (Emlen and Oring, 1977; Kokko and Rankin, 2006). In this situation, the number of males fighting for the possession of a territory may increase and the energy necessary to maintain a territory may surpass the benefits associated with mating success (see, for example, Alcock and O'Neill, 1986). In the same way, if the abundance/synchrony of sexually receptive females is very high, the chances of a male finding a mate by actively searching for females may increase and, consequently, male spatial concentration should not be favored.

Although there are different classifications of territorial mating systems (Emlen and Oring, 1977; Shuster and Wade, 2003), the environmental cues used to define a territory may determine three main categories. When males guard areas with resources used by females, such as reproductive resources, the system is called resource-defense polygyny (e.g., Fischer and Fiedler, 2001). When males directly defend female aggregations and not resources, the system is called female-defense polygyny (e.g., Buzatto and Machado, 2008). Finally, males may hold areas located near landmark points in which the only resource used by females is the male. They visit such places just to copulate, and leave afterwards, determining a system called lek polygyny (e.g., Aspi and Hoffmann, 1998).

Specifically for butterflies, it is thought that the location of larval host plants, the spatial distribution of pupae and their developing time, as well as the place selected by females to become sexually receptive or mate after emergence, should be the best predictors for the occurrence of territorial mating systems (Rutowski, 1991a). If larval host plants are aggregated, the immature stages must also be aggregated. In this situation, if females emerge sexually receptive from the pupae, males that concentrate and defend territories near larval host plants should increase their reproductive success (e.g., Rutowski and Gilchrist, 1988). However, if the larvae do not pupate near their host plant, the host plant is widely distributed, or females need a few days after emergence to become sexually receptive, the territorial defense of host plants should not be favored. If this occurs and females are unpredictably found in space and time, territoriality should not increase male mating success. Although the life history and mating behavior of female butterflies often generate predictions associated with resource-based territoriality, in many species females are attracted to (or select) landmarks that do not contain any resource needed for feeding or reproduction, such as sunspots or hilltops (Davies, 1978; Alcock, 1985). In this situation, males that defend such sites may increase their mating success (Wickman, 1985a; Bergman *et al.*, 2007), favoring the evolution of non-resource-based territorial mating systems.

Finally, even in populations that fulfill the requirements that promote the evolution of territorial mating systems, morphological or physiological differences among males (e.g., size, age, wing shape or color) associated with variation in individual efficiency or performance could reduce the male capacity for establishing a territory. Such individual trait variation may configure conditions for the occurrence of alternative non-defense mating systems (e.g., Van Dyck *et al.*, 1997). This suggests that individuals can adjust their mate-locating behavior to maximize pay-offs, in a way presumably optimized by natural or sexual selection (Van Dyck and Wiklund, 2002; Shuster and Wade, 2003; Mendoza-Cuenca and Macías-Ordóñez, 2005; Oliveira *et al.*, 2008).

WHAT HAS BEEN CONFIRMED FOR BUTTERFLIES, AND IN WHICH PARTS OF THE GLOBE?

To evaluate contemporary empirical knowledge about territorial mating systems in butterflies, we performed searches in the ISI Web of Knowledge (www.isiknowledge.com) and Scopus databases (http://www.scopus.com) using the keywords "Lepidoptera" and "mating system" (we also searched for related references cited in the selected literature obtained from this main search). In the selected papers, we searched for information about the occurrence of territoriality, and the region where the study was developed. Additionally, we plotted the location of study sites on the most recent version of the Köppen–Geiger climate map (Peel *et al.*, 2007) to obtain the climate type of the work populations in the above-mentioned papers (see methodological details in Chapter 1 of this volume). For species in which males were reported as territorial, we also recorded the landmark adopted for territory establishment and the existence of alternative mate-locating tactics, when this information was presented. In many references, the information about mating system was not the main objective of the study. In this sense, when authors provided information that males patrol in search of females without stating that they defend mating sites, we considered the species as non-territorial.

We selected a total of 68 studies comprising 101 species (Table 4.1). Although our selection is not comprehensive, some trends became evident. First, mating systems without territorial defense are much less clearly reported than territorial ones. There are often brief reports describing male patrols in search of females without detailed information about how they behave. Assuming that all reports about male patrolling indicate non-territorial mating systems, they comprised 40 (39%) species, including butterflies that aggregate on hilltops but do not defend territorial sites in such areas. In addition, studies were concentrated in the Nymphalidae family, particularly in species belonging to the subfamilies Heliconiinae (32% of all species), Satyrinae (18% of all species), and Nymphalinae (15% of all species). If these results reflect a phylogenetic signal associated with species or specific subfamilies, we may lack proper comparisons (particularly with phylogenetic controls) to adequately evaluate the selective forces that shape butterfly mating system evolution. On the other hand, there are interesting patterns that need to be evaluated. For example, all publications with Papilionidae species report territorial defense located on hilltops; moreover, 69% of alternative mate-locating tactics appear in the Nymphalidae family (only 14% of the papers report alternative mate-locating tactics), and in 77% of the cases the alternative strategy is patrolling.

Excluding polar climates, there were small differences in the proportion of territorial species for different Köppen climates (A–D). However, it is important to note that in addition to not controlling for phylogenetic relationships among species, there may be some spatial concentration among studies. Considering the Americas and Europe (see details in Chapter 1), the percentage of the total area covered by tropical (A), arid (B), temperate (C), and cold (D) climates corresponds to 25%, 13%, 19%, and 31%, respectively. However, the percentage of studies developed in tropical, arid, temperate, and cold climates corresponds to 32%, 16%, 27%, and 23% of the published papers, respectively. This pattern is somewhat skewed by the tradition and importance of the workgroups involved

TABLE 4.1 Main Features of the Mating Systems in Butterflies*

Family/Species	Territoriality	Territory Landmark	Habitat	AMLT	Climate Classification	Geographic Location
Nymphalidae (Heliconiinae)						
Acraea aganice, A. alcinoe, A. alciopoides, A. althoffi, A. aurivilli, A. epaea, A. humilis, A. jodutta, A. leucographa, A. lycoa, A. maçaria, A. peneleos, A. pentapolis, A. poggei, A. quirinalis, A. semivitrea[1]	No	Hilltops			Af	Africa
Proclossiana eunomia[2]	No		Wet meadow		Cfb	North America
Dryas iulia[3]	No		Forest		Aw	South America
Eudeis lybia, E. vibilia[3]	No		Forest		Aw	South America

Continued

TABLE 4.1 Main Features of the Mating Systems in Butterflies*—cont'd

Family/Species	Territoriality	Territory Landmark	Habitat	AMLT	Climate Classification	Geographic Location
Heliconius erato, H. melpomene, H. walacei[3]	No		Forest		Aw	South America
Heliconius antiochus[3]	Yes		Forest		Af	South America
Heliconius ricini[3]	Yes	Forest canopy	Forest		Am	South America
Heliconius charitonia[4]	Yes	Female pupae	Montane forest	Patrolling	Cfb	North America
Heliconius sara[3,5]	Yes	Forest edges/prominent points in sunlit vegetation corridors	Forest	Territorial defense	Aw	South America
Eueides aliphera[3]	Yes	Larval host plants (?)			Af	Central America
Eueides isabella[3]	Yes	Low vegetation in roadside ravine			Aw/Af	North America
Eueides tales[3]	Yes	Prominent points in sunlit vegetation corridors	Forest		Aw	South America

TABLE 4.1 Main Features of the Mating Systems in Butterflies*—cont'd

Family/Species	Territoriality	Territory Landmark	Habitat	AMLT	Climate Classification	Geographic Location
Heliconius leucadia[3]	Yes	Prominent points in sunlit vegetation corridors	Forest		Aw	South America
Heliconius hewitsoni[6]	Yes	Female pupae	Tropical forest	Pupal mating	Aw	Central America
Nymphalidae (Satyrinae)						
Neominois ridingsii[7]	No	Hills	Forest		Dfb	North America
Geitoneura acantha[8]	No				Cfb	Australia
Geitoneura klugii[8]	No				Cfb	Australia
Aphantopus hyperanthus[9]	No		Meadow		Dfb	Europe
Coenonympha tullia[10]	No		Wet meadow		Dfb	Europe
Erebia epipsodea[11]	No		Forest		Dfc	North America
Hermeuptychia hermes[12]	No		Semideciduous forest		Cwb	South America

Continued

TABLE 4.1 Main Features of the Mating Systems in Butterflies*—cont'd

Family/Species	Territoriality	Territory Landmark	Habitat	AMLT	Climate Classification	Geographic Location
Coenonympha pamphilus[13]	Yes	Beside bushes, trees or brambles	Heath	Patrolling	Dfb	Europe
Pharneuptychia sp.[14]	Yes	Decomposing fruits	Forest		Aw	South America
Oeneis chryxus[15]	Yes	Exposed rocks			Dfb	North America
Melanitis leda[16]	Yes	Forest edge	Open woodland		Am	North America
Moneuptychia soter[17]	Yes	Forest gaps	Forest		Cwb	South America
Lasiommata megera[18]	Yes	Hilltops			Dfb	Europe
Lethe diana[19]	Yes	Sunlit branch tips facing open areas	Open woodland		Cfa	Asia
Paryphthimoides phronius[12,20]	Yes	Sunny clearings in forest edges (eventually with decomposing fruits)	Forest	Satellite	Cwb	South America
Hermeuptychia fallax[17]	Yes	Sunspots at trail entrances in forest edges	Forest	Satellite	Cwb	South America

TABLE 4.1 Main Features of the Mating Systems in Butterflies*—cont'd

Family/Species	Territoriality	Territory Landmark	Habitat	AMLT	Climate Classification	Geographic Location
Pararge aegeria[21,22]	Yes	Sunspots in forest floor	Open woodland	Territory search	Cfb/Dfb	Europe
Erebia aethiops[23]	No		Sparse woodland		Cfb	Europe
Nymphalidae (Nymphalinae)						
Euphydryas chalcedona[24]	No		Desert		BWh	North America
Junonia villida[25]	Yes	Bare ground	Open woodland		Aw	Australia
Junonia orithya[25]	Yes	Bare ground	Open woodland		Aw	Australia
Polygonia comma[26]	Yes	Bare spots or sunlit sides of trees			Dfa	North America
Inachis io[27,28]	Yes	Female route or molehills			Cfb	Europe
Chlosyne californica[29]	Yes	Hilltops	Open woodland	Patrolling	BSh	North America
Vanessa annabella, V. cardui[30]	Yes	Hilltops	Meadow		BWh	North America
Anartia jatrophae[31]	Yes	Larval host plants			Dfb	North America

Continued

TABLE 4.1 Main Features of the Mating Systems in Butterflies*—cont'd

Family/Species	Territoriality	Territory Landmark	Habitat	AMLT	Climate Classification	Geographic Location
Aglais urticae[27]	Yes	Larval host plant			Cfb	Europe
Nymphalis antiopa[32]	Yes	Ravines	Ravines		Dfa	North America
Vanessa atalanta[30,32]	Yes	Sidewalk edges, sunlit western facing walls, open sunlit lawns/hilltops			BWh	North America
Vanessa kershawi[33]	Yes	Sunspots in hilltops or sunny trails	Forest		Csa	Australia
Hypolimnas bolina[34,35]	Yes	Vantage points in forest clearings	Open woodland		Am	Australia
Melitaea cinxia[36,37]	Yes	Vantage points in vegetation/valleys	Open grassland / Meadow	Patrolling	Dfb/BSk	Europe/Asia
Nymphalidae (Melitaeinae)						
Euphydryas editha taylor[38]	Yes	Female pupae	Meadow	Patrolling (?)	Csb	North America
Nymphalidae (Limenitidinae)						
Limenitis arthemis[39]	Yes	Gravel road with hedgerows			Dfb	North America

TABLE 4.1 Main Features of the Mating Systems in Butterflies*—cont'd

Family/Species	Territoriality	Territory Landmark	Habitat	AMLT	Climate Classification	Geographic Location
Limenitis weidemeyerii[40]	Yes	Larval host plants			Dfb	North America
Nymphalidae (Apaturinae)						
Asterocampa leilia[41]	Yes	Larval host plants	Desert		BWh	North America
Nymphalidae (Brassolinae)						
Caligo idomenaeus[42]	Yes	Low vegetation in roadways	Forest		Am	South America
Nymphalidae (Libytheinae)						
Libytheana bachimanii[43]	No		Desert		BWh	North America
Lycaenidae						
Jalmenus evagoras[44]	No	Female pupae			Cfa	North America
Tarucus theophrastus[45]	No				BWh	Africa
Lycaena hippothoe[46]	Yes	Flower concentration		Patrolling	Cfb	Europe

Continued

TABLE 4.1 Main Features of the Mating Systems in Butterflies*—cont'd

Family/Species	Territoriality	Territory Landmark	AMLT	Habitat	Climate Classification	Geographic Location
Lycaena phlaeas daimio[47]	Yes	Flower concentration (?)			Cfa	Asia
Chrysozephyrus smaragdinus[48]	Yes	Forest edges near streams or forest gaps		Deciduous forest	Dfa	Asia
Favonius taxila[49]	Yes	Forest edge		Deciduous forest	Dfa	North America
Callophrys xami[50]	Yes	Low vegetation surrounded by walls in trails			Cwb	North America
Atlides halesus[51]	Yes	Palo verde trees			BWh	North America
Strymon melinus[52]	Yes	Palo verde trees	Patrolling	Desert	BSh	North America
Heodes virgaureae[53]	Yes	Sun exposed forest edges			Aw	Europe
Eumaeus toxea[54]	Yes	Vantage points near host plants			Am	North America
Incisalia iroides[55]	Yes				Csb	North America
Papilionidae						
Battus philenor[56]	Yes	Hilltops			BWh	North America

TABLE 4.1 Main Features of the Mating Systems in Butterflies*—cont'd

Family/Species	Territoriality	Territory Landmark	AMLT	Habitat	Climate Classification	Geographic Location
Papilio indra minor[57]	Yes	Hilltops			BSk	North America
Papilio polyxenes[58]	Yes	Hilltops			Dfb	North America
Battus polydamas[59]	Yes	Hilltops		Savanna	Aw	South America
Eurythides orthosilaus[59]	Yes	Hilltops		Savanna	Aw	South America
Papilio thoas[59]	Yes	Hilltops		Savanna	Aw	South America
Papilio zelicaon[60]	Yes	Hilltops		Freshwater marsh	Csa	North America
Hesperiidae						
Amblyscirites simius[7]	No	Hills		Forest	Dfb	North America
Hesperia pahaska[7]	No	Hills		Forest	Dfb	North America
Ochlodes venata[61]	Yes	Edge or junction sites in woodland margins facing sun		Open and woodland areas	Cfb	North America
Astraptes galesus[62]	Yes	Sunspots near streams		Forest	Am	Central America

Continued

TABLE 4.1 Main Features of the Mating Systems in Butterflies*—cont'd

Family/Species	Territoriality	Territory Landmark	Habitat	AMLT	Climate Classification	Geographic Location
Carterocephalus palaemon[63]	Yes	Vantage points (?)	Open woodland		Cfb	Europe
Thymelicus lineola[64]	Yes				Dfb	North America
Pieridae						
Pieris rapae crucivora[65]	No				Cfa	Asia
Colias eurytheme[66]	No		Alfalfa fields		BWh	North America
Eurema hecabe[67]	No		Poorly drained vacant lots	Pupal mating	Am	North America
Ridionidae						
Charis cadytis[68]	Yes	Moist sunny areas along forest edges	Forest		Cwb	South America

TABLE 4.1 Main Features of the Mating Systems in Butterflies*—cont'd

Family/Species	Territoriality	Territory Landmark	Habitat	AMLT	Climate Classification	Geographic Location
Mesosemia asa asa[62]	Yes	Small plants in stream edges	Forest		Am	Central America

*Territoriality refers to reports of territorial defense by males in the cited reference; territory landmark indicates the main features that determine territory location when the species was reported as territorial; habitat refers to the main habitat type in which the study was performed; AMLT describe the occurrence of alternate mate-locating tactics (when reported); climate classification describes the Köppen climate classification for the area in which the study was developed; geographic location describes the continent where the butterfly species was studied. Question marks (?) indicate uncertainty, and blank spaces a lack of information. When multiple studies described the same mating system for one species, only the first one that presented the description is cited.

References:

1. Jiggins (2002); 2. Baguette et al. (1998); 3. Benson et al. (1989); 4. Mendoza-Cuenca and Macías-Ordóñez (2005); 5. Hernández and Benson (1998); 6. Deinert et al. (1994); 7. Scott (1973); 8. Braby and New (1988); 9. Wiklund (1982); 10. Wickman (1992); 11. Brussard and Ehrlich (1970); 12. Peixoto and Benson (2009a); 13. Wickman (1985a); 14. Kane (1982); 15. Knapton (1985); 16. Kemp (2003); 17. Peixoto and Benson (2011); 18. Wickman (1988); 19. Ide (2002); 20. Peixoto et al. (2012); 21. Davies (1978); 22. Bergman and Wiklund (2009b); 23. Slamova et al. (2011); 24. Rutowski et al. (1988); 25. Rutowski (1991b); 26. Bitzer and Shaw (1983); 27. Baker (1972); 28. Dennis and Sparks (2005); 29. Alcock (1985); 30. Brown and Alcock (1990); 31. Lederhouse et al. (1992); 32. Bitzer and Shaw (1979); 33. Alcock and Gwynne (1988); 34. Rutowski (1992); 35. Kemp (2000); 36. Niitepöld et al. (2011); 37. Zhou et al. (2012); 38. Bennett et al. (2012); 39. Lederhouse (1993); 40. Rosenberg and Enquist (1991); 41. Rutowski and Gilchrist (1988); 42. Freitas et al. (1997); 43. Rutowski et al. (1997); 44. Hughes et al. (2000); 45. Courtney and Parker (1985); 46. Fischer and Fiedler (2001); 47. Suzuki (1976); 48. Takeuchi and Imafuku (2005a); 49. Takeuchi and Imafuku (2005b); 50. Cordero and Soberón (1990); 51. Alcock (1983); 52. Alcock and O'Neill (1986); 53. Dowes (1975); 54. Martinez-Lendech et al. (2007); 55. Powell (1968); 56. Rutowski et al. (1989); 57. Eff (1962); 58. Lederhouse (1982); 59. Pinheiro (2001); 60. Sims (1979); 61. Dennis and Williams (1987); 62. Alcock (1988); 63. Ravenscroft (1994); 64. Pivnick and McNeil (1985); 65. Hirota et al. (2001); 66. Kemp and Macedonia (2007); 67. Kemp (2008); 68. Chaves et al. (2006).

in the study of Lepidoptera, since, for example, 85% of the studies in arid climates were developed in study sites in the USA, particularly in Arizona.

Butterfly territorial mating systems with resource defense are by far less common than territorial mating systems without resources (Table 4.1). The defense of territories associated with host plants is described for species such as *Asterocampa leilia* (Rutowski and Gilchrist, 1988), *Anartia jatrophae* (Lederhouse *et al.*, 1992), *Eueides aliphera* (Benson *et al.*, 1989), *Limenitis weidemeyerii* (Rosenberg and Enquist, 1991), and *Aglais urticae* (Baker, 1972), while in *Heliconius charitonia* (Mendoza-Cuenca and Macías-Ordóñez, 2010) and *Euphydryas editha taylori* (Bennett *et al.*, 2012) males defend female pupae, from which females emerge sexually receptive from the host plant. However, the defense of hilltops, sunspots, forest edges, and open ground are much more common (Table 4.1). In many of these examples the system does not resemble a typical lek, because males are not concentrated in the same place. However, since the defended resources do not have any other utility to females besides being the place to find a mate, some authors call this system *dispersed leks* (e.g., Alcock, 1981; Peixoto and Benson, 2011a).

Experiments testing the effect of female or resource clumping in mating system organization are very rare for butterflies. Consequently, most information concerning how environmental cues affect male territoriality comes from observational studies that lack appropriate controls to discuss causal effects. Although such reports are heavily biased towards temperate species, they indicate that butterfly territoriality may be associated with a great variety of landmark structures (Rutowski, 1991a). The most common landmarks used as territory sites are hilltops, sunspots, and vantage points along forest edges (Table 4.1). All of them are prominent and somewhat scarce structures in the environment, used by females for copulation, but not to lay eggs or feed. There has only been a single study to date that investigated the importance of changing the distribution of such landmarks on territory formation (Merckx and Van Dyck, 2005). In this system, males of the speckled wood butterfly *Pararge aegeria* typically establish territories located in sunspots inside the forest understory. However, in fragmented systems with few sunspots, territoriality was much less frequent. Given that adult butterflies are heliothermic organisms, the selection of territories may be driven by a combined effect of visual detection and thermal aspects of the substrate (Bergman and Wiklund, 2009a; Velde *et al.*, 2011).

Studies with neotropical species reveal similar mating patterns to those of temperate species, but with a few novelties. Neotropical butterflies also defend territories on hilltops and sunspots along forest edges and vantage points (Cordero and Soberón, 1990; Pinheiro, 2001; Chaves *et al.*, 2006; Peixoto and Benson, 2009a, 2011a). However, differently from temperate butterflies, some tropical species seem to show higher intra- and interpopulation diversity in male mating strategies, including both non-resource-based and resource-based territorial mating systems (Peixoto *et al.*, 2012).

The relationship between density and butterfly territoriality is even less documented than the effects of landmark distribution. There is one example reporting territoriality abandonment with increasing density for a temperate species that occurs in a desert (Alcock and O'Neill, 1986). Two other studies, one with temperate (Lederhouse *et al.*, 1992) and another with neotropical (Peixoto and Benson, 2009b) species, documented that males defend territories irrespective of male density. In these populations, the number of territories increased with male abundance, indicating that the availability of territories is not as low as previously thought. Possibly they differed in quality, with poor territories being occupied only when male density increased. Finally, there is just one example reporting that a decrease in the synchrony of female arrival favored the adoption of alternative non-territorial male mate-locating tactics (Mendoza-Cuenca and Macías-Ordóñez, 2010).

The most striking mating system difference between neotropical butterfly species and species from other areas is related to the period in which males are found defending territories during the day and also during the year. Species that occur in lower latitudes typically defend territories throughout most of the year, and for longer periods during the day (Kemp and Rutowski, 2001; Kemp, 2003; Chaves *et al.*, 2006; Peixoto and Benson, 2009a, 2009b, 2011a), while species in higher latitudes defend territories for just a short period during the day and throughout the hotter months of the year (Wickman and Wiklund, 1983; Wickman, 1985b; Bitzer and Shaw, 1995; Rutowski *et al.*, 1996; Fischer and Fiedler, 2001; Ide, 2002, 2004; Takeuchi and Imafuku, 2005a). This indicates that thermal restrictions may play an important role in determining suitable periods for territory defense in butterflies that inhabit colder climates. In the tropics, due to the more constant and warm conditions, males may be able to defend territories for most of the year. However, species that occur under forest canopy or at higher altitudes may be more thermally dependent on temperature or light incidence, and establish territories associated with sunspots (e.g. Peixoto and Benson, 2009a). In fact, of 29 species (Table 4.1) reported as occurring in forested habitats (forests, deciduous forests, or open woodlands), 18 (62%) defend territories associated with sunny conditions such as forest edges, forest gaps, or sunspots. Daily and seasonal changes in male reproductive activities reflect the high sensitivity of butterfly phenology and active time budgets to climate conditions (Illán *et al.*, 2012), and show how selective pressures adjust the timing of male mating behavior to match the availability of receptive females.

WHAT DO WE SUGGEST AFTER LOOKING AT THE ACCUMULATED KNOWLEDGE?

Due to the great diversity of environmental structures used as landmarks to establish butterfly territory (Table 4.1), we suggest that structural environmental traits are not restrictive in determining the occurrence of territoriality in butterflies. When conditions such as OSR, synchrony in sexually receptive females and population density favor the occurrence of territorial behavior, any prominent point in the environment may be adopted as a landmark convention for territory establishment.

In contrast to landmark points, climatic characteristics play an overriding role in butterfly territoriality (Fig. 4.1). In habitats where temperature and/or water availability are more restrictive the breeding period should occur only when climatic conditions are suitable, and this should lead to narrower breeding periods in comparison to areas in which climate is less restrictive (Chapter 1). The shorter time interval for reproduction may increase the synchronization in female sexual receptivity. Consequently, areas with a very restricted period for reproduction should present few cases of territorial mating systems, since it is impossible for males to monopolize many females when they become sexually receptive simultaneously (Mendoza-Cuenca and Macías-Ordóñez, 2010). On the other hand, areas with constant temperature and regular humidity may possess suitable conditions for longer breeding periods, higher survival, and overlapping generations, reducing the synchronization of female sexual receptivity and favoring territorial mating systems. The defense of host plants (a type of resource-based territorial system) may also be possible, but this is unexpected in butterflies because pupae from few species pupate near their host plants (Rutowski, 1991a) and also because females are rarely sexually receptive upon emergence or during oviposition (Wickman, 1992; Kemp, 2001; Peixoto and Benson, 2009a). Consequently, most territories should be located at landmark points and, in a few cases, near feeding resources (Suzuki, 1976; Fischer and Fiedler, 2001; Peixoto et al., 2012).

Although the reproductive window and female synchrony may be important, we still do not have sufficient empirical tests to refute the possibility that density affects territoriality (Alcock and O'Neill, 1986; Lederhouse et al., 1992; Peixoto and Benson, 2009b). In this sense, even when the breeding window is lengthy, if population density is high it is possible that wandering males have high chances of encountering sexual partners. In conditions of low population density, the adoption of landmark points may increase the encounter rate between sexes. Consequently, territorial mating systems should occur more frequently with moderate climatic conditions and low population density. In addition, the types of territories should be less associated with sunspots in lower latitudes due to the low variability in temperature, although species from the understory in forested areas may be more dependent on sunny habitats or warmer periods to establish territorial sites (Jones and Lace, 1992; Peixoto and Benson, 2009a). In addition to the thermal regulation benefits, sunspots under or over the canopy may provide better chances to detect passing females visually (Bergman and Wiklund, 2009a; Velde et al., 2011).

Biotic interactions, although not directly associated with the occurrence of territorial mating systems, may dictate the level of female choosiness and the evolution of traits used to select sexual partners. Biotic interactions are allegedly more intense in the Neotropics (Schemske et al., 2009). If predation and

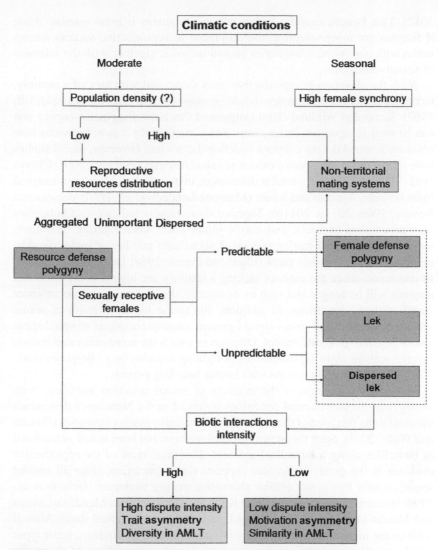

FIGURE 4.1 Causal relationships between environmental restrictions, population density, and intensity of biotic interactions on the occurrence of territorial mating systems and the intensity of male–male interactions for the possession of mating sites. Shaded boxes represent the predictions expected, based on the causal factors outlined above them. Although the variation in all factors is continuous, it has been dichotomized for simplicity. The question mark (?) indicates causal factors for which there are empirical tests. *AMLT* represents alternate mate-locating tactics.

parasitism are higher in the moderate conditions of the Neotropics, it is possible that they dictate some processes related to sexual selection. For example, females that choose males exhibiting cues of low parasitism should benefit if the high resistance to parasitism is genetically determined (Prokop *et al.*,

2012). This benefit may be higher where parasitism is more intense. Also, if females are more selective, the difference in reproductive success among males with and without territories should increase together with the intensity of sexual selection.

With the elevation in reproductive costs associated with lack of a territory, territorial fights between males should be more intense (see, for example, Eff, 1962). Kemp and Wiklund (2001) suggested that butterflies lack weapons that can be used in agonistic interactions, and consequently it is not obvious how costs are accrued during contests to define the winner. However, recent studies have shown that some species exhibit physical disputes (Pinheiro, 2001; Chaves *et al.*, 2006) while others exhibit differences in morphological or physiological traits between winners and losers (Martínez-Lendech *et al.*, 2007; Peixoto and Benson, 2008, 2011a, 2011b). Together these results indicate that butterflies may accumulate energetic and maybe injury costs when fighting for a territory. In fact, in pupal mating species, a significant number of males die during the defense of female pupa (Elgar and Pierce, 1988; Deinert *et al.*, 1994). In this sense, when the costs of lacking a territory are high it is expected that disputes will be longer and with more injury risks in species that occur under steady climatic conditions. In addition, due to the higher strength of sexual selection, winners and losers should present more pronounced morphological and/or physiological differences. Disputes in which the more motivated male is favored without showing differences in fighting capacity (e.g., Bergman *et al.*, 2010) should prevail in species with briefer breeding periods.

The implicit increase in the intensity of sexual selection associated with steady (but diverse) thermal conditions observed in the Neotropics determines the conditions that predict the evolution of alternative mating strategies (Shuster and Wade, 2003). Such theoretical predictions have not been tested or explored in butterflies along a latitudinal gradient. However, most of the reproductive evidence in the genus *Heliconius* supports this expectation, since all studied species within this genus exhibit alternative mating strategies (Deinert *et al.*, 1994; Hernández and Benson, 1998; Klein and Araújo, 2010; Mendoza-Cuenca and Macías-Ordóñez, 2010; L. Mendoza-Cuenca, unpublished data). Also, if females are more selective in areas with longer breeding windows, other types of suboptimal reproductive tactics may evolve, such as the defense of alternative low quality areas instead of the more common non-territorial patrolling or satellite tactics (e.g., Hernández and Benson, 1998; Peixoto *et al.*, 2012).

It is known that other types of reproductive investments may be important in determining the availability of mating partners (Trivers, 1972; Kokko and Jennions, 2008). In some butterfly species males provide a spermathophore with the addition of nutritious secretions, such as proteins and carbohydrates, which can be used by females to increase longevity and fecundity (Gilbert, 1991; Gwynne, 2008). Since the production costs of such secretions are frequently high, investment in each reproductive cell is increased, consequently males may be more selective. In this context, although we expect that disputes should be more intense in territorial species that occur in more constant and less restrictive climates (e.g., the Neotropics), disputes may be less intense in species in which males provide nuptial gifts to females. Spermathophore size and quality should be particularly important in species that breed in areas or periods with low resource availability, such as areas with low productivity or high density.

WHERE SHOULD MORE TERRITORIAL SPECIES OCCUR?

Our basic reasoning is that where territoriality should be rarer, non-territorial mating systems should be more frequent. In this sense, the best way to evaluate how climatic characteristics affect butterfly mating systems is by recording the proportion of species that adopt or do not adopt mating systems with territorial defense. However, since the published investigation of non-territorial mating systems seems to be less frequent and detailed than territorial ones (Table 4.1), we are still unable to test this possibility. It is also important to note that we are treating non-territorial mating systems as any system in which males do not defend a specific mating site against rival males. However, there may be variations among these species. For example, we suggested that high abundance and synchrony of sexually receptive females should favor the evolution of non-territorial mating systems in which males fly through vast areas in search of mates. Nevertheless, if females are concentrated in specific places, males should not search for females through large areas but rather concentrate in sites with high female occurrence (Rutowski, 1991a). In these locations, they should not exhibit agonistic interactions against conspecific rivals due to the high cost associated with territoriality (e.g., Alcock and O'Neill, 1986) and low capacity to monopolize females (Klug *et al.*, 2010).

According to our reasoning, the frequency and diversity of territorial mating systems should be higher in moderate climatic conditions. In this sense, the fact that 91% of studies that report non-territorial mating systems were developed in seasonal habitats strongly supports our theoretical prediction. However, as mentioned above, there is an oversimplification in this argument. For example, one may imagine that moderate climatic conditions should be found in tropical regions. However, Macías-Ordóñez *et al.* (Chapter 1) outlined how different climate regimes are structured in the Neotropics and suggested that such variations are not dictated only by thermal restrictions related to a latitudinal gradient. Consequently, a simple dichotomy between tropical and temperate climates would miss many important characteristics that affect the pay-offs related to the adoption of territorial behaviors. Inside each latitudinal interval, climatic regimes change in response to differences in topography, humidity, and even vegetation types. In this sense, the climatic classification proposed by Köppen may be more precise in representing the variation in climatic conditions that we hypothesize should be important in determining the occurrence of territorial mating systems (Chapter 1).

Climatic regimes near the equator (classifications Af and Am, particularly in the Amazon forest) should be the most moderate in terms of temperature and rainy variations. Therefore, these areas should present a greater incidence of territorial mating systems. Also, since they are dominated by a dense forest cover, territories should be associated with decomposing fruits, sunspots or sunny clearings, emerging trees, and, when available, vantage points in forest edges (Rutowski, 1991a). Arid locations such as steppes and deserts (classifications BSk, BWk, BSh, BWh) should present breeding periods associated with the irregular precipitation regimes typical of these areas (e.g., Rutowski et al., 1996). The more aggregated distribution of vegetation (mainly in deserts) may constrain butterflies to small areas. Consequently, suitable periods for breeding may coincide with high population density and synchrony of female sexual receptivity, leading to the prevalence of non-territorial mating systems. Cold climates (D) should follow a similar pattern, although butterfly population dynamics may be more regulated by temperature because, as for any other helio-thermic insect, cold climates reduce butterfly flight periods. Finally, temperate climates (C) also show a marked variation in temperature during the year, while tropical savanna climates (Aw) show a similar trend related to the rainy period. However, the breeding window in the latter is longer when compared to cold and arid climates, and plants are not restricted to patches, such as in deserts. Consequently, although there may be some synchronization among females, it should not be sufficiently high to favor the prevalence of non-territorial mating systems, and thus territorial males should be found during the breeding periods. In temperate climates, territories should be located in sunspots or other landmarks that improve heating (e.g., Van Dyck et al., 1997; Ide, 2002, 2004). In tropical savannas, on the other hand, trees are often widely spaced, allowing a high incidence of light on the ground. In this scenario, territories associated with vantage points should prevail (e.g., Pinheiro, 2001).

The intensity of disputes and the expression of differences between winners and losers should follow the same trend proposed for the occurrence of territorial mating systems. Where territoriality should be more frequent, disputes should be more intense and winners and losers should present extreme differences in traits such as body mass, parasitic load, immune response, and energetic reserves (e.g., Peixoto and Benson, 2008, 2011a, b). In places where territoriality is less frequent, disputes that follow some convention (e.g., Takeuchi and Honda, 2009) or that are won by more motivated males (e.g., Bergman et al., 2010) should prevail.

WHAT DO WE NEED TO LEARN ABOUT MATING SYSTEMS IN BUTTERFLIES? A NEOTROPICAL PERSPECTIVE

Insect mating systems have been intensively studied, and extensively reviewed (Baker, 1983; Fitzpatrick and Wellington, 1983; Thornhill and Alcock, 1983; Choe and Crespi, 1997). Although many of these studies have been carried out in lower latitudes, studies in temperate regions are over-represented given the relatively higher tropical insect diversity (Table 4.1). Although there are variations in Köppen climatic regimes within each global region, the concentration of studies in some areas may over-represent the effects of particular conditions of such regions that may bias our understanding about how mating systems evolve in response to climatic pressures.

The idea that climatic conditions are important forces in the evolution of animal mating systems is not necessarily new. In fact, it is often implicit in the literature within the concept of ecological factors (see, for example, Emlen and Oring, 1977; Rutowski, 1991a). However, our hypothesis about how climatic conditions determine reproductive patterns over a broad geographic scale affecting the intensity of sexual selection and promoting a greater diversity of mating systems in neotropical butterflies is an issue that deserves further attention. Accordingly, our literature revision unexpectedly shows that many of the processes originally suggested as important determinants of mating system organization (Emlen and Oring, 1977; Thornhill and Alcock, 1983; Rutowski, 1991a; Shuster and Wade, 2003) have not been experimentally tested. The next step needed for a better understanding of the selective forces that may have molded the evolution of territorial mating systems in butterflies, which would also validate our rationale, is to investigate the effects of density, territory availability, female choosiness, and sexual synchronicity on intrasexual male competition. A comparative approach of such effects considering differences in climatic regimes should be particularly illuminating, since it would allow the evaluation of possible differences in the evolutionary pathway among butterfly species. In particular, assuming that butterfly families and subfamilies represent phylogenetic relationships among species, comparisons within the Heliconiinae, Satyrinae, and Nymphalinae and also within poorly known families such as Ridionidae, Hesperiidae, Papilionidae, and Pieridae should be the most effective steps in clarifying how mating systems evolved in this insect group.

ACKNOWLEDGMENTS

We thank Regina Macedo and Ronald Rutowski for the comments that greatly improved the quality of this chapter. We also thank Rogelio Macías-Ordóñez for support in the handling of the Köppen–Geiger Climate Map.

REFERENCES

Alcock, J., 1981. Lek territoriality in the tarantula hawk wasp *Hemipepsis ustulata* (Hymenoptera: Pompilidae). Behav. Ecol. Sociobiol. 8, 309–317.

Alcock, J., 1983. Territoriality by hilltopping males of the great purple hairstreak, *Atlides halesus* (Lepidoptera, Lycaenidae) – convergent evolution with a pompilid wasp. Behav. Ecol. Sociobiol. 13, 57–62.

Alcock, J., 1985. Hilltopping in the nymphalid butterfly *Chlosyne californica* (Lepidoptera). Am. Mid. Nat. 113, 69–75.

Alcock, J., 1988. The mating system of three territorial butterflies in Costa Rica. J. Res. Lepid. 26, 89–97.

Alcock, J., Gwynne, D., 1988. The mating system of *Vanessa kershawi*: males defend landmark territories as mate encounter sites. J. Res. Lepid. 26, 116–124.

Alcock, J., O'Neill, K.M., 1986. Density-dependent mating tactics in the gray hairstreak, *Strymon melinus* (Lepidoptera, Lycaenidae). J. Zool. 209, 105–113.

Andersson, M., 1994. Sexual Selection. Princeton University Press, Princeton.

Aspi, J., Hoffmann, A.A., 1998. Female encounter rates and fighting costs of males are associated with lek size in *Drosophila mycetophaga*. Behav. Ecol. Sociobiol. 42, 163–169.

Baguette, M., Vansteenwegen, C., Convi, I., Nève, G., 1998. Sex-biased density-dependent migration in a metapopulation of the butterfly *Proclossiana eunomia*. Acta Oncol. 19, 17–24.

Baker, R.R., 1972. Territorial behaviour of the nymphalid butterflies, *Aglais urticae* (L.) and *Inachis io* (L.). J. Anim. Ecol. 41, 453–469.

Baker, R.R., 1983. Insect territoriality. Annu. Rev. Entomol. 28, 65–89.

Bateman, A.J., 1948. Intra-sexual selection in *Drosophila*. Heredity 2, 349–368.

Bennett, V.J., Smith, W.P., Betts, M.G., 2012. Evidence for mate guarding behavior in the Taylor's Checkerspot butterfly. J. Insect Behav. 25, 196.

Benson, W.W., Haddad, C.F.B., Zikán, M., 1989. Territorial behavior and dominance in some Heliconiinae butterflies (Nymphalidae). J. Lepid. Soc. 43, 33–49.

Bergman, M., Wiklund, C., 2009a. Visual mate detection and mate flight pursuit in relation to sunspot size in a woodland territorial butterfly. Anim. Behav. 78, 17–23.

Bergman, M., Wiklund, C., 2009b. Differences in mate location behaviours between residents and nonresidents in a territorial butterfly. Anim. Behav. 78, 1161–1167.

Bergman, M., Gotthard, K., Berger, D., Olofsson, M., Kemp, D.J., Wiklund, C., 2007. Mating success of resident versus non-resident males in a territorial butterfly. Proc. R. Soc. B. 274, 1659–1665.

Bergman, M., Olofsson, M., Wiklund, C., 2010. Contest outcome in a territorial butterfly: the role of motivation. Proc. R. Soc. B. 277, 3027–3033.

Bitzer, R.J., Shaw, K.C., 1979. Territorial behavior of the red admiral, *Vanessa atalanta* (L.) (Lepidoptera, Nymphalidae). J. Res. Lepid. 18, 36–49.

Bitzer, R.J., Shaw, K.C., 1983. Territorial behavior of *Nymphalis antiopa* and *Polygonia comma* (Nymphalidae). J. Lepid. Soc. 37, 1–13.

Bitzer, R.J., Shaw, K.C., 1995. Territorial behavior of the red admiral, *Vanessa atalanta* (Lepidoptera, Nymphalidae). 1. The role of climatic factors and early interaction frequency on territorial start time. J. Insect Behav. 8, 47–66.

Blanckenhorn, W.U., Frei, J., Birrer, M., 2003. The effect of female arrivals on mate monopolization in the yellow dung fly. Behav. Ecol. Sociobiol. 54, 65–70.

Braby, M.F., New, T.R., 1988. Adult reproductive biology of adult *Geitoneura klugii* and *G. acantha* (Lepidoptera: Satyrinae) near Melbourne, Australia. Aust. J. Zool. 36, 397–409.

Brown, W.D., Alcock, J., 1990. Hilltopping by the red admiral butterfly: mate searching alongside congeners. J. Res. Lepid. 29, 1–10.

Brussard, P.F., Ehrlich, P.R., 1970. Adult behavior and population structure in *Erebia epipsodea* (Lepidoptera: Satyrinae). Ecology 51, 880–885.

Buzatto, B.A., Machado, G., 2008. Resource defense polygyny shifts to female defense polygyny over the course of the reproductive season of a neotropical harvestman. Behav. Ecol. Sociobiol. 63, 85–94.

Chaves, G.W., Pato, C.E.G., Benson, W.W., 2006. Complex non-aerial contests in the lekking butterfly *Charis cadytis* (Riodinidae). J. Insect Behav. 19, 179–196.

Choe, J.C., Crespi, B.J., 1997. The Evolution of Social Behavior in Insects and Arachnids. Cambridge University Press, Cambridge.

Cordero, C.R., Soberón, J., 1990. Non-resource based territoriality in males of the butterfly *Xamia xami* (Lepidoptera: Lycaenidae). J. Insect Behav. 3, 719–732.

Courtney, S.P., Parker, G.A., 1985. Mating behaviour of the tiger blue butterfly (*Tarucus theophrastus*): competitive mate-searching when not all females are captured. Behav. Ecol. Sociobiol. 17, 213–221.

Davies, N.B., 1978. Territorial defense in the speckled wood butterfly, *Pararge aegeria*: the resident always wins. Anim. Behav. 26, 138–147.

Deinert, E.I., Longino, J.T., Gilbert, L.E., 1994. Mate competition in butterflies. Nature 370, 23–24.

Dennis, R.L.H., Sparks, T.H., 2005. Landscape resources for the territorial nymphalid butterfly *Inachis io*: microsite landform selection and behavioral responses to environmental conditions. J. Insect Behav. 18, 725–742.

Dennis, R.L.H., Williams, W.R., 1987. Mate location behavior of the large skipper butterfly *Ochlodes venata*: flexible strategies and spatial components. J. Lepid. Soc. 41, 45–64.

Dowes, P., 1975. Territorial behaviour in *Heodes virgaureae* L. (Lep., Lycaenidae) with particular reference to visual stimuli. Norw. J. Entomol. 22, 143–154.

Eff, D., 1962. A little about the little-known *Papilio indra minori*. J. Lepid. Soc. 16, 137–143.

Elgar, M.A., Pierce, N.E., 1988. Mating success and fecundity in an ant-tended lycaenid butterfly. In: Clutton-Brock, T.H. (Ed.), Reproductive Success: Studies of Selection and Adaptation in Contrasting Breeding Systems, Chicago University Press, Chicago, pp. 59–75.

Emlen, S.T., Oring, L.W., 1977. Ecology, sexual selection, and the evolution of mating systems. Science 197, 215–223.

Fischer, K., Fiedler, K., 2001. Resource-based territoriality in the butterfly *Lycaena hippothoe* and environmentally induced behavioural shifts. Anim. Behav. 61, 723–732.

Fisher, R.A., 1930. The Genetical Theory of Natural Selection. Oxford University Press, Oxford.

Fitzpatrick, S.M., Wellington, W.G., 1983. Insect territoriality. Can. J. Zool. 61, 471–486.

Freitas, A.V.L., Benson, W.W., Marini-Filho, O.J., Carvalho, R.M., 1997. Territoriality by the dawn's early light: the neotropical owl butterfly *Caligo idomenaeus* (Nymphalidae: Brassolinae). J. Res. Lepid. 34, 14–20.

Gilbert, L.E., 1991. Biodiversity of a Central American *Heliconius* community: patterns, process and problems. In: Price, P.W., Lewinsohn, T.M., Fernandes, G.W., Benson, W.W. (Eds.), Plant-Animal Interactions, John Wiley & Sons, Inc., New York, pp. 403–428.

Gwynne, D.T., 2008. Sexual conflict over nuptial gifts in insects. Ann. Rev. Entomol. 53, 83–101.

Hernández, M.I.M., Benson, W.W., 1998. Small-male advantage in the territorial tropical butterfly *Heliconius sara* (Nymphalidae): a paradoxical strategy? Anim. Behav. 56, 533–540.

Hirota, T., Hamano, K., Obara, Y., 2001. The influence of female post-emergence behavior on the time schedule of male mate-locating in *Pieris rapae crucivora*. Zool. Sci. 18, 475–482.

Höglund, J., Alatalo, R.V., 1995. Leks. Princeton University Press, Princeton.

Hughes, L., Chang, B.S.-W., Wagner, D., Pierce, N.E., 2000. Effects of mating history on ejaculate size, fecundity, longevity, and copulation duration in the ant-tended lycaenid butterfly, *Jalmenus evagoras*. Behav. Ecol. Sociobiol. 47, 119–128.

Ide, J., 2002. Seasonal changes in the territorial behaviour of the satyrine butterfly *Lethe diana* are mediated by temperature. J. Ethol. 20, 71–78.

Ide, J.Y., 2004. Diurnal and seasonal changes in the mate-locating behavior of the satyrine butterfly *Lethe diana*. Ecol. Res. 19, 189–196.

Illán, J.G., Gutiérrez, D., Díez, S.B., Wilson, R.J., 2012. Elevation trends in butterfly phenology: implications for species responses to climate change. Ecol. Entomol. 37, 134–144.

Jiggins, F.M., 2002. Widespread "hilltopping" in *Acraea* butterflies and the origin of sex-role-reversed swarming in *Acraea encedon* and *A. encedana*. Afr. J. Ecol. 40, 228–231.

Jones, M.J., Lace, L.A., 1992. The speckled wood butterflies *Pararge xiphia* and *P. aegeria* (Satyridae) on Madeira – distribution, territorial behavior and possible competition. Biol. J. Linn. Soc. 46, 77–89.

Kane, S., 1982. Notes on the acoustic signals of a neotropical satyrid butterfly. J. Lepid. Soc. 36, 200–206.

Kemp, D.J., 2000. Contest behavior in territorial male butterflies: does size matter? Behav. Ecol. 11, 591–596.

Kemp, D.J., 2001. The ecology of female receptivity in the territorial butterfly *Hypolimnas bolina* (L.) (Nymphalidae): implications for mate location by males. Aus. J. Zool. 49, 203–211.

Kemp, D.J., 2003. Twilight fighting in the evening brown butterfly, *Melanitis leda* (L.) (Nymphalidae): age and residency effects. Behav. Ecol. Sociobiol. 54, 7–13.

Kemp, D.J., 2008. Female mating biases for bright ultraviolet iridescence in the butterfly *Eurema hecabe* (Pieridae). Behav. Ecol. 19, 1–8.

Kemp, D.J., Macedonia, J.M., 2007. Male mating bias and its potential reproductive consequence in the butterfly *Colias eurytheme*. Behav. Ecol. Sociobiol. 61, 415–422.

Kemp, D.J., Rutowski, R.L., 2001. Spatial and temporal patterns of territorial mate locating behaviour in *Hypolimnas bolina* (L.) (Lepidoptera: Nymphalidae). J. Nat. Hist. 35, 1399–1411.

Kemp, D.J., Wiklund, C., 2001. Fighting without weaponry: a review of male–male contest competition in butterflies. Behav. Ecol. Sociobiol. 49, 429–442.

Klein, A.L., Araújo, A.M., 2010. Courtship behavior of *Heliconius erato phyllis* (Lepidoptera, Nymphalidae) towards virgin and mated females: conflict between attraction and repulsion signals? J. Ethol. 28, 409–420.

Klug, H., Heuschele, J., Jennions, M.D., Kokko, H., 2010. The mismeasurement of sexual selection. J. Evol. Biol. 23, 447–462.

Knapton, R.W., 1985. Lek structure and territoriality in the chryxus arctic butterfly, *Oeneis chryxus* (Satyridae). Behav. Ecol. Sociobiol. 17, 389–395.

Kokko, H., Jennions, M.D., 2008. Parental investment, sexual selection and sex ratios. J. Evol. Biol. 21, 919–948.

Kokko, H., Rankin, D.J., 2006. Lonely hearts or sex in the city? Density-dependent effects in mating systems. Phil. Trans. R. Soc. B. 361, 319–334.

Lederhouse, R.C., 1982. Territorial defense and lek behavior of the black swallowtail butterfly, *Papilio polyxenes*. Behav. Ecol. Sociobiol. 10, 109–118.

Lederhouse, R.C., 1993. Territoriality along flyways as mate-locating behavior in male *Limenitis arthemis* (Nymphalidae). J. Lepid. Soc. 47, 22–31.

Lederhouse, R.C., Codella, S.G., Grossmueller, D.W., Maccarone, A.D., 1992. Host plant-based territoriality in the white peacock butterfly, *Anartia jatrophae* (Lepidoptera, Nymphalidae). J. Insect Behav. 5, 721–728.

Martínez-Lendech, N., Córdoba-Aguilar, A., Serrano-Menezes, M.A., 2007. Body size and fat reserves as possible predictors of male territorial status and contest outcome in the butterfly *Eumaeus toxea* Godart (Lepidoptera: Lycaenidae). J. Ethol. 25, 195–199.

Meek, S.B., Herman, T.B., 1991. The influence of oviposition resources on the dispersion and behaviour of calopterygid damselflies. Can. J. Zool. 69, 835–839.

Mendoza-Cuenca, L., Macías-Ordóñez, R., 2005. Foraging polymorphism in *Heliconius charitonia* (Lepidoptera: Nymphalidae): morphological constraints and behavioral compensation. J. Trop. Ecol. 21, 407–415.

Mendoza-Cuenca, L., Macías-Ordóñez, R., 2010. Female asynchrony may drive disruptive sexual selection on male mating phenotypes in a *Heliconius* butterfly. Behav. Ecol. 21, 144–152.

Merckx, T., Van Dyck, H., 2005. Mate location behaviour of the butterfly *Pararge aegeria* in woodland and fragmented landscapes. Anim. Behav. 70, 411–416.

Niitepõld, K., Mattila, A.L.K., Harrison, P.J., Hanski, I., 2011. Flight metabolic rate has contrasting effects on dispersal in the two sexes of the Glanville fritillary butterfly. Oecologia 165, 847–854.

Oliveira, R.F., Taborsky, M., Brockmann, H.J., 2008. Alternative Reproductive Tactics: an Integrative Approach. Cambridge University Press, Cambridge.

Peel, M.C., Finlayson, B.L., McMahon, T.A., 2007. Updated world map of the Köppen–Geiger climate classification. Hydrol. Earth Syst. Sci. 11, 1633–1644.

Peixoto, P.E.C., Benson, W.W., 2008. Body mass and not wing length predicts territorial success in a tropical satyrine butterfly. Ethology 114, 1069–1077.

Peixoto, P.E.C., Benson, W.W., 2009a. Daily activity patterns of two co-occurring tropical satyrine butterflies. J. Insect Sci. 9, 54.

Peixoto, P.E.C., Benson, W.W., 2009b. Seasonal effects of density on territory occupation by males of the satyrine butterfly *Paryphthimoides phronius* (Butler 1867). J. Ethol. 27, 489–496.

Peixoto, P.E.C., Benson, W.W., 2011a. Fat and body mass predict residency status in two tropical satyrine butterflies. Ethology 117, 722–730.

Peixoto, P.E.C., Benson, W.W., 2011b. Influence of previous residency and body mass in the territorial contests of the butterfly *Hermeuptychia fallax* (Lepidoptera: Satyrinae). J. Ethol. 30, 61–68.

Peixoto, P.E.C., Benson, W.W., Muniz, D., 2012. Do feeding resources induce the adoption of resource defence polygyny in a lekking butterfly? Ethology 118, 311–319.

Pinheiro, C.E.G., 2001. Territorial hilltopping behavior of three swallowtail butterflies (Lepidoptera: Papilionidae) in western Brazil. J. Res. Lepid. 29, 134–142.

Pivnick, K.A., McNeil, J.N., 1985. Mate location and mating behavior of *Thymelicus lineola* (Lepidoptera, Hesperiidae). Ann. Entomol. Soc. Am. 78, 651–656.

Powell, J.A., 1968. A study of area occupation and mating behavior in *Incisalia iroides* (Lepidoptera: Lycaenidae). J. NY Entomol. Soc. 76, 47–57.

Prokop, Z.M., Michalczyk, L., Drobniak, S.M., Herdegen, M., Radwan, J., 2012. Meta-analysis suggests choosy females get sexy sons more than "good genes". Evolution 66, 2665–2673.

Ravenscroft, N.O.M., 1994. Environmental influences on mate location in male chequered skipper butterflies, *Carterocephalus palaemon* (Lepidoptera, Hesperiidae). Anim. Behav. 47, 1179–1187.

Rosenberg, R.H., Enquist, M., 1991. Contest behavior in Weidemeyer's admiral butterfly *Limenitis weidemeyerii* (Nymphalidae) – the effect of size and residency. Anim. Behav. 42, 805–811.

Rutowski, R.L., 1991a. The evolution of male mate-locating behavior in butterflies. Am. Nat. 138, 1121–1139.

Rutowski, R.L., 1991b. Temporal and spatial overlap in the mate-locating behavior of the males of two species of Jujonia (Lepidoptera, Nymphalidae). J. Res. Lep. 30, 267–271.

Rutowski, R.L., 1992. Male mate-locating behavior in the common eggfly, *Hypolimnas bolina* (Nymphalidae). J. Lepid. Soc. 46, 24–38.

Rutowski, R.L., Gilchrist, G.W., 1988. Mate-locating behavior of the desert hackberry butterfly, *Asterocampa leilia* (Nymphalidae). J. Res. Lepid. 26, 1–12.

Rutowski, R.L., Gilchrist, G.W., Terkanian, B., 1988. Male mate-locating behavior in *Euphydryas chalcedona* (Lepidoptera: Nymphalidae) related to pupation site preferences. J. Insect Behav. 3, 277–289.

Rutowski, R.L., Alcock, J., Carey, M., 1989. Hilltopping in the pipevine swallowtail butterfly (*Battus philenor*). Ethology 82, 244–254.

Rutowski, R.L., Demlong, M.J., Terkanian, B., 1996. Seasonal variation in mate-locating activity in the desert hackberry butterfly (*Asterocampa leilia*; Lepidoptera: Nymphalidae). J. Insect Behav. 9, 921–931.

Rutowski, R.L., Terkanian, B., Eitan, O., Knebel, A., 1997. Male mate-locating behavior and yearly population cycles in the snout butterfly, *Libytheana bachmanii* (Libytheidae). J. Lepid. Soc. 51, 197–207.

Schemske, D.W., Mittelbach, G.G., Cornell, H.V., Sobel, J.M., Roy, K., 2009. Is there a latitudinal gradient in the importance of biotic interactions? Annu. Rev. Ecol. Evol. Syst. 40, 245–269.

Scott, J.A., 1973. Convergence of population biology and adult behavior in two sympatric butter-flies, *Neominois ridingsii* (Papilionoidea: Nymphalidae) and *Amblyscirtes simius* (Hesperioi-dea: Hesperiidae). J. Anim. Ecol. 42, 663–672.

Shuster, S.M., Wade, M.J., 2003. Mating Systems and Strategies. Princeton University Press, Princeton.

Sims, S.R., 1979. Aspects of mating frequency and reproductive maturity in *Papilio zelicaon*. Am. Mid. Nat. 102, 36–50.

Slamova, I., Klecka, J., Konvicka, M., 2011. Diurnal behavior and habitat preferences of *Ere-bia aethiops*, an aberrant lowland species of a mountain butterfly clade. J. Insect. Behav. 24, 230–246.

Suzuki, Y., 1976. So-called territorial behaviour of the small copper, *Lycaena phlaeas daimio* Seitz (Lepidoptera, Lycaenidae). Kontiû 44, 193–204.

Takeuchi, T., Honda, K., 2009. Early comers become owners: effect of residency experience on territorial contest dynamics in a lycaenid butterfly. Ethology 115, 767–773.

Takeuchi, T., Imafuku, M., 2005a. Territorial behavior of a green hairstreak *Chrysozephyrus smaragdinus* (Lepidoptera: Lycaenidae): site tenacity and wars of attrition. Zool. Sci. 22, 989–994.

Takeuchi, T., Imafuku, M., 2005b. Territorial behavior of *Favonius taxila* (Lycaenidae): territory size and persistency. J. Res. Lepid. 38, 59–66.

Thornhill, R., Alcock, J., 1983. The Evolution of Insect Mating Systems. Harvard University Press, Cambridge.

Trivers, R.L., 1972. Parental investment and sexual selection. In: Campbell, B. (Ed.), Sexual Selec-tion and the Descent of Man 1871–1971, Aldine, Chicago, pp. 136–179.

Van Dyck, H., Wiklund, C., 2002. Seasonal butterfly design: morphological plasticity among three developmental pathways relative to sex, flight and thermoregulation. J. Evol. Biol. 15, 216–225.

Van Dyck, H., Matthysen, E., Dhondt, A.A., 1997. Mate-locating strategies are related to relative body length and wing colour in the speckled wood butterfly *Pararge aegeria*. Ecol. Entomol. 22, 116–120.

Velde, L.V., Turlure, C., Van Dyck, H., 2011. Body temperature and territory selection by males of the speckled wood butterfly (*Pararge aegeria*): what makes a forest sunlit patch a rendezvous site? Ecol. Entomol. 36, 161–169.

Wickman, P.O., 1985a. Territorial defense and mating success in males of the small heath butterfly, *Coenonympha pamphilus* L. (Lepidoptera, Satyridae). Anim. Behav. 33, 1162–1168.

Wickman, P.O., 1985b. The influence of temperature on the territorial and mate locating behavior of the small heath butterfly, *Coenonympha pamphilus* (L) (Lepidoptera, Satyridae). Behav. Ecol. Sociobiol. 16, 233–238.

Wickman, P.O., 1988. Dynamics of mate-searching behavior in a hilltopping butterfly, *Lasiommata megera* (L.): the effects of weather and male density. Zool. J. Linn. Soc. 93, 357–377.

Wickman, P.O., 1992. Mating systems of *Coenonympha* butterflies in relation to longevity. Anim. Behav. 44, 141–148.

Wickman, P.O., Wiklund, C., 1983. Territorial defense and its seasonal decline in the speckled wood butterfly (*Pararge aegeria*). Anim. Behav. 31, 1206–1216.

Wiklund, C., 1982. Behavioural shift from courtship solicitation to mate avoidance in female ringlet butterflies (*Aphantopus hyperanthus*) after copulation. Anim. Behav. 30, 790–793.

Wiklund, C., 2003. Sexual selection and the evolution of butterfly mating systems. In: Boggs, C.L., Wat, W.B., Ehrlich, P.R. (Eds.), Butterflies: Ecology and Evolution Taking Flight, University of Chicago Press, Chicago, pp. 67–90.

Williams, G.C., 1966. Adaptation and Natural Selection. Princeton University Press, Princeton.

Zhou, Y., Cao, Y., Chen, H., Long, Y., Yan, F., Xu, C., Wang, R., 2012. Habitat utilization of the Glanville fritillary butterfly in the Tianshan Mountains, China, and its implication for conservation. J. Insect Conserv. 16, 207–214.

Chapter 5

Macroecology of Harvestman Mating Systems

Bruno A. Buzatto,[1] Rogelio Macías-Ordóñez[2] and Glauco Machado[3]

[1]Centre for Evolutionary Biology, School of Animal Biology, The University of Western Australia, Crawley, WA, Australia, [2]Red de Biología Evolutiva, Instituto de Ecología, A.C., Xalapa, Veracruz, Mexico, [3]Departamento de Ecologia, Instituto de Biociências, Universidade de São Paulo, Brazil

INTRODUCTION

In this chapter, we introduce the reader to a highly diverse group of arachnids with a wide diversity of life histories, reproductive behaviors, and mating systems. Popularly known as harvestmen or daddy longlegs, these organisms are great models for an ecological approach to the study of sexual selection for at least four reasons. First, being large and slow-moving animals with generally low vagility, harvestmen are very suitable for detailed behavioral observations and capture–mark–recapture studies, allowing good estimates of population size, sex ratio, survival rates, and even morph ratio in species with male dimorphism (see, for example, Buzatto *et al.*, 2007, 2011; Zatz *et al.*, 2011; Requena *et al.*, 2012). Second, several species have unique morphological features that make them particularly interesting from a sexual selection perspective, such as cheliceral nuptial glands and exaggerated male weaponry (e.g., Martens, 1969; Willemart *et al.*, 2006, 2009; Zatz *et al.*, 2011). Third, being an ancient group, harvestmen diverged from the classic models of sexual selection (birds or insects) a long time ago, so they can be used to test sexual selection hypotheses in a phylogenetically independent manner (e.g., Macías-Ordóñez *et al.*, 2011). Finally, harvestmen are easily found in all continents except for Antarctica, occurring in a great variety of terrestrial habitats (Curtis and Machado, 2007). Due to the diversity of climate types occupied by harvestman species, it is not surprising that there is also great variation of reproductive strategies and mating systems within the order (e.g., Machado and Macías-Ordóñez, 2007a).

Despite these attractive features, harvestmen have only recently gained attention from behavioral ecologists and evolutionary biologists. Today, a search on *Web of Science* using the keywords "harvestman" and "sexual selection" returns only seven papers. If we replace "harvestman" with "spider", the most intensively studied arachnid group, we find 404 papers. Even considering

Sexual Selection. http://dx.doi.org/10.1016/B978-0-12-416028-6.00005-0

that spiders are almost seven times more diverse than harvestmen, the bias we find in the literature on sexual selection is significantly greater than might be expected considering the relative number of species in each of these two orders ($\chi^2 = 45.62$, $df = 1$, $P < 0.0001$). These results do not change significantly if we use the name of the orders (Opiliones and Araneae) instead of their popular names.

Surprisingly, contrary to nearly every other animal group covered in this book, the reproductive behavior of harvestmen is better studied in the Neotropics than in other parts of the world. An intensive search in the literature, including papers published in 11 languages from 1920 to the present, indicates that 47 of 93 studies published on the reproductive biology of harvestmen were carried out in the Neotropics (Fig. 5.1). Moreover, whereas the increase in the number of papers on harvestmen reproduction in temperate regions of Europe and North America has been fairly constant since the 1930s, there has been an accelerating increase in such studies on neotropical species beginning in the 1990s (Fig. 5.1). This indicates that the reproductive biology of neotropical harvestmen has been receiving growing attention, and it is likely that what we consider to be rules and exceptions in these animals' reproductive biology today has actually been based on species from this region.

In this chapter, we first summarize the most relevant features of the harvestman's reproductive biology and provide the current state of knowledge on the diversity of its mating systems. Then, we adopt the macroecological approach postulated in Chapter 1 and use recent techniques of phylogenetic control in an attempt to understand the influence of climate on some life-history traits in the context of mating systems. Finally, we outline suggested lines of data gathering

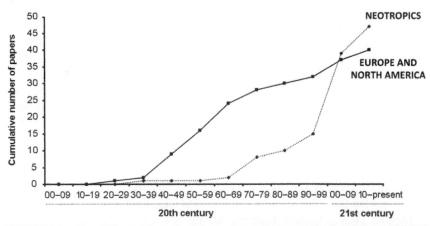

FIGURE 5.1 Cumulative number of articles on harvestmen reproductive biology published from 1901 to 2012. This compilation includes papers published in 11 languages. Note that the increase in the cumulative number of papers in temperate regions of Europe and North America has been fairly constant since the 1930s, whereas there has been an accelerating increase in such studies on neotropical species beginning in the 1990s.

and analyses to build on this approach and obtain a sharper image of the influence of environment at large geographic scales on the evolution of harvestman reproductive strategies.

HARVESTMEN: WHAT THEY ARE AND WHAT THEY DO

Sympathy for the Harvestmen

Harvestmen belong to the order Opiliones, which constitutes the third largest order of arachnids, with about 6500 species distributed worldwide (Kury, 2012). They are divided into four suborders with remarkable differences in morphology and behavior. The mite harvestmen of the suborder Cyphophthalmi comprise a low diversity group composed of small (1–3 mm in body length), short-legged inhabitants of soil and caves (Giribet, 2007). They are a basal lineage of harvestmen whose species still exhibit the primitive form of sperm transfer via spermatophore (Karaman, 2005; Macías-Ordóñez et al., 2011). The remaining three suborders form a clade called Phalangida, whose most conspicuous synapomorphy is the presence of intromittent male genitalia (Shultz and Pinto-da-Rocha, 2007). The suborder Eupnoi is the second most diverse harvestman group, and includes the long-legged forms widely known in the northern hemisphere (Cokendolpher et al., 2007). Not only their legs but also their penises and ovipositors are extremely long – usually as long as their bodies (Fig. 5.2). The suborder Dyspnoi comprises a small group of species exhibiting great diversity of body plans (Gruber, 2007). They also have somewhat intriguing and probably unique reproductive behaviors, such as belly-to-belly copulation in the family Trogulidae (Pabst, 1953), and cheliceral nuptial feeding in the family Ischyropsalididae (Martens, 1969). Finally, the suborder Laniatores is the most diverse lineage, which includes armored species, typically with spiny pedipalps and legs that are often sexually dimorphic and function as weapons in male–male fights (Kury, 2007; Machado and Macías-Ordóñez, 2007a; Willemart et al., 2009; Fig. 5.3).

The order Opiliones is an extremely old group, and fossil specimens from the Devonian period indicate that the basic morphology of the group has not changed since then (Dunlop, 2007). One of the unique traits of the order that is recognizable in the fossil record is intromittent genitalia, suggesting that harvestmen were already transferring gametes using penises about 400 million years ago. In fact, it is likely that they were among the first land animals to evolve internal fertilization, opening the path for the evolution of cryptic female choice, as well as morphological adaptations in the male genitalia derived from sperm competition inside the female's reproductive tract (Macías-Ordóñez et al., 2011).

Another important reproductive feature of the order Opiliones is that their sperm is aflagellate (Macías-Ordóñez et al., 2011). Although this type of sperm has been suggested to be associated with the absence of sperm competition in truly

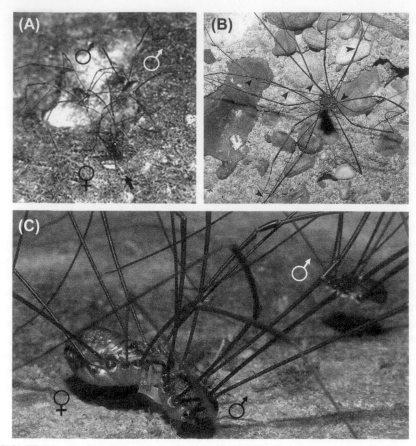

FIGURE 5.2   Sex and violence in *Leiobunum*. (A) A female *Leiobunum vittatum* (below) oviposits and the territory holder (top right) guards her after copulation while engaged in a fight with an intruding male (top left) in a population in Eastern Pennsylvania. The arrow shows the ovipositor inserted in a fissure. (B) Two fighting males of *Leiobunum* sp. close to an oviposition site. One male is pulling the second leg of the rival while a third male approaches him from behind. (C) A mating pair of the same population during intromission and nuptial feeding while a peripheral male (top right) is about to engage in a fight with the copulating male in a likely attempt to obtain a mate. *Photograph (A) courtesy of J. Warfel; photographs (B) and (C) courtesy of H. Wijnhoven.*

monandrous species, there is still no strong evidence supporting this hypothesis (Morrow, 2004). In fact, no harvestman species has been shown to be monandrous, and copulation with multiple partners seems to be the rule in this order for both males and females (Machado and Macías-Ordóñez, 2007a; Table 5.1). The most plausible hypothesis on the evolution of aflagellate sperm in harvestmen relies on the particular morphology of the female reproductive tract. Sperm are transferred into small paired structures called seminal receptacles at the tip of the female's ovipositor. On their way out of the female reproductive tract, just before exiting the ovipositor, eggs come in contact with sperm in the seminal

FIGURE 5.3 **Reproductive biology of *Serracutisoma proximum*.** (A) At the beginning of the reproductive season, males patrol territories on the vegetation and (B) repel other males in seemingly ritualized fights in which territorial males hit each other with their elongated second pair of legs. (C) Females visit the territories, copulate with the territorial males (penis indicated by white arrow) (D), and lay their eggs on the undersurface of the leaves (egg at the tip of the female's ovipositor indicated by white arrow). While the female oviposits, the territorial male guards his mate with the second pair of legs extended towards her (E). Minor males (sneakers) have a short second pair of legs and usually do not defend territories, but instead invade harems and sneak copulations with egg-guarding females (F), sometimes even when the unaware territorial male (top right) is in the vicinity of the female. See color plate at the back of the book.

receptacles, and fertilization occurs. Because the sperm are aflagellate and contained within the seminal receptacles, they have no opportunity to swim toward the eggs, and therefore females seem to have total control over the fate of male gametes (Macías-Ordóñez *et al.*, 2011).

TABLE 5.1 Main Features of the Mating Systems of Harvestmen from the Suborders Cyphophthalmi, Eupnoi, Dyspnoi, and Laniatores[a]

Taxon	Access to Multiple Females	Resource Defense	Female Defense	Mate-Guarding	Male Fights for	Parental Care	Alternative Mating Tactics	Reference(s)
Cyphophthalmi								
Cyphophthalmus serbicus (Sironidae)	No (?)	No (?)	No (?)	?	?	Egg hiding	No (?)*	Karaman (2005)
Stylocellus sp. (Stylocellidae)	No (?)	No (?)	No (?)	?	?	Egg hiding	No (?)	Schwendinger and Giribet (2005)
Dyspnoi								
Anelasmocephalus cambridgei, Trogulus nepaeformis, and T. tricarinatus (Trogulidae)	Yes (?)	No (?)	No	No (?)	Access to females	Egg hiding	No (?)	Pabst (1953)
Ischyropsalis hellwigi, I. luteipes (Ischyropsalididae), and Nemastoma lugubre (Nemastomatidae)	Yes (?)	No (?)	No	No	Access to females	Egg hiding	No (?)	Martens (1969), Meijer (1972)

Eupnoi

Leiobunum sp.* (Sclerosomatidae)	Yes	No	No	Yes	Access to females	Egg hiding	Yes	Wijnhoven (2011)
L. aldrichi* (Sclerosomatidae)	Yes	No	No	Yes	Access to females	Egg hiding	No (?)	Edgar (1971)
L. calcar, L. politum, and *L. vittatum* (Sclerosomatidae)	Yes	No	No	No	No	Egg hiding	No (?)	Edgar (1971)
L. vittatum* (Sclerosomatidae)	Yes	Yes	No	Yes	Access to females and territories	Egg hiding	No (?)	Macías-Ordóñez (1997, 2000)

Laniatores

Acutisoma longipes and *Heteromitobates discolor* (Gonyleptidae)	Yes	Yes	?	Yes	Access to territories	Maternal egg guarding	Yes (?)	Machado and Oliveira (1998), Machado and Macías-Ordóñez (2007a); Buzatto and Machado, unpubl. data

Continued

TABLE 5.1 Main Features of the Mating Systems of Harvestmen from the Suborders Cyphophthalmi, Eupnoi, Dyspnoi, and Laniatores[a]—cont'd

Taxon	Access to Multiple Females	Resource Defense	Female Defense	Mate-Guarding	Male Fights for	Parental Care	Alternative Mating Tactics	Reference(s)
Gonyleptes saprophilus and *Neosadocus* sp. (Gonyleptidae)	Yes	Yes	No	Yes (?)	Access to territories	Paternal egg guarding	?	Machado et al. (2004)
*Iporangaia pustulosa** (Gonyleptidae)	Yes	No	No	No	No	Paternal egg guarding	No (?)	Machado et al. (2004); Requena and Machado, unpubl. data
*Longiperna concolor** (Gonyleptidae)	Yes	Yes	No (?)	No	Access to territories	Egg hiding	Yes	Zatz (2010), Zatz et al. (2011)
*Magnispina neptunus** (Gonyleptidae)	Yes	Yes	No	Yes (?)	Access to territories	Paternal egg guarding	?	Nazareth and Machado (2010)
Neosadocus maximus (Gonyleptidae)	Yes	Yes (?)	No (?)	?	Access to territories	Maternal egg guarding	Yes (?)	Willemart et al. (2009), Chelini and Machado (2012)

Species							
Promitobates ornatus* (Gonyleptidae)	Yes	Yes	No (?)	Access to territories	Egg hiding	Yes (?)	Machado and Macías-Ordóñez (2007a), Zatz (2010)
Serracutisoma proximum* (Gonyleptidae)	Yes	Yes	Yes	Access to territories	Maternal egg guarding	Yes	Buzatto and Machado (2008), Buzatto et al. (2007, 2011)
Zygopachylus albomarginis* (Manaosbiidae)	Yes	Yes	No	Access to territories	Paternal egg guarding	Yes (?)	Mora (1990)

*Information for these species came from studies specifically focused on their mating systems; information for species without the asterisk came from a set of anecdotal studies from which there was enough information to speculate about their mating systems.

aThe names of the families to which each species belongs are in parentheses; (?) indicates uncertainty; and ? indicates lack of information.

In terms of external morphology, harvestmen are typical arachnids with their bodies divided into two main regions: a prosoma, which carries all the appendages (legs, pedipalps, and chelicerae), and a limbless opisthosoma, which bears the genital opening (Shultz and Pinto-da-Rocha, 2007). Although this genital opening is on the ventral part of the opisthosoma, it has shifted anteriorly relative to the dorsal parts, explaining why it is located just below the mouth and facing forward (Shultz and Pinto-da-Rocha 2007; Fig. 5.3C). Despite the widespread presence of a pair of eyes in harvestmen, most species seem incapable of forming images, and most of their incoming information seems to be mechanical or chemical, perceived by their thin and elongated legs (Willemart *et al.*, 2009). Lack of image formation and very limited use of long-range chemical or acoustic stimuli are remarkable features of harvestmen, because our classic model organisms for the study of sexual selection have been those whose sexual behavior stood out in our visual, acoustic, or long-range airborne chemical sensory universe – i.e., species in which long-range contactless information is exchanged in the form of colored feathers, melodious songs, or potent pheromones (for examples, see Andersson, 1994). Sexual cues in animals that rely mainly on tactile or short-range chemical cues have been mostly overlooked, especially in the context of mating systems. In territorial harvestmen, for instance, two males may defend the same territory for a long period of time, unaware of each other's presence, until there is physical contact between them. This has significant implications for the individuals' own evaluation of their status of resident or intruder in a territory, resulting in very long fights between two self-perceived resident males (Macías-Ordóñez, 1997). Such circumstances would be unlikely in organisms in which long-range information exchange is the first step in any interaction, as is the case with vertebrates (such as birds, mammals, or fish) or insects (such as flies, butterflies, or damselflies).

From an ecological perspective, harvestmen are strictly terrestrial (Curtis and Machado, 2007), and their large surface/volume ratio (typical of long-legged arthropods), lack of spiracular control, and low osmotic hemolymph concentration may explain why most species are found in damp and shaded areas (Santos, 2007). Given these general morphological and physiological constraints, several species exhibit behavioral features that reduce water loss, such as strictly nocturnal activity and gregariousness, since close body contact and the intertwining of legs reduces airflow between individuals, thus reducing their net water loss (Machado and Macías-Ordóñez, 2007b). The occurrence of aggregations is sometimes markedly seasonal, generally restricted to the colder and drier periods of the year, when no reproductive activity is observed (Machado and Macías-Ordóñez, 2007b). This pattern suggests that environmental factors, such as temperature and precipitation, may shape the dynamics and evolution of the harvestman's reproductive behavior, either directly or indirectly, as we will suggest in this chapter.

Sex and Violence: The Mating Systems of Harvestmen

The study of mating systems, i.e., the pool of reproductive strategies in a population (Emlen and Oring, 1977), is especially relevant when attempting to link environmental variables to sexual selection. The agents of sexual selection are members of the same population, including members of the same sex competing for mates and thus exerting selection on weapons or any strategies to exclude sexual competitors, and members of the opposite sex discriminating among mates and thus exerting selection on ornaments or any other traits that are potentially attractive to mates. However, the spatial and temporal distribution of resources and other environmental conditions set the stage, and thus modulate sexual selection in a given population (e.g., James and Shine, 1985; Wikelski *et al.*, 2000).

Although there are several studies devoted to harvestmen's reproductive biology, few specific accounts are available regarding their mating systems. Table 5.1 summarizes current information on mating systems in the order. The list is not exhaustive, but it includes all studies specifically focused on mating systems, as well as a selected set of anecdotal studies from which it is possible to extract enough information to speculate about mating systems. Except for the Cyphophthalmi, for which not enough information is available, males from all other suborders seem to have access to multiple females (Table 5.1). As stated before, however, Cyphophthalmi harvestmen exhibit the primitive form of sperm transfer via a spermatophore, which represents a relatively large investment for males, since the entire structure comprises nearly 3% of their body volume (Macías-Ordóñez *et al.*, 2011). Therefore, we suppose that male time-out (Clutton-Brock and Parker, 1992) in species of this suborder is very long because the production of such a large and complex spermatophore probably takes a lot of time and energy. One of the consequences of a long time-out is that the access to multiple females is considerably reduced when compared to males of the other three suborders, which produce free spermatozoids and deliver them inside females through a penis. Moreover, a long time-out would make mating strategies based either on female or resource defense unprofitable, because during most of the time males would pay the costs of fighting rivals without being able to receive the benefits in terms of exclusive access to multiple females. Given these constraints, we hypothesize that the most likely mating system in Cyphophthalmi is a scramble competition with low encounter rates in which males that are ready to mate actively search for receptive females in the population. Furthermore, we may even expect some degree of female courtship if spermatophores contain a significant amount of nutritious resources for females, as observed in many arthropods (Vahed, 1998). However, the study of Cyphophthalmi reproductive ecology is still unexplored and promising territory.

In Dyspnoi, at least in two species of *Ischyropsalis* and also in *Nemastoma lugubre*, males offer a glandular secretion, produced on their chelicerae, to females (Martens, 1969; Meijer, 1972). Besides offering a nuptial gift, males also tap intensively on the females' back as a form of pre-copulatory courtship (Martens, 1969). Males may fight for access to receptive females, but there is no evidence that males defend them (Table 5.1). Unfortunately, we do not have any information regarding the costs of producing glandular secretions, or the time for gland replenishing. Even if males experience some time-out period after copulation due to a high cost of these secretions, we suggest that the mating system of these nuptial gift-giving harvestmen is a scramble competition similar to that observed among nuptial gift-giving spiders (e.g., Albo *et al.*, 2009). Scramble competition also seems to be common among representatives of the family Trogulidae, in which pre-copulatory and copulatory interactions involve mutual cheliceral rubbing and male leg tapping on the female's body (Pabst, 1953). There are reports of males fighting for access to receptive females in two species of the genus *Trogulus*, but no evidence of a female-defense mating system (Table 5.1). Although females oviposit exclusively inside empty snail shells, this reproductive resource is widely scattered on the ground and it is unlikely that males are able to economically defend the resource, especially since each shell is used by a single ovipositing female. Additionally, there is a great time-lag between copulation and oviposition, and in all events of egg deposition witnessed so far there was no male close to the shells (Pabst, 1953). Nevertheless, there are no formal studies of Dyspnoi mating systems or even detailed behavioral observations under field conditions; thus we lack enough elements to speculate further on the potential role of ecological variables on the evolution of their reproductive strategies.

Most reports of mating system in Eupnoi are focused on species of *Leiobunum* (Table 5.1), a seemingly polyphyletic genus that includes a large proportion of the diversity in the suborder (Hedin *et al.*, 2012). Scramble competition is likely to be the mating system of three of these species, but for at least three others there is evidence of resource-defense polygyny (Table 5.1). One of these harvestmen, *L. vittatum*, has been studied in two distantly located sites in the USA, and these studies suggest that environmental conditions are important in influencing mating systems in the group. In the population studied in eastern Pennsylvania, the suitable substrate for oviposition is limited to cracks in rocks (Fig. 5.2A), which males actively patrol and over which males fight (Macías-Ordóñez, 1997, 2000). When a female finds a rock, she slowly searches the whole surface, inserting the ovipositor inside cracks and probing potential sites for egglaying. Searching is impossible when a female encounters a male, since on contact he eagerly attempts to grasp her using his pedipalps. If the female escapes grasping, she usually abandons the rock to avoid the male, thus also abandoning the opportunity to find an oviposition site. If copulation proceeds, a series of short repeated intromissions take place, during which the female seems to obtain some sort of nuptial secretion from the base of the male genitalia using her chelicerae (Macías-Ordóñez *et al.*, 2011). Once copulation is over, the male guards the female by wrapping one of her legs with the terminal tarsi of his own first pair of legs and following her while she walks around (Fig. 5.2A). The female is thus free to probe the rock, undisturbed not only by the guarding male but also by any other male, since the guarding male will aggressively repel any other approaching male (Fig. 5.2A). Mate-guarding stops only when the female abandons the male's territory.

In the population from Pennsylvania, adults of no other harvestman species are present during the mating season. However, in the other study site located in central Michigan, four *Leiobunum* species coexist: *L. aldrichi* (77.4%), *L. calcar* (10.6%), *L. vittatum* (7.5%), and *L. politum* (4.5%). In this site, *L. vittatum* males adopt a completely different mating strategy. No territory defense takes place; neither does any post-copulatory female guarding (Edgar, 1971). The presence of congeneric species probably influences, among other factors, resource availability and sperm competition risk, thus having an effect on mating strategies (Machado and Macías-Ordóñez, 2007a). Moreover, the breeding season in the Michigan population is nearly 1 month shorter when compared with Pennsylvania, and there is no mention of patchy oviposition sites since females lay their eggs inside fissures on fallen trunks, which are widespread and abundant in the study site. Therefore, temporal and spatial distribution of resources may play a role in determining the differences in the mating system of *L. vittatum* between populations.

Unlike the other suborders, resource- or female-defense polygyny seems to be the rule among the species of the suborder Laniatores studied so far (Table 5.1). Male fights for access to territories or nesting sites have been recorded for many species exhibiting different forms of parental care (Table 5.1). In species in which parental care is restricted to egg hiding and parental individuals do not exhibit any additional form of egg protection, males defend territories which are used as oviposition sites by females (Machado and Macías-Ordóñez, 2007a; Zatz *et al.*, 2011). The mating system in these cases is similar to that described for *L. vittatum* in Pennsylvania, except for the fact that males do not exhibit mate-guarding and females remain inside the territories during the entire mating season, forming harems (Table 5.1). Among species with exclusive paternal egg-guarding, males usually defend nesting sites, such as natural cavities in trunks, rocks, or roadside banks (Machado *et al.*, 2004; Nazareth and Machado, 2010). In *Zygopachylus albomarginis*, males build cup-like mud nests, and later are visited, courted, and seemingly selected by females. Females lay their eggs in the nests and leave them after oviposition. Only males brood the eggs, not only by aggressively defending the nest from potential predators (including conspecifics) but also by actively cleaning them of fungal infection (Mora, 1990). A small proportion of females, however, seem also to defend the nest and thus the associated male by staying in close proximity to one of them for several days and fighting off other females (Mora, 1990). Although nests are necessary for males to obtain mates, not all males in the population build nests and, over the course of the mating season, many adult males can be found without a nest. Vagrant males can acquire a nest by occupying an empty one or by displacing the occupant of a defended nest. The proportion of nests whose ownership changes varies during the mating season and also between sites. Unfortunately, there is no information about whether these alternative forms of defending males and nests for females, or acquiring a nest for males, represent alternative mating tactics within each sex. This subject certainly deserves close attention in future field studies with this species.

The best studied harvestman mating system is that of *Serracutisoma proximum*, which exhibits maternal care (Buzatto *et al.*, 2007, 2011; Buzatto and Machado, 2008; Munguía-Steyer *et al.*, 2012; Fig. 5.3). At the beginning of the reproductive season, males fight for territories on the vegetation, as in a typical resource-defense mating system (Figs 5.3A–B). During the fights, territorial males hit each other with their elongated second pair of legs, which are much longer than those of females (Fig. 5.3B). Females visit the territories, copulate with the territorial males, and lay their eggs on the undersurface of the leaves (Figs 5.3C–D). Although nearly 80–90% of the eggs are laid in the first 24 hours after copulation with territorial males, females may take up to 14 days to complete oviposition. While the female oviposits, the territorial male mate guards her with his second pair of legs extended towards her (Fig. 5.3E). Up to six females may mate with a given territory owner, with all females then laying their eggs and guarding them during the period of egg development. After the arrival of several females in a harem, territorial males concentrate their patrolling activity mostly on egg-guarding females. At this stage, the mating system seems to shift to a female-defense polygyny (Buzatto and Machado, 2008). As we show in the next topic, there is also an alternative mating tactic reported for small males which have a short second pair of legs and do not patrol or defend females and territories but instead invade territories and sneak copulations with egg-guarding females (Buzatto *et al.*, 2011; Fig. 5.3F).

Bizarre Love Triangle: Alternative Mating Tactics in Harvestmen

Harvestmen exhibit a remarkable variety of weapons that are manifested in different structures of males, usually in the form of morphologically diverse tubercles or spines that can emerge from many different points on their carapace, legs, chelicerae, or pedipalps (Pinto-da-Rocha and Giribet, 2007). In some species the legs are themselves used as weapons, and might be enormously elongated in the largest males of the population (Buzatto and Machado, 2008; Zatz *et al.*, 2011). In accordance to what would be expected based on the condition dependence of animal weapons (Jennions *et al.*, 2001), harvestman weapons show enormous variation among conspecific males (Fig. 5.4). In the most extreme cases this variation evolved into discontinuous relationships between such traits and body size (i.e., the trait's allometry) in such a way that the weapons of small males are either reduced or completely absent (Fig. 5.4). The phenomenon of male dimorphism (Gadgil, 1972) probably evolved when use of a weapon to fight and/or defend a territory could only be efficient in the very

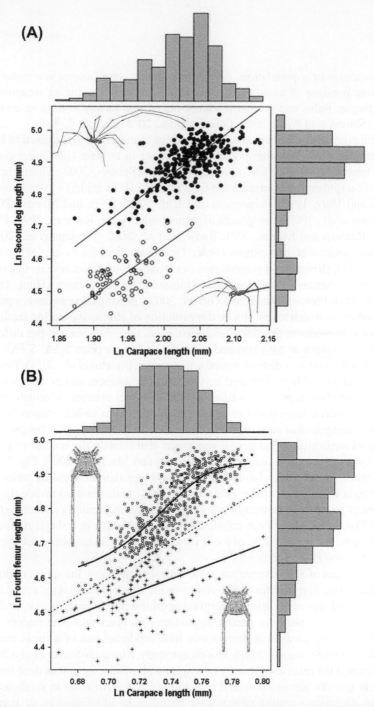

FIGURE 5.4 Allometric relationship between body size and weapon size in males of two male dimorphic harvestmen. (A) *Serracutisoma proximum*: open circles represent minors and full circles represent majors. Note that carapace length (top histogram) exhibits a unimodal distribution whereas second leg length (side histogram) exhibits a clear bimodal distribution. (B) *Longiperna concolor*: the dashed line is the switch line that separates the population into majors (circles) and minors (crosses). The solid lines represent the best models explaining each male morph data set analyzed independently. The solid circles represent males involved in fights. Note that both body size (top histogram) and leg size (side histogram) exhibit unimodal distribution.

largest males in a population. As a consequence, small males are under the selective pressure of avoiding the prohibitive costs of producing weapons or engaging in fights and instead pursuing alternative means to achieve copulations (Shuster and Wade, 2003; Oliveira *et al.*, 2008).

Male dimorphism coupled with alternative mating tactics is known to have evolved repeatedly in arthropods, being present in at least nine insect orders (Brockmann, 2008) and four crustacean orders (Shuster, 2008). In arachnids, male dimorphism has been recorded for a few species of spiders (order Araneae: Clark and Uetz, 1993; Heinemann and Uhl, 2000; Clark and Morjan, 2001; Vanacker *et al.*, 2003), two genera of mites (order Acari: Radwan, 1993, 1995, 2003; Radwan and Klimas, 2001; Radwan *et al.*, 2002; Tomkins *et al.*, 2004), and two species of harvestmen (order Opiliones: Buzatto *et al.*, 2011; Zatz *et al.*, 2011). However, an examination of taxonomic literature reveals that male dimorphism seems to be especially widespread in harvestmen (Hunt, 1979; Taylor, 2004; Pinto-da-Rocha and Giribet, 2007), making these animals a promising group in which to investigate the evolution of alternative mating tactics.

In the harvestman *Serracutisoma proximum*, two male phenotypes differ in the relative length of the elongated second pair of legs (Figs 5.3A, 5.4A), and each morph exhibits a distinct reproductive tactic (Buzatto *et al.*, 2011). Having the second pair of legs elongated is typical of harvestmen, and in most species of the order these legs seem to function like an insect antenna (Machado *et al.*, 2007). However, large males of *S. proximum* (hereafter called "majors") have extreme elongation of such legs, which they use in ritualistic fights for the ownership of territories on patches of vegetation that contain the species of plants used by females as oviposition sites (Buzatto and Machado, 2008; Fig. 5.3B). Meanwhile, males with relatively short second legs (hereafter called "minors") never fight or patrol territories (Fig. 5.4A). These males instead invade the territories of majors and furtively mate with egg-guarding females (Buzatto *et al.*, 2011; Fig. 5.3F). Minors can invade five or more harems in a single reproductive season, and around 70% of these invasions may lead to successful copulations (Buzatto *et al.*, 2011).

As females of *S. proximum* can mate with several males, the sperm of different males may compete for fertilizations inside the female's reproductive tract (Macías-Ordóñez *et al.*, 2011). Sperm competition theory predicts that, in these cases, minors should have greater expenditure on ejaculates than majors, this difference being greatest in populations with moderate risk of a sneak mating (Parker, 1990; Simmons, 2001). In a recent study, Munguía-Steyer *et al.* (2012) investigated the relative frequency of majors and minors, as well as their investment in gonads, across 10 natural populations of *S. proximum* in southeastern Brazil. Alternative mating tactics were found to be ubiquitous in all populations, but there was no relation between gonadal investment, male mating tactic, male size, second leg length, and relative frequency of majors and minors across populations. However, few territories had a very high number of females whereas a high number of territories had no females, and this uneven distribution of females seems to cause minors to preferentially invade territories with more females (Munguía-Steyer *et al.*, 2012). The authors suggested that the high risk of sneak mating in territories with many females generates high sperm competition for both male tactics, which may explain the similarity in gonadal investment between them.

The exaggerated elongation of the second pair of legs is common in other species of the subfamily Goniosomatinae (Machado *et al.*, 2003), and in at least a few species these legs seem to be male-dimorphic (B. A. Buzatto, unpublished data). This elongation of male legs has also evolved in another group of gonyleptids, the subfamily Mitobatinae, but in this case the elongation occurred in another pair of legs (Fig. 5.4B). Males of this subfamily commonly have an extreme exaggeration in the fourth pair of legs (Kury, 2007), in a clearly convergent adaptation to the kind of fights seen in Goniosomatinae. In the seemingly ritualistic fights between males of Goniosomatinae (at least in *S. proximum* and *Acutisoma longipes*), opponents face each other, hold the second pair of legs laterally extended, and repeatedly hit each other with the tips of the second pair of legs, using them as whips (Buzatto and Machado, 2008; Caetano and Machado, 2013). Interestingly, males of at least two species of Mitobatinae, *Longiperna concolor* and *Promitobates ornatus*, show an inverted version of such ritualistic fights: fighting males turn around so that they are back to back, hold their fourth pair of legs laterally extended, and also repeatedly hit each other with the tips of their whip-like legs (Machado and Macías-Ordóñez, 2007a; Zatz *et al.*, 2011). Perhaps not surprisingly, the fourth pair of legs of Mitobatinae harvestmen is not homogenously elongated across males of the same species but rather shows extreme variation; in some species there is nearly 200 times more variance than is seen in their body sizes (variances standardized by the means of the traits; B. A. Buzatto, unpublished data). Moreover, in *L. concolor* the fourth pair of legs is also male dimorphic (Fig. 5.4B), and detailed field observations indicate that the mating tactic of majors is based on resource defense, whereas that of minors seems to rely on sneaking into the territories of majors and furtively copulating with females (Zatz *et al.*, 2011).

For at least one species of Eupnoi – an introduced unidentified *Leiobunum* that has been rapidly invading Europe – males seem to exhibit alternative mating tactics that are not coupled with male dimorphism. The anecdotal description of the mating system of a population from The Netherlands suggests that while some males defend oviposition sites (small crevices in concrete ceilings and rocks), other males do not (Wijnhoven, 2011). Male–male competition for egg-laying females seems to be intense (Fig. 5.2B), and fights between males at the oviposition sites are very common when one or both of them are mate-guarding a partner. Copulating males can be harassed by other males while mating, and if the attacker manages to separate the mating pair he can then repel the previous male, take over the female, and mate (Fig. 5.2C). Male fights in the absence of a receptive female, which would suggest a resource-defense mating system, have never been observed. However, territoriality was inferred by male fidelity to some oviposition sites in the absence of any receptive female (Wijnhoven, 2011). Since populations of this introduced *Leiobunum* are quite abundant, they offer a good opportunity to investigate alternative mating tactics in a species in which males show no evident dimorphism in weaponry.

The current status of knowledge of harvestman mating systems is very fragmentary, and detailed descriptions of the mating systems of these interesting animals are still very scarce. There is no doubt that this is the main reason why, to date, there are only a few described cases of alternative mating tactics and male dimorphism in the order. Future studies on the behavioral ecology and mating systems of harvestmen could reveal the diversity of male weapons that have evolved in the order Opiliones. The few cases of male dimorphism that are currently well understood in Opiliones are probably just the tip of an iceberg of alternative mating tactics waiting to be discovered in the group.

MACROECOLOGY OF HARVESTMAN MATING SYSTEMS

As is true of most arthropods, harvestmen are absent at the lower ends of humidity and temperature ranges, most likely due to the fact that they depend on environmental heat to carry out the most basic physiological processes, and lack adaptations to conserve water in very low humidity (see "Sympathy for the Harvestmen", above). Given that it is hard to obtain water when temperatures are below freezing, water availability is somewhat defined by temperature in very cold environments. In other words, even in climates with high precipitation, if the temperature is below freezing such precipitation will be in the form of ice or snow and thus unavailable for physiological uptake (Chown and Nicolson, 2004). This can explain why harvestmen have not been reported outside latitudes 81°N and 56°S, or at altitudes above 4000 m (Curtis and Machado, 2007). At the other end of the environmental spectrum, high temperature and humidity define tropical climates. Harvestmen are both diverse and abundant in these climates (Curtis and Machado, 2007). As discussed below, abundant resources and somewhat stable conditions seem to favor long breeding seasons and even continuous reproduction.

Arid climates represent another corner of the temperature–humidity environmental space described in Chapter 1. The combination of extremely high temperatures and extremely low humidity, along with drastic temperature contrast in both diurnal and annual cycles, may impose severe restrictions on arthropod physiology (Chown and Nicolson, 2004). Nevertheless, harvestmen may be conspicuous inhabitants of this type of environment, forming large aggregations that may be buffering humidity at the most demanding time of day (Machado and Macías-Ordóñez, 2007b). In fact, the largest aggregation reported for harvestmen, and probably for many arthropods with the exception of social hymenopteran and termitoid insects, was observed in an extremely arid area of northern Mexico in which around 70,000 individuals were gathered in a single columnar cactus (Wagner, 1954). These aggregations may be observed during the day, when temperature is high and humidity is low, but individuals leave the aggregations at night to forage. Although this pattern is common to harvestmen in other environments, it is not unusual to see single individuals during the day in more humid and temperate climates, whereas this is much rarer in arid environments.

Species of the suborder Eupnoi seem to be frequently more diverse and abundant than species of the other suborders in arid environments (e.g., Cokendolpher *et al.*, 1993). Contrary to tropical areas, this is unlikely to be a sampling bias, since the proportion of undescribed species of any arthropod group is most likely lower in arid than in tropical regions. A similar pattern of Eupnoi dominance in terms of diversity may be observed in temperate climates (Giribet and Kury, 2007), whose fauna is probably even better covered than in arid areas. Species of the suborder Laniatores, on the other hand, seem to be more diverse and abundant in tropical climates (Giribet and Kury, 2007), although the data to really test this are probably not available, given the still incipient degree of taxonomic knowledge of harvestmen in the tropics, and especially outside the Neotropics. Species of the suborder Dyspnoi are rarely very abundant (Curtis and Machado, 2007), but they are more diverse in temperate climates (Giribet and Kury, 2007). However, like most species of the suborder Cyphophthalmi, they are mainly found in the litter – a poorly studied environment in terms of behavioral ecology or sexual selection of its inhabitants. This may explain why all studies on the reproductive behavior of species belonging to Dyspnoi have been conducted in the laboratory, and no detailed account in the field is available for the suborder.

Distribution and diversity patterns at large geographic scales are available for many animal groups, and harvestmen are no exception. Reproductive patterns, however, have rarely been subject to analysis at large geographic scales in any animal group, even though theory has long indicated an essential role of key environmental factors defining reproductive strategies (see, for example, Williams, 1966; Emlen and Oring, 1977; Stearns, 1992). Behavioral data accumulate at a much lower rate than presence or abundance data. Additionally, behavioral research tends to be biased toward temperate climates (Chapter 1), which are characterized by less environmental variation. Although the study of harvestman reproduction is biased towards the Neotropics (Fig. 5.1), the great majority of the species studied so far occur in temperate climates prevalent in the southern portion of South America (see discussion on the concept of the Neotropics, in Chapter 1). Tropical harvestmen are still under-represented in the literature, but enough information is available to attempt a formal test of some predictions presented in Chapter 1 for arthropods. Whenever possible, we analyze the data using modern comparative methods to account for possible phylogenetic effects. Where the data available seem to fall short and no analysis can be performed, we discuss the possible patterns and identify the kinds of empirical studies that are lacking.

It's My Life: Environmental Effects on Harvestman Development

The fact that the local richness of harvestmen varies considerably across climate types is the first evidence that climatic factors influence their biology (Curtis and Machado, 2007). In general, our understanding of life-history evolution in the order Opiliones is still very incipient, but some information is available on the reproduction and development of a number of species belonging to various families and living in different climates worldwide. The recently published book *Harvestmen: The Biology of Opiliones* (Pinto-da-Rocha *et al.*, 2007) contains a comprehensive compilation of such information, which should be consulted by readers seeking detailed accounts of the reproduction and development of individual harvestman species. Here we will use the information available in the book, as well as papers published more recently, to make inferences about how climatic factors influence the life cycles and reproduction of harvestmen, and consequently their mating systems.

The very first moment in which climatic factors may start playing a role in the life cycle of a harvestman is the egg stage. As a general rule, climatic factors such as temperature and humidity (or precipitation regime) strongly affect egg survival and time to hatching in several arthropod groups (Zaslavski, 1988). Harvestmen are no exception, and their eggs seem to develop faster in warmer temperatures and experience higher survival in humid conditions, although there are of course upper tolerance limits for both temperature and humidity, after which egg survival decreases sharply (Gnaspini, 2007). Unfortunately, detailed information on the effects of humidity on egg survival and developmental time is extremely rare, and all we can conclude at present is that dry conditions generally cause eggs to die due to dehydration, whereas excessive humidity often decreases egg survival due to increased fungal attack (e.g., Edgar, 1971; Machado and Oliveira, 1998).

On the other hand, somewhat more detailed accounts of the effects of temperature on harvestman eggs are available for a few species. Egg development time in harvestmen is plastic, and usually varies negatively with temperature, at least within the tolerance range between the minimum and maximum lethal temperatures (Gnaspini, 2007). Nonetheless, different species can have very distinct developmental times even when kept at the same temperature, which is evident when comparing, for instance, the two congeneric species *Lacinius ephippiatus*, whose eggs take on average 240 days to develop at 20°C, and *L. horridus*, whose eggs take on average half that time to develop at the same temperature (see Table 13.1 in Gnaspini, 2007). This could reflect species-specific adaptations to sustain optimum development in the different environmental conditions faced by individuals of each species during their evolutionary history. Testing this idea would require a phylogenetic comparative approach using information on a greater number of species than is currently available, and is beyond the scope of this chapter. Nevertheless, it is worth mentioning that egg development time might pose a constraint on the evolution of parental care, which is in itself an extremely important component of the mating system (Trivers, 1972; Emlen and Oring, 1977). The rationale for this constraint is that egg developmental time can translate directly into the duration of the costly behavior of egg-guarding (see "Mother Love: Environmental Effects on Maternal Care", below).

A harvestman hatches from the egg in an ephemeral stage called the larva, and usually molts into the first instar nymph within a few hours. From this point on, the individual goes through a number of molting events (from four to eight depending on the species) until it reaches the adult stage (Gnaspini, 2007). The time needed to reach maturity and the number of molting events have clear effects on adult body size in a number of arthropods, which in turn seems to affect traits that are strongly connected to fitness, such as longevity and female fecundity, for instance (Boggs, 2009). One of the predictions raised in Chapter 1 is that offspring development should be faster in humid and hot environments than in dry and/or cold environments. The rationale behind this prediction is that arid, cold, and temperate climates present a short period of favorable conditions for reproduction, which poses an upper limit on development time, causing fast sexual maturity and small body size. The nymphal phase duration in harvestmen varies across species, from a couple of months to 3 years, but the bulk of this information comes from European species living under similar climatic conditions or from individuals reared in laboratory conditions, making it impossible to test this prediction at the moment.

The nymphal stage ends, and harvestmen reach sexual maturity after the last molting event, which defines the beginning of the adult stage. It is only in this stage that the genital operculum appears and individuals become capable of copulating. The longevity of adults in the order shows great intra- and interspecific variation, from less than 1.5 months in a German population of *Lophopilio palpinalis* (Pfeifer, 1956) to 72 months in some species of the suborder Cyphophthalmi (Juberthie, 1964). This pattern seems to have a strong phylogenetic component: Eupnoi generally live as adults for only a few months (median = 3.3 months, range = 1.5–10), Dyspnoi and Laniatores usually live as adults for a considerably higher number of months (median = 24 months, range = 6–54), and Cyphophthalmi adults live for many years (median = 66 months; range = 54–72) (Gnaspini, 2007). According to another prediction raised in Chapter 1, the longevity of arthropods should be higher in hot and humid environments, due to the lack of a harsh winter that causes species of cold and dry environments to overwinter in the egg stage. Here again, other than emphasizing the phylogenetic effect on adult longevity among harvestmen, we cannot firmly conclude anything about the environmental effects on the evolution of this trait with the information currently available. However, from a sexual selection perspective it is not adult longevity *per se* that directly influences the evolution of a mating system, but rather the length and the number of mating seasons that adult individuals will experience. We do have this kind of information, and it will be discussed in detail in "Summertime: Environmental Effects on the Length of the Mating Season", below.

Despite the paucity of information on the life cycle for a large number of species, it seems that environmental conditions have some influence on the evolution of a variety of developmental traits in species of the order Opiliones. For most of these traits, however, it is impossible at present to investigate how they are affected by such climatic factors while taking the phylogenetic relationships between species into account. However, it is reasonable to expect that the climatic influences on basic features of the developmental biology of harvestmen extend to their mating systems. Below, we analyze and discuss some of the potential connections between environmental conditions and mating system evolution in harvestmen.

Mother Love: Environmental Effects on Maternal Care

The keystone for an ecological perspective of mating system evolution was the study by Emlen and Oring (1977), who emphasized that individuals from the limited sex should attempt to control access to mates of the limiting sex. The authors also stressed that environmental factors determine the degree to which this is possible, and thus play an important role on the intensity of sexual selection through what became known as the environmental potential for polygamy. According to this view, the first step in understanding a population's mating system is identifying which sex is the limiting sex and which sex is the limited one. Females invest more energy into each gamete (and frequently into each offspring), and their reproductive success is hence limited by fecundity, whereas male reproductive success is usually limited by access to receptive females (Bateman, 1948). Harvestmen are no exception to this general pattern, and thus female reproductive success is mainly limited by the rate of food acquisition (Allard and Yeargan, 2005) and the availability of suitable egg-laying substrates (Machado and Oliveira, 2002). On the other hand, males have little investment in sperm and ejaculate accessory material (Juberthie and Manier, 1977), and their reproductive success is less dependent on the rate of food acquisition (Nazareth and Machado, 2010). In some species of *Leiobunum*, males do not seem to be sperm limited and successful individuals may copulate with several females within a few hours (Edgar, 1971; Macías-Ordóñez, 1997; Wijnhoven, 2011). When few or no resources are invested in offspring, male reproductive success will be limited mainly by the number of eggs they can fertilize (Bateman, 1948; Trivers, 1972). As should be expected in the context of mating systems, distribution of harvestman females usually follows resource distribution, whereas male distribution usually follows female distribution, or that of the resources required by females (e.g., Macías-Ordóñez, 1997; Buzatto and Machado, 2008; Zatz *et al.*, 2011).

A key resource to egg production is food, and prey availability is frequently a limiting factor for egg production in predatory arthropods (Wheeler, 1996). However, unlike the great majority of arachnids, which are strictly predatory and are often equipped with costly killing tools such as strong pedipalps, fangs, and/or venom, harvestmen are omnivores or scavengers. Their diet includes a wide range of items, such as rotting fruit, fungi, and live and dead arthropods (Acosta and Machado, 2007). Females thus seem to have a wider option of resources to gather enough protein and fat to produce eggs, and receptive females may be available at any time (Machado and Macías-Ordóñez, 2007a). Given that the number of females that manage to gather enough resources to produce eggs and be receptive also limits the reproductive opportunities of males, individuals of both sexes in harvestmen probably have a much wider potential to mate several times over the course of their lives than do other arachnids, such as spiders and scorpions. Therefore, it is not surprising that, despite the great diversity of mating systems in arachnids (Thomas and Zeh, 1984) and the relatively limited number of mating system studies in harvestmen, all cases of resource-defense polygyny or female-defense polygyny reported so far are restricted to harvestmen (Table 5.1) and some few species of phytophagous mites (e.g., Dimock, 1985; Saito, 1990).

Another important factor that shapes mating systems is parental care, as it has profound implications on the potential reproductive rates of males and females, and thus on the operational sex ratio of natural populations (see detailed discussion in Chapter 8). But what exactly are the implications of parental care for mating systems in harvestmen? In the case of paternal care, the fact that males invest a lot of their time and energy in egg-guarding seems to have caused some degree of sex-role reversal, at least in the harvestman *Zygopachylus albomarginis* (see "Sex and Violence: The Mating Systems of Harvestmen", above). Although paternal care is the rarest form of post-zygotic investment in animals, it is surprisingly common in harvestmen, especially when compared with other arthropod lineages. The evolution of paternal care in arthropods is the central topic of Chapter 8 in this volume, and covers paternal care in harvestmen in depth. Thus, here we will focus our discussion on maternal care.

From the female's perspective, the cost of maternal care in harvestmen can be maintained at its minimum if her clutch develops simultaneously, such that hatching is more or less synchronous, and thus the period of egg-guarding is the shortest possible. Several species of the suborder Laniatores exhibit the most extreme form of post-ovipositional maternal care in harvestmen, with females remaining with the clutch until all nymphs have hatched and dispersed (Machado and Macías-Ordóñez, 2007a). Egg-guarding behavior has been shown to be costly for females of the harvestman *Serracutisoma proximum*, negatively affecting their future fecundity (Buzatto *et al.*, 2007). In this species, eggs develop completely and nymphs disperse in 39.5 days on average, which corresponds to approximately 6.9% of the females' adult life span. It becomes clear, therefore, that a longer egg developmental time would increase the fecundity costs of maternal care, as a greater proportion of the females' adult life span would be invested in egg guarding at the cost of foraging.

From the male's perspective, selection on female egg-laying synchrony makes each copulation with a receptive female an opportunity to sire a much greater number of offspring than if eggs were not all laid at the same time (Machado and Macías-Ordóñez, 2007a). Moreover, because females are not sexually receptive during the egg-guarding period (time-out), the operational sex ratio becomes male-biased during a great part of the mating season. Synchronized egg laying and a male-biased operational sex ratio both lead to stronger male competition for females or for the territories where females lay their eggs, which should produce a female- or resource-defense polygynous mating system. In fact, many examples of polygynous mating systems with female or resource defense in the Laniatores come from species with maternal egg guarding (Table 5.1). However, it is important to stress that in harvestman species with paternal care, nests where eggs are laid probably represent limiting and defendable resources that also select for resource-defense polygyny. In fact, resource defense is observed in at least three species where oviposition sites are limiting, such as costly mud nests or rare natural cavities in trunks, rocks, or roadside banks (Table 5.1). Meanwhile, there is no resource defense in the paternal *Iporangaia pustulosa*, in which females lay eggs on the underside of leaves of abundant plant species (Table 5.1).

An overlooked influence of climate on harvestman mating systems could operate through the types of parental care that occur in different climatic conditions. Understanding the evolution of parental care is not without difficulties (Kokko and Jennions, 2008). The influence of morphological and physiological factors, as well as behavioral preadaptations and phylogenetic constraints on the evolution of maternal care in harvestmen, has already been discussed in detail in Machado and Macías-Ordóñez (2007a). Here, we focus on the possible effects of climatic conditions (more specifically temperature and precipitation regimes) on the occurrence of this trait in species of the suborder Laniatores, the only suborder of harvestmen in which egg guarding has evolved. This influence may occur through an indirect climatic effect on the constraints imposed by egg developmental time and adult longevity, or through more direct effects of climate on egg predation and fungi infection, for instance (Chapter 1). To address this question, we approached the data available for maternal care in Laniatores with phylogenetic comparative methods. We first searched the literature for information on parental care in Laniatores, and then extracted the precise geographic coordinates of where each study was conducted (Appendix 5.1). Next, we classified the climate in each location specified by these coordinates using the Köppen–Geiger climate classification map (Peel *et al.*, 2007; Chapter 1).

We managed to gather data about the presence or absence of maternal egg guarding for 89 species belonging to 13 families, spread across 9 climate subcategories of the Köppen–Geiger map. We used these subcategories to define two major axes of environmental variables: first, a temperature axis in which temperate climates (C) were grouped as *mild*, whereas tropical (A) and arid (B) climates were grouped as *hot*; and second, a precipitation axis in which arid steppes (Bsk) and temperate dry (Csb) climates were grouped as *dry*, temperate seasonal climates (Cwa, Cwb) and tropical savvanas (Aw) as *seasonal*, and tropical rain forests (Af), tropical monsoon forests (Am), and temperate humid climates (Cfa, Cfb) as *humid* (see Peel *et al.*, 2007 for a detailed explanation of each climate regime and the rationale for this grouping criteria). Next, we inferred the phylogenetic relationships among species in our sample based on one molecular and several morphological phylogenies (see Appendix 5.3, Fig. 5.A1). Where these phylogenies disagreed, we conservatively used polytomies to indicate lack of information about the relationship among terminal taxa. We set all branch lengths in this phylogenetic tree to one, as the morphological sources of phylogenetic information provide no data on branch lengths. Finally, we used a phylogenetic multiple logistic regression using the function "compar.gee" (Paradis and Claude, 2002) in the package "ape" (Paradis *et al.*, 2004) in R version 2.14.0 (R Development Core Team, 2011) to investigate the influence of temperature and precipitation (as well as their interaction) on the probability of a species presenting maternal care (the script used for this analysis, with explanatory notes, is available from the authors upon request).

We found a significant relationship between precipitation and probability of egg guarding, which is clearly positive in hot climates, and a marginally non-significant interaction between temperature and precipitation, suggesting a somewhat inverse pattern in mild climates (Table 5.2, Fig. 5.5). The first result is in accordance with a prediction presented in Chapter 1, which states that maternal egg guarding should be more frequently found in hot and humid climates, probably to prevent intense predation and parasitism on the eggs. However, the fact that egg guarding in mild climates is significantly more common in dry and mild climates when compared with humid and mild climates was a

TABLE 5.2 Results of the Phylogenetic Multiple Logistic Regression Used to Model the Probability that a Harvestman Species Will Present Maternal Egg Guarding (Only Representatives of the Suborder Laniatores Were Included in the Analysis)

Coefficients	Estimate	SE	t	$P(\chi^2)$
Intercept	11.9969	6.0731	1.9754	0.0821
Humidity	−5.6757	2.2232	−2.5530	0.0329
Temperature	−5.0523	2.7260	−1.8534	0.0994
Humidity : Temperature	2.2454	0.9880	2.2727	0.0513

Independent variables in the model were the temperature axis (ordered as temperate < hot) in Figure 5.6 and the precipitation axis (ordered as dry < seasonal < humid) in the same figure. The phylogenetic distances between the 89 species in the sample are depicted in Figure 5.A1, and the phylogenetic degrees of freedom of the model were estimated to be 12.348.

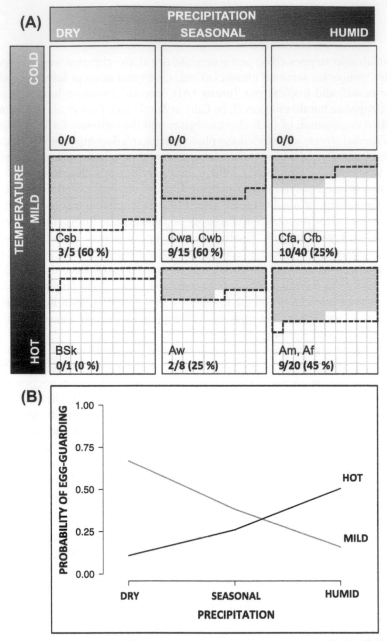

FIGURE 5.5 **Environmental effects on maternal care in harvestmen.** (A) The distribution of maternal egg-guarding in 89 harvestman species of the suborder Laniatores spread across two axes of environmental variables: temperature in the Y-axis and precipitation in the X-axis. The proportion of shaded cells indicates the proportion of species with maternal egg-guarding in each combination of these environmental variables in the data gathered. The broken contour in each cell shows the proportion of cases of maternal egg-guarding predicted by the statistical model used in the analyses. The number of instances of maternal care was divided by sample size and the corresponding percentage are represented. Climate types (from the Köppen–Geiger climate classification map; Chapter 1, Fig. 1.1) are indicated for each category. (B) An alternative representation of the proportion of cases of maternal care predicted by the statistical model used in the analyses illustrating the statistical interaction between temperature and precipitation regime.

surprising result. We had no *a priori* hypothesis for this pattern, and we can only speculate that egg guarding in mild climates may be related to protecting eggs in dry environments. In order to further investigate this pattern, we should look at the details of female brooding behavior in egg-guarding species in dry and mild areas. However, in the few reports we gathered of egg-guarding species in these climates there is no mention of any female behavior that could be linked to protecting the eggs against the potential risk of dehydration.

There are two central messages from these results. First, the effects of temperature and precipitation should be simultaneously taken into account when trying to understand climate influences on the occurrence of maternal care or any other trait of potential relevance within a reproductive strategy. Secondly, the evolution of harvestman mating systems might be indirectly affected by climatic conditions, through the influence of such conditions on the occurrence of maternal egg guarding in the order. The next step toward testing this hypothesis would be to investigate the types of mating systems in a greater number of Laniatores with and without maternal egg guarding. We predict that strong male–male competition and resource- or female-defense polygyny should be more common in species with maternal egg guarding.

We would also like to emphasize that there is a great sampling bias toward species of mild (rather than hot) and of humid (rather than seasonal or dry) environments. As an extreme example, our knowledge of maternal care in harvestmen from arid areas is restricted to a single species (Fig. 5.5). This is clearly not surprising, as harvestmen are much more abundant and diverse in humid environments, especially among representatives of the suborder Laniatores (Curtis and Machado, 2007). There is a great need for studies on parental care in harvestman species that are capable of living in dry areas. This information, as well as any information about maternal care in species from cold climates (for which we have no data), could greatly enhance our ability to understand the effects of climate on the occurrence of maternal care and, consequently, its influence on mating systems in the order.

Summertime: Environmental Effects on the Length of the Mating Season

As we argued earlier, the length and the number of mating seasons that adult individuals experience could directly influence the evolution of a mating system. This rationale has already been developed in Chapter 1, and we will only summarize it here. In general, short mating seasons select for reproductive synchrony, leading more males to be seeking copulations at the same time, hence making territoriality much more costly and unprofitable (Thornhill and Alcock, 1983). Reproductive synchrony also means that mature individuals of both sexes are reproductively active at the same time, causing an even operational sex ratio. These two factors should select for scramble competition polygynous mating systems (Shuster and Wade, 2003). Meanwhile, long mating seasons do not select for reproductive synchrony, so that the population is composed of a mixture of females in time-in and time-out, and consequently the operational sex ratio is generally male-biased. This potentially leads to stronger sexual selection (Shuster and Wade, 2003), fierce male–male competition, and can select for resource- or female-defense polygynous mating systems (Thornhill and Alcock, 1983).

Climatic conditions universally affect the length of reproductive seasons in arthropods (Wolda, 1988). Reproductive activities are usually constrained to the part of the year with optimum environmental conditions, and the period with such conditions might last for only a few months or even weeks in cold and dry environments, or almost the whole year in hot and humid environments. Here, we tested the idea that mating season length in harvestmen is influenced by temperature and precipitation – an influence that might extend to their mating systems. To address that question, we approached the data available for the length of mating seasons in harvestmen with phylogenetic comparative methods. Just as we analyzed the maternal egg-guarding data, we first searched the literature extensively for information on the length of mating seasons in harvestmen, and then extracted the precise geographic coordinates of where each study was conducted (Appendix 5.2). Again, we classified the climate in each location using the Köppen–Geiger climate classification map.

This time we gathered information about the length of mating seasons in 55 species of 14 families of Opiliones. Additionally, in four of these species we were able to find information for two or three different populations, creating a total of 61 populations. These populations were studied in locations spread across 13 climate subcategories of the Köppen–Geiger map. These subcategories were again used to define two axes of environmental variables: first, a temperature axis in which *cold* (D) climate types were grouped as *cold*, temperate (C) climates as *mild*, and tropical (A) and arid (B) climates as *hot*; and second, a precipitation axis in which arid steppes (Bsk) and temperate dry (Csa, Csb) climates were grouped as *dry*, temperate seasonal climates (Cwa, Cwb) and tropical savannas (Aw) as *seasonal*, and tropical rain forests (Af), tropical monsoon forests (Am), temperate humid (Cfa, Cfa), and cold humid (Dfa, Dfb, Dfc) climates as *humid* (see Peel *et al.*, 2007). We then inferred the phylogenetic relationships between the species in our sample, based on two molecular and several morphological phylogenies (see Appendix 5.3, Fig. 5.A2). We conservatively used polytomies to indicate lack of information about the relationship between some species, and especially between different populations of the same species. In this later case, we also set the branch length of conspecific populations to 0.1 (rather than 1) to account for the fact that such populations have certainly diverged from each other more recently than any pair of sister species in our sample. Setting the branch lengths of these conspecific populations to 0.01 or 0.001 returned similar results (not shown).

Finally, we log-transformed the data on the proportion of the year in which each species is reproductively active, and investigated the likely influence of the temperature and precipitation regime (as well as their interaction) on this variable with multiple regression using a phylogenetic generalized least-squared (PGLS) approach (Nunn, 2011). We did this with the function "pgls" (Freckleton *et al.*, 2002) from the package "caper" (Orme *et al.*, 2012) in R version 2.14.0 (R Development Core Team, 2011). This function calculates the value of a parameter "lambda" (λ) through maximum likelihood, allowing an estimation of the degree of phylogenetic dependence exhibited by the data (Pagel, 1999; Ives and Garland, 2010), and then uses the obtained value of λ to fit a linear model taking into account the degree of phylogenetic non-independence between data points (the script used for this analysis with explanatory notes is available from the authors upon request). This approach is preferable to the multiple logistic regression that we used earlier with the function "compar.gee" because it adjusts the amount of phylogenetic control according the degree of phylogenetic dependence of the data, instead of assuming that the phylogeny accurately describes such non-independence. However, this could not be employed in our analysis of the maternal care data (see above) because binomial response variables are currently not developed for the "pgls" function.

Our results suggest that both temperature and precipitation may have a significant positive influence on the length of harvestman mating seasons, and there was a significant interaction effect between these two variables (Table 5.3, Fig. 5.6). In dry environments, extremely hot temperatures of dry steppes

TABLE 5.3 Results of the Multiple Regression Used to Model the Length of Mating Seasons in Harvestmen (in the Log-Transformed Proportion of the Year in Reproductive Activity), Fit with Phylogenetic Generalized Least-Squares (Nunn, 2011).

Coefficients	Estimate	SE	t	$P(\chi^2)$
Intercept	1.7807	1.6905	1.0534	0.2966
Humidity	−1.1343	0.5757	−1.9704	0.0537
Temperature	−1.4110	0.6099	−2.3137	0.0243
Humidity : Temperature	0.5808	0.2152	2.6991	0.0091

Independent variables in the model were the temperature axis (ordered as cold < temperate < hot) in Figure 5.6 and the precipitation axis (ordered as dry < seasonal < humid) in the same figure. The phylogenetic distances between the 55 species (and 61 populations) in the sample are depicted in Figure 5.A2. Lambda (λ), which reflects the degree of phylogenetic dependence exhibited by the data, was estimated to be 0.363 (95% CI: 0.060–0.758), significantly different from 0 ($P=0.009$) and 1 ($P< 0.0001$). The degrees of freedom in the model were 57.

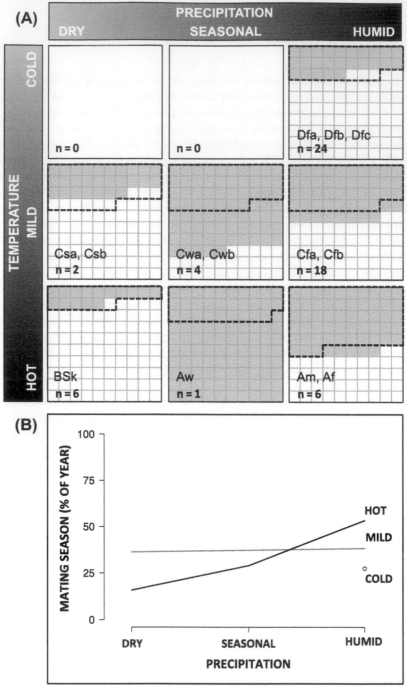

FIGURE 5.6 Environmental effects on the length of the mating season in harvestmen. (A) The average mating season length in 61 populations (of 55 species) of harvestmen spread across two axes of environmental variables: temperature on the Y-axis and precipitation on the X-axis. The proportion of shaded cells indicates the average proportion of the year in which adults are reproductively active in each combination of these environmental variables in the data gathered. The broken contour in each cell shows the average proportion of the year in which adults would be reproductively active predicted by the statistical model used in the analyses. Sample size and climate types (from the Köppen–Geiger climate classification map; Chapter 1, Fig. 1.1) are indicated for each category. (B) An alternative representation of the proportion of cases of maternal care predicted by the statistical model used in the analyses illustrating the statistical interaction between temperature and precipitation regime.

(BSk climate) result in a high risk of desiccation, and harvestman from these areas can probably only breed in the few "least dry" months. Therefore, such regions have one of the shortest periods of favorable conditions for harvestmen reproduction (Fig. 5.6). On the other hand, a higher temperature in humid environments always corresponds to longer periods of favorable conditions for harvestmen reproduction. These patterns strongly suggest some influence of climate on the type of harvestman mating systems. High temperatures and humidity, when acting together, seem to increase the length of mating seasons. As predicted in Chapter 1, this should lead to male-biased operational sex ratios, intense male–male competition, and a higher frequency of resource- or defense-polygynous mating systems. Unfortunately, the references we gathered for hot and humid environments do not include enough information on mating systems to test this hypothesis.

Once again, the central message here is that the effects of temperature and precipitation need to be simultaneously taken into account when trying to understand climate influences on the length of mating seasons. There is no doubt that our understanding of these climatic effects on harvestmen mating seasons will benefit a lot from information about species from cold and dry or from cold and seasonal environments, for which we currently have no data. However, we predict that such environments will present very short periods of favorable conditions for breeding harvestmen, and their mating seasons will hence be very brief in such conditions, leading frequently to extreme breeding synchrony and potentially to scramble mating polygyny.

CONCLUDING REMARKS

Although the reproductive ecology of harvestmen has received much less attention than other taxa, the recent surge in research on this area has turned this group into a promising opportunity to test macroecological hypotheses such as those presented or outlined in this chapter and in Chapter 1. This opportunity will be enhanced if we identify where most work is required. First, as in other taxa, the study of mating systems in extreme climates is under-represented. Few or no studies have been carried out in really extreme climates, i.e., the actual corners of the two dimensional environmental space presented in Chapter 1. We can never say there is enough done in temperate climates, but too much theory has been developed around mating systems only in this subset of environmental conditions and not really tested in a full environmental range. Whenever possible, these studies should include detailed behavioral observations of individually tagged and frequently re-sighted individuals, allowing good estimates of population parameters. Comparative studies of conspecific or congeneric populations in contrasting climates, when available, would be particularly valuable. In all cases, the importance of phylogenetically controlled analyses when comparing different species can never be overstated.

In terms of major harvestman taxa, formal research on Cyphophthalmi mating strategies is badly needed. Field studies may be especially challenging, but probably little effort would be required to have laboratory colonies that could provide valuable information. Although we know more about Dyspnoi reproductive behavior, no studies are available in the field with a mating system approach. Some behavioral and morphological traits of Dyspnoi mentioned above anticipate extremely interesting mating systems in this suborder. We have some detailed information on the reproductive strategies of some Eupnoi species, but, given their diversity, abundance, global distribution, how easy they may be to handle, and thus their potential generosity in providing data, a whole array of fascinating hypotheses could be tested using Eupnoi only. Last but not least, Laniatores themselves are abundant and diverse enough to provide ample opportunities to explore the effect of environment on the evolution of mating strategies. More studies on tropical Eupnoi and temperate Laniatores would balance our somewhat environmentally biased knowledge of these suborders, and thus enhance their joint potential to further explore some of the ideas presented in this chapter.

Even when we are able to gather enough information to carry out analyses such as those presented here for maternal egg guarding and length of mating season, we still need enough detailed information to explain correlates between climatic variables and mating system traits. In these two cases, for instance, we need to study the details of female egg brooding in dry temperate climates in order to fully understand our results, and the frequency of resource- or defense-polygynous mating systems in hot and humid environments to test our hypotheses. For a long time the empirical exploration of the macroecology of reproductive strategies, and thus the role of large-scale environmental variation shaping traits under sexual selection, have appeared to be an unreachable goal. We hope to have shown in this chapter that, although the data are far from perfect, an attempt to explore the relationship between climatic variation and sexual selection should be made for harvestmen and many other groups.

ACKNOWLEDGMENTS

We are grateful to John Alcock for helpful comments on an early version of the manuscript, and to Joe Warfel and Hay Wijnhoven for providing the photos presented here as Figures 5.2A and 5.2B–C, respectively. BAB is supported by the University of Western Australia through an International Postgraduate Research Scholarship, GM has research grants from FAPESP (2012/50229-1) and Conselho Nacional de Desenvolvimento Cientifico e Tecnológico, and RMO is supported by INECOL.

APPENDICES

Appendix 5.1: Maternal Care and Habitat Climate for the Harvestmen Species Used in the Comparative Analyses Reported in Table 5.2

Species	Climate Type	Temperature	Precipitation	Maternal care (1 = present)	Coordinates
Lepchana spinipalpis	Cwb	2	2	0	27°43′N, 88°3′E
Holoscotolemon querilhaci	Cfb	2	3	0	42°50′N, 1°30′E
Cryptopoecilaema almipater	Aw	3	2	0	10°55′15.79″N, 85°28′3.53″W
Cynortoides cubanus	Aw	3	2	0	23°07′N, 82°23′W
Eucynortula lata	Am	3	3	0	4°4′S, 59°7′W
Erginulus clavotibialis	Am	3	3	1	17°8′N, 88°38′W
Metalibitia paraguayemsis	Cfa	2	3	0	34°35′S, 58°22′W
Vonones sayi	Cfa	2	3	0	29°45′N, 95°22′W + 35°28′N, 97°32′W
Gryne coccineloides	Cwb	2	2	0	22°4941.06″S, 47°4′51.71″W
Gryne orensis	Cfa	2	3	0	34°35′S, 58°22′W
Phalangodus sp.	Csb	2	1	1	6°47′N, 73°15′W
Santinezia serratotibialis	Am	3	3	1	10°39′N, 61°14′W + 10°46′N, 61°14′W
Santinezia sp.	Af	3	3	1	03°28′N, 76°37′W
Pseudobiantes japonicus	Cfa	2	3	0	32°58′N, 129°48′E
Bourguyia albiornata	Cfa	2	3	1	25°18′S, 48°05′W
Ampheres leucopheus	Cfa	2	3	0	25°43′S, 48°50′W + 24°14′S, 48°04′W
Acutisoma hamatum	Cwa	2	2	1	22°21′S, 44°44′W
Acutisoma longipes	Cwa	2	2	1	23°10′S, 46°25′W
Goniosoma roridum	Aw	3	2	1	22°29′S, 43°4′W
Goniosoma venustum	Af	3	3	1	22°28′S, 42°27′W
Heteromitobates albiscriptum	Cwa	2	2	1	23°38′S, 46°22′W

Continued

Appendix 5.1: Maternal Care and Habitat Climate for the Harvestmen Species Used in the Comparative Analyses Reported in Table 5.2—cont'd

Species	Climate Type	Temperature	Precipitation	Maternal Care (1 = present)	Coordinates
Heteromitobates discolor	Cwa	2	2	1	23°26'S, 45°04''W
Mitogoniella indistincta	Am	3	3	1	20°26'S, 41°52'W
Mitogoniella taquara	Cwb	2	2	1	20°23'S, 45°42'W
Mitogoniella unicornis	Aw	3	2	1	4°31'S, 40°05'W
Serracutisoma molle	Cfa	2	3	1	24°28'S, 48°39'W
Serracutisoma aff. proximum	Cfa	2	3	1	25°18'S, 48°05'W
Serracutisoma guaricana	Cfa	2	3	1	25°43'S, 48°50'W
Serracutisoma catarina	Cfa	2	3	1	27°41'S, 48°46'W
Serracutisoma proximum	Cfa	2	3	1	24°14'S, 48°04'W
Serracutisoma pseudovarium	Cfa	2	3	1	24°14'S, 48°04'W
Serracutisoma spelaeum	Cfa	2	3	1	24°14'S, 48°04'W
Gonyleptes saprophilus	Cwa	2	2	0	23°10'S, 46°25'W + 23°17'S, 47°00'W
Liogonyleptoides tetracanthus	Am	3	3	1	19°06'S, 39°45'W
Megapachylus grandis	Cwb	2	2	1	21°43'S, 48°01'W
Mischonyx cuspidatus	Cwb	2	2	0	18°53'S, 48°15'W
Neosadocus maximus	Cfa	2	3	1	24°14'S, 48°04'W
Neosadocus sp2	Cfa	2	3	0	24°14'S, 48°04'W
Parampheres bimaculatus	Cfa	2	3	0	34°36'S, 55°53'W
Parampheres ronae	Cfa	2	3	0	34°36'S, 55°53'W
Hernandaria scabricula	Cfa	2	3	0	34°35'S, 58°22'W
Pseudotrogulus funebris	Cfa	2	3	0	23°46'S, 46°18'W
Chavesincola inexpectabilis	Am	3	3	0	19°58'S, 40°32'W
Magnispina neptunus	Am	3	3	0	19°06'S, 39°45'W
Longiperna concolor	Cfa	2	3	0	24°14'S, 48°04'W
Promitobates ornatus	Cfa	2	3	0	24°14'S, 48°04'W

Species						
Acanthopachylus aculeatus	Cfa	2	3	34°53'S, 56°10'W	1
Discocyrtus dilatatus	Cwa	2	2	31°25'S, 64°12'W	0
Discocyrtus oliverioi	Cwb	2	2	18°53'S, 48°15'W	1
Discocyrtus pectinifemur	Cwb	2	2	22°24'39"S, 47°33'39"W	1
Discocyrtus prospicuus	Cfa	2	3	34°35'S, 58°22'W + 34°36'S, 55°53'W	0
Pachyloidellus goliath	Cwa	2	2	31°43'S, 64°58'W	1
Pachyloides thorelli	Cfa	2	3	34°35'S, 58°22'W + 34°36'S, 55°53'W	0
Pachylus paessleri	Csb	2	1	33°39'S, 70°50'W	1
Pachylus quinamavidensis	Csb	2	1	35°47'S, 71°26'W	1
Parapachyloides fontanensis	Cfa	2	3	34°35'S, 58°22'W	0
Pygophalangodus canalsi	Cfa	2	3	34°35'S, 58°22'W	0
Pachylospeleus strinatii	Cfa	2	3	24°14'S, 48°04'W	0
Cadeadoius niger	Cfa	2	3	23°42'S, 51°38'W	0
Iguapeia melanocephala	Cfa	2	3	24°14'S, 48°04'W	0
Iporangaia pustulosa	Cfa	2	3	24°14'S, 48°04'W	0
Progonyleptoidellus striatus	Cfa	2	3	23°46'S, 46°18'W	0
Progonyleptoidellus orguensis	Aw	3	2	22°29'S, 43°4'W	0
Camarana flavipalpi	Cwa	2	2	23°26'S, 45°04'W	0
Poassa limbata	Aw	3	3	9°55'N, 84°4'W	0
Saramacia lucasae	Am	3	3	02°25'S, 59°45'W	0
Zygopachylus albomarginis	Am	3	3	9°09'N, 79°51'W	0
Scotolemon doriae	Cfb	2	3	42°50'N, 1°30'E	0
Scotolemon lespesi	Cfb	2	3	42°50'N, 1°30'E	0
Scotolemon lucasi	Cfb	2	3	42°50'N, 1°30'E	0
Ibalonoius sp.	Af	3	3	9°28'S, 159°49'E	0
Leytpodoctis oviger	Af	3	3	10°50'N, 124°50'E	0
Hoplobunus boneti	Am	3	3	21°59'N, 99°1'W	1
Eutimesius sp.	Af	3	3	0°38'S, 76°08'W	1
Stenostygnellus flavolimbatus	Aw	3	2	10°31'5.95"N, 66°48'21.87"W	0
Stenostygnellus aff. flavolimbatus	Aw	3	2	10°24'56.30"N, 67°17'12.64"W	0

Continued

Appendix 5.1: Maternal Care and Habitat Climate for the Harvestmen Species Used in the Comparative Analyses Reported in Table 5.2—cont'd

Species	Climate Type	Temperature	Precipitation	Maternal Care (1 = present)	Coordinates
Auranus parvus	Am	3	3	0	4°4'S, 59°7'W
Protimesius longipalpis	Am	3	3	0	02°25'S, 59°45'W
Stygnus sp.	Am	3	3	0	2°37'33''S, 60°56'37''W
Peltonychia clavigera	Cfb	2	3	0	42°50'N, 1°30'E
Karamea spp.	Cfb	2	3	0	42°0'S, 174°0'E
Soerensenella spp.	Cfb	2	3	0	42°0'S, 174°0'E
Hendea myersi	Cfb	2	3	0	42°0'S, 174°0'E
Ceratomontia argentina	Cfa	2	3	0	38°9'S, 61°48'W
Araucanobunus juberthiei	Csb	2	1	0	36°46'S, 73°3'W
Cynorta sp.	Am	3	3	0	02°25'S; 59°45'W
Phareicranaus calcariferus	Am	3	3	1	10°28'15''N, 61°11'50''W + 10°47'39''N, 61°13'33''W
Sclerobunus robustus	Bsk	3	1	0	32°43'N, 108°22'W
Stygnomma sp.	Csb	2	1	0	6°47'N, 73°15'W

Appendix 5.2: Mating Season Length and Habitat Climate for the Harvestmen Species Used in the Comparative Analyses Reported in Table 5.3

Species	Mating Season Length (months)	Mating Season Length (proportion of the year)	Climate Type	Temperature	Precipitation	Coordinates
Parogovia pabsgarmoni	7	0.58	Am	3	3	8°29′32″N, 13°12′34″W
Dicranolasma scabrum	6	0.5	Dfb	1	3	47°58′N, 16°36′E
Ischyropsalis luteipes	6.5	0.54	Dfc	1	3	43°1′N, 1°6′E
Ischyropsalis pyrenaea	6.5	0.54	Dfc	1	3	43°1′N, 1°6′E
Mitostoma chrysomelas	3	0.25	Cfb	2	3	53°15′N, 6°44′E
Nemastoma lugubre1	1	0.08	Dfb	1	3	48°9′N, 14°1′E
Nemastoma lugubre2	5	0.42	Cfb	2	3	53°15′N, 6°44′E
Sabacon imamurai	2	0.17	Dfa	1		38°55′N, 140°1′E + 43°5′N, 141°32′E
Sabacon makinoi	5	0.42	Dfa	1	3	38°55′N, 140°1′E + 43°5′N, 141°32′E
Dalquestia formosa	2	0.17	Bsk	3	1	29°16′N, 103°18′W
Lophopilio palpinalis	3	0.25	Dfb	1	3	51°32′N, 9°56′E
Mitopus morio1	3	0.25	Cfb	2	3	54°46′33.96″N, 1°34′23.88″W
Mitopus morio2	3.5	0.29	Dfb	1	3	51°32′N, 9°56′E
Mitopus morio3	2	0.17	Dfc	1	3	63°20′N, 10°25′E
Odiellus aspersus	2	0.17	Dfa	1	3	38°55′N, 140°1′E + 43°5′N, 141°32′E
Oligolophus hanseni	3	0.25	Cfb	2	3	54°46′33.96″N, 1°34′23.88″W
Oligolophus tridens	2	0.17	Cfb	2	3	54°46′33.96″N, 1°34′23.88″W

Continued

Appendix 5.2: Mating Season Length and Habitat Climate for the Harvestmen Species Used in the Comparative Analyses Reported in Table 5.3—cont'd

Species	Mating Season Length (months)	Mating Season Length (proportion of the year)	Climate Type	Temperature	Precipitation	Coordinates
Phalangium opilio	3	0.25	Dfb	1	3	51°32'N, 9°56'E
Platybunus bucephalus	2.5	0.21	Dfb	1	3	51°32'N, 9°56'E
Rilaena triangularis1	3	0.25	Dfb	1	3	51°32'N, 9°56'E
Rilaena triangularis2	1	0.08	Dfb	1	3	48°18'N, 14°17'E
Rilaena triangularis3	1	0.08	Cfb	2	3	54°46'33.96"N, 1°34'23.88"W
Protolophus singularis	4	0.33	Bsk	3	1	32°43'N, 108°22'W
Globipes sp.	1	0.08	Bsk	3	1	32°43'N, 108°22'W
Leiobunum calcar	2	0.17	Dfb	1	3	45°28'N, 84°30'W
Leiobunum longipes	3	0.25	Dfb	1	3	45°28'N, 84°30'W
Leiobunum japonicum	1	0.08	Dfa	1	3	38°55'N, 140°1'E
Leiobunum politum	1	0.08	Dfb	1	3	45°28'N, 84°30'W
Leiobunum rotundum	2	0.17	Cfb	2	3	54°46'33.96"N, 1°34'23.88"W
Leiobunum vittatum1	3	0.25	Dfa	1	3	40°35'N, 75°21'W
Leiobunum vittatum2	2	0.17	Dfb	1	3	45°28'N, 84°30'W
Leiobunum townsendii	1.5	0.13	Bsk	3	1	32°43'N, 108°22'W
Nelima genufusca	3	0.25	Dfa	1	3	38°55'N, 140°1'E + 43°5'N, 141°32'E
Nelima suzuki	2	0.17	Dfa	1	3	38°55'N, 140°1'E + 43°5'N, 141°32'E
Trachyrhinus marmoratus	1	0.08	Bsk	3	1	30°45'N, 105°0'W
Paroligolophus agrestis	3	0.25	Cfb	2	3	54°46'33.96"N, 1°34'23.88"W

Appendix 5.3 Phylogenetic Relationships Among the Harvestmen Species Used in the Comparative Analyses

Species						Coordinates
Acanthopachylus aculeatus	6	0.5	Cfa	2	3	34°53'S, 56°10'W
Acutisoma longipes	11	0.92	Cwa	2	2	23°10'S, 46°25'W
Discocyrtus oliverio	12	1	Cwb	2	2	18°53'S, 48°15'W
Iporangaia pustulosa	12	1	Cfa	2	3	24°14'S, 48°04'W
Longiperna concolor	8	0.67	Cfa	2	3	24°14'S, 48°04'W
Neosadocus maximus	6	0.5	Cfa	2	3	24°14'S, 48°04'W
Pachyloidellus goliath	5	0.42	Cwa	2	2	31°43'S, 64°58'W
Promitobates ornatus	8	0.67	Cfa	2	3	24°14'S, 48°04'W
Serracutisoma proximum	8	0.67	Cfa	2	3	24°14'S, 48°04'W
Serracutisoma spelaeum	12	1	Cfa	2	3	24°14'S, 48°04'W
Sclerobunus robustus	1	0.08	Bsk	3	1	32°43'N, 108°22'W
Dicranopalpus ramosus	4	0.33	Csa	2	1	41°24'N, 2°6'E
Paraumbogrella pumilio	4	0.33	Dfa	1	3	43°4'N, 141°20'E
Auranus parvus	6	0.5	Am	3	3	4°4'S, 59°7'W
Bourguyia trochanteralis	8	0.67	Cfa	2	3	25°18'S, 48°05'W
Cynortoides cubanus	12	1	Aw	3	2	23°07'N, 82°23'W
Erginulus clavotibialis	8	0.67	Am	3	3	17°8'N, 88°38'W
Eucynortula lata	6	0.5	Am	3	3	4°4'S, 59°7'W
Gonyleptes saprophilus	8	0.67	Cwa	2	2	23°10'S, 46°25'W
Magnispina neptunus	9	0.75	Af	3	3	19°06'S, 39°45'W
Pachylus quinamavidensis	2.5	0.21	Csb	2	1	35°47'S, 71°26'W
Phalangodus sp.	12	1	Cfb	2	3	6°47'N, 73°15'W
Vonones ornatus	3	0.25	Dfa	1	3	40°25'N, 86°52'W
Vonones sayi	2	0.17	Cfa	2	3	29°45'N, 95°22'W
Zygopachylus albomarginis	6	0.5	Am	3	3	9°09'N, 79°51'W

Appendix 5.3: Phylogenetic Relationships Among the Harvestmen Species Used in the Comparative Analyses

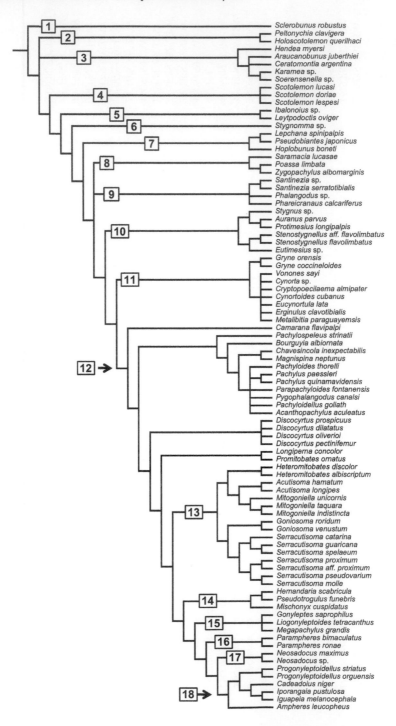

FIGURE 5.A1   Phylogenetic relationships among species used in our analysis of the effects of temperature and precipitation on the probability of occurrence of maternal care among harvestmen of the suborder Laniatores (see Table 5.2 for results). The topology of this tree was based on one molecular (Giribet *et al.*, 2010) and several morphological phylogenies (Pinto-da-Rocha, 1997, 2002; Giribet and Kury, 2007; DaSilva and Gnaspini, 2009; Mendes, 2009). The basal position of *Sclerobunus robustus* (**1**), the positions of the clade Travuniidae + Cladonychiidae (**2**) and Triaenonychidae (**3**), and the internal topology of the latter were based on Mendes (2009). The position of Phalangodidae (**4**), Podoctidae (**5**), Stygnommatidae (**6**), and Assamiidae (**7**) were based on Giribet *et al.* (2010). The relationships between Manaosbiidae (**8**), Cranaidae (**9**), and the clade composed of Stygnidae (**10**), Cosmetidae (**11**), and Gonyleptidae (**12**) were depicted as a polytomy because molecular (Giribet *et al.*, 2010) and morphological (Kury, 1993) phylogenies show completely different topologies. Stygnidae (**10**), however, was placed as the sister group to the clade Cosmetidae (**11**) + Gonyleptidae (**12**) because a recent morphological phylogeny strongly supports this relationship (Yamaguti and Pinto-da-Rocha, 2009). The internal topology of the Stygnidae (**10**) was based on Pinto-da-Rocha (1997). Within the Gonyleptidae, the relationship among subfamilies was based on Pinto-da-Rocha (2002). For all those subfamilies represented here by more than two species and that have already been revised, the internal topology is based on the most recent phylogenetic studies (Goniosomatinae (**13**), DaSilva and Gnaspini, 2009; Progonyleptoidellinae (**18**), Pinto-da-Rocha, unpub. data). Gonyleptinae is polyphyletic and the clade *Mischonyx cuspidatus* + Hernandariinae (**14**), as well as the position of other representatives of this artificial group (**15, 16, 17**), are based on Caetano (2011).

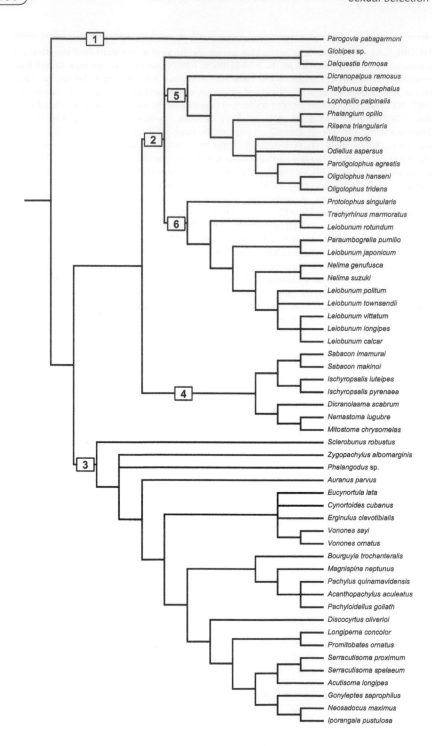

FIGURE 5.A2   **Phylogenetic relationships among species used in our analysis of the effects of temperature and precipitation on the length of mating seasons in harvestmen (see Table 5.3 for results).** The topology of this tree was based on two molecular (Giribet *et al.*, 2010; Hedin *et al.*, 2012) and several morphological phylogenies (reviewed in Giribet and Kury, 2007). The relationships among the suborders (numbers **1, 2, 3, 4**), as well as the internal topology of Dyspnoi (**4**), were based on Giribet *et al.* (2010). The polytomy in the base of Eupnoi (**2**), the internal topology of the clade Sclerosomatidae + Protolophidae (**6**), and the position of Phalangiidae (**5**) were based on Hedin *et al.* (2012). The internal topology of the Phalangiidae (**5**) was based merely on taxonomic information available in Kury's web page (http://www.museunacional.ufrj.br/mndi/Aracnologia/opiliones.html). The internal topology of Laniatores (**3**) follows the same general scheme as Figure 5.A1.

REFERENCES

Acosta, L.E., Machado, G., 2007. Diet and foraging. In: Pinto-da-Rocha, R., Machado, G., Giribet, G. (Eds.), Harvestmen: The Biology of Opiliones, Harvard University Press, Cambridge, pp. 309–338.

Albo, M.J., Costa-Schmidt, L.E., Costa, F.G., 2009. To feed or to wrap? Female silk cues elicit male nuptial gift construction in a semiaquatic trechaleid spider. J. Zool. 277, 284–290.

Allard, C.M., Yeargan, K.V., 2005. Effect of diet on development and reproduction of the harvestman *Phalangium opilio* (Opiliones: Phalangiidae). Environ. Entomol. 34, 6–13.

Andersson, M., 1994. Sexual Selection. Princeton University Press, Princeton.

Bateman, A.J., 1948. Intra-sexual selection in *Drosophila*. Heredity 2, 349–368.

Boggs, C.L., 2009. Understanding insect life histories and senescence through a resource allocation lens. Funct. Ecol. 23, 27–37.

Brockmann, H.J., 2008. Alternative reproductive tactics in insects. In: Oliveira, R.F., Taborsky, M., Brockmann, H.J. (Eds.), Alternative Reproductive Tactics: An Integrative Approach, Cambridge University Press, Cambridge, pp. 177–223.

Buzatto, B.A., Machado, G., 2008. Resource defense polygyny shifts to female defense polygyny over the course of the reproductive season of a neotropical harvestman. Behav. Ecol. Sociobiol. 63, 85–94.

Buzatto, B.A., Requena, G.S., Martins, E.G., Machado, G., 2007. Effects of maternal care on the lifetime reproductive success of females in a neotropical harvestman. J. Anim. Ecol. 76, 937–945.

Buzatto, B.A., Requena, G.S., Lourenço, R.S., Machado, G., 2011. Conditional male dimorphism and alternative reproductive tactics in a neotropical arachnid (Opiliones). Evol. Ecol. 25, 331–349.

Caetano, D.S., Machado, G., 2013. The ecological tale of Gonyleptidae (Arachnida, Opiliones) evolution: phylogeny of a Neotropical lineage of armored harvestmen using ecological, behavioral, and chemical characters. Cladistics, http://dx.doi:org/101111/cla.12009.

Chelini, M.C., Machado, G., 2012. Costs and benefits of temporary brood desertion in a neotropical harvestman (Arachnida: Opiliones). Behav. Ecol. Sociobiol. 66, 1619–1627.

Chown, S., Nicolson, S.W., 2004. Insect Physiological Ecology: Mechanisms and Patterns. Oxford University Press, Oxford.

Clark, D.L., Morjan, C.L., 2001. Attracting female attention: the evolution of dimorphic courtship displays in the jumping spider *Maevia inclemens* (Araneae: Salticidae). Proc. R. Soc. B 268, 2461–2465.

Clark, D.L., Uetz, G.W., 1993. Signal efficacy and the evolution of male dimorphism in the jumping spider *Maevia inclemens*. Proc. Natl. Acad. Sci. U. S. A. 90, 11954–11957.

Clutton-Brock, T.H., Parker, G.A., 1992. Potential reproductive rates and the operation of sexual selection. Q. Rev. Biol. 67, 437–456.

Cokendolpher, J.C., MacKay, W.P., Muma, M.H., 1993. Seasonal population phenology and habitat preferences of montane harvestmen (Arachnida: Opiliones) from southwestern New Mexico. S. West. Nat. 38, 236–240.

Cokendolpher, J.C., Tsurusaki, N., Tourinho, A.L., Taylor, C.K., Gruber, J., Pinto-da-Rocha, R., 2007. Taxonomy: Eupnoi. In: Pinto-da-Rocha, R., Machado, G., Giribet, G. (Eds.), Harvestmen: The Biology of Opiliones, Harvard University Press, Cambridge, pp. 108–131.

Curtis, D.J., Machado, G., 2007. Ecology. In: Pinto-da-Rocha, R., Machado, G., Giribet, G. (Eds.), Harvestmen: The Biology of Opiliones, Harvard University Press, Cambridge, pp. 280–308.

DaSilva, M.B., Gnaspini, P., 2009. A systematic revision of Goniosomatinae (Arachnida: Opiliones: Gonyleptidae), with a cladistic analysis and biogeographical notes. Invertebr. Syst. 23, 530–624.

Dimock, R.V., 1985. Population dynamics of *Unionicola formosa* (Acari, Unionicolidae), a water mite with a harem. Am. Midl. Nat. 114, 168–179.

Dunlop, J.A., 2007. Paleontology. In: Pinto-da-Rocha, R., Machado, G., Giribet, G. (Eds.), Harvestmen: The Biology of Opiliones, Harvard University Press, Cambridge, pp. 247–265.

Edgar, A.L., 1971. Studies on the biology and ecology of Michigan Phalangida (Opiliones). Misc. Pub. Mus. Zool., Univ. Mich. 144, 1–64.

Emlen, S.T., Oring, L.W., 1977. Ecology, sexual selection, and evolution of mating systems. Science 197, 215–223.

Freckleton, R.P., Harvey, P.H., Pagel, M., 2002. Phylogenetic analysis and comparative data: a test and review of evidence. Am. Nat. 160, 712–726.

Gadgil, M., 1972. Male dimorphism as a consequence of sexual selection. Am. Nat. 106, 574–580.

Giribet, G., 2007. Taxonomy: Cyphophthalmi. In: Pinto-da-Rocha, R., Machado, G., Giribet, G. (Eds.), Harvestmen: The Biology of Opiliones, Harvard University Press, Cambridge, pp. 89–108.

Giribet, G., Kury, A.B., 2007. Phylogeny and biogeography. In: Pinto-da-Rocha, R., Machado, G., Giribet, G. (Eds.), Harvestmen: The Biology of Opiliones, Harvard University Press, Cambridge, pp. 62–87.

Giribet, G., Vogta, L., Pérez-González, A., Sharma, P., Kury, A.B., 2010. A multilocus approach to harvestman (Arachnida: Opiliones) phylogeny with emphasis on biogeography and the systematics of Laniatores. Cladistics 26, 408–437.

Gnaspini, P., 2007. Development. In: Pinto-da-Rocha, R., Machado, G., Giribet, G. (Eds.), Harvestmen: The Biology of Opiliones, Harvard University Press, Cambridge, pp. 455–472.

Gruber, G., 2007. Taxonomy: Dyspnoi. In: Pinto-da-Rocha, R., Machado, G., Giribet, G. (Eds.), Harvestmen: The Biology of Opiliones, Harvard University Press, Cambridge, pp. 131–159.

Hedin, M., Tsurusaki, N., Macías-Ordóñez, R., Shultz, J.W., 2012. Molecular systematics of sclerosomatid harvestmen (Opiliones, Phalangioidea, Sclerosomatidae): geography is better than taxonomy in predicting phylogeny. Mol. Phylogenet. Evol. 62, 224–236.

Heinemann, S., Uhl, G., 2000. Male dimorphism in *Oedothorax gibbosus* (Araneae, Linyphiidae): A morphometric analysis. J. Arachnol. 28, 23–28.

Hunt, G.S., 1979. Male dimorphism and geographic variation in the genus *Equitius* Simon (Arachnida, Opiliones). PhD Thesis, University of New South Wales, Sydney, Australia.

Ives, A.R., Garland Jr, T., 2010. Phylogenetic logistic regression for binary dependent variables. Syst. Biol. 59, 9–26.

James, C., Shine, R., 1985. The seasonal timing of reproduction: a tropical–temperate comparison in Australian lizards. Oecologia 67, 464–474.

Jennions, M.D., Møller, A.P., Petrie, M., 2001. Sexually selected traits and adult survival: a meta-analysis. Q. Rev. Biol. 76, 3–36.

Juberthie, C., 1964. Recherches sur la biologie des opilions. Ann. Spéléol. 19, 1–244.

Juberthie, C., Manier, J.F., 1978. Étude ultrastructurale de la spermiogenèse de deux Opilions Laniatores: *Cynorta cubana* Banks (Cosmetidae) et *Strisilvea cavicola* Roewer (Phalangodidae). Rev. Arachnol. 1, 103–115.

Karaman, I.M., 2005. Evidence of spermatophores in Cyphophthalmi (Arachnida, Opiliones). Rev. Suisse Zool. 112, 3–11.

Kokko, H., Jennions, M.D., 2008. Parental investment, sexual selection and sex ratios. J. Evol. Biol. 21, 919–948.

Kury, A.B., 1993. Análise filogenética de Gonyleptoidea (Arachnida, Opiliones, Laniatores). PhD thesis. Universidade de São Paulo. São Paulo, Brazil.

Kury, A.B., 2007. Taxonomy: Laniatores. In: Pinto-da-Rocha, R., Machado, G., Giribet, G. (Eds.), Harvestmen: The Biology of Opiliones, Harvard University Press, Cambridge, pp. 159–246.

Kury, A.B., 2012. A synopsis of catalogs and checklists of harvestmen (Arachnida, Opiliones). Zootaxa 3184, 35–58.

Machado, G., Macías-Ordóñez, R., 2007a. Reproduction. In: Pinto-da-Rocha, R., Machado, G., Giribet, G. (Eds.), Harvestmen: The Biology of Opiliones, Harvard University Press, Cambridge, pp. 414–454.

Machado, G., Macías-Ordóñez, R., 2007b. Social behavior. In: Pinto-da-Rocha, R., Machado, G., Giribet, G. (Eds.), Harvestmen: The Biology of Opiliones, Harvard University Press, Cambridge, pp. 400–413.

Machado, G., Oliveira, P.S., 1998. Reproductive biology of the neotropical harvestman *Goniosoma longipes* (Arachnida, Opiliones: Gonyleptidae): mating and oviposition behaviour, brood mortality, and parental care. J. Zool. 246, 359–367.

Machado, G., Oliveira, P.S., 2002. Maternal care in the neotropical harvestman *Bourguyia albiornata* (Arachnida, Opiliones): oviposition site selection and egg protection. Behaviour 139, 1509–1524.

Machado, G., Requena, G.S., Buzatto, B.A., Osses, F., Rossetto, L.M., 2004. Five new cases of paternal care in harvestmen (Arachnida: Opiliones): implications for the evolution of male guarding in the neotropical family gonyleptidae. Sociobiology 44, 577–598.

Machado, G., Pinto-da-Rocha, R., Giribet, G., 2007. What are harvestmen? In: Pinto-da-Rocha, R., Machado, G., Giribet, G. (Eds.), Harvestmen: The Biology of Opiliones, Harvard University Press, Cambridge, pp. 1–13.

Machado, S.F., Ferreira, R.L., Martins, R.P., 2003. Aspects of the population ecology of *Goniosoma* sp. (Arachnida Opiliones Gonyleptidae) in limestone caves in southeastern Brazil. Trop. Zool. 16, 13–31.

Macías-Ordóñez, R., 1997. The mating system of *Leiobunum vittatum* Say 1821 (Arachnida: Opiliones: Palpatores): resource defense polygyny in the striped harvestman. PhD Thesis, Lehigh University, USA.

Macías-Ordóñez, R., 2000. Touchy harvestmen. Nat. Hist. 109, 58–67.

Macías-Ordóñez, R., Machado, G., Pérez-González, A., Shultz, J.W., 2011. Genitalic evolution in Opiliones. In: Leonard, J., Córdoba-Aguilar, A. (Eds.), The Evolution of Primary Sexual Characters in Animals, Oxford University Press, New York, pp. 285–306.

Martens, J., 1969. Die Sekretdarbietung während des Paarungsverhaltens von *Ischyropsalis* C. L. Koch (Opiliones). Z. Tierpsychol. 26, 513–523.

Meijer, J., 1972. Some data on the phenology and the activity patterns of *Nemastoma lugubre* (Müller) and *Mitostoma chrysomelas* (Herman) (Nemastomatidae: Opilionida: Arachnida). Neth. J. Zool. 22, 105–118.

Mendes, A.C., 2009. Avaliação do status sistemático dos táxons supra-genéricos da infra-ordem Insidiatores Loman, 1902 (Arachnida, Opiliones, Laniatores). PhD thesis, Museu Nacional do Rio de Janeiro, Rio de Janeiro, Brazil.

Mora, G., 1990. Parental care in a neotropical harvestman, *Zygopachylus albomarginis* (Arachnida, Opiliones: Gonyleptidae). Anim. Behav. 39, 582–593.

Morrow, E.H., 2004. How the sperm lost its tail: the evolution of aflagellate sperm. Biol. Rev. 79, 795–814.

Munguía-Steyer, R., Buzatto, B.A., Machado, G., 2012. Male dimorphism of a neotropical arachnid: harem size, sneaker opportunities, and gonadal investment. Behav. Ecol. 23, 827–835.

Nazareth, T.M., Machado, G., 2010. Mating system and exclusive postzygotic paternal care in a neotropical harvestman (Arachnida: Opiliones). Anim. Behav. 79, 547–554.

Nunn, C., 2011. The Comparative Approach in Evolutionary Anthropology and Biology. University of Chicago Press, Chicago.

Oliveira, R.F., Taborsky, M., Brockmann, H.J., 2008. Alternative Reproductive Tactics: an Integrative Approach. Cambridge University Press, Cambridge.

Orme, D., Freckleton, R., Thomas, G., Petzoldt, T., Fritz, S., Isaac, N., Pearse, W., CAPER: Comparative Analyses of Phylogenetics and Evolution in R. R package version 0.5. http://CRAN.R-project.org/package=caper2012.

Pabst, W., 1953. Zur Biologie der mitteleuropäischen Troguliden. Zool. Jb, Abt. Syst. Ökol. u. Geog. Tiere 82, 1–156.

Pagel, M., 1999. Inferring the historical patterns of biological evolution. Nature 401, 877–884.

Paradis, E., Claude, J., 2002. Analysis of comparative data using generalized estimating equations. J. Theor. Biol. 218, 175–185.

Paradis, E., Claude, J., Strimmer, K., 2004. APE: analyses of phylogenetics and evolution in R language. Bioinformatics 20, 289–290.

Parker, G.A., 1990. Sperm competition games – sneaks and extra-pair copulations. Proc. R. Soc. Lond. B 242, 127–133.

Peel, M.C., Finlayson, B.L., McMahon, T.A., 2007. Updated world map of the Köppen–Geiger climate classification. Hydrol. Earth Syst. Sci. 11, 1633–1644.

Pfeifer, H., 1956. Zur Ökologie und Larvalsystematik der Weberknechte. Mitt. naturh. Mus. Berlin 32, 59–104.

Pinto-da-Rocha, R., 1997. Systematic review of the neotropical family Stygnidae (Opiliones, Laniatores, Gonyleptoidea). Arq. Zool. 33, 163–342.

Pinto-da-Rocha, R., 2002. Systematic review and cladistic analysis of the Brazilian subfamily Caelopyginae (Opiliones: Gonyleptidae). Arq. Zool. 36, 357–464.

Pinto-da-Rocha, R., Giribet, G., 2007. Taxonomy. In: Pinto-da-Rocha, R., Machado, G., Giribet, G. (Eds.), Harvestmen: The Biology of Opiliones, Harvard University Press, Cambridge, pp. 88–246.

Pinto-da-Rocha, R., Machado, G., Giribet, G., 2007. Harvestmen: The Biology of Opiliones. Harvard University Press, Cambridge.

Radwan, J., 1993. The adaptive significance of male polymorphism in the acarid mite *Caloglyphus berlesei*. Behav. Ecol. Sociobiol. 33, 201–208.

Radwan, J., 1995. Male morph determination in two species of acarid mites. Heredity 74, 669–673.

Radwan, J., 2003. Heritability of male morph in the bulb mite, *Rhizoglyphus robini* (Astigmata, Acaridae). Exp. Appl. Acarol. 29, 109–114.

Radwan, J., Klimas, M., 2001. Male dimorphism in the bulb mite, *Rhizoglyphus robini*: fighters survive better. Ethol. Ecol. Evol. 13, 69–79.

Radwan, J., Unrug, J., Tomkins, J.L., 2002. Status-dependence and morphological trade-offs in the expression of a sexually selected character in the mite *Sancassania berlesei*. J. Evol. Biol. 15, 744–752.

R Development Core Team, 2011. R: A Language and Environment for Statistical Computing. R Foundation for Statistical Computing, Vienna, Austria, ISBN 3-900051-07-0, URL http://www.R-project.org/.

Requena, G.S., Buzatto, B.A., Martins, E.G., Machado, G., 2012. Paternal care decreases foraging activity and body condition, but does not impose survival costs to caring males in a neotropical arachnid. PLoS ONE 7, e46701.

Saito, Y., 1990. Harem and non-harem type mating systems in two species of subsocial spider-mites (Acari, Tetranychidae). Res. Popul. Ecol. 32, 263–278.

Santos, F.H., 2007. Ecophysiology. In: Pinto-da-Rocha, R., Machado, G., Giribet, G. (Eds.), Harvestmen: The Biology of Opiliones, Harvard University Press, Cambridge, pp. 473–488.

Schwendinger, P.J., Giribet, G., 2005. The systematics of the south-east Asian genus *Fangensis* Rambla, 1994 (Opiliones: Cyphophthalmi: Stylocellidae). Invertebr. Syst. 19, 297–323.

Shultz, J.W., Pinto-da-Rocha, R., 2007. Morphology and functional anatomy. In: Pinto-da-Rocha, R., Machado, G., Giribet, G. (Eds.), Harvestmen: The Biology of Opiliones. Harvard University Press, Cambridge, pp. 14–61.

Shuster, S.M., 2008. The expression of crustacean mating strategies. In: Oliveira, R.F., Taborsky, M., Brockmann, H.J. (Eds.), Alternative Reproductive Tactics: An Integrative Approach. Cambridge University Press, Cambridge, pp. 224–250.

Shuster, S.M., Wade, M.J., 2003. Mating Systems and Strategies. Princeton University Press, New Jersey.

Simmons, L.W., 2001. Sperm Competition and its Evolutionary Consequences in the Insects. Princeton University Press, Princeton.

Stearns, S.C., 1992. The Evolution of Life Histories. Oxford University Press, Oxford.

Taylor, C.K., 2004. New Zealand harvestmen of the subfamily Megalopsalidinae (Opiliones: Monoscutidae) – the genus Pantopsalis. Tuhinga 15, 53–76.

Thomas, R.H., Zeh, D.W., 1984. Sperm transfer and utilization strategies in arachnids: ecological and morphological constraints. In: Smith, R.L. (Ed.), Sperm Competition and the Evolution of Animal Mating Systems, Academic Press, London, pp. 179–221.

Thornhill, R., Alcock, J., 1983. The Evolution of Insect Mating Systems. Harvard University Press, Cambridge.

Tomkins, J.L., LeBas, N.R., Unrug, J., Radwan, J., 2004. Testing the status-dependent ESS model: population variation in fighter expression in the mite *Sancassania berlesei*. J. Evol. Biol. 17, 1377–1388.

Trivers, R.L., 1972. Parental investment and sexual selection. In: Campbell, B. (Ed.), Sexual Selection and the Descent of Man, 1871-1971, Aldine Publishing Co., Chicago, pp. 136–179.

Vahed, K., 1998. The function of nuptial feeding in insects: a review of empirical studies. Biol. Rev. 73, 43–78.

Vanacker, D., Maes, L., Pardo, S., Hendrickx, F., Maelfait, J.P., 2003. Is the hairy groove in the gibbosus male morph of *Oedothorax gibbosus* (Blackwall 1841) a nuptial feeding device? J. Arachnol. 31, 309–315.

Wagner, H.O., 1954. Massenansammlungen von Weberknechten. Z. Tierpsychol. 11, 348–352.

Wheeler, D., 1996. The role of nourishment in oogenesis. Annu. Rev. Entomol. 41, 407–431.

Wijnhoven, H., 2011. Notes on the biology of the unidentified invasive harvestman *Leiobunum* sp. (Arachnida: Opiliones). Arachnol. Mitt. 41, 17–30.

Wikelski, M., Hau, M., Wingfield, J.C., 2000. Seasonality of reproduction in a neotropical rain forest bird. Ecology 81, 2458–2472.

Willemart, R.H., Farine, J.P., Peretti, A.V., Gnaspini, P., 2006. Behavioral roles of the sexually dimorphic structures in the male harvestman, *Phalangium opilio* (Opiliones, Phalangiidae). Can. J. Zool. 84, 1763–1774.

Willemart, R.H., Osses, F., Chelini, M.C., Macías-Ordóñez, R., Machado, G., 2009. Sexually dimorphic legs in a neotropical harvestman (Arachnida, Opiliones): ornament or weapon? Behav. Process. 80, 51–59.

Williams, G.C., 1966. Adaptation and Natural Selection. Princeton University Press, Princeton.

Wolda, H., 1988. Insect seasonality: why? Annu. Rev. Ecol. Syst. 19, 1–18.

Yamaguti, U.Y., Pinto-da-Rocha, R., 2009. Taxonomic review of Bourguyiinae, cladistic analysis, and a new hypothesis of biogeographic relationships of the Brazilian Atlantic Rainforest (Arachnida: Opiliones, Gonyleptidae). Zool. J. Linn. Soc. 156, 319–362.

Zaslavski, V.A., 1988. Insect Development, Photoperiodic and Temperature Control. Springer, New York.

Zatz, C., 2010. Seleção sexual e evolução do dimorfismo sexual em duas espécies de opiliões (Arachnida: Opiliones). Master Thesis, University of São Paulo, Brazil.

Zatz, C., Werneck, R.M., Macías-Ordóñez, R., Machado, G., 2011. Alternative mating tactics in dimorphic males of the harvestman *Longiperna concolor* (Arachnida: Opiliones). Behav. Ecol. Sociobiol. 65, 995–1005.

Chapter 6

Adventurous Females and Demanding Males: Sex Role Reversal in a Neotropical Spider

Anita Aisenberg

Laboratorio de Etología, Ecología y Evolución, Instituto de Investigaciones Biológicas Clemente Estable, Montevideo, Uruguay

INTRODUCTION

Traditionally, as a consequence of their lower expenditure in gamete production and maintenance, males are considered as the competitive sex (Darwin, 1871; Andersson, 1994). Due to males' higher reproductive rate, they try to maximize the number of matings, while females are the choosy sex with higher parental investment (Darwin, 1871; Bateman, 1948; Trivers, 1972; Andersson, 1994). Darwin (1871) defined sexual selection as the evolutionary process that provides advantages to some individuals over others of the same species in terms of reproductive success. He distinguished the different forms this selection could take through intrasexual and intersexual competition and discussed how they modeled sex roles. Darwin provided elegant descriptions and explanations of why males of most animal species possess conspicuous weapons and display elaborate ornaments during courtship, while females seem relatively shy and disinterested. However, he recognized some cases in which the competitive-and-courting male role and choosy female role seem to operate differently from the generalized and expected fashion.

A century later, Trivers (1972) stated that anisogamy and differences in parental care were the main factors determining the direction and intensity of sexual selection acting on each sex. These arguments were later revised by many authors (Emlen and Oring, 1977; Gwynne, 1991; Owens and Thompson, 1994; Queller, 1997; Clutton-Brock, 2007; Kokko and Jennions, 2008; Perry and Rowe, 2012) who realized that determining the causes of sex role determination in the animal kingdom was not as straightforward as it seemed. Sex roles are the result of a combination of many factors acting simultaneously or sequentially, such as female and male contributions in gamete production, courtship and copulatory effort, delivery of nuptial gifts or other

Sexual Selection. http://dx.doi.org/10.1016/B978-0-12-416028-6.00006-2

resources associated with reproduction, and parental care (Bonduriansky, 2001). The relative investment of each sex in reproduction will define their potential reproductive rates and, consequently, the operational sex ratio of the population over the course of the mating season (Gwynne, 1991). If females provide most parental investment, this sex will delay future mating opportunities and their potential reproductive rate will be low. In this context, if males continue their mate search while females invest more in parental effort, the operational sex ratio will turn male-biased and females will become the scarce resource for which males compete. Nevertheless, this argument is much too simplified and the direction and intensity of sexual selection on each sex is not constrained only by potential reproductive rates (Kokko and Jennions, 2008; Karlsson Green and Madjidian, 2011). Factors such as primary sex ratios, age of adulthood, and longevity and mortality rates of each sex should also be considered.

We should always keep in mind that during an animal's lifetime and according to its age, reproductive status, size, body condition, and timing of the breeding period, the requirements and interests related to reproduction can fluctuate. Thus, sex roles of certain species can change and may be flexible and not as gender-specific as expected. Furthermore, characteristics of the habitat, such as temperature, humidity, availability of refuges, predictability of environmental conditions, predation risk, and/or prey abundance, may also be modeling the evolution of the mating system, sexual size dimorphism, and sex roles of a species (Gwynne and Simmons, 1990; Karlsson et al., 1997). As an example, when males provide a resource valuable to females during courtship or copulation (i.e., a nuptial gift), females will probably be willing to mate and re-mate to obtain the nuptial gift, and males may have low reproductive potential due to the costs of the production of this item. These conditions can transform males into the scarce resource for females, which can drive the occurrence of mutual or exclusive male mate choice and affect the mating system of the species (Svensson et al., 1989; Gwynne and Simmons, 1990; Jia et al., 2000).

During recent decades, the conventional sex role assignment of competitive active males and selective coy females has been widely discussed and reinterpreted (Eberhard, 1985, 1996; Andersson, 2005; Arnqvist and Rowe, 2005; Roughgarden et al., 2006; Clutton-Brock, 2007, 2009; Kokko and Jennions, 2008; Karlsson Green and Madjidian, 2011). The historical masculine bias in scientific research at the origins of the seminal research on sexual selection may have influenced our perceptions of female and male functions (Eberhard, 1996). Detailed observations of female sexual behavior and its consequences on male behavior and reproductive success have shown that females are not at all sexually passive, as was originally perceived (Eberhard, 1996). Furthermore, examples of exceptions to the traditional sex roles concept are widespread in various animal groups and could be more common than initially expected (Gwynne, 1991).

EXCEPTIONS THAT TEST GENERAL RULES: SEX ROLE REVERSAL

In species with high male contributions to reproduction, traditional sex roles and sexual dimorphism can reverse from the expected patterns leading to choosy males and active females that search for males, compete for their access, and perform courtship (Gwynne, 1991; Andersson, 1994; Bonduriansky, 2001). So, sex role reversal does not imply a change in sex, but in the female and male behavioral roles predicted for species in certain taxa. Sex role reversal is considered total when only males are selective when making mating decisions, or partial when both sexes select their mating partners (Andersson, 1994). It has been described or suggested in vertebrates, including fish, frogs, and birds, and in invertebrates such as crustaceans, insects, and arachnids (Gwynne, 1991; Eens and Pinxten, 2000; Bonduriansky, 2001; Huber, 2005; Aisenberg et al., 2007; Aisenberg and Costa, 2008). Most examples from vertebrates come from species with high paternal investment, and both courtship and parental roles are reversed. However, several arthropods that lack parental care show a reversal in typical mate searching, competing, and courtship roles. In these cases, high male reproductive costs are associated with intensive courtship or copulatory effort, which frequently includes the delivery of expensive nuptial gifts to their mating partners (Vahed, 2007). The reversal in sex roles can also involve changes in traditional sexual dimorphism (Gwynne, 1991; Andersson, 1994). In these cases females, and not males, will display bright and colorful ornaments while seducing their potential mates.

The main causes driving these unexpected sexual behaviors are still obscure and debated. Nevertheless, we can sustain that sex role reversal is associated with high male reproductive investment that equals or exceeds female reproductive costs, reducing male mating opportunities and making them the scarce resource for which females compete (Gwynne, 1991). However, high male reproductive costs do not determine sex role reversal *per se*, and possibly many factors related to the natural history, behavioral, ecological, and phylogenetic characteristics of the species should also be considered. The inevitable dilemma arises regarding what is cause and what is consequence of sex role reversal, leading to the question: if high male reproductive costs drive sex role reversal, then what is driving male reproductive investment?

Nuptial gift delivery and the evolution of paternal investment and sex role reversal have been associated with harsh and unpredictable environments (Karlsson et al., 1997; Lorch, 2002). Examples of quick shifts in sex roles suggest that responses to changes in the environment can occur in brief periods considering the life of an animal. An example of this is the inspiring study by Gwynne and Simmons (1990) with katydids. During mating, males of this katydid species provide females with a nutritious nuptial gift that is part of the spermatophore, the spermatophylax, which increases female fecundity. These authors were able to experimentally reverse sex roles in one direction or the other – from conventional to reversed sex roles – by manipulating food abundance and, consequently, the value and cost of male nuptial gifts. An additional factor that can play a role in determining sex roles in a short period is parasite infection. Another experiment with bush crickets showed that when females were free of parasites the typical sex roles prevailed; conversely, when females were infected with gut parasites they actively solicited matings, while males adopted the choosy role (Simmons, 1994).

Summarizing, these behavioral exceptions could be more widespread than previously thought, and much future research is needed to decide if these exceptions are truly exceptions (see Chapter 2). Despite the fact that sex role reversal has been observed in many animal orders, few cases have been exhaustively studied. Indeed, various authors have highlighted the importance of studying these atypical cases towards establishing a more robust theory of sexual roles in the animal kingdom (Tallamy, 2000; Bonduriansky, 2001; Roughgarden et al., 2006; Karlsson Green and Madjidian, 2011).

SEX ROLES AND SEXUAL DIMORPHISM IN SPIDERS

Looking for potential mates and seducing mating partners is not a trivial issue in spiders (Foelix, 2011; Schneider and Andrade, 2011). After reaching adulthood, small and agile males rove around looking for females (Schneider and Andrade, 2011). Females are the choosy sex that imposes sexual selection on certain behavioral or morphological characteristics of males. Males follow airborne or contact pheromones released by the females (Gaskett, 2007), and once they find a potential mate they expend a great deal of effort to be accepted as mates by frequently aggressive, large, and carnivorous females. The risk of injury for males occurs not only during mate search (Schneider and Lubin, 1997; Andrade, 2003) but also during courtship and copulation. Sexual cannibalism, when the male is attacked and consumed by the female at some stage during courtship and/or mating, is widely distributed in arachnids (Elgar, 1992; Elgar and Schneider, 2004; Wise, 2006).

In most spider species females are larger than males, leading to extreme sexual size dimorphism in some cases (Vollrath and Parker, 1992; Hormiga et al., 2000; Moya-Laraño et al., 2002; Foellmer and Moya-Laraño, 2007). Sexual size dimorphism is driven by differences in the evolutionary history of each sex imposed by sexual selection, natural selection, or both. Many hypotheses have been proposed regarding the origin and maintenance of sexual size dimorphism in spiders. One of these hypotheses proposes that male dwarfism is a consequence of the relaxation of sexual selection for large size in this sex due to high mortality during mate search (Vollrath and Parker, 1992). Conversely, female gigantism has been discussed as a response to selection for higher fecundity (Coddington et al., 1997; Prenter et al., 1999; Hormiga et al., 2000). Smaller size in males could also increase agility during mate search (Framenau, 2005) or during climbing in orb web spiders (Moya-Laraño et al., 2002). There are very few reports of male spiders of a larger body size compared to females

(Alderweireldt and Jocqué, 1991; Prenter *et al.*, 1995; Lang, 2001; Gasnier *et al.*, 2002; Schutz and Taborsky, 2005; Aisenberg and Costa, 2008).

Though sex role reversal seems to be widespread in insects, there are very few reports or exhaustive studies on these topics for arachnids. We could expect males to be choosy when they are the sex that provides parental care (as, for example, in several harvestman species; see Chapter 8), when they are cannibalized during or after mating, or when they suffer permanent damage to their genitalia after copulation (Huber, 2005). Male mate assessment has been observed in several spider species (Herberstein *et al.*, 2002; Rypstra *et al.*, 2003; Andrade and Kasumovic, 2005; Roberts and Uetz, 2005; Baruffaldi and Costa, 2009; Schulte *et al.*, 2010; Aisenberg and González, 2011), and in general the preference is biased towards virgin females. Female courtship and pursuit of mating has been described for several orb web spiders, theraphosids, and wolf spiders (see, for example, Robinson and Robinson, 1980; Knoflach, 1998; Costa and Pérez-Miles, 2002; Huber, 2005; Aisenberg and Eberhard, 2009). Nonetheless, previous reports have not scrutinized their observations considering sex role reversal hypotheses.

One of the few studies of sex role reversal in arachnids comes from the tropical harvestman *Zygopachylus albomarginis* (Mora, 1990). Males of this species build mud nests, or usurp them from other males, and perform all their maintenance. On the other hand, females are wanderers that search and compete with other females for access to males offering good nests for ovipositing. Females actively court males, oviposit inside their nests, and then leave. Males perform exclusive parental care by guarding the nests and providing protection to eggs laid by different females. Male investment in nest construction and egg-guarding could explain the reversal in typical mate searching, competing, and courting roles. Nevertheless, it is unclear whether males of *Z. albomarginis* are choosy when deciding on mating partners. Paternal care has been reported in several harvestmen (Machado *et al.*, 2004; Machado and Macías-Ordóñez, 2007), so these arthropods remain challenging models for determining whether the costs imposed by paternal care are enough to turn males into the scarce resource for which females compete (see discussion in Chapter 8).

The other example of sex role reversal in arachnids comes from the Mediterranean wolf spider *Lycosa tarantula* (Moya-Laraño *et al.*, 2003; Huber, 2005). Females of this burrowing tarantula are highly territorial, and males that visit them during the breeding season are frequently cannibalized (Moya-Laraño *et al.*, 2003). At the end of the mating season, males become a limited resource for females due to cannibalism and high mortality during mate search. Though males of *L. tarantula* are always the mobile and courting sex, towards the end of the breeding period they become choosy and prefer small females with good body condition. Females show conspicuous colorations and have high variance in mating success. These characteristics, in addition to the female-biased sex ratio towards the end of the breeding period, suggest that sex role reversal could be occurring (Moya-Laraño *et al.*, 2003). Nonetheless, further studies are needed to determine if sex roles are really reversed in this system, or if males are just mating more frequently with small and well-fed females that are more willing to mate compared with large and hungry females.

A NEOTROPICAL WOLF SPIDER AS A MODEL OF SEX ROLE REVERSAL

Wolf spiders or lycosids are distributed all over the world. They are wandering spiders that do not build webs for capturing prey, but hunt them actively (Foelix, 2011). They exhibit four small anterior and four large posterior eyes (Barth, 2002). Wolf spiders usually have low to moderate sexual size dimorphism, but, as is general for spiders, females are larger than males (Walker and Rypstra, 2003; Logunov, 2011). One distinctive feature of wolf spiders is that females carry their egg sacs attached to spinnerets, and when spiderlings emerge they move to their mother's dorsum where they stay until it is time to disperse (Foelix, 2011).

Allocosa brasiliensis also named the white spider of the sand dunes, is a medium-sized South American sand-dwelling wolf spider (Capocasale, 1990). Individuals of this species build burrows where they stay during the day and in the coldest months, becoming very active during summer nights (Costa, 1995; Costa *et al.*, 2006) (Fig. 6.1). Their whitish cryptic coloration makes them inconspicuous in their sandy habitat. The Uruguayan coastline, where most of the studies regarding the biology of *A. brasiliensis* have been carried out, is characterized by the occurrence of reflective or dissipative beaches of variable width, with sand dunes of variable heights and dynamics (Gómez-Pivel, 2006; Gómez and Martino, 2008; Aisenberg *et al.*, 2011a). Vegetation is scarce, and mainly composed of *Panicum racemosum*

FIGURE 6.1   A male *A. brasiliensis* walking on the sand dunes during sunset. See color plate at the back of the book. *Photograph courtesy of M. Casacuberta.*

(Poaceae), *Hydrocotyle bonariensis*, *Senecio crassiflorus* (Apiaceae), and the exotic shrub *Acacia longifolia* (Leguminosae) (Costa, 1995; Alonso-Paz and Bassagoda, 2006) (Fig. 6.2).

The white spider of the sand dunes shows a reversal in the typical sex roles and sexual size dimorphism expected for spiders (Aisenberg *et al.*, 2007). Males are larger than females (carapace width: females, 4.6 ± 0.5 mm; males, 5.8 ± 0.6 mm) and they remain sedentary inside their burrows for long periods (Costa *et al.*, 2006; Aisenberg *et al.*, 2009). They build long burrows, while females build just temporary silk capsules for refuge during the day (Aisenberg *et al.*, 2007). Contrary to the generalized pattern, females are the mobile sex that searches for sexual partners (Fig. 6.3). The male white spider emits volatile pheromones that help the female locate his burrow (Aisenberg *et al.*, 2010a). After finding it, the female initiates courtship by waving her forelegs asynchronously while leaning into the male burrow entrance. The male can respond to female courtship by performing body shakes at the bottom of his burrow. Females are selective in their mating partners, and they prefer males that have long burrows (Aisenberg *et al.*, 2007). Possibly, when the female follows the male to the bottom of his burrow during courtship, she inspects its length.

After this, if courtship continues, the sexes exchange positions inside the burrow and the male stays at the top and the female at the bottom of the burrow (Aisenberg *et al.*, 2007). Then, mounting occurs in the typical lycosid mating position (Foelix, 2011) but vertically, with the male on the female's dorsal surface but facing in the opposite direction. After mating, the male exits the burrow and blocks the entrance from outside (Aisenberg *et al.*, 2007). He departs only after completely closing the entrance and scattering sand around the area with palps, chelicerae, and forelegs, camouflaging the burrow entrance. The female stays inside the male's burrow, where she oviposits, and exits 1 month later

FIGURE 6.2   Typical sand dune habitat of *A. brasiliensis*. See color plate at the back of the book. *Photograph courtesy of M. Casacuberta.*

FIGURE 6.3　Traces left on the sand by a white sand dune spider. See color plate at the back of the book. *Photograph courtesy of M. Casacuberta.*

when it is time for spiderling dispersal (Costa *et al.*, 2006; Postiglioni *et al.*, 2008) (Fig. 6.4). Females can lay up to four consecutive egg sacs during one breeding period, but the first egg sac is the most successful in number of eggs (Postiglioni *et al.*, 2008). Because females are not good diggers they need to re-mate to obtain a new burrow for subsequent ovipositions. On the other hand, males need to build a new burrow to obtain a refuge and to generate new mating opportunities.

　　Summarizing, females of *A. brasiliensis* are the mobile sex that look for males and start courtship, whereas males are sedentary and wait for their potential mating partners inside their burrows. The male donation of a burrow to the female as a nuptial gift limits male mating chances and, as we will see later, burrow construction in the sand is energetically expensive. In the context of sex role reversal hypotheses, we should expect males to be choosy. The next section describes how males perform extreme mate choice, and discusses the consequences of this behavior for the sexual strategies of the species.

FIGURE 6.4   A female with spiderlings riding on her abdomen. See color plate at the back of the book. *Photograph courtesy of M. Casacuberta.*

MALE'S REVENGE: REVERSED SEXUAL CANNIBALISM IN *A. BRASILIENSIS*

Sexual cannibalism occurs when one sex is attacked and consumed by the other sex at some stage during courtship or mating (Elgar, 1992). In general, in arachnids, males are the victims due to their smaller size and higher vulnerability when approaching their aggressive mating partners (Wilder *et al.*, 2009). But what happens when sex roles and sexual size dimorphism are reversed?

As we have already seen, *A. brasiliensis* shows a reversal in sex roles and males are larger than females, contrary to the expected sexual size dimorphism pattern in spiders. High male reproductive cost and a consequently low mating rate could be driving male mate assessment in this species. We found that males preferred to mate with virgin females in good body condition (Aisenberg *et al.*, 2011b). Females with these characteristics are expected to maximize male paternity expectations. First of all, the first egg sac of *A. brasiliensis* is the most successful, and since after their first mating females remain in their burrows until spiderling emergence and apparently do not re-mate during this period, they are ideal mates. Good body condition and body weight are positively associated with the number of eggs in spiders (Wise and Wagner, 1992). However, the most surprising finding for *A. brasiliensis* was not that males are choosy, but that rejected females could be cannibalized (Aisenberg *et al.*, 2011b). The attacks always occurred during mating and after both sexes had performed courtship, with the male on top of the female and before palpal insertion. Male sexual cannibalism in this species was confirmed by field studies (Aisenberg *et al.*, 2009).

Thus, in this case we must reinterpret the hypotheses used to explain female sexual cannibalism of males. We can discard Gould's hypothesis (1984) of mistaken identity, since females and males always perform courtship before the attacks occur. The adaptive foraging hypothesis (Newman and Elgar, 1991) postulates that copulation or sexual cannibalism occur according to the nutritional status of the strongest and selective sex. However, we did not find that males with lower body condition and/or weight attacked more frequently (Aisenberg *et al.*, 2011b). Arnqvist and Henriksson (1997) proposed that aggressiveness, adaptive in juvenile phases, could result in a runaway process out of control in adults, triggering non-adaptive attacks on the opposite sex. However, this maladaptive runaway process that expects indiscriminate aggressiveness directed towards mating partners cannot explain male selective choice of virgin females in good body condition or the higher frequencies of male sexual cannibalism of mated females of *A. brasiliensis*. Male cannibalistic behavior in this wolf spider best agrees with the extreme mate choice hypothesis (Elgar and Nash, 1988), which proposes that the choosy sex evaluates the quality of the potential mate during pre-inseminatory phases and decides to mate with those of higher quality, cannibalizing less than satisfactory individuals. This coincides with our results that males more frequently attacked mated females with low body condition.

Male sexual cannibalism in *A. brasiliensis* is thus an exception to the general biological rules within this context, constituting the first case of reversed sexual cannibalism in spiders and one of the very few examples in the animal kingdom (Aisenberg *et al.*, 2011b) (Fig. 6.5). These spiders inhabit an environment that exhibits wide fluctuations in prey quality and abundance (Aisenberg *et al.*, 2009), and males perform extreme sexual choice and obtain mates or food according to the quality of the potential partner. Males can be certain that they have not recently mated with the visiting female because females remain in their burrows after mating (Costa *et al.*, 2006; Aisenberg *et al.*, 2007). Males are larger than females and the attack takes place inside the burrow, with the female placed at the bottom and the male covering the burrow entrance (Aisenberg *et al.*, 2011b). Consequently, females are vulnerable to male attacks and expose

FIGURE 6.5 **Male cannibalizing a female.** *Photograph courtesy of L. Watson.*

themselves to this risk because they need new burrows to oviposit and are not themselves good diggers.

FORAGING OPPORTUNITIES AND CANNIBALISM

The habitat occupied by *A. brasiliensis* in Uruguay has been dramatically reduced and modified during recent decades due to urbanism (Costa *et al.*, 2006). Populations of the white spider have been affected by these changes, suffering habitat fragmentation and processes of local extinctions (Aisenberg *et al.*, 2011a). Prey abundance on the Uruguayan coastline is very variable and entirely dependent on weather conditions (A. Aisenberg, unpublished data). As previously stated, harsh environments with unpredictable climatic fluctuations, scarce refuges, and periods of scarcity of prey possibly promote the evolution of paternal care and sex role reversal (Karlsson *et al.*, 1997; Lorch, 2002). All these characteristics could be modeling the sex role reversal and reversed sexual cannibalism described for *A. brasiliensis*.

The white spider of the sand dunes forages intensively during summer nights and is a highly opportunistic predator, capable of varying its diet according to prey availability (Aisenberg *et al.*, 2009). Their diet consists mainly of Diptera, Hymenoptera, Coleoptera, and other *Allocosa* spiders (Fig. 6.6). Intraguild cannibalism is frequent, and always follows the pattern that the larger eats the smaller (Polis, 1981; Wise, 2006). Males forage when they leave their burrows after mating, having spent a long period buried without feeding, or only feeding sporadically on mating partners. Thus, males are truly voracious when they venture out of their burrows, and attack whatever large prey they find, including conspecific adult females. Females need to forage intensively before mating, because later they will remain in male burrows for at least 1 month, until they exit the burrow with the spiderlings. The consumption of ants, infrequent prey for wolf spiders, suggests food limitation (Nentwig, 1987). Ants are very abundant

FIGURE 6.6   A female hunting a bug. *Photograph courtesy of M. Casacuberta.*

in coastal areas (Costa *et al.*, 2006), and though they are small prey for a spider of the size of a white sand dune spider, they are the most abundant and predictable prey in these areas. This latter argument may explain the high levels of cannibalism in areas where this spider is the most abundant species (Costa *et al.*, 2006).

BURROW DIGGING, SPATIAL DISTRIBUTION, AND MALE ADAPTATIONS AS NEST PROVIDER

One way of avoiding the extreme heat during daylight in sandy habitats and preventing dehydration is to burrow in the sand. However, burrow digging in the sand can be an energetically costly activity for spiders, mainly due to the costs related to the production and deposition of multiple layers of silk for maintaining a stable burrow in this substrate, in addition to the digging activities *per se* (Henschel and Lubin, 1992; Aisenberg and Peretti, 2011a).

Strong selective pressures shape male burrow digging in *A. brasiliensis*. Males need to build stable burrows, long enough to accommodate two adult individuals during mating, and with suitable temperature and humidity conditions for adequate development of the future progeny (Fig. 6.7). We also know that females prefer males with longer burrows (Aisenberg *et al.*, 2007), so both natural and sexual selection should be driving burrow length in this species. Male burrows are approximately 10 cm long, whereas female burrows, which operate as transitory refuges, are less than 2 cm in length (Capocasale 1990; Aisenberg *et al.*, 2011a). Male burrows function as temperature buffers, and buffering increases with burrow length (Aisenberg *et al.*, 2011c). In this way, males create a refuge with a more stable environment for egg sac care and spiderlings' development. Also, burrow spatial distribution is not uniform, and

FIGURE 6.7   A male inside his burrow. See color plate at the back of the book. *Photograph courtesy of M. Casacuberta.*

adult burrows are dug near the base of the dunes (Aisenberg *et al.*, 2011c). This preference for the base of the dunes could be related to higher protection from strong winds and higher humidity levels relative to the top of the dunes.

What makes males good diggers and females bad ones? First of all, the larger body size could predispose males to construct longer burrows. Second, individuals of *A. brasiliensis* use their palps, front legs, and chelicerae for digging. Males have higher muscle mass on their legs and palps, and larger chelicerae, compared to females (Aisenberg *et al.*, 2010b; Aisenberg and Peretti, 2011b). Furthermore, the tip of the male palpal tarsi are shaped like spades, instead of the typical claw of juveniles and females (Aisenberg *et al.*, 2010b), and this probably helps to remove sand when digging.

Most burrowing activities take place at night, and individuals reduce their time outside the burrow possibly to avoid attracting predators. To build a stable 10-cm burrow in the sand, males need to dig for the whole night and sometimes for two nights (A. Aisenberg, unpublished data). While they are building their burrows, males perform very fast movements, sporadically by hiding inside the burrow they are digging, presumably to reduce predation risk (Aisenberg and Peretti, 2011b).

WHEN HUNTERS ARE HUNTED

One of the most dangerous enemies of the white spider of the sand dunes is the wasp *Anoplius bicintus*, a frequent parasitoid of *A. brasiliensis*. This wasp is very abundant on the Uruguayan coastline, and shares the same habitat of *A. brasiliensis* (Costa, 1995; Costa *et al.*, 2006). During daytime in the summer of the southern hemisphere, female *A. bicinctus* wander along the sand dunes searching for *A. brasiliensis* burrows. When the wasp finds a burrow, she enters or forces the spider to exit from the burrow, and then stings it. The paralyzed spider is dragged for several meters along the sand, until the wasp finds an adequate place for digging. The wasp builds a new burrow to bury the paralyzed spider, and deposits one egg on the spider's abdomen. When the wasp larva emerges it feeds on the anesthetized spider body, which is kept alive until wasp pupation (Stanley *et al.*, 2013).

Once again, we have an exception to a general rule. Recent studies have found that *A. brasiliensis* females are more frequently parasitized compared with males of this species (Stanley *et al.*, 2013). In systems with sex role reversal, the risk of predation is expected to reverse so that females will suffer more attacks than males (Gwynne and Bussière, 2002). In *A. brasiliensis*, females are the mobile sex (Costa *et al.*, 2006; Aisenberg *et al.*, 2007), which could explain why they build transitory and shallow silk refuges. Also, males are larger than females and show positive allometry in chelicerae length (Aisenberg *et al.*, 2010b), presumably a trait used to ward off wasps when they invade their burrows. All these characteristics result in males being more capable of defending themselves, while females are more susceptible to wasp attacks.

WHAT ARE THE CAUSES DRIVING SEX ROLE REVERSAL IN THIS SPECIES?

As we have seen, the sex role reversal described for the white spider of the sand dunes has multiple consequences on the behaviors and adaptations of females and males. It is difficult to determine the specific causes driving the atypical behaviors described for this species. Wolf spiders are abundant in neighboring areas, but *Allocosa* are the only lycosids that inhabit Uruguayan coastal sand dunes (Costa *et al.*, 2006). Similarly to other wolf spiders (Capocasale and Costa, 1975; Wagner, 1995), females need to forage intensively before mating. Afterwards, they remain buried without foraging for approximately 1 month, during egg sac deposition and egg sac care, and until spiderling dispersal (Costa *et al.*, 2006; Aisenberg *et al.*, 2007). Because prey abundance and quality can fluctuate during the reproductive period (Aisenberg *et al.*, 2009), it would appear that *A. brasiliensis* females need to be mobile to forage intensively before finding a safe and adequate refuge to mate and oviposit. Strong pressures related to the habitat and wolf spider life history could be driving the need for high male reproductive investment, as a way to compensate female foraging needs and scarcity of available refuges in such a harsh and dynamic environment as sand dunes.

Two species of *Allocosa* inhabit the coastline in Uruguay (Costa *et al.*, 2006; Aisenberg and Costa, 2008), *A. brasiliensis* and *A. alticeps*, and both of them show sex role reversal (Aisenberg *et al.*, 2007; Aisenberg and Costa, 2008). Though these two species share characteristics such as the reversal in sexual size dimorphism expected for spiders, females being the mobile sex, and males being providers of the mating refuge and breeding nest, they show significant differences regarding longevity, operational sex ratio, and frequency of sexual cannibalism. In *A. alticeps*, males survive for only one breeding period, the primary sex ratio is female biased, and, though males are selective, sexual cannibalism of females is absent or very rare (Aisenberg and Costa, 2008; Aisenberg and González, 2011). Conversely, in *A. brasiliensis*, males survive for two breeding seasons as adults, the primary sex ratio is 1 : 1, and male mate choice and sexual cannibalism of females is frequent (Aisenberg and Costa, 2008; Aisenberg *et al.*, 2009). Therefore, these two sex role reversed wolf spiders that inhabit the same coastal habitat show divergences in their solutions to similar selective pressures. These divergences could be a consequence of differences in body size (*A. alticeps* is smaller: carapace width, females 2.9 ± 0.3 mm; males 3.3 ± 0.5 mm), foraging opportunities, inter- and intraspecific competition for access to refuges or other resources, or different predation risk, among other factors.

Allocosa brasiliensis is an outstanding model to test hypotheses of sex role reversal, sexual dimorphism, and sexual cannibalism, and although a great deal has been uncovered relative to the biology of the species, much remains to be done. There are several *Allocosa* species in South America and also in the northern hemisphere. However, northern species seem to have typical mobility and sexual dimorphism patterns (Stratton, 2005). This could reflect variations in the direction and intensity of selective forces acting on neotropical species. This emerges as an interesting question that requires further exploration, and that may have multiple explanations associated with the environmental conditions experienced by the different *Allocosa* species (see Chapter 1). However, the sexual behavior of northern *Allocosa* remains almost unstudied, and much further research is needed on other species of the genus that inhabit the coasts of the Atlantic Ocean and internal rivers and lakes of South America. Comparisons of the habitat and mating systems of *Allocosa* species will undoubtedly shed light on the pressures driving sex role reversal and on whether this occurs incidentally or is associated to neotropical members of this genus.

Studying the behavior of other members of the sand-dwelling fauna of the coastline would also provide answers to some broader ecological questions. Some tropical burrowing crabs also show active and mobile females that search for male burrows in the sand, similarly to *A. brasiliensis* (Christy *et al.*, 2002). Data on behavioral strategies of other arthropods that inhabit similar habitats to this wolf spider could help us understand the convergence of behavioral and morphological solutions to the same environmental problem at a broader ecological scale. Furthermore, a more global knowledge on species inhabiting the sandy Uruguayan coastline is essential for elaborating adequate management plans for these areas. Finally, as explained in Chapter 1 of this book, the heterogeneity of landscapes and climatic conditions in the Neotropics, with its high diversity in species and behavioral strategies, offers ideal scenarios for sexual selection studies and for the emergence of new challenging models, with more exceptions to test general and traditional biological rules.

ACKNOWLEDGMENTS

I thank Maydianne Andrade, Luiz Ernesto Costa Schmidt, and editors Regina H. Macedo and Glauco Machado for their reviews and suggestions on the text. Marcelo Casacuberta provided the pictures included in the present chapter.

REFERENCES

Aisenberg, A., Costa, F.G., 2008. Reproductive isolation and sex role reversal in two sympatric sand-dwelling wolf spiders of the genus *Allocosa*. Can. J. Zool. 86, 648–658.

Aisenberg, A., Eberhard, W.G., 2009. Female cooperation in plug formation in a spider: effects of male copulatory courtship. Behav. Ecol. 20, 1236–1241.

Aisenberg, A., González, M., 2011. Male mate choice in *Allocosa alticeps* (Araneae, Lycosidae): a sand-dwelling spider with sex role reversal. J. Arachnol. 39, 444–448.

Aisenberg, A., Peretti, A.V., 2011a. Sexual dimorphism in immune response, fat reserves and muscle mass in a sex role reversed spider. Zoology 114, 272–275.

Aisenberg, A., Peretti, A.V., 2011b. Male burrow digging in a sex role reversed spider inhabiting water-margin environments. Bull. Br. Arachnol. Soc. 15, 201–204.

Aisenberg, A., Viera, C., Costa, F.G., 2007. Daring females, devoted males and reversed sexual size dimorphism in the sand-dwelling spider *Allocosa brasiliensis* (Araneae, Lycosidae). Behav. Ecol. Sociobiol. 62, 29–35.

Aisenberg, A., González, M., Laborda, Á., Postiglioni, R., Simó, M., 2009. Reversed cannibalism, foraging and surface activities of *Allocosa alticeps* and *Allocosa brasiliensis* (Lycosidae), two wolf spiders from coastal sand dunes. J. Arachnol. 37, 135–138.

Aisenberg, A., Baruffaldi, L., González, M., 2010a. Behavioural evidence of male volatile pheromones in the sex-role reversed wolf spiders *Allocosa brasiliensis* and *Allocosa alticeps*. Naturwissenschaften 97, 63–70.

Aisenberg, A., Costa, F.G., González, M., Postiglioni, R., Pérez-Miles, F., 2010b. Sexual dimorphism in chelicerae, forelegs and palpal traits in two burrowing wolf spiders (Araneae: Lycosidae) with sex-role reversal. J. Nat. Hist. 44, 1189–1202.

Aisenberg, A., Simó, M., Jorge, C., 2011a. Spider as a model towards the conservation of coastal sand dunes in Uruguay. In: Murphy, J.A. (Ed.), Sand Dunes: Conservation, Types and Desertification, NOVA Science Publishers, USA, pp. 75–93.

Aisenberg, A., Costa, F.G., González, M., 2011b. Male sexual cannibalism in a sand-dwelling wolf spider with sex role reversal. Biol. J. Linnean Soc. 103, 68–75.

Aisenberg, A., González, M., Laborda, Á, Postiglioni, R., Simó, M., 2011c. Spatial distribution, burrow depth and temperature: implications for the sexual strategies in two *Allocosa* wolf spiders. Stud. Neotrop. Fauna Environ. 46, 147–152.

Alderweireldt, M., Jocqué, R., 1991. A remarkable new genus of wolf spider from southwestern Spain (Araneae, Lycosidae). Bull. Inst. R. Nat. Belg., Entomol. 61, 103–111.

Alonso-Paz, E., Bassagoda, M.J., 2006. Flora y vegetación de la costa platense atlántica. In: Menafra, R., Rodríguez-Gallego, L., Scarabino, F., Conde, D. (Eds.), Bases para la Conservación y el Manejo de la Costa Uruguaya, Vida Silvestre, Uruguay, pp. 21–34.

Andersson, M., 1994. Sexual selection. Princeton University Press, Princeton.

Andersson, M., 2005. Evolution of classical polyandry: three steps to female emancipation. Ethology 111, 1–23.

Andrade, M.C.B., 2003. Risky mate search and male self-sacrifice in redback spiders. Behav. Ecol. 14, 531–538.

Andrade, M.C.B., Kasumovic, M.M., 2005. Terminal investment strategies and male mate choice: extreme tests of Bateman. Integr. Comp. Biol. 45, 838–847.

Arnqvist, G., Henriksson, S., 1997. Sexual cannibalism in the fishing spider and a model for the evolution of sexual cannibalism based on genetic constraints. Evol. Ecol. 11, 255–273.

Arnqvist, G., Rowe, L., 2005. Sexual Conflict. Princeton University Press, Princeton.

Barth, F.G., 2002. A Spider's World. Senses and Behavior. Springer, Berlin.

Baruffaldi, L., Costa, F.G., 2009. Changes in male sexual responses from silk cues of females at different reproductive states in the wolf spider *Schizocosa malitiosa*. J. Ethol. 28, 75–85.

Bateman, A.J., 1948. Intra-sexual selection in *Drosophila*. Heredity 2, 349–368.

Bonduriansky, R., 2001. The evolution of male mate choice in insects: a synthesis of ideas and evidence. Biol. Rev. 76, 305–339.

Capocasale, R.M., 1990. Las especies de la subfamilia Hipassinae de América del Sur (Araneae, Lycosidae). J. Arachnol. 18, 131–141.

Capocasale, R.M., Costa, F.G., 1975. Descripción de los biotopos y caracterización de los habitats de *Lycosa malitiosa* Tullgren (Araneae: Lycosidae) en Uruguay. Viet et Milieu Série C 25, 1–15.

Christy, J.H., Backwell, P.R.Y., Goshima, S., Kreuter, T.J., 2002. Sexual selection for structure building by courting male fiddler crabs: an experimental study of behavioral mechanisms. Behav. Ecol. 13, 366–374.

Clutton-Brock, T., 2007. Sexual selection in males and females. Science 318, 1882–1885.

Clutton-Brock, T., 2009. Sexual selection in females. Anim. Behav. 77, 3–11.

Coddington, J.A., Hormiga, G., Scharff, N., 1997. Giant females or dwarf males? Nature 385, 687–688.

Costa, F.G., 1995. Ecología y actividad diaria de las arañas de la arena *Allocosa* spp. (Araneae, Lycosidae) en Marindia, localidad costera del sur del Uruguay. Rev. Brasil. Biol. 55 (3), 457–466.

Costa, F.G., Perez-Miles, F., 2002. Reproductive biology of Uruguayan theraphosids (Araneae, Mygalomorphae). J. Arachnol. 30, 571–587.

Costa, F.G., Simó, M., Aisenberg, A., 2006. Composición y ecología de la fauna epígea de Marindia (Canelones, Uruguay) con especial énfasis en las arañas: un estudio de dos años con trampas de intercepción. In: Menafra, R., Rodríguez-Gallego, L., Scarabino, F., Conde, D. (Eds.), Bases para la Conservatión y el Manejo de la Costa Uruguaya, Vida Silvestre Uruguay, pp. 427–436.

Darwin, C., 1871. The Descent of Man, and Selection in Relation to Sex. Murray, London.

Eberhard, W.G., 1985. Sexual Selection and Animal Genitalia. Harvard University Press, Cambridge.

Eberhard, W.G., 1996. Female Control: Sexual Selection by Cryptic Female Choice. Princeton University Press, Princeton.

Eens, M., Pinxten, R., 2000. Sex-role reversal in vertebrates: behavioural and endocrinological accounts. Behav. Processes 51, 135–147.

Elgar, M.A., 1992. Sexual cannibalism in spiders and other invertebrates. In: Elgar, M.A., Crespi, B.J. (Eds.), Cannibalism: Ecology and Evolution Among Diverse Taxa. Oxford University Press, Oxford, pp. 128–155.

Elgar, M.A., Nash, D.R., 1988. Sexual cannibalism in the garden spider *Araneus diadematus*. Anim. Behav. 36, 1511–1517.

Elgar, M.A., Schneider, J.M., 2004. Evolutionary significance of sexual cannibalism. Adv. Stud. Behav. 34, 135–163.

Emlen, S.T., Oring, L.W., 1977. Ecology, sexual selection, and the evolution of mating systems. Science 197, 215–223.

Foelix, R.F., 2011. Biology of Spiders. Oxford University Press, New York.

Foellmer, M., Moya-Laraño, J., 2007. Sexual size dimorphism in spiders: patterns and processes. In: Fairbairn, D.J., Blanckenhorn, W., Székely, T. (Eds.), Sex, Size and Gender Roles: Evolutionary Studies of Sexual Size Dimorphism. Oxford University Press, Oxford, pp. 71–81.

Framenau, V.W., 2005. Gender specific differences in activity and home range reflect morphological dimorphism in wolf spiders (Araneae, Lycosidae). J. Arachnol. 33, 334–346.

Gaskett, A.C., 2007. Spider sex pheromones: emission, reception, structures, and functions. Biol. Rev. 82, 27–48.

Gasnier, T.R., Azevedo, C.S., Torres Sanchez, M.P., Höfer, H., 2002. Adult size of eight hunting spider species in Central Amazonia: temporal variations and sexual dimorphism. J. Arachnol. 30, 146–154.

Gómez, M., Martino, D., 2008. GEO Uruguay, Informe del Estado del Ambiente. CLAES/PNUMA/DINAMA, Montevideo.

Gómez-Pivel, M.A., 2006. Geomorfología y procesos erosivos en la costa atlántica Uruguaya. In: Menafra, R., Rodríguez-Gallego, L., Scarabino, F., Conde, D. (Eds.), Bases para la Conservación y el Manejo de la Costa Uruguaya, Vida Silvestre, Uruguay, pp. 35–43.

Gould, S.J., 1984. Only his wings remained. Nat. Hist. 93, 10–18.

Gwynne, D.T., 1991. Sexual competition among females: what causes courtship role-reversal? Trends Ecol. Evol. 6, 118–121.

Gwynne, D.T., Bussière, L.F., 2002. Female mating swarms increase predation risk in a "role reversed" dance fly (Diptera: Empididae: *Rhamphomyia longicauda* Loew). Behaviour 139, 1425–1430.

Gwynne, D.T., Simmons, L.W., 1990. Experimental reversal of courtship roles in an insect. Nature 346, 172–174.

Henschel, J.R., Lubin, Y.D., 1992. Environmental factors affecting the web and activity of a psammophilous spider in the Namib Desert. J. Arid Environ. 22, 173–189.

Herberstein, M.E., Schneider, J.M., Elgar, M.A., 2002. Costs of courtship and mating in a sexually cannibalistic orb-web spider: female mating strategies and their consequences for males. Behav. Ecol. Sociobiol. 51, 440–446.

Hormiga, G., Scharff, N., Coddington, J.A., 2000. The phylogenetic basis of sexual size dimorphism in orb-weaving spiders (Araneae, Orbiculariae). Syst. Biol. 49, 435–462.

Huber, B., 2005. Sexual selection research on spiders: progress and biases. Biol. Rev. 80, 363–385.

Jia, Z., Jiang, Z., Sakaluk, S.K., 2000. Nutritional condition influences investment by male katydids in nuptial food gifts. Ecol. Entomol. 25, 115–118.

Karlsson, B., Leimar, O., Wiklund, C., 1997. Unpredictable environments, nuptial gifts and the evolution of size dimorphism in insects: an experiment. Proc. R. Soc. B 64, 475–479.

Karlsson Green, C., Madjidian, J.A., 2011. Active males, reactive females: stereotypic sex roles in sexual conflict research? Anim. Behav. 81, 901–907.

Knoflach, B., 1998. Mating in *Theridion varians* Hahn and related species (Araneae: Theridiidae). J. Nat. Hist. 32, 545–604.

Kokko, H., Jennions, M.D., 2008. Parental investment, sexual selection and sex ratios. J. Evol. Biol. 21, 919–948.

Lang, G.H., 2001. Sexual size dimorphism and juvenile growth rate in *Linyphia triangularis* (Linyphiidae, Araneae). J. Arachnol. 29, 64–71.

Logunov, D.V., 2011. Sexual size dimorphism in burrowing wolf spiders (Araneae: Lycosidae). Proc. Zoolog. Inst. RAS 315, 274–288.

Lorch, P., 2002. Understanding reversals in the relative strength of sexual selection on males and females: a role for sperm competition? Am. Nat. 6, 645–657.

Machado, G., Macías-Ordóñez, R., 2007. Reproduction. In: Pinto-da-Rocha, R., Machado, G., Giribet, G. (Eds.), Harvestmen: The Biology of Opiliones. Harvards University Press, Cambridge, pp. 414–454.

Machado, G., Requena, G.S., Buzatto, B.A., Osses, F., Rossetto, L.M., 2004. Five new cases of paternal care in harvestmen (Arachnida: Opiliones): implications for the evolution of male guarding in the neotropical family Gonyleptidae. Sociobiology 44, 577–598.

Mora, G., 1990. Paternal care in a neotropical harvestman, *Zygopachylus albomarginis* (Arachnida, Opiliones:Gonyleptidae). Anim. Behav. 39, 582–593.

Moya-Laraño, J., Halaj, J., Wise, D.H., 2002. Climbing to reach females: Romeo should be small. Evolution 56, 420–425.

Moya-Laraño, J., Pascual, J., Wise, D.H., 2003. Mating patterns in late-maturing female Mediterranean tarantulas may reflect the costs and benefits of sexual cannibalism. Anim. Behav. 66, 469–476.

Nentwig, W., 1987. The prey of spiders. In: Nentwig, W. (Ed.), Ecophysiology of Spiders. Springer-Verlag, Berlin, pp. 249–263.

Newman, J.A., Elgar, M.A., 1991. Sexual cannibalism in orbweaving spiders: an economic model. Am. Nat. 138, 1372–1395.

Owens, I.P.F., Thompson, D.B.A., 1994. Sex differences, sex ratios and sex roles. Proc. R. Soc. B 258, 93–99.

Perry, J.C., Rowe, L., 2012. Sex role stereotyping and sexual conflict theory. Anim. Behav. http://dx.doi.org/10.1016/j.anbehav.2012.01.030.

Polis, G.A., 1981. The evolution and dynamics of intraspecific predation. Annu. Rev. Ecol. Syst. 12, 225–251.

Postiglioni, R., González, M., Aisenberg, A., 2008. Permanencia en la cueva masculina y producción de ootecas en dos arañas lobo de los arenales costeros. Proc. XI Jornadas de Zoología del Uruguay. Uruguay, Montevideo 145.

Prenter, J., Montgomery, W.I., Elwood, R.W., 1995. Multivariate morphometrics and sexual dimorphism in the orb-web spider *Metellina segmentata* (Clerck, 1757) (Araneae, Metidae). Biol. J. Linn. Soc. 55, 345–354.

Prenter, J., Elwood, R.W., Montgomery, W.I., 1999. Sexual size dimorphism and reproductive investment by female spiders: a comparative analysis. Evolution 53, 1987–1994.

Queller, D.C., 1997. Why do females care more than males? Proc. R. Soc. B 264, 1555–1557.

Roberts, J.A., Uetz, G.W., 2005. Information content of chemical signals in the wolf spider, *Schizocosa ocreata*: male discrimination of reproductive state and receptivity. Anim. Behav. 70, 217–223.

Robinson, M.H., Robinson, B., 1980. Comparative studies of the courtship and mating behavior of tropical araneid spiders. Pac. Insects Monogr. 36, 1–218.

Roughgarden, J., Oishi, M., Akcay, E., 2006. Reproductive social behavior: cooperative games to replace sexual selection. Science 311, 965–969.

Rypstra, A.L., Wieg, C., Walker, S.E., Persons, M.H., 2003. Mutual mate assessment in wolf spiders: differences in the cues used by males and females. Ethology 109, 315–325.

Schneider, J.M., Andrade, M., 2011. Mating behaviour and sexual selection. In: Herberstein, M.E. (Ed.), Spider Behaviour. Flexibility and Versatility. Cambridge University Press, Cambridge, pp. 215–274.

Schneider, J.M., Lubin, Y., 1997. Does high adult mortality explain semelparity in the spider *Stegodyphus lineatus* (Eresidae)? Oikos 79, 92–100.

Schulte, K.F., Uhl, G., Schneider, J.M., 2010. Mate choice in males with one-shot genitalia: limited importance of female fecundity. Anim. Behav. 80, 699–706.

Schutz, D., Taborsky, M., 2005. Mate choice and sexual conflict in the size dimorphic water spider *Argyroneta aquatica* (Araneae, Argyronetidae). J. Arachnol. 33, 767–775.

Simmons, L.W., 1994. Courtship role reversal in bush crickets: another role for parasites? Behav. Ecol. 5, 259–266.

Stanley, E., Toscano-Gadea, C., Aisenberg, A., 2013. Spider hawk in sand dunes: *Anoplius bicinctus* (Hymenoptera: Pompilidae), a parasitoid wasp of the sex-role reversed spider *Allocosa brasiliensis* (Araneae: Lycosidae). J. Insect Behav. 26, 514–524.

Stratton, G.E., 2005. Evolution of ornamentation and courtship behavior in *Schizocosa*: insights from a phylogeny based on morphology (Araneae, Lycosidae). J. Arachnol. 33, 347–376.

Svensson, B.G., Petersson, E., Forsgren, E., 1989. Why do males of the dance fly *Empis borealis* refuse to mate? The importance of female age and size. J. Insect Behav. 2, 387–395.

Tallamy, D.W., 2000. Sexual selection and the evolution of exclusive paternal care in arthropods. Anim. Behav. 60, 559–567.

Trivers, R.L., 1972. Parental investment and sexual selection. In: Campbell, B. (Ed.), Sexual Selection and the Descent of Man 1871–1971, Aldine, Chicago, pp. 136–179.

Vahed, K., 2007. All that glisters is not gold: sensory bias, sexual conflict and nuptial feeding in insects and spiders. Ethology 113, 105–127.

Vollrath, F., Parker, G.A., 1992. Sexual dimorphism and distorted sex ratios in spiders. Nature 360, 156–159.

Wagner, J.D., 1995. Egg sac inhibits filial cannibalism in the wolf spider *Schizocosa ocreata*. Anim. Behav. 50, 555–557.

Walker, S.A., Rypstra, A.L., 2003. Sexual dimorphism and the differential mortality model: is behavior related to survival? Biol. J. Linn. Soc. 78, 97–103.

Wilder, S.M., Rypstra, A.L., Elgar, M.A., 2009. The importance of ecological and phylogenetic conditions for the occurrence and frequency of sexual cannibalism. Annu. Rev. Ecol. Evol. Syst. 40, 21–39.

Wise, D.H., 2006. Cannibalism, food limitation, intraspecific competition, and the regulation of spider copulations. Annu. Rev. Entomol. 51, 441–465.

Wise, D.H., Wagner, J.D., 1992. Evidence of exploitative competition among young stages of the wolf spider *Schizocosa ocreata*. Oecologia 91, 7–13.

Chapter 7

Sexual Selection, Ecology, and Evolution of Nuptial Gifts in Spiders

Maria J. Albo,[1,2] Søren Toft[2] and Trine Bilde[2]

[1]*Laboratorio de Etología, Ecología y Evolución, Instituto de Investigaciones Biológicas Clemente Estable, Montevideo, Uruguay, [2]Department of Bioscience, Aarhus University, Aarhus, Denmark*

THE SUBLIME TREASURES: NUPTIAL GIFTS

Nuptial gifts, where the male offers a gift that is often nutritious to the female in order to mate, are an intriguing sexually selected trait. As with most secondary sexual traits, it is most commonly found in males and only exceptionally in females. The nuptial gift-giving trait has evolved independently many times and in a wide range of forms among different animal taxa. Studies on nuptial gifts appeared at the beginning of the 1900s with the first descriptions of males offering food to females in Diptera, Mecoptera, and Orthoptera (*cf.* Vahed, 1998). In recent decades, an increasing number of studies have examined how sexual selection shapes gift-giving behavior in birds, snails, and spiders, but with a particular focus on insects (Austad and Thornhill, 1986; Vahed, 1998, 2007; Mougeot *et al.*, 2006; Burela and Martín, 2007; Gwynne, 2008; Lewis and South, 2012). Nuptial gifts comprise an extensive diversity of donations, such as oral food gifts given during courtship, and seminal substances transferred with the sperm during copulation (Vahed, 1998, 2007; Gwynne, 2008; Lewis and South, 2012). A recent classification, which is followed by Lewis and South (2012), distinguishes gifts depending on the method of gift production and on how gifts are absorbed by females. Gifts produced by males themselves are called endogenous gifts, such as glandular and salivary secretions or seminal fluids, while items that males collect from the environment, such as prey, seeds, or inedible items, are called exogenous gifts. Females receive these donations orally when eating food items or regurgitations, or via the genital tract when males transfer substances together with the sperm.

Nuptial gifts can be given before, during, and after copulation, and have positive, neutral, or negative effects on male and female reproductive success (Vahed, 1998, 2007; Gwynne, 2008; Lewis and South, 2012). In scorpionflies,

Sexual Selection. http://dx.doi.org/10.1016/B978-0-12-416028-6.00007-4

for instance, the male's salivary secretion is important during courtship, and males that produce more secretions increase their mating rate. Females adjust mating duration to the number of secretions they receive, which in turn may influence the number of eggs fertilized by the male (Sauer *et al.*, 1998; Engels and Sauer, 2006). Among bushcrickets and crickets, females consume a gelatinous spermatophylax associated with the spermatophore during mating, which functions to prevent females from removing the spermatophore before sperm transfer is complete (Gwynne *et al.*, 1984; Sakaluk, 1984). Seminal fluids of some fruit fly species contain substances, produced in the male accessory glands, that promote acceleration of female egg production (Wolfner, 1997; Heifetz *et al.*, 2001).

The aim of this chapter is to discuss and compare the function and evolution of nuptial gifts in two spider species. *Paratrechalea ornata* is a Neotropical species from the Trechaleidae family, while *Pisaura mirabilis* is a Palearctic species belonging to the Pisauridae family. These species are not only distributed on different continents, but also inhabit different environments. We present similarities and differences in sexual behavior, and discuss how ecological conditions seem to shape variation in nuptial gift-giving behavior of males and females.

NUPTIAL GIFTS IN SPIDERS

Nuptial gifts appear to be an uncommon sexual trait in spiders. Of 43,678 extant species described in the world (Platnick, 2012), only a few species are known to have males feeding females during courtship and/or mating (Bristowe and Locket, 1926; Andrade, 1996; Huber, 1997; Costa-Schmidt *et al.*, 2008; Uhl and Maelfait, 2008). Nuptial gift-giving in spiders takes three forms that comprise both endogenous and exogenous gifts, all of which are orally transferred: the male's body, glandular secretions, and wrapped prey items. Seminal gifts have so far not been documented in spiders.

Male Body

Sexual cannibalism confers direct benefits to females, and may also allow reproductive advantages for males (Buskirk *et al.*, 1984; Elgar, 1998; Schneider and Elgar, 2001; Herberstein *et al.*, 2011). If sexual cannibalism occurs before mating (pre-copulatory), it is obviously maladaptive for males but may confer fecundity benefits for females. Sexual cannibalism that occurs syn- (during) or post- (after) copulation may also be adaptive for males. Andrade (1996) showed the adaptive value of male suicidal sexual behavior in redback spiders (*Latrodectus hasselti*). During mating, the male performs a "somersault" by which he presents his abdomen directly to the female's chelicerae. Sperm transfer continues while the female consumes the male's abdomen, which results in higher paternity for cannibalized males than for non-cannibalized males. The benefit to males does not derive from the min-ute nutritional contribution from consumption of their bodies; instead, male sacrifice prolongs sperm transfer and confers an advantage in sperm compe-tition. Male sacrifice evolved as an extreme form of nuptial gift to enhance the male's abilities in sperm competition. However, male sacrifice is unlikely to be the general explanation for sexual cannibalism in spiders (Schneider *et al.*, 2000; Schneider and Elgar, 2001; Elgar and Schneider, 2004).

Glandular Secretions

Another nuptial gift form known in spiders is the external glandular secretions described from species of the families Theridiidae, Linyphiidae, and Pholcidae (Lopez, 1987; Huber, 1997; Uhl and Maelfait, 2008). Males of these species often have differentiated head protuberances – knobs, turrets, pits, humps – that are connected to exocrine secretory glands (Vanacker *et al.*, 2003; Michalick and Uhl, 2011). During courtship and/or mating the female inserts her chelicerae into the male's protuberance, excreting digestive saliva, and afterwards ingests the fluids (Uhl and Maelfait, 2008; Kunz *et al.*, 2012). By experimentally cover-ing the male protuberance it was shown that males increase mating probability and coupling efficiency when the glandular secretion is consumed by the female (Kunz *et al.*, 2012). So far, there is no indication that these secretions supply females with nutritional or other benefits, but further research is needed. This type of nuptial gift may be much more common than presently documented.

Wrapped Prey Items

Nuptial gifts as wrapped prey items are known in two spider families: Pisauridae and Trechaleidae (Bristowe, 1958; Costa-Schmidt *et al.*, 2008). Interestingly, fewer than 10 species displaying this behavior have been described, but the behavior may be more widespread in both families. Behavioral research has focused on only two species: the neotropical *Paratrechalea ornata* (Trechalei-dae), and the Palearctic *Pisaura mirabilis* (Pisauridae). In both species, the male captures a prey item (typically an arthropod) and wraps it in silk. The gift is thus a mixed exogenous (prey) and endogenous (silk) one. The result is a round, white package, which the male carries in his chelicerae while searching for a female (Bristowe and Locket, 1926; Costa-Schmidt *et al.*, 2008). After find-ing a female, the male actively courts her by vibrating the first pair of legs and pedipalps while offering the nuptial gift. Female mating acceptance happens when she grasps the gift with her chelicerae and begins to feed while the male inserts the pedipalps and initiates sperm transfer. Several possible functions of the nuptial gift are suggested for these species, including avoidance of sexual cannibalism, paternal investment, and mating effort (see below).

Like other trechaleid species, *P. ornata* is a semi-aquatic spider associated with freshwater courses in southern Brazil, northern Argentina, and Uruguay

(Carico, 2005). These spiders have crepuscular/nocturnal habits, and are able to walk on the water surface. In the field, adults and large juveniles are usually observed during the night perching on stones and pebbles emerging from the water and capturing flying prey (e.g., Ephemeroptera) (Silva *et al.*, 2006; Costa-Schmidt *et al.*, 2008). In contrast, *P. mirabilis* occupies terrestrial habitats, typically living in meadows, and has a Palearctic distribution (Bristowe and Locket, 1926). It has diurnal habits, generally perching on vegetation approximately 25 cm above the ground, from where it captures prey such as Diptera, Hemiptera, and Araneae (Nitzsche, 1988).

Trechaleidae and Pisauridae together with Lycosidae, Ctenidae, and eight additional families make up the superfamily Lycosoidea (Griswold, 1993; Coddington, 2005), which comprises wandering spiders with global distributions. The classification and phylogeny of the species belonging to these families have been extensively discussed, so far based on morphological and some behavioral data (Dondale, 1986; Sierwald, 1990; Griswold, 1993; Carico, 2005). Phylogenetic analyses indicate that Trechaleidae is a sister group of Lycosidae (Griswold, 1993), which suggests that the nuptial gift is an independent trait in Pisauridae and Trechaleidae. If the nuptial gift is a convergent trait, it has appeared at least twice in the evolution of the Lycosoidea lineage. Sexual selection combined with particular ecological conditions may thus shape the evolution of adaptive sexual behaviors.

WRAPPED PREY GIFTS IN AN ECOLOGICAL AND EVOLUTIONARY CONTEXT

A broad understanding of selective pressures acting separately on females and males needs substantial ecological and evolutionary frameworks. In fact, the way in which environmental constraints shape reproductive behaviors, aside from phylogenetic history, is a central subject of research. Our discussion is based upon the assumption that courtship and mating behaviors often are species-specific and highly ritualized, to the point of being useful systematic characters, while other aspects of reproductive behavior vary individually and are influenced by environmental factors.

Phylogenetic Constraints: Ritualized Behaviors

From a phylogenetic perspective, courtship and mating behaviors are ritualized and conserved traits (Stratton *et al.*, 1996). Spiders show a wide diversity of courtship and mating behaviors, but patterns are largely consistent within families (Foelix, 2011). In the particular case of gift-giving species, the male needs to present the prey gift to the female during courtship, and patterns of gift construction are very similar within and among species (Bristowe, 1958; Costa-Schmidt *et al.*, 2008; Albo *et al.*, 2009). Once the male has captured a prey, it attaches silk threads to the substrate and starts spinning a thin basal plate, then the male deposits the prey onto the basal plate and covers it with silk. The male finally removes the package from the substrate with the chelicerae, and shapes it into a round form by rolling it with the legs. The whole silk-wrapping process can be repeated several times, while adding more silk.

Gift-offering postures differ slightly between *Pisaura mirabilis* and *Paratrechalea ornata*. In the latter species, males lift and fold their legs above the cephalothorax when presenting the gift (Costa-Schmidt *et al.*, 2008; Fig. 7.1A). *Pisaura mirabilis* males present the gifts while waving the pedipalps and vertically raising the whole body (Bristowe and Locket, 1926; Fig. 7.1B). In both species, the female then grasps the gift and the sexes remain in a face-to-face position pulling the gift from either side (Figs 7.2A, 7.2B). The female may steal the gift at this stage and run off with it, in which case no mating will follow (Andersen *et al.*, 2008). If the female accepts the gift (and the male), the spiders adopt the mating position, which is also stereotyped and differs between species (Bristowe and Locket, 1926; Costa-Schmidt *et al.*, 2008). The typical mating position of *P. ornata* is similar to the one performed by lycosid spiders: the male mounts the female by first climbing over her prosoma (male and female in opposite directions), then turning towards one side of the female's abdomen and performing a pedipalp insertion. Once the insertion ends, the male returns to the face-to-face position and again grasps the gift with his chelicerae. These behavioral sequences may be repeated up to four times, with changing use of the pedipalps (Costa-Schmidt *et al.*, 2008). The mating position in *P. mirabilis* differs from the one in *P. ornata* in that the male does not climb on top of the female cephalothorax; instead, he pushes the female body upwards from the ventral side to perform pedipalp insertions (Bristowe and Locket, 1926). Similarly to *P. ornata*, once the pedipalp insertion ends the male returns to the face-to-face position and again grasps the gift with his chelicerae. Again, up to four pedipalp insertions can occur (Albo *et al.*, 2011a).

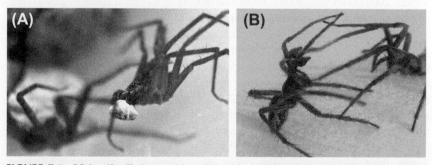

FIGURE 7.1   Male gift-offering position: (A) male (right) and female (left) *Paratrechalea ornata*; (B) male (left) and female (right) in *Pisaura mirabilis*. See color plate at the back of the book. *Photograph (A) courtesy of M. C. Trillo; photograph (B) courtesy of M. J. Albo.*

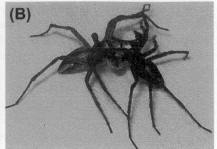

FIGURE 7.2   Female (left) and male (right) in the face-to-face position grasping the gift, during mating in (A) *Paratrechalea ornata***, and (B)** *Pisaura mirabilis***.** See color plate at the back of the book. *Photographs courtesy of M. J. Albo.*

Ecological Constraints: Variable Behaviors

Some aspects of sexual behavior are quite variable and are likely shaped by ecological factors that influence the degree of sexual selection (Emlen and Oring, 1977). One of the important drivers of this type of selection is the spatial and temporal distribution of resources. For instance, when food is scarce in the habitat individuals are forced into heavier competition, increasing the selective forces acting on both males and females. For females, the amount of food not only determines their life span but also the number and quality of eggs and offspring, hence directly influencing reproductive success (Wise, 1975, 2006). In contrast, although males in poor feeding condition may suffer a reduction in reproduction (Mappes *et al.,*1996; Andrade and Mason, 2000; Ahtiainen *et al.*, 2002; Kotiaho, 2002; Engqvist and Sauer, 2003; Hunt *et al.*, 2004; Engels and Sauer, 2006; Lomborg and Toft, 2009), acquiring food during the reproductive season is less essential for them. For example, sexually mature males of many spider species stop feeding to search for mates (Foelix, 2011).

The relative costs and benefits of providing and receiving nuptial gifts may be experienced differently by males and females (Arnqvist and Nilsson, 2000; Vahed, 2007; Gwynne, 2008). Food availability is a central issue in gift-giving mating systems, where males have a particular interest in acquiring nutrients to offer females. Females may not only obtain valuable nutrients from nuptial gifts, but can also be released from the time and energetic costs as well as the dangers of foraging. They are therefore likely to prefer males offering large and high-quality gifts, thus imposing strong selection for nuptial gifts (Leimar *et al.*, 1994; Arnqvist and Nilsson, 2000; Vahed, 2007). Hence, females may eventually benefit from multiple matings with males that offer nutritive gifts, as this may increase their feeding rate and fecundity, and the hatching success of their eggs (Arnqvist and Nilsson, 2000). Differences in food intake by females during their lifetime (resulting either from changes in prey availability or differences in individual foraging skills) can determine mating frequencies in gift-giving species, so that food-limited females engage in more matings

(Boggs, 1990; Gwynne, 1990; Simmons and Bailey, 1990). In the classic example of katydids, Gwynne (1981, 1984, 1990) showed that when food is scarce, females compete intensively for gifts and males become the choosy sex, thus leading to a sex role-reversed mating system. There is evidence to believe that *P. mirabilis* males exploit females' foraging motivation, since food-deprived females are more willing to accept mates than satiated ones (Bilde *et al.*, 2007; Prokop and Maxwell, 2009). Thus, as in other gift-giving species (Arnqvist and Nilsson, 2000), direct benefits obtained by females via gift consumption could lead to the evolution and maintenance of polyandry. Unfortunately, little is known from the Neotropical *P. ornata*, except that females do mate multiple times (I. Pandulli, unpublished data) and likely benefit from this if males offer nutritive gifts (but see below).

Sexual conflict over mating rate may occur when the evolutionary interests diverge between the sexes (Arnqvist and Rowe, 2002, 2005). Most commonly, males increase reproductive success by increasing the number of mating partners and by being successful in sperm competition, whereas increased mating rates may not benefit females. This can lead to an evolutionary arms race where males evolve traits to lure females into mating, while females evolve resistance to these traits. In such cases, adaptations in one sex are disadvantageous for the other sex, and thus may drive selection towards counter-adaptations to minimize these costs. Differences in male investment in nuptial gifts ultimately create a co-evolutionary scenario in which females' and males' interests diverge. Although males of many species benefit from gift-giving by luring females into mating, they may also experience costs of gift production (Gwynne, 1990; Engqvist and Sauer, 2003; Engels and Sauer, 2006; Immonen *et al.*, 2009). There is evidence of condition-dependence in the production of nuptial gifts, which therefore may function as honest indicators of male quality for female choice (Zahavi, 1975; Zahavi and Zahavi, 1997). In both *P. mirabilis* and *P. ornata* gift construction is condition-dependent, since males in poor feeding condition perform silk wrapping inefficiently and present badly wrapped gifts (Albo *et al.*, 2011b; M. J. Albo, unpublished data). This means that the actual silk-wrapping quality, indicated by the amount of silk deposited, may potentially function as an honest indicator of male quality. Contrary to this prediction, *P. mirabilis* females seem to ignore this information and instead evaluate males directly by their body condition (Albo *et al.*, 2012). This makes sense if the silk itself facilitates male interests, and not female interests, for example by disguising the gift or making it more difficult to consume.

Under food-limited conditions, male gift-giving behavior may evolve in the direction of reducing the costs by decreasing the time of production and therefore gift quality. This becomes evident when males offer exogenous prey gifts, as they may try to avoid the costs of searching and capturing prey by re-using gifts after mating, or by using inedible items (Thornhill, 1976; Preston-Mafham, 1999; LeBas and Hockham, 2005). Indeed, worthless gifts have been suggested as an alternative mating tactic used by males in some insect species

(Preston-Mafham, 1999; LeBas and Hockham, 2005). Males from gift-giving spider species may also offer worthless gifts (Albo and Costa, 2010; Albo *et al.*, 2011a). In these species, it is likely that gift content – i.e., whether the male presents a genuine or a worthless gift – is influenced by prey availability, male condition, or their interaction. If males also depend on food, they must decide between eating or wrapping the prey, and when food is scarce they may be tempted to produce worthless gifts by eating it first. On the other hand, females would prefer genuine nutritive gifts, and favor matings with males offering such gifts.

In *P. mirabilis*, males may wrap prey leftovers or parts of plants, which are without nutritive value for females (Albo *et al.*, 2011a). At least two facts indicate strong female preference for nutritive gifts: first, females penalize males that present a worthless gift by interrupting matings earlier compared to those with genuine gifts; second, worthless donations are a minority in the field (about one-third), supporting the idea that they are in fact an alternative male mating tactic. Alternatively, in conditions of high food abundance females may be more tolerant of worthless gifts, imposing less selection on males, which seems to be the case in *P. ornata*. Worthless donations also occur in this species (Albo and Costa, 2010; M. J. Albo unpublished data). Contrary to *P. mirabilis* and to the alternative male tactic hypothesis, field studies suggest that *P. ornata* males use prey that have already been sucked dry, and occasionally parts of plants, more often than genuine prey (about two-thirds). Furthermore, females appear not to penalize males offering worthless gifts in terms of reduced mating success or copulation duration (M. J. Albo, unpublished data). Thus, *P. mirabilis* males seem to hold the upper hand in the co-evolutionary cycle, since, despite female preference for genuine gifts, they are only able to discover male deception after copulation is initiated, probably due to the silk wrapping. However, the balance between female and male interests is even more biased towards males in *P. ornata*, since females seem neither to have preferences for genuine gifts nor to penalize male deception. This difference could be due to abundant insect prey in *P. ornata* habitats. Further studies evaluating the potential reproductive costs of worthless gifts are needed to discuss the evolution of worthless gifts in this species.

FUNCTIONAL EXPLANATIONS OF NUPTIAL GIFTS

The selective forces that favor origin and maintenance of the gift-giving trait may vary in response to ecological conditions or co-evolutionary responses. The function of the gift-giving behavior may consequently also change in relation to the context under which the trait is maintained. The origin of male donations must have been due to strong female selection for nutrients, with the resulting fitness benefit to both partners. Since spiders are often food limited in nature (Wise, 1993), fluctuations in food availability during the female lifespan and especially during adulthood would drive these preferences and the further evolution of male and female behaviors coupled with inevitable costs of mating.

Thus, to maximize their own reproductive success males would originally have benefited by investing in nutritive nuptial gifts that enhanced female interests. Later, however, they may have potentially changed their investment by reducing the costs of gift production to enhance their own interests. In the following sections, we present functional hypotheses for the origin and maintenance of nuptial gifts.

Sexual Cannibalism Avoidance

Females of many species kill and consume males before, during, or after mating. This phenomenon, known as sexual cannibalism, is relatively common in spiders, and its evolutionary significance depends on the timing of male consumption (Elgar and Schneider, 2004; Herberstein *et al.*, 2011). Males from some spider species are consumed during or after mating, but this usually benefits those males since they can prolong matings and hence increase their fertilization success (Schneider and Elgar, 2001). However, if males are killed before mating, when no sperm transfer has occurred, they obviously lose all potential fitness and only females gain benefits (food). Courtship behavior is particularly important under the risk of pre-copulatory sexual cannibalism. Spider males have evolved different strategies to avoid female attacks, such as approaching females when they are feeding or molting – so called "opportunistic matings" (Lubin, 1986; Fromhage and Schneider, 2004). In empidid flies, nuptial gifts are suggested to function to prevent males from being eaten by females during courtship (Kessel, 1955).

Sexual cannibalism occurs at low frequency in the gift-giving spider species, *P. mirabilis* and *P. ornata* (Bilde *et al.*, 2006; Albo and Costa, 2010). Nevertheless, pre-copulatory cannibalism was observed in a small percentage of staged matings in *P. mirabilis*, which may impose strong selection on cannibalism avoidance in males (Bilde *et al.*, 2006). It is possible that nuptial gift-giving evolved as an anticannibalism precaution in this species, while other functions may occur in concert. When food is scarce and females suffer from low feeding regimes, pre-copulatory gift stealing by females increases dramatically, and sexual cannibalism to some extent (S. Toft and M. J. Albo, unpublished data). Thus, it can also be imagined that the gift may create the opportunity for gift stealing as a substitute for pre-copulatory cannibalism – obviously a better option for the males.

Paternal Investment

Parental investment was defined by Trivers (1972) as any investment from the parents that increases offspring survival and fitness at a cost to parents' ability to invest in a future reproduction. Paternal investment is very important in bird species, for example, where males supply food and guard the progeny until they are independent (Stokes and Williams, 1971; Mougeot *et al.*, 2006). Ecological conditions such as the level of predation risk or food abundance may determine the relative parental investment of females and males (Gwynne, 1990; Simmons and Bailey, 1990). It is predicted that males would supply females with resources if food is scarce or females are under high risk of predation when foraging, thus increasing the likelihood of females succeeding in reproduction and securing the male's own paternity. Alternatively, paternal investment would be low when prey availability is high. Some interesting examples arise from gift-giving species (reviewed in Boggs, 1995, and Vahed, 1998). Nutrients from the male gift (spermatophylax) are incorporated by female bushcrickets into the developing eggs (Simmons, 1990; Simmons and Gwynne, 1993), while nuptial gifts also improve female fecundity and longevity in other insect species (Thornhill, 1976; Gwynne, 1984; Simmons and Parker, 1989; Wiklund et al., 1993; Karlsson, 1998; Lewis and Cratsley, 2008).

In the gift-giving spiders P. ornata and P. mirabilis, most previous studies have been unable to demonstrate any effect of the food gift on female fecundity or egg-hatching success (Austad and Thornhill, 1986; Stålhandske, 2001; Albo and Costa, 2010). However, it can be argued that these results are a consequence of the experimental designs that were insufficient to detect any effects from male nutrients (Vahed, 1998). In fact, recent evidence suggests that females gain direct benefits from mating with males with nutritive gifts, such as accelerated egg production and oviposition in both P. ornata and P. mirabilis (Albo and Costa, 2010; Tuni et al., 2013). Furthermore, experiments where females received food only through matings, and thus engaged in high levels of polyandry, showed an enhanced hatching success and only slightly reduced fecundity in P. mirabilis (S. Toft and M. J. Albo, unpublished data). Because female spiders simultaneously develop a large number of eggs and lay them in a single clutch (Foelix, 2011), the gift nutrients will be divided among all of the female's eggs. Thus, all males providing a nutritive gift and succeeding in fertilizing part of the eggs actually contribute some paternal investment. This differs from the situation in insects, which present a more gradual maturation of eggs, which allows a direct correspondence between nutrient allocation to and fertilization of specific eggs by each male (cf. Simmons and Parker, 1989).

Mating Effort

The alternative but non-exclusive mating effort hypothesis suggests that by using gifts males can improve their reproductive success through increasing the number of matings or prolonging the time of copulation, regardless of any nutritive function. In insects, nuptial gifts appear to be maintained by selection to maximize ejaculate transfer and therefore reduce the risk of sperm competition from other males (Simmons and Gwynne, 1991; Eady et al., 2000; Wolfner, 1997; Heifetz et al., 2001; Sakaluk et al., 2006). Similarly, it has been shown that the nuptial gift functions in the context of male mating effort in both P. ornata and P. mirabilis (Stålhandske, 2001; Bilde et al., 2007; Albo and Costa, 2010). Although males may obtain matings without a gift, the chance of acceptance dramatically increases when a gift is offered. Furthermore, the presence of a gift possibly facilitates the mating position, and a large gift keeps the female occupied for longer and consequently prolongs the mating (Stålhandske, 2001; Albo and Costa, 2010; A. Klein, M. Trillo, F. G. Costa and M. J. Albo, unpublished data). In *P. mirabilis*, it is known that by prolonging matings males increase the number of sperm transferred (M. J. Albo, G. Uhl, and T. Bilde, unpublished data) and hence the number of fertilized eggs (Drengsgaard and Toft, 1999; Stålhandske, 2001). Empirical evidence is limited for *P. ornata*, but recent studies have revealed very similar results (M. J. Albo, unpublished data).

In accordance with the male mating effort hypothesis, males of *P. mirabilis* often perform a unique behavior during mating: so called "death feigning" or "thanatosis". This remarkable behavior usually occurs when the pair is disturbed and the female moves away with the gift, possibly attempting to end copulation. Then, while grasping the gift with his chelicerae, the male "feigns death" (with stretched-out legs) and is dragged through the vegetation until the female stops. Subsequently, the male "revives" and resumes copulation (Bilde *et al.*, 2006; Hansen *et al.*, 2008). Thanatosis increases male mating success, and gives males the opportunity of prolonging copulation and hence increasing sperm transfer (Hansen *et al.*, 2008). So far, there is no evidence of thanatosis in *P. ornata* males, suggesting it is a unique innovation in *P. mirabilis*.

SILK-WRAPPING: POTENTIAL FUNCTIONS

The phenomenon of the male gift-wrapping prey in white silk is in itself remarkable. Wrapping of prey caught for their own consumption is common in spiders, and it usually functions to immobilize active and dangerous prey, or to fix it in some place for later consumption (Barrantes and Eberhard, 2007). Since the males capture and kill the gift prey before wrapping it in silk, they could easily offer it without investing energy, time, and silk material on wrapping. Despite this, silk wrapping seems to be an important trait in both species. So why do males wrap gift-prey in silk? Silk wrapping is not obligatory for mating to occur, and sometimes males offer unwrapped prey to females. However, usually these males wrap the prey after contact with females, in particular if they were initially rejected by the female, subsequently succeeding in mating (Bilde *et al.*, 2007; Albo and Costa, 2010). The function of the gift-wrapping trait is intriguing, and needs to be analyzed from both male and female perspectives. In the following sections, we discuss possible hypotheses. So far, there is no evidence that silk wrapping may favor the female's interests; instead, we conclude that all advantages favor the male.

Silk-Wrapping Triggered by Female Silk Cues

As in most spider species, contact sex pheromones associated with female silk seem to be important stimuli that elicit male courtship and silk wrapping in *P. ornata* and *P. mirabilis* (Nitzsche, 1988; Lang, 1996; Albo *et al.*, 2009, 2011b). In *P. ornata*, males initiate nuptial gift construction only after contact with female silk, and indeed female silk seems to be as important a stimulus as is the female herself (Albo *et al.*, 2009, 2011b). In contrast, male *P. mirabilis* may produce gifts without the presence of female sexual stimuli, as in this species gift production may be a spontaneous behavior associated with sexual maturation, although it is still enhanced by female stimuli (Nitzsche, 1988; Lang, 1996; Albo *et al.*, 2011b). In the field, it is common to find males of both species carrying a wrapped gift (Lang, 1996; Albo *et al.*, 2011a; M. J. Albo, personal observation). Searching for females with a ready-wrapped gift is an advantageous male strategy, because the males can avoid delays in prey capture or wrapping, court the female immediately on encounter, and thus minimize the risk of a lost mating opportunity.

Female Attraction

Sensory Exploitation Hypothesis

Stålhandske (2002) performed the first experiments attempting to verify the role of the wrapped gift as a visual signal during courtship in *P. mirabilis*. Since the wrapped gift has a visual resemblance to the egg sac that females carry in their chelicerae, Stålhandske's hypothesis was that gift wrapping evolved through male sensory exploitation of the female maternal instinct. Thus, if wrapped gifts mimic egg sacs, females are sensory-biased to respond to any item resembling an egg sac (Ryan *et al.*, 1990; Christy, 1995; Sakaluk, 2000). By mimicking the female's egg sac with the white-wrapped gift, males would attract females' attention and increase the chances of mating. Stålhandske (2002) manipulated the gift color (brown, natural white, and extra-white) and found that females were more attracted to extra-white gifts and accepted them faster. Subsequent studies using wrapped and unwrapped prey were unable to verify female preference for silk wrapping, however (Bilde *et al.*, 2007; Andersen *et al.*, 2008; Albo *et al.*, 2012). In particular, experiments that varied the gift type (using an egg sac, a silk-wrapped fly, or an unwrapped fly) revealed no female preference for egg sacs or wrapped gifts. Instead, female hunger level (starved or satiated) predicted mating acceptance (Bilde *et al.*, 2007). Thus, it was instead suggested that males offering food gifts exploit the females' foraging motivation in *P. mirabilis* (Bilde *et al.*, 2007; Prokop and Maxwell, 2009).

So far, the function of silk wrapping has not been well studied in *P. ornata*. Evidence suggests many similarities with *P. mirabilis* (M. J. Albo, unpublished data). For instance, males' silk investment is independent of prey size (indicated in *P. mirabilis* by Lang, 1996), thus with similar silk-wrapping duration small prey become white whereas big prey become grey. In spite of this, white gift color seems to have a positive effect on female acceptance (A. Klein, M. Trillo, F. G. Costa, and M. J. Albo, unpublished data).

Phagostimulants

The theory that gift silk might be a potential source of protein for females has been tested and refuted (Nitzsche, 1988). Silk has also been suggested as a source of chemical substances used to attract females in *P. mirabilis* (Lang, 1996; Bilde *et al.*, 2007). In fact, recent studies indicate that male pheromones associated with gift silk play a major role in female attraction in *P. ornata* (Brum *et al.*, 2012). The role of chemical substances as phagostimulants enclosed in nuptial gifts has been reported in insects (Warwick *et al.*, 2009). Whether substances with such effects are involved in wrapped gifts needs to be studied in spiders.

Mating Control and Hiding of the Gift Content

In accordance with the male mating effort hypothesis, silk wrapping increases the time the female spends feeding on the gift and therefore the time in copula (Lang, 1996). In addition, since males maintain a hold on the gift with the claws of the third pair of legs during pedipalp insertion, silk wrapping allows the male to secure the mating position and prevents the female from escaping with the gift. Indeed, *P. mirabilis* females have higher success in stealing the prey when it is unwrapped than if wrapped (Andersen *et al.*, 2008). Similarly, in *P. ornata* it has been observed that during the face-to-face position unwrapped prey may be split in two and each sex remains with one prey piece, consequently ending the mating (M. J. Albo and F. G. Costa, unpublished data). These results indicate that silk wrapping has a significant function in maintaining male mating control.

Silk wrapping also functions to preserve prey, or to facilitate prey handling and transport, in many spiders (Barrantes and Eberhard, 2007). Recently, it has been suggested that the silk-wrapping trait facilitates the evolution of worthless gifts by hiding the gift content (Albo *et al.*, 2011a). Silk wrapping allows the male to offer prey leftovers, which would fall apart and are impossible to carry without being compacted together by silk. It also prevents the female from properly assessing the gift content during courtship. Indeed, the female can only assess the gift content when she consumes the gift, providing clear advantages to males in transferring some sperm before this happens.

SEXUAL SELECTION: COOPERATION AND CONFLICT

Reproduction is an outcome of cooperation between females and males, and encompasses common interests for encountering and recognizing potential mates. However, differences in gamete investment, continuing through interactions during courtship, mating, and parental investment, arising from diverse selective pressures, ultimately shape the differences in sex roles (Bateman, 1948; Trivers, 1972; Arnqvist and Rowe, 2005). As discussed in this chapter, ecological fluctuations may be important factors influencing whether cooperation or conflict dominates, and may lead to co-evolutionary responses that affect mating systems. Direct benefits leading to reproductive advantages for females arise when males offer nutritive gifts, while males benefit by potentially increasing their share of paternity. However, conflicts of interest over mating rate may lead males to exploit the female foraging motivation to gain additional copulations. Fluctuations in food availability could, for example, favor female multiple mating and polyandrous mating systems if females are under strong selection to accept a nutritious gift. If prey availability is low, males may be selected to reduce the cost of producing a nuptial gift, which can favor the evolution of worthless gifts. Females may counteract male deception by restricting the male's paternity when the costs of receiving inedible items are high. If food availability and the costs and benefits of mating change over the course of the mating season, this may ultimately explain the occurrence of polymorphism (genuine and worthless gifts) in the nuptial gift-giving trait. Differences in food resource availability among seasons, habitats, and regions might explain differences in sexually selected traits between *P. ornata* and *P. mirabilis*, and focused studies are needed to test this idea.

ACKNOWLEDGMENTS

We would like to thank Glauco Machado and Regina H. Macedo for inviting us to contribute to this book. Regina H. Macedo and Fernando G. Costa for their helpful comments on this chapter. MJA is supported by the Agencia Nacional de Investigación e Innovación (ANII), Uruguay.

REFERENCES

Ahtiainen, J., Alatalo, R.V., Kotiaho, J.S., Mappes, J., Parri, S., Vertainen, L., 2002. Sexual selection in the drumming wolf spider *Hygrolycosa rubrofasciata*. In: Toft, S., Scharff, N. (Eds.), European Arachnology 2000, Aarhus University Press, Aarhus, pp. 129–137.

Albo, M.J., Costa, F.G., 2010. Nuptial gift giving behaviour and male mating effort in the neotropical spider *Paratrechalea ornata* (Trechaleidae). Anim. Behav. 79, 1031–1036.

Albo, M.J., Costa-Schmidt, L.E., Costa, F.G., 2009. To feed or to wrap? Female silk cues elicit male nuptial gift construction in the spider *Paratrechalea ornata* (Trechaleidae). J. Zool. 277, 284–290.

Albo, M.J., Winther, G., Tuni, C., Toft, S., Bilde, T., 2011a. Worthless donations: male deception and female counter play in a nuptial gift-giving spider. BMC Evol. Biol. 11, 329.

Albo, M.J., Toft, S., Bilde, T., 2011b. Condition dependence of male nuptial gift construction in the spider *Pisaura mirabilis* (Pisauridae). J. Ethol. 29, 473–479.

Albo, M.J., Toft, S., Bilde, T., 2012. Female spiders ignore condition-dependent information from nuptial gift wrapping when choosing mates. Anim. Behav. 84, 907–912.

Andersen, T., Bollerup, K., Toft, S., Bilde, T., 2008. Why do males of the spider *Pisaura mirabilis* wrap their nuptial gifts in silk: female preference or male control? Ethology 114, 775–781.

Andrade, M.B.C., 1996. Sexual selection for male sacrifice in the Australian redback spider. Science 271, 70–72.

Andrade, M.C.B., Mason, A.C., 2000. Male condition, female choice, and extreme variation in repeated mating in a scaly cricket, *Ornebius aperta* (Orthoptera: Gryllidae: Mogoplistinae). J. Insect Behav. 13, 483–497.

Arnqvist, G., Nilsson, T., 2000. The evolution of polyandry: multiple mating and female fitness in insects. Anim. Behav. 60, 145–164.

Arnqvist, G., Rowe, L., 2002. Antagonistic coevolution between the sexes in a group of insects. Nature 415, 787–789.

Arnqvist, G., Rowe, L., 2005. Sexual Conflict. Princeton University Press, Princeton.

Austad, S.N., Thornhill, R., 1986. Female reproductive variation in a nuptial-feeding spider, *Pisaura mirabilis*. Bull. Br. Arachnol. Soc. 7, 48–52.

Barrantes, G., Eberhard, W.G., 2007. The evolution of prey-wrapping behaviour in spiders. J. Nat. Hist. 41, 1631–1658.

Bateman, A.J., 1948. Intra-sexual selection in *Drosophila*. Heredity 2, 349–368.

Bilde, T., Tuni, C., Elsayed, R., Pekar, S., Toft, S., 2006. Death feigning in the face of sexual cannibalism. Biol. Lett. 2, 23–35.

Bilde, T., Tuni, C., Elsayed, R., Pekar, S., Toft, S., 2007. Nuptial gifts of male spiders: sensory exploitation of female's maternal care instinct or foraging motivation? Anim. Behav. 73, 267–273.

Boggs, C.L., 1990. A general model of the role of male-donated nutrients in female insects' reproduction. Am. Nat. 136, 598–617.

Boggs, C.L., 1995. Male nuptial gifts: behaviour consequences and evolutionary implications. In: Leather, S.R., Haerdie, J. (Eds.), Insect Reproduction, CRC Press, Boca Raton, pp. 215–242.

Bristowe, W.S., 1958. The World of Spiders. Collins, London.

Bristowe, W.S., Locket, G.H., 1926. The courtship of British lycosid spiders, and its probable significance. Proc. Zool. Soc. 22, 317–347.

Brum, P.E.D., Costa-Schmidt, L.E., Araújo, A.M., 2012. It is a matter of taste: chemical signals mediate nuptial gift acceptance in a neotropical spider. Behav. Ecol. 23, 442–447.

Burela, S., Martín, P.R., 2007. Nuptial feeding in the freshwater snail *Pomacea canaliculata* (Gastropoda: Ampullariidae). Malacologia 49, 465–470.

Buskirk, R.E., Frohlich, C., Ross, K.G., 1984. The natural selection of sexual cannibalism. Am. Nat. 123, 612–625.

Carico, J.E., 2005. Descriptions of two new spider genera of Trechaleidae (Araneae, Lycosoidea) from South America. J. Arachnol. 33, 797–812.

Christy, J.H., 1995. Mimicry, mate choice, and the sensory trap hypothesis. Am. Nat. 146, 171–181.

Coddington, J.A., 2005. Phylogeny and classification. In: Ubick, D., Paquin, P., Cushing, P.E., Roth, V. (Eds.), Spiders of North America: An Identification Manual, American Arachnological Society, Columbia, pp. 18–24.

Costa-Schmidt, L.E., Carico, J.E., Araújo, A.M., 2008. Nuptial gifts and sexual behaviour in two species of spider (Araneae, Trechaleidae, *Paratrechalea*). Naturwissenschaften 95, 731–739.

Dondale, C.D., 1986. The subfamilies of wolf spiders (Araneae: Lycosidae). *Actas X Congreso de Aracnología*, Jaca. España 1, 327–332.

Drengsgaard, I.L., Toft, S., 1999. Sperm competition in a nuptial feeding spider *Pisaura mirabilis*. Behaviour 136, 877–897.

Eady, P.E., Wilson, N., Jackson, M., 2000. Copulating with multiple mates enhances female fecundity but not egg-to-adult survival in the bruchid beetle *Callosobruchus masculatus*. Evolution 54, 2161–2165.

Elgar, M.A., 1998. Sperm competition and sexual selection in spiders and other arachnids. In: Birkhead, T.R., Moller, A.P. (Eds.), Sperm Competition and Sexual Selection, Academic Press, London, pp. 307–337.

Elgar, M.A., Schneider, J.M., 2004. Evolutionary significance of sexual cannibalism. In: Brockmann, H.J., Roper, T.J., Naguib, M., Mitani, J.C., Simmons, L.W. (Eds.), Adv. Stud. Behav. Elsevier, New York, pp. 133–163.

Emlen, S.T., Oring, L.W., 1977. Ecology, sexual selection, and evolution of mating systems. Science 197, 215–223.

Engels, S., Sauer, K.P., 2006. Resource-dependent nuptial feeding in *Panorpa vulgaris*: an honest signal for male quality. Behav. Ecol. 17, 628–632.

Engqvist, L., Sauer, K.P., 2003. Influence of nutrition on courtship and mating in the scorpionfly *Panorpa cognata* (Mecoptera, Insecta). Ethology 109, 911–928.

Foelix, R.F., 2011. Biology of Spiders, 3rd edn. Oxford University Press, New York.

Fromhage, L., Schneider, J.M., 2004. Safer sex with feeding females: sexual conflict in a cannibalistic spider. Behav. Ecol. 16, 377–382.

Griswold, C.E., 1993. Investigations into the phylogeny of the lycosoid spiders and their kin (Arachnida, Araneae, Lycosoidea). Smithson. Contrib. Zool. 539, 1–39.

Gwynne, D.T., 1981. Sexual difference theory: Mormon crickets show role reversal in mate choice. Science 213, 779–780.

Gwynne, D.T., 1984. Courtship feeding increase female reproductive success in bushcrickets. Nature 307, 361–363.

Gwynne, D.T., 1990. Testing parental investment and the control of sexual selection in katydids: the operational sex ratio. Am. Nat. 136, 474–484.

Gwynne, D.T., 2008. Sexual conflict over nuptial gifts in insects. Annu. Rev. Entomol. 53, 83–101.

Gwynne, D.T., Bowen, B.J., Codd, C.G., 1984. The function of the katydid spermatophore and its role in fecundity and insemination (Orthoptera: Tettigoniidae). Aust. J. Zool. 32, 15–22.

Hansen, L.S., Fernández González, S., Toft, S., Bilde, T., 2008. Thanatosis as an adaptive male mating strategy in the nuptial gift-giving spider *Pisaura mirabilis*. Behav. Ecol. 19, 546–551.

Heifetz, Y., Tram, U., Wolfner, M.F., 2001. Male contributions to egg production: the role of behaviour gland products and sperm in *Drosophila melanogaster*. Proc. R. Soc. B 268, 905–908.

Herberstein, M.E., Schneider, J.M., Harmer, A.M.T., Gaskett, A.C., Robinson, K., Shaddick, K., Soetkamp, D., Wilson, P.D., Pekar, S., Elgar, M.A., 2011. Sperm storage and copulation duration in a sexually cannibalistic spider. J. Ethol. 29, 9–15.

Huber, A.B., 1997. Evidence for gustatorial courtship in a haplogyne spider *Hedypsilus culicinus* (Pholcidae: Araneae). Neth. J. Zool. 47, 95–98.

Hunt, J., Brooks, R., Jennions, M.D., Smith, M.J., Bentsen, C.L., Bussiere, L.F., 2004. High-quality male field crickets invest heavily in sexual display but die young. Nature 432, 1024–1027.

Immonen, E., Hoikkala, A., Kazem, A.J.N., Ritchie, M.G., 2009. When are vomiting males attractive? Sexual selection on condition-dependent nuptial feeding in *Drosophila subobscura*. Behav. Ecol. 20, 289–295.

Karlsson, B., 1998. Nuptial gifts, resource budgets, and reproductive output in a polyandrous butterfly. Ecology 78, 2931–2940.

Kessel, E.L., 1955. The mating activities of balloon flies. Syst. Zool. 4, 97–104.

Kotiaho, J.S., 2002. Sexual selection and condition dependence of courtship display in three species of horned dung beetles. Behav. Ecol. 13, 791–799.

Kunz, K., Garbe, S., Uhl, G., 2012. The function of the secretory cephalic hump in males of the dwarf spider *Oedothorax retusus* (Linyphiidae: Erigoninae). Anim. Behav. 83, 511–517.

Lang, A., 1996. Silk investments in gifts by males of the nuptial feeding spider *Pisaura mirabilis* (Araneae: Pisauridae). Behaviour 133, 697–716.

LeBas, N.R., Hockham, L.R., 2005. An invasion of cheats: The evolution of worthless nuptial gifts. Curr. Biol. 15, 64–67.

Leimar, O., Karlsson, B., Wiklund, C., 1994. Unpredictable food and sexual size dimorphism in insects. Proc. R. Soc. B 258, 121–125.

Lewis, S.M., Cratsley, C.K., 2008. Flash behaviour, mate choice, and predation in fireflies. Annu. Rev. Entomol. 53, 293–321.

Lewis, S.M., South, A., 2012. The evolution of animal nuptial gifts. In: Brockmann, H.J., Roper, T.J., Naguib, M., Mitani, J.C., Simmons, L.W. (Eds.), Adv. Stud. Behav. Elsevier, New York, pp. 53–97.

Lomborg, J.P., Toft, S., 2009. Nutritional enrichment increase courtship intensity and improves mating success in male spiders. Behav. Ecol. 20, 700–708.

Lopez, A., 1987. Glandular aspects of sexual biology. In: Nentwig, W. (Ed.), Ecophysiology of Spiders, Springer Verlag, Heidelberg, pp. 121–132.

Lubin, Y.D., 1986. Courtship and alternative mating tactics in a social spider. J. Arachnol. 14, 239–257.

Mappes, J., Alatalo, R.V., Kotiaho, J., Parri, S., 1996. Viability costs of condition-dependent sexual male display in a drumming wolf spider. Proc. R. Soc. B 263, 785–789.

Michalik, P., Uhl, G., 2011. Cephalic modifications in dimorphic dwarf spiders of the genus *Oedothorax* (Erigoninae, Linyphiidae, Araneae). J. Morphol. 272, 814–832.

Mougeot, F., Arroyo, B.E., Bretagnolle, V., 2006. Paternity assurance responses to first-year and adult male territorial intrusions in a courtship-feeding raptor. Anim. Behav. 71, 101–108.

Nitzsche, R.O.M., 1988. Brautgeschenk' und Umspinnen der Beute bei *Pisaura mirabilis, Dolomedes fimbriatus* und *Thaumasia uncata* (Arachnida, Araneida, Pisauridae). Verh. Naturwiss. Ver. Hamburg. 30, 353–393.

Platnick, N.I., 2012. The World Spider Catalog, version 8.5. American Museum of Natural History, online at http://research.amnh.org/entomology/spiders/catalog/index.html.

Preston-Mafham, K.G., 1999. Courtship and mating in *Empis (Xanthempis) trigramma* Meig., *E. tesselata* F., and *E. (Polyblepharis) opaca* F. (Diptera: Empididae) and the possible implications of "cheating" behaviour. J. Zool. 247, 239–246.

Prokop, P., Maxwell, M.R., 2009. Female feeding and polyandry in the nuptially feeding nursery web spider *Pisaura mirabilis*. Naturwissenschaften 96, 259–265.

Ryan, M.J., Fox, J.H., Wilczynski, W., Rand, A.S., 1990. Sexual selection for sensory exploitation in the frog *Physalaemus pustulosus*. Nature 343, 66–67.

Sakaluk, S.K., 1984. Male crickets feed females to ensure complete sperm transfer. Science 223, 609–610.

Sakaluk, S.K., 2000. Sensory exploitation as an evolutionary origin to nuptial food gifts in insects. Proc. R. Soc. Lond. B. 267, 339–343.

Sakaluk, S.K., Avery, R.L., Weddle, C.B., 2006. Cryptic sexual conflict in gift-giving insects: chasing the chase away. Am. Nat. 167, 94–104.

Sauer, K.P., Lubjuhn, T., Sindern, J., Kullmann, H., Kurtz, J., 1998. Mating system and sexual selection in the scorpionfly *Panorpa vulgaris* (Mecoptera: Panorpidae). Naturwissenschaften 85, 219–228.

Schneider, J.M., Elgar, M.A., 2001. Sexual cannibalism and sperm competition in the golden orb-web spider *Nephila plumipes* (Araneoidea): female and male perspectives. Behav. Ecol. 5, 547–452.

Schneider, J.M., Herberstein, M.E., Champion de Crespigny, F., Ramamurthy, S., Elgar, M.A., 2000. Sperm competition and small size advantage for males of the golden orb-web spider *Nephila edulis*. J. Evol. Biol. 13, 939–946.

Sierwald, P., 1990. Phylogenetic analysis of Pisaurine nursery web spiders, with revsions of *Tetragonophthalma* and *Perenethis* (Araneae, Lycosoidea, Pisauridae). J. Arachnol. 25, 361–407.

Silva, E.L.C., Lise, A.A., Buckup, E.H., Brescovit, A.D., 2006. Taxonomy and new records in the neotropical spider genus *Paratrechalea* (Araneae, Lycosoidea, Trechaleidae). Biociências 14, 71–82.

Simmons, L.W., 1990. Nuptial feeding in tettigoniids: male costs and the rates of fecundity increase. Behav. Ecol. Sociobiol. 27, 43–47.

Simmons, L.W., Bailey, W.J., 1990. Resource influenced sex roles of Zaprochiline Tettigoniids (Orthoptera: Tettigoniidae). Evolution 44, 1853–1868.

Simmons, L.W., Gwynne, D.T., 1991. The refractory period of female katydids (Orthoptera: Tettigoniidae): sexual conflict over the mating interval? Behav. Ecol. 12, 691–697.

Simmons, L.W., Gwynne, D.T., 1993. Reproductive investment in bushcrickets: the allocation of male and female nutrients to offspring. Proc. R. Soc. B 252, 1–5.

Simmons, L.W., Parker, G.A., 1989. Nuptial feeding in insects: mating effort versus paternal investment. Ethology 81, 332–343.

Stålhandske, P., 2001. Nuptial gift in the spider *Pisaura mirabilis* maintained by sexual selection. Behav. Ecol. 6, 691–697.

Stålhandske, P., 2002. Nuptial gifts of male spiders function as sensory traps. Proc. R. Soc. B 269, 905–908.

Stokes, A.W., Williams, H.W., 1971. Courtship feeding in gallinaceous birds. The Auk 88, 543–559.

Stratton, G., Hebets, E.A., Miller, P.R., Miller, G.L., 1996. Pattern and duration of copulation in wolf spiders (Araneae, Lycosidae). J. Arachnol. 24, 186–200.

Thornhill, R., 1976. Sexual selection and nuptial feeding behaviour in *Bittacus apicalis* (Insecta: Mecoptera). Am. Nat. 110, 529–548.

Trivers, R.L., 1972. Parental investment and sexual selection. In: Campbell, B. (Ed.), Sexual Selection and the Descent of Man, Aldine, Chicago, pp. 136–1791871-1971.

Tuni, C., Albo, M.J., Bilde, T., 2013. Polyandrous females acquire indirect benefits in a nuptial-feeding species. J. Evol. Biol. May 3, http://dx.doi.org/10.1111/jeb.12137 (Epub ahead of print).

Uhl, G., Maelfait, J.P., 2008. Male head secretion triggers copulation in the dwarf spider *Diplocephalus permixtus*. Ethology 114, 760–767.

Vahed, K., 1998. The function of nuptial feeding in insects: review of empirical studies. Ethology 113, 105–127.

Vahed, K., 2007. All that glisters is not gold: sensory bias, sexual conflict and nuptial feeding in insects and spiders. Ethology 113, 105–127.

Vanacker, D., Maes, L., Pardo, S., Hendrickx, F., Maelfait, J.P., 2003. Is the hairy groove in the gibbosus male morph of *Oedothorax gibbosus* (Blackwall 1841) a nuptial feeding device? J. Arachnol. 31, 309–315.

Warwick, S., Vahed, K., Raubenheimer, D., Simpson, S.J., 2009. Free amino acids as phagostimulants in cricket nuptial gifts: support for the "Candymaker" hypothesis. Biol. Lett. 5, 194–196.

Wiklund, C., Kaitala, A., Wedell, N., 1993. Decoupling of reproductive rates and parental expenditure in a polyandrous butterfly. Behav. Ecol. 9, 20–25.

Wise, D.H., 1975. Food limitation of the spider *Linyphia marginata*: experimental field studies. Ecology 56, 637–646.

Wise, D.H., 1993. Spiders in Ecological Webs. Cambridge University Press, Cambridge.

Wise, D.H., 2006. Cannibalism, food limitation, intraspecific competition, and the regulation of spider populations. Annu. Rev. Entomol. 51, 441–465.

Wolfner, M.F., 1997. Tokens of love: functions and regulation of *Drosophila* male accessory gland products. Insect Biochem. Molec. 27, 179–192.

Zahavi, A., 1975. Mate selection – a selection for a handicap. J. Theor. Biol. 53, 205–214.

Zahavi, A., Zahavi, A., 1997. The Handicap Principle: a Missing Piece of Darwin's Puzzle. Oxford University Press, Oxford.

Chapter 8

Paternal Care and Sexual Selection in Arthropods

Gustavo S. Requena,[1] Roberto Munguía-Steyer[2] and Glauco Machado[1]

[1]*Departamento de Ecologia, Instituto de Biociências, Universidade de São Paulo, São Paulo, Brazil*, [2]*Departamento de Ecología Evolutiva, Instituto de Ecología, Universidad Nacional Autónoma de México, México DF, Mexico*

INTRODUCTION

Exclusive paternal care is probably the rarest form of post-zygotic parental investment in nature. According to the current literature, this behavior is known to have independently evolved in 15 lineages of arthropods, including nearly 1500 species (Table 8.1; Figs 8.1 and 8.2, below). This is a tiny fraction of the arthropod diversity, which encompasses nearly 1.2 million described species (Hammond *et al.*, 1995). Although many arthropod species exhibiting exclusive paternal care are easily observed and manipulated both in the field and in the laboratory, thus offering unique opportunities to test hypotheses on parental investment and evolution of sex roles, only recently have researchers started to pay attention to such interesting biological systems (Fig. 8.1B). In fact, there has been a rapid increase in the recognition of new and independently evolved cases of paternal care in arthropods during the past three decades (Fig. 8.1A). This accumulation of basic biological information has revealed a great diversity of forms in which paternal care is expressed, involving specific male traits and behaviors that enhance offspring survival. At the same time, recent advances in the theory of sex roles evolution have incorporated co-evolutionary feedbacks between parental investment and sexual selection into the models, challenging the classical foundations of behavioral ecology and proposing new hypotheses to be tested (see, for example, Manica and Johnstone, 2004; Kokko and Jennions, 2008; McNamara *et al.*, 2009; Alonzo, 2012).

In this chapter, we review the theoretical background for the evolution of parental investment and sex roles, contrasting classical views with the most recent criticisms and advances proposed by new mathematical models. Next, we introduce the cases in which males exclusively care for the offspring in arthropods, stressing both particularities and general patterns. We explore the

Sexual Selection. http://dx.doi.org/10.1016/B978-0-12-416028-6.00008-6

TABLE 8.1 Basic Biological Information and Geographic Distribution of the Arthropod Species Exhibiting Exclusive Paternal Care

Each Line Represents a Lineage in Which Male Care Has Independently Evolved, Except for Water Bugs (Belostomatidae), Presented by Subfamilies Separately Due to Differences in Reproductive and Parental Biology. (?) Indicates Uncertainty and ? Indicates Lack of Information

Taxa*	Oviposition Site	Multiple Clutches	Number of Eggs	Brood Adoption	Additional Investment	Geographic Distribution
Hexapoda: Hemiptera						
BELOSTOMATIDAE: Lethocerinae (24 spp.)[1,2]	Vegetation[3]	Yes[3]	150–300[4–5]	?	?	Cosmopolitan[2]
BELOSTOMATIDAE: Belostomatinae (143 spp.)[1,2]	Males' dorsum[3]	Yes[3]	ca. 100[6]	Not possible	?	Cosmopolitan[2]
COREIDAE: Plunentis porosus[7,8] and P. yurupucu[9]	Males' venter[7–9]	?	20–257[7–9]	Not possible	?	SE Brazil[7–9]
PENTATOMIDAE: Edessa nigropunctata[10] and Lopadusa augur[10]	Vegetation[10]	Yes[10]	16–48[10]	?	?	SE Brazil[10]
REDUVIIDAE: Rhinocoris albopilosus[11], R. albopunctatus[12], and R. tristis[13]	Vegetation[11–13]	Yes[11–13]	4–40[11–13]	Yes[14]	Hard eggshell[15] (?)	Sub-Saharan Africa[11–13]
Diplopoda: Platydesmida						
ANDROGNATHIDAE: Brachycybe genus (8 spp.)[16–30] and Yamasinaium noduligerum[21]	Under rotten logs[17, 19, 21]	Yes[19] (?)	10–100[19]	?	?	Japan, Korea, Taiwan, and SE and SW United States[20]
Pycnogonida						
SEVEN FAMILIES[22] (ca. 1,100 spp.)[23]	Ovigerous legs[22]	Yes[22]	Dozens to hundreds[22]	Not possible	?	Cosmopolitan (marine)[22–23]

Arachnida: Opiliones

ASSAMIIDAE: *Lepchana spinipalpis*[24]	Yes[24]	Under rocks or rotten logs[24]	20–130[24]	Yes[24] (?)	?	Himalayas, Nepal[24]
COSMETIDAE: *Cryptopoecilaema almipater*[25]	Yes[25]	Vegetation[25]	20–150[25]	?	Mucus coat[25]	Costa Rica[25]
GONYLEPTIDAE (Gonyleptinae): *Gonyleptes saprophilus* and *Neosadocus* sp.[26]	Yes[26]	Natural cavities[26]	8–560[26]	Yes[26]	Eggs covered with debris[26]	SE Brazil[26]
GONYLEPTIDAE (Heteropachylinae): *Chavensicola inexpectabilis*[27] and *Magnispina neptunus*[28]	Yes[27,28]	Natural cavities[27,28]	110–160[27,28]	Yes[27,28]	Eggs covered with debris[27,28]	NE Brazil [27,28]
GONYLEPTIDAE (Caelopyginae + Progonyleptoidellinae): *Ampheres leucopheus*[29], *Cadeadoius niger*[30], *Iguapeia melanocephala*[26], *Iporangaia pustulosa*[26], *Progonyleptoidellus striatus*[26], *P. orguensis*[31]	Yes[26,29]	Vegetation[26,29–31]	14–420[26,29,32]	No[32]	Mucus coat[26,29–33]	SE Brazil[26,29–31]
MANAOSBIIDAE: *Zygopachylus albomarginis*[34]	Yes[34–35]	Mud nests[34–35]	1–100[35]	No[35]	Eggs covered with debris[34–35]	Panama[34]
PODOCTIDAE: *Leytpodoctis oviger*[24] and *Ibalonius* sp.[31]	No[24,31] (?)	Males' legs[24,31]	4–17[24,31]	Not possible	?	Philippines and Melanesia[24,31]

Continued

TABLE 8.1 Basic Biological Information and Geographic Distribution of the Arthropod Species Exhibiting Exclusive Paternal Care—cont'd

Taxa*	Oviposition Site	Multiple Clutches	Number of Eggs	Brood Adoption	Additional Investment	Geographic Distribution
STYGNIDAE: Stenostygnellus flavolimbatus and S. aff. flavolimbatus[36]	Vegetation or under rotten logs[36]	Yes[36]	ca. 100[36]	?	?	Venezuela[36]
TRIAENONYCHIDAE (Soerensenellinae): Karamea genus (6 spp.)[37,38] and Soerensenella genus (10 spp.)[37,38]	Under rocks or trunks[37,38]	Yes[37,38]	ca. 50–100[37,38]	?	Eggs covered with debris[37,38]	New Zealand[37,38]

References: 1. Lanzer-de-Souza (1980); 2. Polhemus and Polhemus (2008); 3. Smith (1997); 4. Hoffman (1933); 5. Smith (1979a); 6. Smith (1979a); 7. Lima (1940); 8. Morgado and Monteiro (2012); 9. Brailovsky (1989); 10. Requena et al. (2010); 11. Odhiambo (1959); 12. Nyiira (1970); 13. Thomas (1994); 14. Thomas and Manica (2005); 15. Gilbert et al. (2010); 16. Murakami (1962); 17. Kaestner (1968); 18. Gardner (1974); 19. Kudo et al. (2011); 20. Brewer et al. (2012); 21. Kudo et al. (2009); 22. King (1973); 23. León (1999); 24. Martens (1993); 25. Proud et al. (2011); 26. Machado et al. (2004); 27. Nazareth and Machado (2009); 28. Nazareth and Machado (2010); 29. Hara et al. (2003); 30. Stefanini-Jim (1985); 31. Machado and Macías-Ordóñez (2007); 32. Requena and Machado (unpublished data); 33. Requena et al. (2009); 34. Rodríguez and Guerrero (1976); 35. Mora (1990); 36. Villareal Manzanilla and Machado (2011); 37. Forster (1954); 38. Machado (2007).

*We are not considering two arthropod lineages that have been included in the review on paternal care by Tallamy (2001): the leaf-footed bugs of the genus Scolopocerus (Coreidae), and the thrips of the genera Hoplothrips, Sporothrips, and Idolothrips (Thysanoptera). Although there are records of eggs laid on the body of several species of phytophagous heteropterans, there is no convincing demonstration based on adequate sample sizes to show that only males carry the eggs. On the contrary, there are at least three species in which eggs are laid both on males and females (Panizzi and Santos, 2001), including the well-studied Phyllomorpha laciniata (see discussion in Kaitala et al., 2001). In relation to the colonial thrips, the available data indicate that males fiercely defend territories where females lay eggs (Crespi, 1986, 1988). Despite intensive behavioral studies, there is no record of males defending the egg masses against predators. According to the author of the papers, the mating system is territory-based without any evidence of parental care (B.J. Crespi, personal communication). However, we would like to stress that the inclusion of these two lineages does not change any of our conclusions.

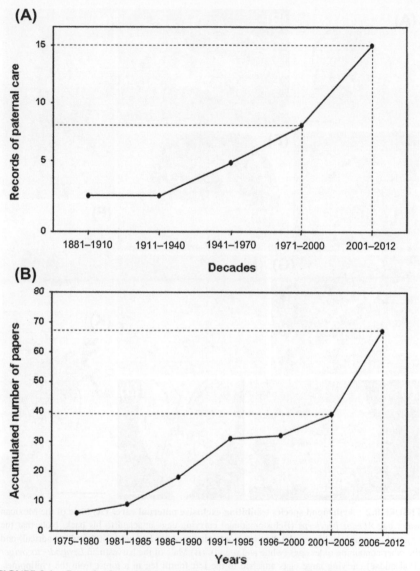

FIGURE 8.1   Growth of empirical research on paternal care in arthropods. (A) Records of independently evolved cases of paternal care in arthropods since the first formal description in 1881. Although the most recent time interval is almost three times shorter than the previous ones, we can see that there is an accentuated increase in the reports of new records in the past 12 years. Most of these records concern species from the Neotropics (see Table 8.1). (B) Accumulated number of empirical papers on paternal care published since 1975. Information was obtained from the *Web of Science* database using first the key words "paternal care" or "male brooding", and then conducting a search on the output for papers containing at least one of the following terms in the title, abstract, or key words: arthropod, insect, Belostomatidae, water bug, Opiliones, harvestman, Diplopoda, millipede, Pycnogonida, or sea spider. Although more than 1600 papers were found using this procedure, only 67 were selected because the remaining focused primarily on taxonomy or other aspects not related to behavior. Similarly to graph (A), there is a clear increase in the number of empirical papers on exclusive paternal care in the past 12 years.

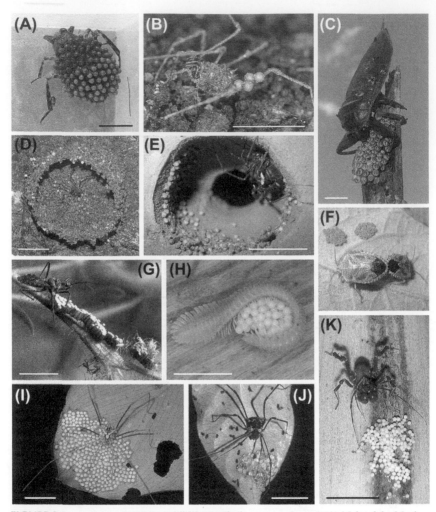

FIGURE 8.2 Arthropod species exhibiting exclusive paternal care. (A) Male of the Mexican
water bug *Abedus breviceps* (Belostomatinae) carrying eggs attached to his back. Note that the
egg-pad is composed of two clutches, one containing recently laid eggs (dark and cylindrical) and
the other containing older eggs (white and round). (B) Male of the harvestman *Leytpodoctis oviger*
(Podoctidae) carrying large eggs attached to the left fourth leg in a forest from the Philippines.
(C) Male of the giant water bug *Lethocerus* sp. (Lethocerinae) guarding a clutch on the emergent
vegetation of a lake in the Brazilian Amazon forest. (D) Male of the harvestman *Zygopachylus
albomarginis* (Manaosbiidae) inside his mud nest built on a fallen trunk in Panama. Some white
eggs are visible on the nest floor just below the guarding male. (E) Inside view of an artificial
nest of the harvestman *Magnispina neptunus* (Gonyleptidae) in the laboratory. The nest is full of
eggs, which were covered by debris by the ovipositing females. (F) Mating pair of the stink bug
Edessa nigropunctata (Pentatomidae) on the host plant in southeastern Brazil. Under the male (individual on the left) there is a recently laid clutch, and close to the mating pair there are two other
clutches. (G) Male of the assassin bug *Rhinocoris tristis* (Reduviidae) guarding a multiple clutch
on the host plant in Uganda. Note that the clutch is composed of eggs and hatched nymphs, which

empirical information in the context of modern theoretical models, focusing on the costs and benefits of male egg-guarding behavior, and the possible role of paternal care for male attractiveness and sex role reversal. In the last section of this chapter, we analyze the possible effects of temperature and rainfall on several ecological aspects that are likely to influence the evolution and maintenance of paternal care, and also derive specific macroecological predictions to be tested in future studies.

HISTORICAL PERSPECTIVE ON SEX ROLES AND PARENTAL INVESTMENT

Those 1970s Theories

In 1972, Trivers published his seminal work on the relationship between sexual selection and parental investment, which provided the first theoretical framework for our understanding of sex roles. According to his argument, anisogamy lies at the heart of sexual differences between males and females because gametes are a form of pre-zygotic parental investment. Females, which produce large and costly eggs, would make a greater parental investment prior to mating when compared to males, which produce cheap and abundant sperm. Indeed, in the great majority of species, male ejaculates represent a smaller total investment per mating than the eggs produced by females. Males therefore would be able to replenish their gamete supply and return to the mating pool sooner than females. Consequently, males should intensively compete for access to females and, assuming that male–male competition and/or active searching for other females prevents any investment in the offspring, males should also not exhibit any sort of parental care. On the other hand, given that females already invest more than males in gametes, they should be selected to provide care in order to minimize breeding failure for lack of additional investment. Therefore, highest past investment in gametes would select females to be choosy in relation to potential partners, and to show high future investment in offspring protection. This causal link between anisogamy and parental investment has led to the notion of *conventional sex roles*. In this sense, sex role reversal, in

---

remain under male protection for a few days. (H) Male of the Japanese millipede *Bachycybe nodulosa* (Andrognathidae) curled around a mass of eggs in the laboratory. (I) Male of the harvestman *Iporangaia pustulosa* (Gonyleptidae) guarding a multiple clutch on the abaxial surface of a leaf in southeastern Brazil. Eggs are covered with a mucus coat that confers additional protection (see Fig. 8.3). (J) Male of the harvestman *Cryptopoecilaema almipater* (Cosmetidae) guarding eggs in Costa Rica. The oviposition substrate and the presence of mucus around the eggs are very similar to patterns found for the harvestman *I. pustulosa*, suggesting that both features are convergent in these two species. (K) Male of the harvestman *Stenostygnellus* aff. *flavolimbatus* guarding a multiple clutch laid on the petiole of a palm leaf in Venezuela . Scale bars ~1 cm. See color plate at the back of the book. *Photographs courtesy of J. Martens (B), R. Macías-Ordóñez (C), T. M. Nazareth (E), L. K. Thomas (G), S. Kudo (H), C. Víquez (J), and O. Villareal Manzanilla (K).*

which females compete for mates more strongly than males, which in turn are the choosy sex, should only occur when parental investment by males is somehow greater than by females, as observed when males invest in expensive nuptial donations to females or, in some cases, when they care for the offspring (Trivers, 1972).

Although the association between pre- and post-zygotic parental investment provided a possible explanation for the evolution of sex roles, a central question remained unanswered: why should males care for the offspring at all? Until the late 1970s, the great majority of known examples of species exhibiting paternal care occurred among fishes and frogs (see review in Ridley, 1978). This led researchers to propose a causal link between external fertilization, widespread in these two groups, and paternal care. Therefore, the first two explanations for the evolution of exclusive paternal care were based on the mode of fertilization. According to the *certainty of paternity* (Trivers, 1972) and the *order of gamete release* hypotheses (Dawkins and Carlisle, 1976), internal fertilization decreases confidence of paternity in polyandrous species, increasing the costs of parental care for males and promoting a time lag between copulation and oviposition, which allows male desertion and predisposes females towards parental care. In species with external fertilization, on the other hand, females spawn earlier than males – a situation that may increase confidence of paternity, but also allows females to leave males in possession of the zygotes, forcing them into the so called "cruel bind" (Trivers, 1972). Several empirical studies, however, have reported the presence of sneaker males close to guarding males in fishes, leading to a high proportion of cuckolded broods in nature (reviewed in Coleman and Jones, 2011). Moreover, males from some fish families remain in their nests and take care of the eggs even when they supposedly have the opportunity to desert first, as reported for some species with simultaneous release of gametes (Gross and Shine, 1981) or some catfishes, in which males build foam nests wherein they release sperm before female spawning (Hostache and Mol, 1998).

A third explanation for the evolution of paternal care, known as the *association with offspring* or *territoriality* hypothesis, postulates that males would monopolize suitable oviposition sites in an attempt to attract females and acquire mates (Williams, 1975). Males defending a territory would further increase their fitness because they would indirectly defend eggs against conspecific predators. In this case, paternal care does not necessarily decrease the probability of a caring male to acquire additional mates, because several females may visit his territory. The territoriality hypothesis predicts that resource-defense polygyny mating systems are required to promote the evolution of exclusive paternal care. Although this hypothesis does not account for the evolution of paternal care in all animal groups, it has been proposed as the primary explanation for two groups in which this behavior is widely distributed, namely fish and anurans (Ridley, 1978; Gross and Shine, 1981; Coleman and Jones, 2011).

In 1977, Maynard Smith proposed a model to explain sexual differences in post-zygotic parental investment in which individuals of each sex have the option to desert or to care, so that the best tactic adopted for individuals of one sex depends on the tactics adopted by individuals of the other sex. According to his model, ecological and physiological parameters should determine the costs and benefits of each tactic. The direct benefits of caring were incorporated into the model as the probability of offspring survival when protected by zero, one, or two parents. The fecundity costs associated with parental care were also included, such as females producing fewer eggs when providing care than when deserting. The final parameter is the probability of males acquiring an additional mate when they desert, which should depend on the population's operational sex ratio (OSR), defined as the relative proportion of sexually active males to receptive females. As a result, only a specific combination of parameters favors the evolution of exclusive paternal care, i.e., when: (1) offspring survival strongly depends on the presence of at least one parent, but two parents do not significantly increase offspring protection; (2) females enjoy a great increase in fecundity if they desert; and (3) males have a low probability of acquiring additional mates after desertion. Maynard Smith's (1977) model was useful to highlight conditions under which each sex is expected to care for or desert the offspring, and it was the cornerstone on which subsequent models were built.

Challenging Classic Theories

During the following decades, Trivers' (1972) theory remained the most accepted explanation for the evolution of sex roles. However, his original argument that sexual differences in past parental investment condemn females to keep investing in costly post-zygotic parental care hid a logical failure known as the *Concorde fallacy*; i.e., the false impression that past investment in a costly activity makes it more profitable *per se* to continue with rather than abandon such activity (Dawkins and Carlisle, 1976). Furthermore, in the early 1990s Clutton-Brock and Vincent (1991) proposed that the potential reproductive rate (PRR), defined as the maximum number of progeny that adults of each sex could produce when there is no limitation in the amount of food or sexual partners, should be used as a more general factor determining sex roles (see also Clutton-Brock and Parker, 1992; Parker and Simmons, 1996). The PRR concept leads to the same general pattern predicted by Trivers (1972): due to anisogamy, males generally require fewer nutrients than females to produce their cheap gametes, so they have a higher PRR than females and, consequently, the OSR would be male-biased. This scenario would force males to compete more intensely for access to mates and also, rarely, to adopt the caring role.

The argument provided by the PRR concept reformulated the causal link between past and future parental investment, extending the acceptance of classical verbal models on sex roles until the late 1990s, when Queller (1997) stressed the importance of the so called *Fisher consistency*. In any diploid sexually reproducing species, each offspring has one mother and one father, with the consequence that the mean reproductive rate of males and females in natural populations with even adult sex ratios should be the same. Fisher consistency challenges the predictions of the classical models, pointing out a natural constraint: even with a potentially higher PRR, deserting males cannot reproduce faster if there are no receptive females with which to mate (Queller, 1997). This situation intensifies reproductive competition among males, and different outcomes are possible. If there is high mating success variance among males, a subset of males would mate many times, whereas all other males would mate a few times or not at all. Therefore, the higher reproductive rate of successful males would make them less prone to care, if there is a trade-off between caring and acquiring new mates (Queller, 1997). Existing differences in mortality rates between the sexes would lead to biases in sex ratio, and individual strategies would change dynamically: frequency-dependent selection would favor parental investment among individuals of the more common sex until OSR reaches 1 : 1 (Kokko and Jennions, 2008). Finally, if reproductive variance among males is not high and receptive females are a limiting resource (i.e., the sex ratio is male-biased), paternal care could evolve as a strategy to increase males' fitness by improving the viability of their offspring (Kokko and Jennions, 2008).

Once Fisher consistency is taken into account, sex role divergence is the expected scenario and, contrary to past theoretical models, male care should be a widespread behavior in nature (Kokko and Jennions, 2008) – which is clearly not the case. In an attempt to explain the rarity of male care in nature, it became necessary to generate a new theoretical framework that accounted for additional variables. Recent mathematical models for the evolution of sex roles have considered not only fecundity costs and direct benefits for offspring survival associated with parental care, but also population parameters such as adult density, OSR, and sex differences in mortality and mate searching rates, which ultimately influence mate encounter rate (e.g., Kokko and Jennions, 2008; McNamara et al., 2009; Alonzo, 2012). In all these models, parental care in general is favored when there are high benefits in terms of improvement in offspring survival. Paternal care, in particular, should evolve when the mate encounter rate is high, since it minimizes both the mating cost of caring and the probability of dying without having achieved any mating. Another general conclusion of the model proposed by Kokko and Jennions (2008) is that the reduced relatedness to any given set of offspring due to multiple mating by females (i.e., sperm competition) would select against male care.

Although Kokko and Jennions' (2008) model proposes that the most likely strategy for successful males with greater reproductive opportunities is deserting the offspring, the authors recognize that their conclusions would be reversed if paternal care itself were a sexually selected trait. However, their model assumed an inherent trade-off between parental and mating efforts, which seems to be the rule among birds. Moreover, they treated the variance in mating success as a fixed characteristic, which was not allowed to evolve in response to changes in other parameters. Subsequent studies, however, have argued that models on sex role evolution should not only demonstrate the inter-actions between sexual selection and parental investment, but also investigate how sex differences arise from a starting scenario that does not assume any asymmetry between sexes in mating or parental decisions (Stiver and Alonzo, 2009; Alonzo, 2012).

Trade-offs between mating effort and parental effort may arise due to variations in life-history traits, and can be broadly classified into three main categories (Stiver and Alonzo, 2009): (1) resource limitations, since individuals have to invest their limited energy and resources either to attract sexual part-ners and compete for access to mates, or to care for the offspring; (2) temporal limitations, since behaviors related to mating and parenting usually cannot be performed simultaneously; and (3) mechanistic limitations, since morphologi-cal, physiological, or behavioral traits involved mainly in the sexual selection context are at the same time frequently detrimental to effective offspring care. Theoretical models, however, have already proposed that paternal care does not necessarily conflict with male mating effort (Manica and Johnstone, 2004), and empirical studies with fishes (Stiver and Alonzo, 2009) and arthropods (Tallamy, 2001) have shown that males may care for multiple clutches composed of eggs laid by several females. Once exclusive paternal care has evolved, males exhib-iting guarding behaviors could also provide an honest signal of their quality as offspring defenders, and thus female preference for caring males could be responsible for maintaining the trait (e.g., Hoelzer, 1989; Tallamy, 2001; Wagner, 2011; Alonzo, 2012).

PATERNAL CARE IN ARTHROPODS

Overview of Cases

The first case of paternal care in arthropods was reported at the end of the 19th century for a sea spider of the family Phoxichilidiidae (Pycnogonida), in which males carry egg masses attached to a specialized pair of legs (Sars, 1891; see also Cole, 1901). During the 20th century, similar behavior was described for species in seven of the eight families belonging to the class Pycnogonida (Bain and Govedich, 2004; Table 8.1). Some years before the first description of pater-nal care in sea spiders, Dimmock (1887) had already reported water bugs of the subfamily Belostomatinae carrying eggs attached to their backs (Fig. 8.2A). However, these described cases of parental behavior were originally attributed to females, probably because paternal care was totally undescribed in arthro-pods until that moment, and also because there is no clear sexual dimorphism among the Belostomatinae. The mistake was only corrected 12 years later, when the caring behavior was attributed to the male sex (Slater, 1899). Later, four additional cases of species with males carrying eggs attached to their bodies were described, two among leaf bugs of the family Coreidae and two among harvestmen of the family Podoctidae (Fig. 8.2B, Table 8.1).

Although water bugs of the subfamily Lethocerinae are the sister group to the Belostomatinae, they show some behavioral differences regarding the form of parental care (Smith, 1997). Lethocerinae males care for the offspring on the emergent vegetation at the edges of lakes and ponds (Fig. 8.2C), which does not impose a spatial limit to the number of cared eggs. Despite this important difference, paternal care has probably evolved only once in the family Belostomatidae, and oviposition on emergent vegetation is regarded as the ancestral state (Smith, 1997). Back brooding has probably evolved as a response to a restriction or even absence of appropriate oviposition sites outside the water (Smith, 1997). Moreover, males in both subfamilies perform specific parental behaviors devoted to protect the eggs from stressful abiotic conditions. In the Belostomatinae, males may either remain at the air–water interface or perform different types of underwater movements that promote egg aeration (Smith, 1997; Munguía-Steyer et al., 2008). In the Lethocerinae, males constantly water the clutch exposed on the vegetation to avoid egg dehydration (Smith, 1997).

Another form of paternal care is found among species in which the mating system and male reproductive success are associated with the possession of a nest. The first example of this type of paternal care was described for males of the neotropical harvestman Zygopachylus albomarginis, which build cup-like mud nests used to attract females, and where they copulate and guard the eggs (Table 8.1, Fig. 8.2D). Males actively invest in repairing the nests and protecting them from invasion of conspecific males, as well as in defending the eggs against fungal attack and predation (Rodriguez and Guerrero, 1976; Mora, 1990). In two other lineages of neotropical harvestmen (Gonyleptinae and Heteropachylinae), males do not build nests but rather defend natural cavities (in rocks, trunks, and roadside banks) against conspecific intruders and egg predators (Table 8.1, Fig. 8.2E). Although males of both Z. albomarginis and Magnispina neptunus (=Pseudopucrolia sp.) may attempt to take over nests by fighting the original male owners, the behavior of successful intruders in these two species are markedly different: Z. albomarginis males invariably eat all the eggs inside the mud nest (Mora, 1990), while M. neptunus males usually adopt the unrelated eggs, even repelling intruders (Nazareth and Machado, 2010; see also discussion in "Caring Male Attractiveness, a Neglected Benefit", below).

Finally, male egg-guarding behavior on exposed substrates is found in a couple of species of stink bugs (Fig. 8.2F), some species of assassin bugs of the genus Rhinocoris (Fig. 8.2G), millipedes of the genera Brachycybe and Yamasinaium (Fig. 8.2H), and harvestmen of five different families (Table 8.1, Figs 8.2I–K). For at least two of these species, namely Iporangaia pustulosa and R. tristis, there is experimental evidence that males defend the eggs against predators (Requena et al., 2009) and parasitoids (Gilbert et al., 2010). Males of the millipede B. nodulosa protect the eggs mainly from fungus infection, and orphaned clutches are promptly covered by hyphae so that no egg hatches (Kudo et al., 2011). Eggs in at least three harvestman families are additionally covered either by mucus or a debris coat, both deposited on the eggs by females just after oviposition

(Table 8.1, Figs 8.2D, E, I, J). Experimental evidence shows that the mucus coat that surrounds the eggs of the harvestman *I. pustulosa* works as an efficient physical barrier, providing additional protection to the embryos by hampering predator access to them (Requena *et al.*, 2009; see also "Costs and Benefits of Paternal Care", below).

Despite the diversity of forms of paternal care found in arthropods, some general patterns seem to occur (Table 8.1). Given that males are in charge of egg care, females are free to forage after oviposition and to allocate the additional energy they acquire in the continuous production of eggs over the course of the breeding season. This would be particularly important in predatory species, in which individuals have less regular access to food than do herbivorous species (Tallamy, 2001). In fact, females from most of the arthropod species known to exhibit exclusive paternal care are iteroparous, and it has been suggested that female mate preferences towards caring males may have been favored due to the direct, fitness-enhancing gift of cost-free care of the offspring (Maynard Smith, 1977; Tallamy, 2001). Additionally, another general pattern observed among almost all arthropods with paternal care is that males are able to copulate sequentially with several females and usually care for multiple clutches simultaneously (Table 8.1, Fig. 8.2). However, due to the asynchronous development of the eggs laid in the multiple clutches, the total caring period is inevitably prolonged, imposing additional energetic costs for caring individuals in species in which males are prevented from foraging while caring for the offspring (see, for example, Tallamy, 1994; Gilbert *et al.*, 2010; Nazareth and Machado, 2010, Requena *et al.*, 2012). In the following sections we focus on the trade-offs sustained by arthropod males while caring for the offspring, and discuss how sexual selection affects and is affected by paternal care in well-studied biological systems.

Costs and Benefits of Paternal Care

Virtually all theoretical models proposed to explain the evolution and maintenance of parental care take into account, either explicitly or implicitly, the costs and benefits of this behavior. The benefits of caring are generally related to increasing offspring survival, and all experimental studies with arthropods exhibiting exclusive paternal care have obtained the same general result: the absence of caring males condemns the offspring to death (Mora, 1990; Smith, 1997; Munguía-Steyer *et al.*, 2008; Requena *et al.*, 2009; Gilbert *et al.*, 2010; Kudo *et al.*, 2011; Fig. 8.3). However, males from different lineages exhibit different behaviors that increase offspring survival. In giant water bugs, males actively reduce the risk of dehydration and/or increase egg aeration, as stated previously (Smith, 1997; Munguía-Steyer *et al.*, 2008). In other lineages, males protect the eggs against predators (as observed in two harvestman species; Mora, 1990; Requena *et al.*, 2009), parasitoids (as reported in one assassin bug species; Gilbert *et al.*, 2010), and even fungal attack (as described for one harvestman and one millipede; Mora, 1990; Kudo *et al.*, 2011).

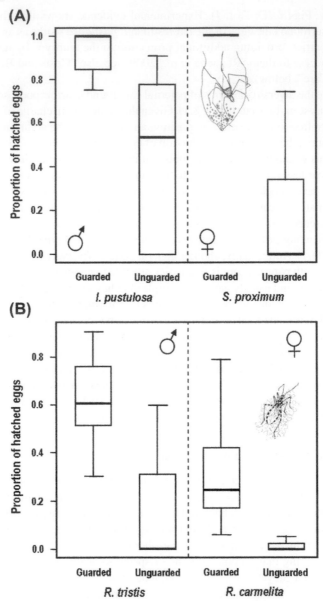

FIGURE 8.3 Efficiency of paternal care when compared to maternal care in two arthropod species. (A) Proportion of surviving eggs in an experiment in which parental individuals of the syntopic harvestmen *Iporangaia pustulosa* (exhibiting male care) and *Serracutisoma proximum* (exhibiting female care) were removed from their clutches in the field for 2 weeks. The median values for guarded clutches in both species are 1.0, indicating that male care is as efficient as female care in preventing egg predation. However, the median proportion of surviving eggs in unguarded clutches is considerably higher for *I. pustulosa*, in which the eggs are additionally protected by a thick mucus coat that reduces predation even in the absence of the guarding males. (B) Proportion of hatched eggs in a similar experiment in which parental individuals of the syntopic assassin bugs

In addition to the effort from caring males to minimize offspring mortality, females from several species cover the eggs with debris or a thick mucus coat, which may confer additional protection and also prevent filial cannibalism (Table 8.1, Figs 8.2D, E, I, J). Eggs experimentally manipulated to remove the mucus coat in the harvestman *Iporangaia pustulosa*, for instance, suffered higher predation than eggs covered by mucus deposited by females after oviposition (Requena *et al.*, 2009). Furthermore, when compared to a syntopic harvestman species with uniparental female care (*Serracutisoma proximum*), in which there is no deposition of mucus on the eggs, mortality of unattended clutches of *I. pustulosa* was lower (Requena *et al.*, 2009; Fig. 8.3A). A comparable pattern has been reported in a pair of syntopic sub-Saharan assassin bug species: *Rhinocoris tristis* with paternal care, and *R. carmelita* with maternal care (Gilbert *et al.*, 2010). Although the median proportion of hatched eggs in unattended clutches of both *R. tristis* and *R. carmelita* is zero, the number of clutches in which a great proportion of unattended eggs survived was much higher in the former than in the latter (Fig. 8.3B). The authors proposed the existence of an additional investment by *R. tristis* females in a hard eggshell to avoid filial cannibalism, which is commonly observed among caring males (Thomas and Manica, 2003). The protection against filial cannibalism could also provide protection against other natural enemies, which may explain why unattended eggs of *R. tristis* had higher chances of hatching when compared to *R. carmelita*. Moreover, the additional protection that would be afforded by the hard eggshell could explain why the efficiency of paternal care was higher when compared to maternal assistance in this pair of species (Fig. 8.3B). These suggestions, however, remain as open and interesting questions to be investigated in *Rhinocoris* and perhaps other arthropod species exhibiting paternal care.

The costs paid by caring males are generally classified into three main categories (Gross and Sargent, 1985; Clutton-Brock, 1991): (1) increases in mortality risk due to predator, parasite, and parasitoid attacks during the caring period; (2) decreases in body condition as a consequence of either decreasing foraging activity or increasing metabolic expense associated to parental behaviors; and (3) decreases in mating rate due to a trade-off between parental and mating efforts. Empirical data on mortality risks of paternal care are restricted to a few species, with no common pattern. In the assassin bug *R. tristis* (Fig. 8.2G), caring males have higher mortality rates when compared to non-caring males (Gilbert *et al.*, 2010), because, according to the authors, caring males suppress their escape behavior while guarding the eggs, thus becoming more vulnerable

---

Rhinocoris tristis (exhibiting male care) and *R. carmelita* (exhibiting female care) were removed from their clutches in the field until all eggs had hatched or disappeared (modified from Gilbert *et al.*, 2010). In this pair of species, male care seems to be more efficient than female care. Although the median proportion of hatched eggs in unguarded clutches is zero in both species, the number of clutches in which a great proportion of unattended eggs survives is much higher in the species with paternal care. In both graphics, thick horizontal lines represent the median, boxes represent the interval between the first and third quartiles, and vertical lines indicate 90% confidence intervals.

to predators. Alternatively, no evidence of survival costs was observed in two species of water bugs. In a laboratory experiment without predators, *Belostoma flumineum* males that had their egg pads removed showed a similar life span to either virgin or brooding males (Gilg and Kruse, 2003). The same pattern was also observed in a mark–recapture study conducted under field conditions with *Abedus breviceps* (Fig. 8.2A), in which parental status of adult males did not influence their survival probabilities (Munguía-Steyer and Macías-Ordóñez, 2007). Even more surprising are the results obtained in the field for the harvestman *I. pustulosa* (Fig. 8.2I), in which caring males showed higher survival probabilities than non-caring individuals (both males and females). This pattern is interpreted as resulting from differences in movement activity, since non-caring individuals are constantly moving on the vegetation and are more likely than caring males to be caught by sit-and-wait predators, which are the most important natural enemies of the adults (Requena *et al.*, 2012).

Results reported for energetic costs of paternal care in arthropods are as rare and controversial as data on mortality risks. While egg carrying allows males to move and forage while caring, the extra weight gained with the eggs may negatively affect the movement, foraging efficiency, and food intake of brooding males. Although adults of the sea spider *Achelia simplissima* seek food (sessile worms) and non-brooding males cover larger areas and move more frequently than caring males, the presence of the eggs does not affect male foraging efficiency (Burris, 2010). On the other hand, empirical results for a Belostomatinae water bug showed that males carrying eggs on their back swim slower than both non-caring males and females (Crowl and Alexander, 1989; Kight *et al.*, 1995), which is likely to affect their attack response and, consequently, their foraging efficiency. Additionally, water bug males exhibit expensive parental behaviors that may incur additional energetic costs to parental individuals (see "Overview of Cases", below).

Egg-guarding behavior, on the other hand, would be expected to impose relatively lower energetic costs to caring males because parental activities involve basically staying on the clutch and protecting the eggs against natural enemies. However, the higher potential for multiple and asynchronous mating opportunities in this case may lead to prolonged caring periods with limited foraging opportunities. Indeed, caring males of one assassin bug (Thomas and Manica, 2005) and two harvestman species (Nazareth and Machado, 2010; Requena *et al.*, 2012) feed less frequently than non-caring individuals. The consequences for male body condition may vary widely: paternal care erodes caring male body condition over the course of the caring period in the harvestman *I. pustulosa* (Requena *et al.*, 2012), while it has no effect on caring male weight in the assassin bug *R. tristis* (Thomas and Manica, 2005). A plausible explanation for this difference is that males of *R. tristis* usually engage in filial cannibalism during caring (Thomas and Manica, 2003), while males of *I. pustulosa* have never been observed consuming eggs from their own clutches (Requena *et al.*, 2012).

Finally, the potential loss of additional mating opportunities while caring, as well as the uncertainty of genetic relatedness with the offspring under protection, have always been pointed out as the strongest selective pressures against the evolution of paternal care (Magrath and Komdeur, 2003). The available data for arthropod species with exclusive paternal care, however, challenge these prohibitive mating costs associated with paternal effort. In most of the reported cases at least some caring males in the population (if not all of them) are able to care for multiple clutches simultaneously, even when they carry the eggs attached to their body (Table 8.1). This evidence strongly suggests that males are usually not constrained to care for eggs from just one female at a time, but instead can mate with more than one female during the caring period. Therefore, caring for the offspring and acquiring new mates are not mutually exclusive activities for males from most arthropod species, which usually have many mating opportunities while guarding eggs (Table 8.1). This pattern is similar to that described for fishes with exclusive paternal care, in which males also care for multiple clutches (Gross and Shine, 1981; Gross and Sargent, 1985; Stiver and Alonzo, 2009, Coleman and Jones, 2011). In some fishes, females preferentially mate with males that exhibit high quality care, which promotes a fast increase in offspring number inside the nests of these males (e.g. Östlund and Ahnesjö, 1998; Pampoulie *et al.*, 2004). Given that offspring number can be directly assessed by other females during mating decisions, preference for males caring for large clutches may have driven the evolution of specialized male traits in fishes, such as egg mimicry, egg thievery, and nest takeovers (Porter *et al.*, 2002). Could paternal care also play an important role in mate decisions among arthropods? If so, should we keep interpreting male post-zygotic parental investment as a general constraint in future reproductive events? These questions are explored in more detail in the following two sections.

Caring Male Attractiveness, a Neglected Benefit

The evidence that males have multiple mates while caring is not strong enough to refute the existence of a trade-off between parental and mating efforts, which is the underlying assumption of the so called reproductive costs associated with paternal care (Clutton-Brock, 1991). In theory, non-caring males may assess more females and copulate more frequently than caring males, without paying the costs of caring for the offspring. However, for at least two harvestman species there is strong observational evidence indicating that females exclusively mate inside nests and leave their eggs only with territorial males owning such nests (Mora, 1990; Nazareth and Machado, 2010). It is worth noting, though, that observational data alone cannot disentangle the effect of male individual traits, nest-associated traits, or paternal care quality. Only experimental manipulations are able to isolate these three factors, and such experiments have already been performed using three species with marked differences in their reproductive biology: the harvestman *Magnispina neptunus* (Nazareth and Machado,

2010), the assassin bug *Rhinocoris tristis* (Gilbert *et al.*, 2010), and the sea spider *Achelia simplissima* (Burris, 2011). In these three species, the presence of eggs indeed influenced the probability of males receiving additional eggs, but the specific effect differed among species, and will be discussed in detail in the following paragraphs.

Females of the harvestman *M. neptunus* were given the opportunity to simultaneously choose between two males with their respective nests in a paired mate choice experiment. First, each female was allowed to select either a caring male (with eggs inside his nest) or a non-caring male (with an empty nest). After female decision and oviposition, individuals (the female and both males) were isolated from one another for a few days. The same triplet was allowed to interact again, but the parental state of the males was reversed: offspring of caring males in the first round had hatched and were removed from their nests, while non-caring males in the first round had copulated with other females and received eggs in their nests. The results obtained showed that females consistently preferred to lay eggs inside nests already containing eggs, regardless of nest or individual male traits (Nazareth and Machado, 2010).

The experimental design and the results obtained are slightly different for the assassin bug *R. tristis*, in which females sequentially assessed males: females indiscriminately mated with caring and non-caring males, but they preferred to lay eggs in already established broods. Furthermore, caring males were more likely to receive eggs in their broods if females had interacted with a non-caring male first (Gilbert *et al.*, 2010). Observational data of the sea spider *Ammothella biunguiculata* are in accordance with this pattern, since male mate acquisition seems to be random, but once a male encounters a mate the number of eggs received depends on whether he is already carrying eggs (Barreto and Avise, 2011). Therefore, there are cases in which caring males are clearly more attractive than non-caring males, and the latter, in turn, can be expected to exhibit other mating tactics to increase their reproductive success. Examples among arthropods include non-caring males of the harvestman *M. neptunus* taking over nests containing eggs (Nazareth and Machado, 2010), and non-caring males of the assassin bug *R. tristis* adopting eggs from unattended clutches or from clutches whose owners have been displaced (Thomas and Manica, 2005). These tactics may be profitable to non-caring males only if the benefits in terms of increased attractiveness are expected to be higher than the costs of caring for unrelated offspring (Tallamy, 2001).

The last example regards a simultaneous mate choice experiment conducted with the sea spider *Achelia simplissima* to investigate female mating decisions, which revealed a pattern opposite to that described above: females clearly preferred non-caring males. Furthermore, when allowed to select one of two parental males, females preferred to lay their eggs with males carrying fewer egg masses (Burris, 2011). The author argued that active aeration of egg masses, the extra weight conferred by the eggs, and the extra resistance to avoid dislodgement by water flow imposed by the egg masses (Burris, 2010) would deplete caring males' energetic reserves and, consequently, decrease the quality of male care. Given that females are expected to select males based on their parental effort (Hoelzer, 1989; Alonzo 2012; Fig. 8.4A), the negative effect of the number of eggs on the quality of paternal care may explain the pattern of female preference reported for *A. simplissima*.

To have a broad understanding of the influence of clutch size on caring male attractiveness, it may be useful to recognize two types of care, which represent extremes of a continuum (Clutton-Brock, 1991): *depreciable care*, in which the quality of male care declines as the number of eggs in the clutch increases (Fig. 8.4B), and *non-depreciable care*, in which the quality of male care does not depend on the number of eggs in the clutch (Fig. 8.4B). Regardless of the type of paternal care, caring male attractiveness should increase as the quality of care provided increases (Fig. 8.4A). However, when paternal care is non-depreciable, male attractiveness should not be directly influenced by the number of eggs because the quality of male care is not affected by offspring number (Fig. 8.4B). Assuming that more attractive males accumulate eggs faster, a positive relationship between offspring number and male attractiveness should be interpreted only as a consequence of female mate choice based on male traits, not implying any causality between these two variables. On the other hand, when paternal care is depreciable, caring male attractiveness should decrease as the number of eggs in the clutch increases (Fig. 8.4B). According to this rationale, males caring for small clutches would be more attractive because the quality of care per egg is better when compared with males caring for large clutches (Fig. 8.4B). Therefore, the more strongly depreciable paternal care is, the more negative

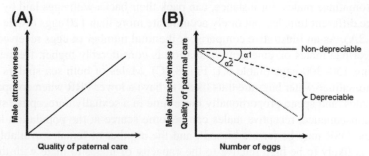

FIGURE 8.4   Theoretical predictions for the relationship among clutch size, quality of paternal care per egg, and caring male attractiveness. (A) In species with exclusive paternal care, females should select males based on their parental effort. Thus, the quality of parental care per egg provided by males should positively affect their attractiveness. (B) The influence of clutch size on the quality of paternal care or on caring male attractiveness, however, depends on the type of care. When parental care is depreciable, its quality declines as clutch size increases; when parental care is non-depreciable, the quality of care is independent of clutch size. How depreciable paternal care is depends on how intensely the quality of care decreases with an increase in clutch size (higher values of α). For the sake of simplicity, the relationship between variables in both graphs is represented linearly. Regardless of the functional form of the relationships, the tendencies predicted here should be the same.

should be the relationship between the number of eggs in the clutch and male attractiveness (Fig. 8.4B). In extreme cases, females should prefer non-caring males even over males caring for small clutches, as reported for the sea spider *A. simplissima*.

Sex Role Reversal

In the first section of this chapter, we provided a historical perspective on parental investment and the evolution of sex roles. Regardless of the proposed theoretical model, extreme cases of sex role reversal, characterized in terms of competitive females and choosy males, are expected to evolve only under particular conditions (see Chapter 6). Male mate choice, for instance, is more likely to occur when: (1) males have large parental investment or suffer from sperm depletion, so that they are unable to mate with many available females; (2) females exhibit great variation in quality; and (3) mate searching effort is relatively inexpensive (see, for example, Bateman and Fleming, 2006; Barry and Kokko, 2010; Edward and Chapman, 2011). Bonduriansky (2001) provides a review on this subject among arthropods, emphasizing species in which males offer nuptial gifts, and here we explore the available empirical evidence of sex role reversal among species exhibiting paternal care.

Sex role reversal in paternally caring arthropods has been described for some water bugs, sea spiders, and at least one harvestman species. In Belostomatinae water bugs and sea spiders, males carry eggs attached to their own body, which limits the number of eggs they can receive due to the available space either on their backs (water bugs) or on their legs (sea spiders) (Table 8.1). Belostomatinae males, for instance, can pack their backs with eggs laid by one to three different females, but rarely accumulate more than 120 eggs (Table 8.1; Fig. 8.2A). As an illustrative comparison, the total number of eggs received by Lethocerinae males on emergent vegetation is considerably higher, frequently reaching 150–300 eggs (Table 8.1; Fig. 8.2C). Males of both sea spiders and Belostomatinae water bugs are thus likely to have a lower PRR when compared to females, and spend proportionally more time in a sexually unreceptive state. As a consequence, receptive males can become scarce at the population level (i.e., the OSR may be female-biased), and the number of females available as mates is likely to be high relative to the capacity of males to mate with them. If the benefit of mating with specific high-quality females exceeds the cost of assessing them, male choice is expected to evolve (Edward and Chapman, 2011).

Saturation of male back space in the wild, however, is rare (Krause, 1989), and detailed records of sex role reversal in Belostomatinae are scarce and inconclusive (e.g., Ichikawa, 1989). Although there is no record of female rejection prior to oviposition, male mate choice expressed as premature termination of parental care has been reported for the Belostomatinae *Belostoma flumineum* (Kight and Kruse, 1992; Kight *et al.*, 2000, 2011). Data obtained

under captive conditions show that males discard small egg pads, condemn-
ing the eggs to submerge and die. Extensive fieldwork on *Abedus breviceps*
(Belostomatinae), however, shows that oviposition of a second egg pad without
removing the first one is the rule, and male abortion has never been recorded
in nature (R. Munguía-Steyer, unpublished data). Differences between these
two species suggest that male abortion may be influenced by the conditions
under which data are taken. Captivity experiments maintaining *B. flumineum*
adults in a predator-free environment greatly reduce the risks of mate searching,
increasing the residual reproductive value of parental males in comparison with
natural conditions (Kight and Kruse, 1992; Kight *et al.*, 2000, 2011). Under
natural conditions, the mortality risk experienced by *A. breviceps* males is high
(Munguía-Steyer and Macías-Ordóñez, 2007). As predicted by theory (Barry
and Kokko, 2010; Edward and Chapman, 2011), males experiencing low mate
search costs could afford to selectively discard small clutches as a reproductive
tactic to make their back space available to a new and potentially larger clutch.
On the other hand, males living in a risky environment may have their choosiness constrained by the high probability of dying before encountering a new
female. Additional data and experiments specifically designed to test the adaptive role of male abortion are needed, and the influence of mate search costs on
the probability of premature termination of parental care remains an open and
interesting line of investigation.

Given that the cost of carrying unrelated eggs is probably very high for
males of sea spiders and Belostomatinae water bugs, males are also expected
to exhibit paternity assurance strategies. In fact, alternating repeated copulations and ovipositions, and post-copulatory female guarding, are common male
mating behaviors in water bugs, reported for representatives of all genera of
back-brooders and three species of emergent-brooders (Smith, 1979a, 1997),
suggesting an ancient origin probably related to mate guarding (Alcock, 1994).
Although some level of cuckoldry has been detected in a Japanese back-brooder
(Inada *et al.*, 2011), extremely high levels of paternity have been reported for
a back-brooder from North America (Smith, 1979b), which led the author to
interpret the above-mentioned mating behaviors as male tactics that diminish
the risk of sperm competition (Smith, 1997). Unfortunately, paternity data on
Lethocerinae are lacking, thus it is not possible to compare the levels of cuckoldry between back-brooders and emergent-brooders. Among sea spiders, fertilization is external, but the gonopores of males and females are in close contact
during mating interactions and fertilization usually occurs immediately after
the transfer of eggs onto males' ovigerous legs (Wilhelm *et al.*, 1997; Bain and
Govedich, 2004). This tactic could explain the lack of registered cuckoldry in
some species (Barreto and Avise, 2008, 2011).

Some degree of sex role reversal is also observed among female water bugs,
especially among representatives of Lethocerinae. Given that appropriate oviposition sites on emergent vegetation may be in short supply in some places,
females may aggressively compete for access to them. Competition among

gravid females of some species of *Lethocerus* occurs in the form of cannibalism and destruction of non-filial eggs, which is probably a tactic that opens space for their own offspring and also allows females to acquire nutritional benefits from the consumed eggs (Ichikawa, 1990; Smith, 1997). Although males attempt to chase attacking females away from the eggs, in all observed cases they did not succeed and the eggs were eaten and/or dropped into the water. After destruction of the clutch, attacking females always mated with the guarding male following the same general pattern described above, and replaced the destroyed clutch with their own eggs. Males, in turn, always cared for the new clutch (Ichikawa, 1990). The best explanation for why clutch cannibalism and destruction are not widespread among *Lethocerus* species probably relies on the fact that size dimorphism is relatively rare in the genus. According to this hypothesis, infanticide should occur in species in which females are considerably larger than males and thus able to displace the males and destroy the clutch they are caring for, but not in species monomorphic in size, where multiple clutches should be the rule (Smith, 1997).

Among sea spiders, information on either intra or intersexual interactions prior to mating is very scarce and no general pattern of sex role reversal is evident. Females of *Propallene saengeri* are the exclusive initiators of courtship, with instances of female–female aggression also described during mating interaction, which strongly suggests sex role reversal (Bain and Govedich, 2004). Curiously, males but not females exhibit courtship behavior in two closely related species, *P. longiceps* and *Parapallene avida*, with no reported instance of competition or aggressive interactions between females (Nakamura and Sekiguchi, 1980; Hooper, 1980). Finally, in another well-studied species of sea spider, *Achelia simplissima*, most of the observed mating interactions were initiated by females, which exhibited specific courtship behaviors (Burris, 2011), but they do not seem to aggressively compete for mates. Under captive conditions, however, it has been demonstrated that females reject some mates, and in fact their mating decisions are based on the number of egg masses that males are already carrying, with a clear preference for non-caring males (Burris, 2011). This finding suggests that males are unlikely to accumulate eggs fast enough to fill up the space on their ovigerous legs, which may explain why nearly all males in the population are carrying egg masses below their maximum capacity (Burris, 2010). Therefore, there are receptive males available to mate in the population at any time, implying that *A. simplissima* may have conventional sex roles (Burris, 2011).

Among harvestmen, there is unequivocal observational evidence of sex role reversal for *Zygopachylus albomarginis*, in which males build cup-like mud nests that are visited by females (Mora, 1990; Fig. 8.2D). Females actively court males when they enter a nest, but males do not copulate with all courting females. Males aggressively reject 14% of the visiting females, biting their legs and chasing them out of the nest. Attacked females never respond to the male's bites, but rather remain quiet on the nest floor or leave the nest without fighting

(Mora, 1990). The reasons why some females are accepted while others are rejected remain unknown, but it is likely that males are able to evaluate visiting females through chemical information based on tegumentary hydrocarbons. Given that the great majority (90%) of females in the population visit only one or two nests during the mating season (Mora, 1990), males may recognize their frequent partners and repel newcomer females that visit many nests and thus represent a high risk of sperm competition or egg cannibalism. Field observations also showed that some females spend several days in the vicinity of one or two nests, and sometimes show aggressive behaviors toward newcomer females that approach one of these nests. Given that females are egg predators, the risk of cannibalism may explain female–female aggression (Mora, 1990).

In the harvestman *Magnispina neptunus*, territorial males can reject females visiting their nests, which suggests some degree of sex role reversal (Nazareth and Machado, 2010; R. Werneck and G. Machado, unpublished data). Males of this species do not build a mud nest, but rather occupy natural cavities on roadside banks that are used as oviposition sites (Nazareth and Machado, 2010). Contrary to sea spiders and Belostomatinae water bugs, the number of eggs that males of both *Z. albomarginis* and *M. neptunus* can receive in their nests is not limited and is similar to other harvestman species that do not have nests (Table 8.1). However, nests are in short supply, either because only good quality males are able to pay the costs of building and maintaining a mud nest (*Z. albomarginis*) or because natural cavities of appropriate dimensions are extremely rare in nature (*M. neptunus*). Given that females of both harvestman species copulate only with males holding a nest (Mora, 1990; Nazareth and Machado, 2010), the OSR may be slightly or strongly female-biased, which may partially explain sex role reversal in these two species.

MACROECOLOGY OF PATERNAL CARE

Chapter 1 lists several predictions regarding the influence of biotic and abiotic conditions on numerous reproductive traits of arthropods, and Chapter 5 provides a formal test of some of them using harvestmen as a model system. In this section, we explore in detail the potential effects of biotic and abiotic conditions on the costs and benefits of exclusive paternal care, as well as on population parameters affecting sexual selection dynamics. Given that the evolution and maintenance of paternal care are likely to be influenced by both natural and sexual selection (see "Historical Perspective on Sex Roles and Parental Investment", above), we take into account these two selective pressures when deriving macroecological predictions. Unfortunately, hard data are scarce and detailed studies are restricted to a few species, which prevents any formal test of our predictions. We expect, however, that the ideas presented here will stimulate researchers to gather basic information on a wider set of arthropod species exhibiting exclusive paternal care, and also take into account the role of environmental conditions on the costs and benefits of this behavior.

Temperature and rainfall have an important role in modulating the intensity and diversity of biotic interactions, including predation and parasitism (Schemske, 2009; Moya-Laraño, 2010; Fig. 8.5). Therefore, temperature and rainfall may have indirect effects on the benefits of parental care in terms of offspring protection, regardless of the parental sex (Fig. 8.5). In hot and humid climates, the benefits of offspring protection, for instance, should be higher than in cold and dry climates, because eggs and early hatched juveniles are expected to be under more intense threat from predators, parasitoids, and fungal attack. In fact, the importance of biotic interactions on the benefits of parental care in general was recognized long ago by Wilson (1971), who stated that intense predation on eggs by ants and also by conspecifics (in predatory species) as well as the high risk of fungal attack in tropical rain forests, may have been the major forces favoring the evolution of parental care in arthropods. Although this hypothesis does not provide an explanation for why parental care has evolved in some species and not in others, it may explain why this behavior is comparatively so frequent among tropical arthropods (see Chapter 5, and Table 8.1).

For the same reasons that hot and humid climates may increase the benefits of parental care in terms of offspring protection, the costs to parents in terms of mortality risks may also be affected by more frequent encounters with natural enemies (Schemske, 2009; Moya-Laraño, 2010; Fig. 8.5). If offspring protection makes parents more conspicuous, increasing their susceptibility to predator and parasitoid attacks, mortality rates of caring individuals (either males or females) should be higher in tropical climates. However, high temperatures accelerate embryonic development of arthropods (Zaslavski, 1988; Fig. 8.5),

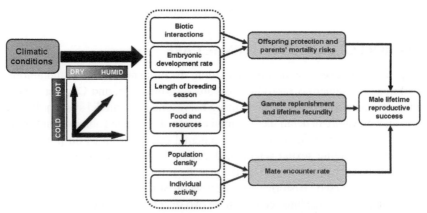

FIGURE 8.5 **Macroecology of paternal care.** The scheme depicts predictions for the influence of climatic conditions on several features of arthropod ecology and life history. Temperature, rainfall, or their combination, positively affect all ecological and life-history features, as depicted by the arrows connecting these features of the figure. Through each one of these features, climatic conditions may also indirectly influence the costs and benefits of paternal care, as well as the intensity and direction of sexual selection. See text for a detailed explanation of the causal links among variables.

thus decreasing the time parents are exposed to natural enemies. Short caring periods promoted by increases in temperature may also diminish the energetic/ foraging costs of parental behavior in species in which parents are prevented from seeking food while caring, or have their foraging activities severely constrained. Therefore, the net effect of climatic conditions on the costs of parental activities will depend on how biotic and abiotic conditions influence the rate of attacks suffered by parents and the rate of egg development (Fig. 8.5). Variations in these rates among species that occur in different climate types may partially explain the contradictory results for both mortality and energetic costs reported for different arthropod species exhibiting exclusive paternal care (see "Costs and Benefits of Paternal Care", above).

Due to the positive effect of temperature and rainfall on ecosystem productivity and resource availability (Hawkins *et al.*, 2003; Moya-Loraño, 2010; Fig. 8.5), individuals are likely to have access to more food in tropical climates. Therefore, abundant food in hot and humid climates may reduce the time parents need to recover their body condition after the caring period (especially among predatory species), and may also accelerate the rate of egg replenishing (Fig. 8.5). Particularly among species with exclusive paternal care, we predict that female lifetime fecundity and female PRR will be higher in hot and humid climates than in cold and dry climates (Fig. 8.5) because females are free to forage just after oviposition and produce additional eggs relatively faster (Tallamy, 2001). Moreover, if the negative effect of low temperature and little rainfall on female egg replenishing and female PRR is strong, the OSR may approach 1 : 1 or even become female-biased (Fig. 8.5). According to recent mathematical models for the evolution of parental care, these conditions would at least partly favor male care (see "Historical Perspective on Sex Roles and Parental Investment", above).

Temperature and rainfall may also influence the length of the breeding season (Fig. 8.5), so that in hot and humid climates it should be longer than in cold and dry climates. This pattern has already been supported by empirical data on harvestmen, and is explored in Chapter 5. Moreover, increases in temperature and rainfall are also known to positively influence individual activity patterns in arthropods (Chown and Nicolson, 2004). The combination of a short period for males and females to return to the mating pool (promoted by higher food supply, faster gamete replenishment, and shorter caring period), high population density, and intense activity of individuals in tropical climates should increase encounter rates between receptive adults (Fig. 8.5). Under these conditions, recent mathematical models predict that both males and females would exhibit some degree of mate choice (Bonduriansky, 2001; Manica and Johnstone, 2004; Barry and Kokko, 2010; Edward and Chapman, 2011). Moreover, since the benefits of paternal care in terms of offspring survival may be affected by climatic conditions (see above), females that discriminate among mates based on the quality of their care should enjoy higher fitness benefits in hot and humid climates than in dry and cold regions.

Finally, given that fecundity, egg replenishment, and mate encounter rate are all positively affected by temperature and rainfall (Fig. 8.5), another prediction we derive is that the number of eggs in broods of species with paternal care should increase at higher rates in hot and humid climates than in cold and dry climates (Fig. 8.5). This pattern may, however, depend on some specific features of the systems of interest. For example, such high rates might decrease through the breeding season if there is any limitation of the oviposition site to receive additional eggs, as is the case for Belostomatinae water bugs and sea spiders (Table 8.1). On the other hand, the rate of increase in clutch size should be relatively high and constant over time if there is no spatial limit to the number of eggs males can care for.

CONCLUDING REMARKS

In this chapter we have assembled scattered biological information for a wide set of arthropod species exhibiting exclusive paternal care, and contextualized the available data relative to the most recent theoretical frameworks for the evolution of sex roles and parental investment. Several important messages emerge from this synthesis. First, theoretical models for the evolution of parental care have been traditionally based on avian systems in which there is a strong trade-off between parental and mating efforts and bi-parental care is widespread. Among arthropods and fishes, however, exclusive paternal care is relatively common and the trade-off between parental and mating effort is considerably relaxed or even non-existent because males are able to care for many eggs and simultaneously have access to several mates (Table 8.1). This finding has profound theoretical implications because these trade-offs should be mathematically incorporated into the models instead of assumed *a priori*. Not surprisingly, recent attempts at modeling the evolution of sex roles and parental investment using arthropods and fishes have indeed adopted this approach (Manica and Johnstone, 2004; Alonzo, 2012).

The second important message is that basic information on the costs and benefits of paternal care in arthropods is extremely scarce and the results on the few well-studied species have only rarely been convergent. Whereas male care may benefit offspring, protecting the eggs from natural enemies and/or stressful abiotic conditions, the costs of egg-guarding markedly differ among species. Divergent results on the costs of male care can be partially explained by methodological differences among studies. Survival costs, in particular, were estimated both in the field and in the laboratory using either an experimental approach or mark–recapture protocols. We strongly suggest that future studies on this subject use a mark–recapture modeling approach because it dissociates survival from recapture through maximum likelihood techniques using the encounter histories of individuals under natural field conditions (Lebreton *et al.*, 1992). Regarding energetic costs, whenever possible researchers should avoid proxies of body condition correlated with structural body size, such as body mass. To remove the effect of body size and accurately assess body condition, estimates controlling for the effect of structural body size or based on body density should be taken into account because they increase the power to detect condition differences in terms of nutrient storage (Moya-Laraño *et al.*, 2008).

The third message is that exclusive paternal care does not necessarily lead to sex role reversal in arthropods. In fact, the available empirical data show unequivocal evidence of sex role reversal only in a few species. However, given the paucity of detailed behavioral studies of most species known to exhibit exclusive paternal care in arthropods, it would not be surprising if additional cases of sex role reversal were described as our knowledge increases. Future studies should pay attention to population parameters, such as density and sex ratios, because recent theoretical models have stressed their importance for the evolution of mate choice by males and females (Bonduriansky, 2001; Barry and Kokko, 2010; Edward and Chapman, 2011). Moreover, when focusing on species with exclusive paternal care, researchers should not neglect observations on female behavior and morphology. Field observations for at least one harvestman species suggest that females may exhibit alternative mating tactics (Mora, 1990), which seems an exciting and promising avenue of investigation. Detailed behavioral data coupled with paternity analyses in this and other species would be extremely welcome to better understand the role of sexual selection in the evolution and maintenance of paternal care and sex roles in arthropods.

Finally, the last message of this chapter is that environmental (climatic) conditions may play an important yet unexplored role affecting the costs and benefits of paternal care in arthropods. In the future, when more information on paternal care is available, most of the predictions presented in the section "Macroecology of Paternal Care" could be tested. There are two equally interesting ways to approach these predictions: (1) interspecific comparisons that take into account the phylogenetic relationships among species occurring in different climate types worldwide (see Chapter 5); and (2) interpopulation comparisons using individual species with wide geographic or altitudinal distribution and thus with populations occurring in different climate types. For both approaches, three arthropod groups exhibiting exclusive paternal care seem to be especially appropriate as model systems. The first two are the water bugs of the family Belostomatidae and millipedes of the genus *Brachycybe*, whose species are distributed worldwide in markedly different climatic conditions (Table 8.1), and thus could be used for interspecific comparisons. Moreover, some species of these groups are distributed over a wide area encompassing a great variety of climate types (Perez-Goodwyn, 2006; Brewer *et al.*, 2012), which makes them particularly suitable for interpopulation comparisons. The third group is the harvestmen, in which paternal care has evolved at least nine times independently (Table 8.1). There are phylogenies for many groups in the suborder Laniatores, which concentrates all cases of egg-guarding in Opiliones, and basic data on reproduction are accumulating fast (see Chapter 5). The group may therefore offer unique opportunities to test macroecological predictions using comparative methods.

ACKNOWLEDGMENTS

We are deeply grateful to the following colleagues for discussions and helpful comments on early versions of the manuscript: James Gilbert, Rogelio Macías-Ordóñez, Regina H. Macedo, Paulo Inácio Prado, and Leornardo Wedekin. We also thank the following researchers, who kindly helped by providing literature as well as behavioral and phylogenetic information about some taxa: Harry Brailovsky and Adalberto J. Santos (leaf bugs), Bernard Crespi (thrips), Shin-ichi Kudo (millipedes), and José Ricardo I. Ribeiro (water bugs). The composition of Figure 8.2 was only possible because photographs were generously provided by Jochen Martens, Rogelio Macías-Ordóñez, Lisa Thomas, Taís Nazareth, Shin-ichi Kudo, Carlos Víquez, and Osvaldo Villareal Manzanilla. GSR was supported by a fellowship from Fundação de Amparo à Pesquisa do Estado de São Paulo (FAPESP, 08/50466-8), GM has research grants from FAPESP (2012/50229-1) and Conselho Nacional de Desenvolvimento Científico e Tecnológico and RMS is supported by a postdoctoral fellowship from Dirección General de Asuntos de Personal Académico, Universidad Nacional Autónoma de México (DGAPA-UNAM).

REFERENCES

Alcock, J., 1994. Post-insemination associations between males and females in insects: the male-guarding hypothesis. Annu. Rev. Entomol. 39, 1–21.

Alonzo, S.H., 2012. Sexual selection favours male parental care, when females can choose. Proc. R. Soc. B 279, 1784–1790.

Bain, B.A., Govedich, F.R., 2004. Mating behavior, female aggression, and infanticide in *Propallene saengeri* (Pycnogonida: Callipallenidae). Vic. Nat. 121, 168–171.

Barreto, F.S., Avise, J.C., 2008. Polyandry and sexual size dimorphism in the sea spider *Ammothea hilgendorfi* (Pycnogonida: Ammotheidae), a marine arthropod with brood carrying males. Mol. Biol. 17, 4164–4175.

Barreto, F.S., Avise, J.C., 2011. The genetic mating system of a sea spider with male-biased sex size dimorphism: evidence for paternity skew despite random mating success. Behav. Ecol. Sociobiol. 65, 1595–41604.

Barry, K.L., Kokko, H., 2010. Male mate choice: why sequential choice can make its evolution difficult. Anim. Behav. 80, 163–4169.

Bateman, P.W., Fleming, P.A., 2006. Males are selective too: mating, but not courtship, with sequential females influences choosiness in male field crickets (*Gryllus bimaculatus*). Behav. Ecol. Sociobiol. 59, 577–581.

Bonduriansky, R., 2001. The evolution of male mate choice in insects: a synthesis of ideas and evidence. Biol. Rev. 76, 305–339.

Brailovsky, H., 1989. Nuevos arreglos tribales dentro de la familia Coreidae y descripción de dos especies nuevas sudamericanas (Hemiptera–Heteroptera). Anales Inst. Biol. Univ. Nal. Autón. México 59, 159–180.

Brewer, M.S., Spruill, C.L., Rao, N.S., Bond, J.E., 2012. Phylogenetics of the millipede genus *Brachycybe* Wood, 1864 (Diplopoda: Platydesmida: Andrognathidae): Patterns of deep evolutionary history and recent speciation. Mol. Phylogenet. Evol. 64, 232–242.

Burris, Z.P., 2010. Costs of exclusive male parental care in the sea spider *Achelia simplissima* (Arthropoda: Pycnogonida). Mar. Biol. 158, 381–390.

Burris, Z.P., 2011. The polygamous mating system of the sea spider *Achelia simplissima*. Inv. Rep. Dev. 55, 162–167.

Chown, S.L., Nicolson, S., 2004. Insect Physiological Ecology: Mechanisms and Patterns. Oxford University Press, Oxford.

Clutton-Brock, T.H., 1991. The Evolution of Parental Care. Princeton University Press, Princeton.

Clutton-Brock, T., Parker, G.A., 1992. Potential reproductive rates and the operation of sexual selection. Q. Rev. Biol. 67, 437–456.

Clutton-Brock, T., Vincent, A.A.A., 1991. Sexual selection and the potential reproductive rates of males and females. Nature 351, 58–60.

Cole, L.J., 1901. Notes on the habits of pycnogonids. Biol. Bull. 2, 195–207.

Coleman, S.T., Jones, A.G., 2011. Patterns of multiple paternity and maternity in fishes. Biol. J. Linn. Soc. 103, 735–760.

Crespi, B.J., 1986. Territoriality and fighting in a colonial thrips, *Hoplothrips pedicularius*, and sexual dimorphism in Thysanoptera. Ecol. Entomol. 11, 119–130.

Crespi, B.J., 1988. Risks and benefits of lethal male fighting in the polygynous, colonial thrips *Hoplothrips karnyi*. Behav. Ecol. Sociobiol. 22, 293–301.

Crowl, T.A., Alexander, J.E., 1989. Parental care and foraging ability in male water bugs (*Belostoma flumineum*). Can. J. Zool. 67, 513–515.

Dawkins, R., Carlisle, T.R., 1976. Parental investment and mate desertion: a fallacy. Nature 262, 131–133.

Dimmock, G., 1887. Belostomatidae and other fish destroying bugs. Zoologist 11, 101–105.

Edward, D.A., Chapman, T., 2011. The evolution and significance of male mate choice. Trends. Ecol. Evol. 26, 647–654.

Forster, R.R., 1954. The New Zealand harvestmen (sub-order Laniatores). Canterbury Mus. Bull. 2, 1–329.

Gardner, M.R., 1974. Revision of the millipede family Andrognathidae in the Nearctic region. Mem. Pac. Coast Entomol. Soc. 5, 1–61.

Gilbert, J.D.J., Thomas, L.K., Manica, A., 2010. Quantifying the benefits and costs of parental care in assassin bugs. Ecol. Entomol. 35, 639–651.

Gilg, M.R., Kruse, K., 2003. Reproduction decrease life span in the giant waterbug (*Belostoma flumineum*). Am. Mid. Nat. 149, 306–319.

Gross, M.R., Sargent, R.C., 1985. The evolution of male and female parental care in fishes. Am. Zool. 25, 807–822.

Gross, M.R., Shine, R., 1981. Parental care and mode of fertilization in ectothermic vertebrates. Evolution 35, 775–793.

Hammond, P., Aguirre-Hudson, B., Dadd, M., Groombridge, B., Hodges, J., Jenkins, M., Mengesha, M.H., Stewart Grant, W., 1995. The current magnitude of biodiversity. In: Heywood, V.H. (Ed.), Global Biodiversity Assessment, Cambridge University Press, Cambridge, pp. 113–138.

Hara, M.R., Gnaspini, P., Machado, G., 2003. Male guarding behavior in the neotropical harvestman *Ampheres leucopheus* (Mello-Leitão, 1922) (Opiliones, Laniatores, Gonyleptidae). J. Arachnol. 31, 441–444.

Hawkins, B.A., Field, R., Cornell, H.V., Currie, D.J., Guegan, J.F., Kaufman, D.M., Kerr, J.T., Mittelbach, G.G., Oberdorff, T., O'Brien, E.M., Porter, E.E., Turner, J.R.G., 2003. Energy, water, and broad-scale geographic patterns of species richness. Ecology 84, 3105–3117.

Hoelzer, G.A., 1989. The good parent process of sexual selection. Anim. Behav. 38, 1067–1078.

Hoffman, W.E., 1933. Additional data on the life history of *Lethocerus indicus* (Hemiptera, Belostomatidae). Ling. Sci. J. 12, 595–601.

Hooper, J.N.A., 1980. Some aspects of the reproductive biology of *Parapallene avida* Stock (Pycnogonida: Callipallenidae) from northern New South Wales. Aust. Zool. 20, 473–483.

Hostache, G., Mol, J.H., 1998. Reproductive biology of the neotropical armored catfish *Hoplosternum littorale* (Siluriformes: Callichthyidae): a synthesis stressing the role of the floating bubble nest. Aquat. Living Resour. 11, 173–185.

Ichikawa, N., 1989. Breeding strategy of the male brooding water bug, *Diplonychus major* Esaki (Heteroptera: Belostomatidae): is male back space limiting? J. Ethol. 7, 133–140.

Ichikawa, N., 1990. Egg mass destroying behavior of the female giant water bug *Lethocerus deyrollei* Vuillefroy (Heteroptera: Belostomatidae). J. Ethol. 8, 5–11.

Inada, K., Kitade, O., Morino, H., 2011. Paternity analysis in an egg-carrying aquatic insect *Appasus major* (Hemiptera: Belostomatidae) using microsatellite DNA markers. Entomol. Sci. 14, 43–48.

Kaestner, A., 1968. Invertebrate Zoology – Volume II: Arthropod Relatives, Chelicerata, Myriapoda (trans. and adapted by Levi, H. W., and Levi, L. R.). Wiley, New York.

Kaitala, A., Härdling, R., Katvala, M., Macías-Ordóñez, R., Miettinen, M., 2001. Is nonparental egg carrying parental care? Behav. Ecol. 12, 367–373.

Kight, S.L., Kruse, K., 1992. Factors affecting the allocation of paternal care in waterbugs (*Belostoma flumineum* Say). Behav. Ecol. Sociobiol. 30, 409–414.

Kight, S.L., Sprague, J., Kruse, K., Johnson, L., 1995. Are egg-bearing male waterbugs, *Belostoma flumineum* Say (Hemiptera: Belostomatidae), impaired swimmers? J. Kans. Entomol. Soc. 68, 486–470.

Kight, S.L., Batino, M., Zhang, Z., 2000. Temperature-dependent parental investment in the giant waterbug *Belostoma flumineum* (Heteroptera: Belostomatidae). Ann. Entomol. Soc. Am. 93, 340–342.

Kight, S.L., Tanner, A.W., Coffey, G.L., 2011. Termination of parental care in male giant waterbugs, *Belostoma flumineum* Say (Heteroptera: Belostomatidae) is associated with breeding season, egg pad size, and presence of females. Inv. Rep. Dev. 55, 197–204.

King, P.E., 1973. Pycnogonids. St Martin's Press, New York.

Kokko, H., Jennions, M.D., 2008. Parental investment, sexual selection and sex ratios. J. Evol. Biol. 21, 919–948.

Krause, W.F., 1989. Is male backspace limiting? An investigation into the reproductive demography of the giant water bug, *Abedus indentatus* (Heteroptera: Belostomatidae). J. Insect Behav. 2, 623–648.

Kudo, S., Koshio, C., Tanabe, T., 2009. Male egg-brooding in the millipede *Yamasinaium noduligerum* (Diplopoda: Andrognathidae). Entomol. Sci. 12, 346–347.

Kudo, S., Akagi, Y., Hiraoka, S., Tanabe, T., Morimoto, G., 2011. Exclusive male egg care and determinants of brooding success in a millipede. Ethology 117, 19–27.

Lanzer-de-Souza, M.E., 1980. Inventário da distribuição geográfica da família Belostomatidae Leach, 1815, (Hemiptera – Heteroptera) na região Neotropical. Iheringia 55, 43–86.

Lebreton, J.D., Burnham, K.P., Clobert, J., Anderson, D.R., 1992. Modeling survival and testing biological hypotheses using marked animals: a unified approach with case studies. Ecol. Monograph. 62, 67–118.

León, T.M., 1999. Evolución y filogenia de los picnogónidos. Bol SEA 26, 273–279.

Lima, A.C., 1940. Insetos do Brasil, Tomo 2, Hemipteros. Escola Nacional de Agronomia, Rio de Janeiro.

Machado, G., 2007. Maternal or paternal egg guarding? Revisiting parental care in trianonychid harvestmen (Opiliones). J. Arachnol. 35, 202–204.

Machado, G., Macías-Ordóñez, R., 2007. Reproduction. In: Pinto-da-Rocha, R., Machado, G., Giribet, G. (Eds.), Harvestmen: the Biology of Opiliones, Harvard University Press, Cambridge, pp. 414–454.

Machado, G., Requena, G.S., Buzatto, B.A., Osses, F., Rossetto, L.M., 2004. Five new cases of paternal care in harvestmen (Arachnida: Opiliones): implications for the evolution of male guarding in the neotropical family Gonyleptidae. Sociobiology 44, 577–598.

Magrath, M.J.L., Komdeur, J., 2003. Is male care compromised by additional mating opportunity? Trends Ecol. Evol. 18, 424–430.

Manica, A., Johnstone, R.A., 2004. The evolution of paternal care with overlapping broods. Am. Nat. 164, 517–530.

Martens, J., 1993. Further cases of paternal care in Opiliones (Arachnida). Trop. Zool. 6, 97–107.

Maynard Smith, J., 1977. Parental investment: a prospective analysis. Anim. Behav. 25, 1–9.

McNamara, J.M., Fromhage, L., Barta, Z., Houston, A.I., 2010. The optimal coyness game. Proc. R. Soc. Lond. B. 276, 953–960.

Mora, G., 1990. Parental care in a neotropical harvestman, Zygopachylus albomarginis (Arachnida: Gonyleptidae). Anim. Behav. 39, 582–593.

Morgado, B.M., Monteiro, R.F., 2012. Parasitismo em Plunentis porosus (Hemiptera: Coreidae), uma espécie de percevejo com cuidado parental. I Workshop do Instituto Nacional de Ciência e Tecnologia dos Hymenoptera Parasitóides da Região Sudeste Brasileira. São Carlos, São Paulo, Brazil.

Moya-Laraño, J., 2010. Can temperature and water availability contribute to the maintenance of latitudinal diversity by increasing the rate of biotic interactions? Open Ecol. J. 3, 1–13.

Moya-Laraño, J., Macías-Ordóñez, R., Blanckenhorn, W.U., Fernández-Montraveta, C., 2008. Analysing body condition: mass, volume or density? J. Anim. Ecol. 77, 1099–1108.

Munguía-Steyer, R., Macías-Ordóñez, R., 2007. Is it risky to be a father? Survival assessment depending on sex and parental status in the waterbug Abedus breviceps (Hemiptera: Belostomatidae) using multistate modeling. Can. J. Zool. 85, 49–55.

Munguía-Steyer, R., Favila, M.E., Macías-Ordóñez, R., 2008. Brood pumping modulation and the benefits of paternal care in Abedus breviceps (Hemiptera: Belostomatidae). Ethology 114, 693–700.

Murakami, Y., 1962. Postembryonic development of the common Myriopoda of Japan. XI. Life history of Bazillozonium nodulosum (Corobognatha, Platydesmidae). Zool. Mag. 71, 250–255.

Nakamura, K., Sekiguchi, K., 1980. Mating behavior and oviposition in the pycnogonid Propallene longiceps. Mar. Ecol. Prog. Ser. 2, 163–168.

Nazareth, T.M., Machado, G., 2009. Reproductive behavior of Chavesincola inexpectabilis (Opiliones: Gonyleptidae), with the description of a new and independently evolved case of paternal care in harvestman. J. Arachnol. 37, 127–134.

Nazareth, T.M., Machado, G., 2010. Mating system and exclusive postzygotic paternal care in a neotropical harvestman (Arachnida: Opiliones). Anim. Behav. 79, 547–554.

Nyiira, Z.M., 1970. The biology and behaviour of Rhinocoris albopunctatus. Ann. Entomol. Soc. Am. 63, 1224–1227.

Odhiambo, T.R., 1959. An account of parental care in Rhinocoris albopilosus Signoret (Hemiptera – Heteroptera: Reduviidae) with notes on its life history. Proc. R. Entomol. Soc. Lond. Ser. A. 34, 175–187.

Östlund, S., Ahnesjö, I., 1998. Female fifteen-spined sticklebacks prefer better fathers. Anim. Behav. 56, 1177–1183.

Pampoulie, C., Lindström, K., St Mary, C.M., 2004. Have your cake and eat it too: male sand gobies show more parental care in the presence of female partners. Behav. Ecol. 15, 199–204.

Panizzi, A.R., Santos, C.H., 2001. Unusual oviposition on the body of conspecifics by phytophagous heteropterans. Neotrop. Entomol. 30, 471–472.

Parker, G.A., Simmons, L.W., 1996. Parental investment and the control of sexual selection: predicting the direction of sexual competition. Proc. R. Soc. B 263, 315–321.

Perez-Goodwyn, P.J., 2006. Taxonomic revision of the subfamily Lethocerinae Lauck & Menke (Heteroptera: Belostomatidae). Stuttgarter. Beitr. Naturkd. Ser. A. 695, 1–71.

Polhemus, J.T., Polhemus, D.A., 2008. Global diversity of true bugs (Heteroptera; Insecta) in freshwater. Hydrobiologia 595, 379–391.

Porter, B.A., Fiumera, A.C., Avise, J.C., 2002. Egg mimicry and allopaternal care: two mate-attracting tactics by which nesting striped darter (*Etheostoma virgatum*) males enhance reproductive success. Behav. Ecol. Sociobiol. 51, 350–359.

Proud, D.N., Víquez, C., Townsend Jr., V.R., 2011. Paternal care in a neotropical harvestman (Opiliones: Cosmetidae) from Costa Rica. J. Arachnol. 39, 497–499.

Queller, D.C., 1997. Why do females care more than males? Proc. R. Soc. B 264, 1555–1557.

Requena, G.S., Buzatto, B.A., Munguía-Steyer, R., Machado, G., 2009. Efficiency of uniparental male and female care against egg predators in two closely related syntopic harvestmen. Anim. Behav. 78, 1169–1176.

Requena, G.S., Nazareth, T.M., Schwertner, C.F., Machado, G., 2010. First cases of exclusive paternal care in stink bugs (Heteroptera: Pentatomidae). Zoologia 27, 1018–1021.

Requena, G.S., Buzatto, B.A., Martins, E.G., Machado, G., 2012. Paternal care decreases foraging activity and body condition, but does not impose survival costs to caring males in a neotropical arachnid. PLoS One 7, e46701.

Ridley, M., 1978. Paternal care. Anim. Behav. 26, 904–932.

Rodriguez, C.A., Guerrero, S., 1976. La historia natural y el comportamiento de *Zygopachylus albomarginis* (Chamberlin) (Arachnida, Opiliones: Gonyleptidae). Biotropica 8, 242–247.

Sars, G.O., 1891. Pycnogonidea, Norwegian North-Atlantic expedition (1876–1878). Zoology 6, 1–163.

Schemske, D.W., Mittelbach, G.G., Cornell, H.V., Sobel, J.M., Roy, K., 2009. Is there a latitudinal gradient in the importance of biotic interactions? Annu. Rev. Ecol. Syst. 40, 245–269.

Slater, F.W., 1899. The egg-carrying habit of *Zaitha*. Am. Nat. 33, 931–933.

Smith, R.L., 1979a. Repeated copulation and sperm precedence: paternity assurance for a male brooding water bug. Science 205, 1029–1031.

Smith, R.L., 1979b. Paternity assurance and altered roles in the mating behaviour of a giant water bug, *Abedus herberti* (Heteroptera: Belostomatidae). Anim. Behav. 27, 716–725.

Smith, R.L., 1997. Evolution of paternal care in the giant water bugs (Heteroptera: Belostomatidae). In: Crespi, B.J., Choe, J.C. (Eds.), The Evolution of Social Behavior in Insects and Arachnids, Cambridge University Press, Cambridge, pp. 116–149.

Stefanini-Jim, R.L., Soares, H.E.M., Jim, J., 1987. Notas sobre a biologia de *Cadeadoius niger* (Mello-Leitão, 1935) (Opiliones, Gonyleptidae, Progonyleptoidellinae). Anais do XX Encontro Brasileiro de Etologia Botucatu, São Paulo, Brazil.

Stiver, K.A., Alonzo, S.H., 2009. Parental and mating effort: is there necessarily a trade-off? Ethology 115, 1101–1126.

Tallamy, D.W., 1994. Nourishment and the evolution of paternal care in subsocial arthropods. In: Hunt, J.H., Nalepa, C.A. (Eds.), Nourishment and Evolution in Insect Societies, Westview Press, Boulder, pp. 21–55.

Tallamy, D.W., 2001. Evolution of exclusive paternal care in arthropods. Annu. Rev. Entomol. 46, 139–165.

Tawfik, M.F.S., 1969. The life history of the giant water-bug *Lethocerus niloticus* Stael (Hemiptera: Belostomatidae). Bull. Soc. Entomol. Egypte 53, 299–310.

Thomas, L., 1994. The evolution of paternal care in assassin bugs. PhD thesis, University of Cambridge, Cambridge.

Thomas, L.K., Manica, A., 2003. Filial cannibalism in an assassin bug. Anim. Behav. 66, 205–210.

Thomas, L.K., Manica, A., 2005. Intrasexual competition and mate choice in assassin bugs with uniparental male and female care. Anim. Behav. 69, 275–281.

Trivers, R.L., 1972. Parental investment and sexual selection. In: Campbell, B. (Ed.), Sexual Selection and the Descent of Man, Aldine Press, Chicago, pp. 136–179.

Villarreal-Manzanilla, O., Machado, G., 2011. First record of paternal care in the family Stygnidae (Opiliones: Laniatores). J. Arachnol. 39, 500–502.

Wagner Jr., W.E., 2011. Direct benefits and the evolution of female mating preferences: conceptual problems, potential solutions, and a field cricket. Adv. Stud. Behav. 43, 273–319.

Wilhelm, E., Buckmann, D., Tomaschko, K.H., 1997. Life cycle and population dynamics of *Pycnogonum litorale* (Pycnogonida) in a natural habitat. Mar. Biol. 129, 601–606.

Williams, G.C., 1975. Sex and Evolution. Princeton University Press, Princeton.

Wilson, E.O., 1971. The Insect Societies. Harvard University Press, Cambridge.

Zaslavski, V.A., 1988. Insect Development, Photoperiodic and Temperature Control. Springer, New York.

Chapter 9

Underestimating the Role of Female Preference and Sexual Conflict in the Evolution of ARTs in Fishes: Insights from a Clade of Neotropical Fishes

Molly R. Morris[1] and Oscar Ríos-Cardenas[2]

[1]*Department of Biological Sciences, Ohio University, Athens, Ohio, USA,* [2]*Departamento de Biología Evolutiva, Instituto de Ecología, A.C., Xalapa, Veracruz, Mexico*

INTRODUCTION

Alternative Reproductive Tactics (ARTs) refer to alternative ways of obtaining fertilizations by same-sex individuals (within a species). We examine how the initial description of ARTs in fishes with external fertilization and parental care may have impacted our current understanding of their evolution. While both intra- and intersexual selection are considered important in the evolution of traits that make up the bourgeois tactic, the parasitic tactic is thought to have evolved mainly under the influence of male–male competition. To what extent has the importance of circumventing female mate preference and sexual conflict in the evolution of ARTs been overlooked? To what extent does one of the tactics (i.e., parasitic) rely on the presence of the other (i.e., bourgeois) to gain reproductive success? Has the bourgeois-parasitic paradigm led to the assumption that the bourgeois tactic is the ancestral state? We consider these questions, using a clade of neotropical live-bearing fishes (*Xiphophorus*) as an example. More inclusive models of the evolution of ARTs that consider the full roles of female mate preference and sexual conflict will help reveal new types of ARTs as well as a better understanding of the evolution of the common ARTs. The extent to which individuals with different ARTs are dependent on one another to reproduce will determine the extent to which ARTs may have played a role in speciation. Finally, we argue that a more complete and less biased data set is necessary to conduct phylogenetic analyses of ancestral states.

Sexual Selection. http://dx.doi.org/10.1016/B978-0-12-416028-6.00009-8

Our understanding of the causes and consequences of variation within indi-viduals of the same sex in relation to reproduction has grown extensively over the past 30 years. ARTs are the discrete variation in reproductive morphology and/or behaviors within a sex, and are quite common in teleost fishes (for review, see Knapp and Neff, 2008; Oliveira *et al.*, 2008). In one of the first descriptions of variation in mating behaviors (Perrill *et al.*, 1978), it was shown that smaller male frogs sitting next to calling males were not just waiting for the territory to be vacated, but would attempt to intercept females attracted to the calling males. The behavior of these smaller males was described as "sexual parasit-ism", as the smaller males were taking advantage of the calling males' ability to attract females. Taborsky (1994, 1998) proposed the use of bourgeois and parasitic as unifying functional terms to describe alternative reproductive tac-tics, and today the bourgeois–parasitic paradigm is in widespread use. The bour-geois–parasitic paradigm is based on the concept that animals gain access to a resource either directly (bourgeois males) or by exploiting the investment of oth-ers (parasitic males). As such, the bourgeois males compete for monopolization of females, investing in a suite of traits that function in defense and courtship, while the parasitic males exploit this investment and invest in sperm competition (e.g., spermatogenesis and testicular mass) (Taborsky, 1998).

In considering the type of selection that drives the evolution of ARTs, there appears to be an overall agreement that ARTs are more likely to evolve when differential reproduction among members of the same sex is high, or sexual selec-tion is strong (Shuster and Wade, 2003; Taborsky *et al.*, 2008). And yet, while the bourgeois–parasitic paradigm identifies both male–male competition and female mate preference as being important in the evolution of the bourgeois tactic, only male–male competition is considered in the selection acting on the parasitic tactic (Taborsky *et al.*, 2008). Fertilization is external in most fishes, and therefore the bourgeois males not only attract females to an area but also coax the females to lay eggs, thereby making it possible for the competitively inferior parasitic males to gain access to the eggs (Taborsky, 1998). The paradigm has been expanded, how-ever, to apply to organisms with internal fertilization, with male–male competition moving into the realm of sperm competition (Taborsky, 1998), and the parasitic males exploiting the bourgeois males' investment in sexually selected traits that attract females (Oliveira *et al.*, 2008; Butts *et al.*, 2011). The only context in which female preference is acknowledged to produce disruptive selection within this paradigm is in the case of polymorphisms within the females (Taborsky, 2008), which select for polymorphisms in the males (e.g., Kawase and Nakazono, 1996).

We argue below why it is important to fully incorporate selection due to female mate preference and sexual conflict in considering the evolution of ARTs. We have identified three assumptions associated with this paradigm that need to be addressed: (1) the disruptive selection acting on ARTs comes primar-ily from male–male competition, not female preference; (2) the parasitic tactic is dependent on the bourgeois tactic to reproduce; (3) the bourgeois tactic is the ancestral state. We describe some of the consequences associated with each assumption, and then evaluate them in light of the available evidence in a clade of live-bearing fishes with internal fertilization.

CONSEQUENCES OF NOT CONSIDERING THE ROLE OF FEMALE MATE PREFERENCE IN THE BOURGEOIS–PARASITIC PARADIGM AND ASSOCIATED ASSUMPTIONS

It may be useful to consider females as a resource, with males competing either directly or indirectly to gain access to females. However, unlike abiotic resources, females can also evolve under the influences of sexual selection and sexual conflict with males, and therefore it is important to consider the dynamic nature of this evolutionary relationship. Sexual conflict, which occurs when the genetic interests of males and females diverge (Chapman *et al.*, 2003), is an important mechanism that can drive the evolution of alternative male mating tactics (Westneat and Sih, 2009). Traits will evolve that allow unpreferred males to gain matings by circumventing female choosiness through sneaking and forced matings, and the arms race between males and females over reproduction will select for males with yet more extreme tactics of not only attracting females but also circumventing female preferences. Even though it may be difficult to separate selection due to female mate choice and male–male competition (Murphy, 1998), the influence of these different components of sexual selection can be quite different and therefore worth considering. For example, morphological and behavioral traits that allow males to avoid male–male competition may or may not be the same traits that evolve to circumvent female mate preference (see, for example, Morris *et al.*, 2008; Ríos-Cardenas *et al.*, 2010). In addition, without consideration of female mate preference, some of the different types of ARTs, as well as female mate preference for parasitic males, could be overlooked.

When considering selection on ARTs, we suggest use of a two-axes selection space (Fig. 9.1): ARTs can fall along a continuum from "coaxing" to "coercing"

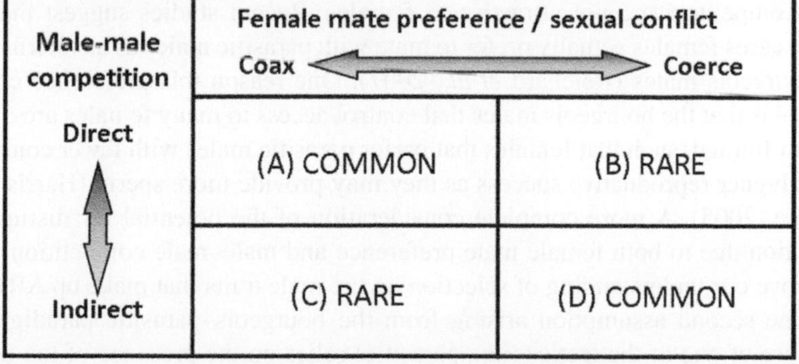

Male-male competition	Female mate preference / sexual conflict	
	Coax ⟵⟶	Coerce
Direct ↑	(A) COMMON	(B) RARE
↓ **Indirect**	(C) RARE	(D) COMMON

FIGURE 9.1   **Both male–male competition and female mate preference can produce disruptive selection in males resulting in the evolution of ARTs.** While the ARTs in most species are likely to fall into selection space (A) and (D), exceptions are expected.

in relation to female choice and sexual conflict, and along a continuum from "direct" male–male competition (e.g., guarding females, territories, fighting with competitors) to "indirect" male–male competition (e.g., sneaking into dominant males' territories, producing more sperm in sperm competition). We expect that most ARTs will fall into one of two corners of the selection space: A, direct competition and coaxing, or D, indirect competition and coercing. We make this prediction because traits that allow a male to avoid male–male competition are often useful in circumventing female mate preference as well. Even in the case of three tactics, such as the rock–paper–scissors example in the side-blotched lizards *Uta stansburiana*, two of the three tactics fall into space A, and the third into space D (Corl *et al.*, 2010). The orange-throated and blue-throated males use direct competition and coax females, thus falling into selection space "A", with the orange throats occupying a selection space slightly higher on the competition axis than the blue throats, and the blue throats slightly higher than the orange throats on the coaxing axis. The yellow-throated males in this system use indirect competition and circumvent female mate preference (coercing), and would therefore occupy selection space "D" (Corl *et al.*, 2010). However, exceptions to tactics falling into one of these two areas of the selection space are expected. One possibility is that traits used to dominate males (direct competition) are also used to dominate females (coercing), leading to ARTs in selection space "B" (Fig. 9.1). In addition, traits used to avoid male–male competition could make it possible for males to invest more in attracting females (selection space "C", Fig. 9.1). Finally, traits that allow males to avoid male–male competition may not necessarily require males to circumvent female mate preference as well. The male ART identified as "floaters" in the mouth-brooding cichlid *Ophthalmotilapia ventralis* (Haesler *et al.*, 2009) sneak into in the territory of the dominant males (indirect male–male competition), but once on the territory will court the females (coaxing; Haesler *et al.*, 2009). Therefore, floaters in this system would fall into selection space "C" in Figure 9.1.

In addition to assuming that indirect competitors (parasites) will always be coercing and never coaxing females to mate, it is also assumed that these indirect competitors are not attractive to females. Recent studies suggest that in some cases females actually prefer to mate with parasitic males as an alternative to bourgeois males (Reichard *et al.*, 2007). One reason this preference could evolve is that the bourgeois males that control access to many females are often sperm-limited, such that females that prefer parasitic males with fewer consorts have higher reproductive success as they may provide more sperm (Harris and Moore, 2005). A more complete consideration of the potential for disruptive selection due to both female mate preference and male–male competition will improve our understanding of selection on the male traits that make up ARTs.

The second assumption arising from the bourgeois–parasitic paradigm is that use of one of the tactics (i.e., parasitic) relies on the presence of the other (i.e., bourgeois) to gain reproductive success. Oliveira *et al.* (2001: 267) state that "In any scenario, without the initial bourgeois investment the alternative tactic would not be functional". This is an important consideration in light of the role of ARTs in speciation. Alternative phenotypes that are "dependent" on one another to reproduce (queens and workers of social insects, males and females of the same species) are not as likely to lead to speciation as "independent" phenotypes (West-Eberhard, 1989). Therefore, the extent to which ARTs are independent needs to be addressed. The third assumption often follows from the second. By assuming that without the initial bourgeois investment the alternative tactic would not be functional (Oliveira *et al.*, 2008), it is clear how one might also assume that the bourgeois tactic is the ancestral state. The question of ancestral state has been addressed in a phylogenetic comparative analysis of the evolution of ARTs in teleost fishes (Mank and Avise, 2006). However, this study highlights the influence of the bourgeois–parasitic paradigm on the interpretation of the ancestral states in ARTs. In the numerous cases where the reconstruction of ancestral states was equivocal, Mank and Avise (2006) cite Taborsky (1994, 2001) when inferring that the ancestral state was mate monopolization in each of these cases (i.e., bourgeois is assumed to be the "requisite ancestral state"; Mank and Avise, 2006: 1312).

Below we address these assumptions, primarily within the swordtail fishes (*Xiphophorus*). This genus is restricted to the Neotropics, with various native species distributed in Belize, Guatemala, Honduras, and especially in Mexico. We consider the evidence that there is selection on the parasitic ARTs to circumvent female mate preference, and that the parasitic ART is dependent on the bourgeois to reproduce, and discuss the data that will be necessary to examine the ancestral state of ARTs in this group in a phylogenetic context. Given the theme of this book, a comparison of the ARTs between species in the Neotropics and the temperate regions would be fitting. However, as we will discuss in the following sections, due to the lack of reliable information regarding the mating system of swordtails, in addition to the larger more widely distributed Poeciliinae clade (Parenti and Rauchenberger, 1989) to which swordtail fishes belong, this approach would need to be applied in future studies.

ARTS IN THE SWORDTAIL FISHES: AN EXAMPLE

Swordtails belong to the larger Poeciliinae clade (Parenti and Rauchenberger, 1989), consisting of approximately 200 species (Evans *et al.*, 2011) with internal fertilization, and we will refer to the members of this clade as poeciliids from here on. The natural distribution of the clade is fresh, brackish, and salt waters of the New World temperate and tropical zones, ranging from the northeastern United States to the Río de la Plata estuary in northern Argentina (Rosen, 1973). Males have a modified anal fin (gonopodium) that is used as an intromittent organ to transfer sperm to the females, which give birth to live young. A male size polymorphism was first detected in *X. hellerii* (Peters, 1964) with early-maturing males remaining small, as male growth strongly declines at sexual maturity in most poeciliid fish, with late-maturing males growing to a larger adult size (Kallman, 1989). Of the 22 species of *Xiphophorus* examined by Kallman (1989), 10 were shown to have a size polymorphism that is genetically influenced. Genetic polymorphisms for male size have been detected in other poeciliid fish (*Poecilia*, *Limia perugiae*; Erbelding-Denk *et al.*, 1994), suggesting that this could be the ancestral state for *Xiphophorus* (Kallman, 1989). The size polymorphism is an important component of the alternative life-history strategies that make up the ARTs in these fishes (Franck *et al.*, 2003).

Two of the species that have been studied extensively in this group in relation to ARTs are *Xiphophorus nigrensis* (Ryan and Causey, 1989; Ryan *et al.*, 1990, 1992; Zimmerer and Kallman, 1989) and *Xiphophorus multilineatus* (Luo *et al.*, 2005; Ríos-Cardenas *et al.*, 2007; Morris *et al.*, 2010; Bono *et al.*, 2011). Both species have genetically-influenced ARTs associated with early- and late-maturing males. *Xiphophorus nigrensis* is endemic to the Río Choy and *X. multilineatus* to the Río Coy and its tributaries in the State of San Luis Potosi, northeast Mexico. In *X. multilineatus* there are four genetically-influenced size classes of males (Y-s, Y-I, Y-II, and Y-L; see Fig. 9.2). Based on a combination of male adult size and associated pigment patterns (Kallman, 1989; Zimmerer and Kallman, 1989), it is possible to identify these size classes. The "parasitic" ART in this species are the small Y-s "sneaker" males. As the smallest size class (Y-s), these males use a sneak-chase behavior in addition to courtship to acquire fertilizations, mature at the same size as females, and have reduced or no vertical bars. Counter males are the "bourgeois" ART, and correspond to the three larger-size classes of males that only use courtship behavior. The Y-I and Y-L size classes both have yellow caudal margins but do not overlap in size. Males in the Y-II size class overlap in size with both the smaller Y-I and the larger Y-L males, but their entire caudal fin is yellow (Zimmerer and Kallman, 1989). The male size polymorphism in *X. multilineatus* has recently been attributed to variation in the copy number of the melanocortin-4 receptor (Mc4r) gene on the Y chromosome, which is involved in the regulation of body weight and appetite (Lampert *et al.*, 2010).

In species with genetically-influenced ARTs, theory suggests that tactics should have equal average fitness maintained by negative frequency-dependent selection. In both *X. nigrensis* and *X. multilineatus* sneakers and courters seem to have equal average fitness, as the mating advantage for the courter males is countered by a higher probability of reaching sexual maturity for the sneakers, which reach sexual maturity sooner (Ryan *et al.*, 1992; Bono *et al.*, 2011). Current studies with *X. multilineatus* are evaluating the role of negative frequency dependence as a mechanism to maintain this polymorphism.

Evidence that Circumventing Female Mate Preference is Important for "Parasitic" Males

Xiphophorus multilineatus females have an overall significant preference for the larger "courter" males as compared to the smaller "sneaker" males; however, the strength of this preference is positively related to female size, with smaller

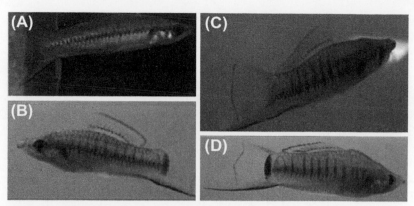

FIGURE 9.2   Males from the four genetically influenced size classes of *Xiphophorus multilineatus*. (A) The smallest size class males (Y-s) are reversible in their use of mating behaviors (parasitic tactic), using both sneak-chase behavior and courtship depending on social context. The three larger size classes ((B) Y-I, (C) Y-II, and (D) Y-L) are irreversible in their use of courtship behavior (bourgeois tactic). See color plate at the back of the book. *Photographs courtesy of K. de Queiroz.*

females having a weaker preference (Ríos-Cardenas *et al.*, 2007). Further studies have shown that in a wild population the relationship between female size and strength of preference results in smaller females being more likely to have mated with sneakers, and larger females with courters (Morris *et al.*, 2010), and in laboratory mesocosms and the field the frequency of "sneakers" to "courters" influences the strength of this preference (Tudor and Morris, 2011). Therefore, the strength of preference for courter males will vary across populations, and the use of association time in the laboratory is a good indicator of the mate choices females make in the field.

There are several morphological differences between the sneaker and courter males that have been shown to be influenced by female mate preference in *X. multilineatus* specifically (e.g., body size, Morris *et al.*, 1995a; vertical bars, Morris *et al.*, 2007) or in the swordtails in general (e.g., sword length, Basolo, 1990). The question here, however, is the extent to which selection due to female mate choice has not only driven courter males to exaggerate these traits to attract females, but also driven sneaker males to reduce these traits to circumvent mate preference. Some traits are reduced in the sneaker males primarily due to their smaller size (e.g., sword; Robinson and Morris, unpublished data), and therefore the differences can be attributed to the selection on sneaker males to reach sexual maturity earlier, and thus at a smaller size. However, this is not the case for the differences between courters and sneakers in expression of the sexually selected trait of vertical bars (Fig. 9.3). In *X. multilineatus*, some sneaker males have a suppressor gene for vertical bars, and in those sneaker males that do express bars there is a reduced number of vertical bars relative to expectations due to body size (Zimmerer and Kallman, 1988). Vertical bars function both to

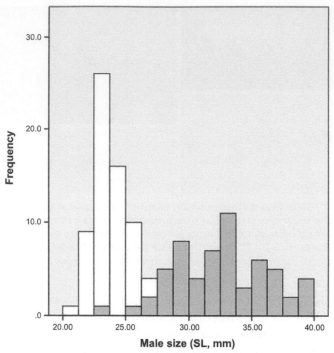

FIGURE 9.3 Size distribution and vertical bar number for "sneaker" males (white) that use courtship and sneak-chase behavior, and "courter" males (gray) that only use courtship behavior in *X. multilineatus*. All males were wild-caught from the Río Oxitipa, Mexico.

attract females and to deter rival males (Morris *et al.*, 1995b). The reduced bar expression in sneaker males could reflect selection to circumvent male–male competition, as courter males fade their bars when they lose contests, which reduces male–male aggression towards them (Morris *et al.*, 1995b). One way to determine the extent to which male–male competition and female mate preference have played a role in the reduction of bars in the sneaker males would be to examine current selection on the variation in bar expression within the sneaker males. Experiments that examine the relationship between mating success and bar expression within the sneaker males when male–male competition is included (mating contests including both sneaker and courter males) as well as when competition with the larger courter males has been removed (mating contests with only sneaker males) have yet to be conducted.

Another way to examine this question is to use comparative studies as natural experiments. Both *X. pygmaeus* and *X. continens* are northern sword-tail species in which the larger courter males have been evolutionarily lost (Fig. 9.4), setting up natural mating contests where the larger competitors have been removed. If we assume that both species were derived from an ancestral

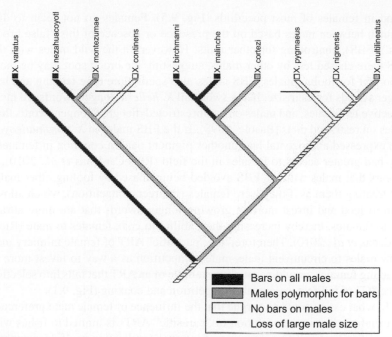

FIGURE 9.4   **The phylogenetic relationships among the northern swordtails as supported by Rauchenberger _et al._ (1990) and Morris _et al._ (2001).** The pigment pattern "vertical bars" has been mapped onto the tree using parsimony, implemented with the program MacClade version 3.0 (Maddison and Maddison, 1992). Black, bars present on all males; gray, males polymorphic for bars; white, males do not have bars. The loss of large males (>35.0 mm) is denoted by a line on the branches where large male size was lost. _From Morris et al., 2005._

species with a polymorphism in vertical bars (Fig. 9.4), what happened to the bars when competition with the "bourgeois" males was removed? In both of these species, males do not court females but use only a behavior that has been called "gonopodial thrusts" (use of the gonopodium to force copulation with females; Farr, 1984) during mating (Ryan and Causey, 1989; Morris _et al._, 2005). In both these species, males also have no vertical bars. We hypothesize that the lack of bars is an indication of selection against bars in males that are trying to circumvent female mate preference with gonopodial thrusting as their primary mating behavior.

A common parasitic ART is female mimicry, in which males evolve traits that fool conspecifics into treating them as females. To determine if sexual conflict is important in the evolution of males with this tactic, one needs to ask if the traits have evolved to fool males (circumventing male–male competition) or to fool females (circumventing female mate choice). This question was addressed in the swordtail _X. nezahualcoyotl_ (Ríos-Cardenas _et al._, 2010). Some males in this species have a permanent pigment pattern that resembles the brood spot (pigmentation of tissue surrounding the female reproductive organs) found in females of most poeciliids (Fig. 9.5). Females do not seem to distinguish between males based on the presence or absence of this "false brood spot" (FBS), controlling for other traits. However, in the wild males with the FBS were chased less by other males, suggesting the brood spot may be more important for fooling males. FBS males also spent more time feeding and had longer swords for their size. In the swordtail *X. hellerii*, longer swords are more attractive to females, and males on an unrestricted diet grew longer swords than males on restricted diets (Basolo, 1998). If the FBS males in *X. nezahualcoyotl* also expressed a horizontal bar (another pigment pattern common in females), they had greater access to females in the field (Ríos-Cardenas *et al.*, 2010). It appears that males with the FBS avoided being chased by fooling other males into treating them as if they were females (indirect competition), which allows them to feed and invest more in growing longer swords that are more attractive to females, thereby increasing their ability to coax females to mate (Ríos-Cardenas *et al.*, 2010). Therefore, this "parasitic" ART of female mimicry may allow males to circumvent male–male competition as a way to invest more in attracting females, providing another example of an ART that falls into selection space "C" of indirect male–male competition and coaxing (Fig. 9.1).

To what extent could one argue that the influence of female mate preference or sexual conflict in the evolution of "parasitic" ARTs is limited to fishes with internal fertilization? The evolution of internal fertilization itself does suggest an important role for sexual conflict, and sexual conflict could be less important in species with external fertilization. However, we argue that circumventing female mate preference is important in fishes with external fertilization as well. Taborsky (2008) proposes that in fishes with external fertilization, there is ample evidence that females prefer to spawn with bourgeois males rather than "parasitic" males – a similar preference to that prevailing in fishes with internal fertilization (Farr, 1980; Bisazza, 1993). Such preference is thought to occur when females refuse to spawn while "parasitic" males are present (Taborsky, 2008). Examples of this type of mate preference occur in the Mediterranean wrasse *Symphodus ocellatus* (Taborsky, 1994) and the bluegill sunfish *Lepomis macrochirus* (Neff, 2008). Therefore, it appears that being less conspicuous to females would be advantageous for "parasitic" males in fishes with external fertilization as well. Even if circumventing female preferences proves to be more important in species with internal fertilization, we suggest that it will be important to examine the potential influence of female preference and sexual conflict in species with external fertilization and parental care as well.

Do "Parasitic" Sneaker Males Rely On "Bourgeois" Courter Males to Reproduce?

Rosen and Tucker (1961) reviewed the sexual behaviors of poeciliid fishes, describing variation across species. They noted that while in some species males used a stereotypic courtship display, in many species there was no pre-copulatory

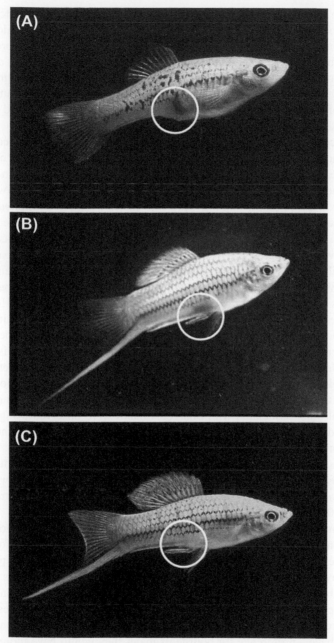

FIGURE 9.5   Circles indicate (A) true brood spot on *X. nezahualcoyotl* female; (B) false brood spot (FBS) on a male; and (C) location of where FBS would be, on a male without FBS. See color plate at the back of the book. *Modified from Ríos-Cardenas et al., 2010. Photographs courtesy of K. de Queiroz.*

courtship but males used gonopodial thrusting (Farr, 1984; also referred as "copulatory thrusts" by Bisazza, 1993). Of the 45 species of poeciliid fishes reported in Bisazza (1993), 24 were scored as using only copulatory thrusts, 18 as using both courtship and thrusts, and 2 (*X. gordoni* and *X. variatus*) as using only courtship. The argument that without the initial bourgeois investment the alternative tactic would not be functional is questionable, given the numerous species of poeciliid fishes in which all males use gonopodial thrusts and employ no courtship during mating.

One could argue that the mating behaviors of the species described above are not homologous to the behaviors used by the "parasitic" sneaker males in those species that have both bourgeois and parasitic males, such as in *X. nigrensis* and *X. multilineatus*. While we do not believe this is the case, we present some evidence below that the "parasitic" males within *X. multilineatus* can reproduce independent of the bourgeois males as well. Members of the smaller size class of males in *X. multilineatus* are more likely to use "sneak-chase" mating behavior in the presence of the larger courter males, and many "parasitic" males will switch to courtship when alone with a female (Zimmerer and Kallman, 1989). This behavioral plasticity alone would suggest that the "parasitic" size class of males can reproduce independent of the bourgeois males. However, we suggest that even the behavior of sneak-chase could be successful independent of bourgeois males. Not all sneaker males switch to courtship when alone with a female, and the propensity to use sneak-chase in this context is highly repeatable (Robinson and Morris, unpublished data). Behavioral observations in the field suggest that the parasitic sneaker males are often found alone with females, and while some were observed to switch to courtship in this context, other males not associated with bourgeois males were observed using sneak-chase and attempting to force copulation with the females they encountered as they swam through the stream (Morris and Lyons, unpublished data). Finally, the two species within the northern swordtails that have lost large male-size classes (Fig. 9.4) have also lost courtship behavior. All the males in *X. pygmaeus* and *X. continens* use the sneak-chase mating behavior exclusively (Morris *et al.*, 1996, 2005).

Again, we can ask if the extent to which the presence of the bourgeois males is necessary for the function of the parasitic males is different for fishes with external fertilization. Here we argue that the function of the parasitic ART may be more reliant on the presence of the bourgeois males in systems with external fertilization, and even more so if the bourgeois males also provide parental care. One would assume that at some point females would need to release their eggs, even if a bourgeois male is not present, but that if females have a preference for bourgeois males they would have more opportunity to express that preference with external fertilization than in species with internal fertilization and forced copulations. Indeed, if it is true that the ARTs in externally fertilized fish are more "dependent" on one another than in internally fertilizing fishes, one might expect a relationship between ARTs and speciation to be more prevalent once internal fertilization evolves, as independent alternative phenotypes can more easily lead to speciation (West-Eberhard, 1989). This would be an interesting hypothesis to test in a phylogenetic comparative analysis. The family Poeciliidae is a monophyletic group (Parenti and Rauchenberger, 1989) that includes certain external-fertilizing fish (the Amazonian *Fluviphylax*, and the African and Madagascan lampeyes or procatopines and *Pantanodon*) in addition to the live-bearing poeciliids. However, the relationships within the Poeciliidae are not well supported, and the hypothesis that poeciliids are closely related to these external fertilizer fishes requires further testing. We suggest that once the phylogenetic relationships in this group have stronger support, and enough information on the mating system of these species has been collected, a comparative analysis could test the effect of the mode of fertilization on the dependency of ARTs and its subsequent effect on rates of speciation.

Is Bourgeois the Ancestral State?

The most common mating behavior in poecillid fishes is gonopodial thrusts, involving no territorial defense or traits to coax the female to mate (thus, the "parasitic" ART). However, a phylogenetic reconstruction of the ancestral state for ARTs in this group is still needed. For many of the questions concerning the evolution of alternative mating behaviors, a broader analysis across the teleosts would be necessary, as poecillids are similar in terms of having internal fertilization and no parental care. Behavioral ecologists have been frustrated in the past by the lack of support for some of the phylogenic hypotheses used in comparative analyses; however, newer methods for testing hypotheses across several trees are making this less of a problem. In addition, work underway on the evolutionary relationships among Teleosts (Lundberg, http://tolweb.org/Teleostei) will be providing a well-supported tree for these types of studies. We suggest that the primary obstacle that stands in the way of conducting an analysis to reconstruct the ancestral state for the ARTs within poeciliid fishes is that the data set on ARTs is currently incomplete, even for some of the smaller, well-studied clades within this group such as *Xiphophorus* and *Limia*. The mating behaviors for numerous species of poeciliids have not been described. In addition, for those species where some description is available, the behaviors are often not scored in a way that allows determination of whether or not a species is polymorphic.

We contend that the current data set underestimates species that are polymorphic for mating behaviors. This will be particularly true if observations are not made across a wide range of males and in more than one context, and if the ARTs are flexible. Determination of which males to test and in which contexts should depend on the particular system that is being examined. For *Xiphophorus* fishes that do not defend a nest or a territory (only groups of females) it will be important to observe male mating behaviors across the size range of males in a species, testing each male in at least two contexts: (1) alone with a variety of females to determine male behavior in the absence of male–male competition; and (2) with a female and a larger male competitor, to determine if males change their behaviors in response to male–male competition. These tests will determine if males use courtship exclusively, use a sneaking behavior such as gonopodial thrusting exclusively, or use both depending on the presence of a larger competitor (see Morris *et al.*, 2008). However, it will also be important to examine the extent to which female preference and/or sexual conflict is important. Otherwise, as we have argued throughout this chapter, the focus on male–male competition will not adequately describe the extent of variation in ARTs. Finally, field experiments will be valuable in many instances to validate the variation observed in the laboratory, as well as to uncover the situations in which the different behaviors are most commonly used (see Haesler *et al.*, 2009).

West-Eberhard (1986, 2003) has hypothesized that in groups where ARTs have contributed to speciation, polyphenism will be the ancestral state. In a recent analysis of the evolution of ARTs in side-blotched lizards, Corl *et al.* (2010) determined from a phylogenetic reconstruction that the polymorphism was ancestral, in addition to showing that it had been independently lost eight times, often giving rise to morphologically distinct subspecies/species. The ability to test this hypothesis in fishes will require that we correct the current bias against detecting polymorphisms/polyphenisms, by examining the mating behaviors of more species across a range of males and in a diversity of contexts.

CONCLUDING REMARKS

It is clear that there is a bias towards considering selection due to male–male competition and not female mate preference or sexual conflict in the evolution of ARTs, which means we are missing an important component of selection in their evolution. Even if circumventing female preferences proves to be more important in species with internal fertilization, it is possible that female preference and sexual conflict still play a role among species with external fertilization and parental care. We argue that while male–male competition selects for males that are bourgeois and parasitic, female mate preference also produces disruptive selection, favoring males that coax females to mate as well as those that coerce, resulting in the evolution of traits that help them circumvent mate preference. In addition, it is important to consider the possibility that not all ARTs will fall into the two extremes of direct competition/coaxing and indirect competition/coercing. The terminology we currently use to describe ARTs may be applicable to some systems, but could have been partially responsible for the incorrect assumptions about ARTs in general. Both "bourgeois" and "parasitic" are loaded terms that not only mislead us into ignoring the role of female mate preference on the parasitic males, but also to mistaken assumptions about their independence and the ancestral state. A more inclusive view of ARTs will require that these assumptions be tested across fishes, describing the mating behaviors of more species, and ultimately analyzing the evolution of ARTs in a phylogenetic context.

ACKNOWLEDGMENTS

We thank Donelle Robinson and Susan Lyons for access to unpublished data, and Kevin de Queiroz and Jane Brockmann for thoughtful discussions regarding alternative mating tactics.

REFERENCES

Basolo, A.L., 1990. Female preference for male sword length in the green swordtail, *Xiphophorus helleri* (Pisces: Poeciliidae). Anim. Behav. 40, 332–338.

Basolo, A.L., 1998. Shift in investment between sexually selected traits: tarnishing of the silver spoon. Anim. Behav. 55, 665–671.

Bisazza, A., 1993. Male competition, female mate choice and sexual size dimorphism in poeciliid fishes. Marine Behav. Physiol. 23, 257–286.

Bono, L.M., Ríos-Cardenas, O., Morris, M.R., 2011. Alternative life histories in *Xiphophorus multilineatus*: evidence for different ages at sexual maturity and growth responses in the wild. J. Fish Biol. 78, 1311–1322.

Butts, I.A.E., Love, O.P., Farwell, M., Pitcher, T.E., 2011. Primary and secondary sexual characters in alternative reproductive tactics of Chinook salmon: associations with androgens and the maturation-inducing steroid. Gen. Comp. Endocrinol. 175, 449–456.

Chapman, T., Arnqvist, G., Bangham, J., Rowe, L., 2003. Sexual conflict. Trends Ecol. Evol. 18, 41–47.

Corl, A., David, A.R., Kutcha, S.R., Sinervo, B., 2010. Selective loss of polymorphic mating types is associated with rapid phenotypic evolution during morphic speciation. Proc. Natl. Acad. Sci. U. S. A. 107, 4254–4259.

Erbelding-Denk, C., Schroder, J.H., Schartl, M., Nanda, I., Schmid, M., Epplen, J.T., 1994. Male polymorphism in *Limia perugiae* (Pisces: Poeciliidae). Behav. Genet. 24, 95–101.

Evans, J.P., Pilastro, A., Schlupp, I., 2011. Ecology and Evolution of Poeciliid Fishes. University of Chicago Press, Chicago.

Farr, J.A., 1980. Social behaviour patterns as determinants of reproductive success in the guppy, *Poecilia reticulata* Peters (Pisces, Poeciliidae). An experimental study of the effects of inter-male competition, female choice, and sexual selection. Behaviour 74, 38–91.

Farr, J.A., 1984. Premating behavior in the subgenus *Limia* (Pisces: Poeciliidae): Sexual selection and the evolution of courtship. Z. Tierpsychol. 65, 152–165.

Franck, D., Müller, A., Rogmann, N., 2003. A colour and size dimorphism in the green swordtail (population Jalapa): female mate choice, male–male competition, and male mating strategies. Acta Ethol. 5, 75–79.

Haesler, M.P., Lindeyer, C.M., Taborsky, M., 2009. Reproductive parasitism: male and female responses to conspecific and heterospecific intrusions at spawning in a mouth-brooding cichlid *Ophthalmotilapia ventralis*. J. Fish Biol. 75, 1845–1856.

Harris, E.W., Moore, P.J., 2005. Female mate preference and sexual conflict: females prefer males that had fewer consorts. Am. Nat. 165, S64–S71.

Kallman, K.D., 1989. Genetic control of size at maturity in *Xiphophorus*. In: Meffe, G.K., Snelson Jr., F.F. (Eds.), Ecology and Evolution of Livebearing Fishes (Poeciliidae), Prentice Hall, Englewood Cliffs, NJ, pp. 163–184.

Kawase, H., Nakazono, A., 1996. Two alternative female tactics in the polygynous mating system of the threadsail filefish, *Stephanolepis cirrhifer* (Monacanthidae). Ichthyol. Res. 43, 315–323.

Knapp, R., Neff, B.D., 2008. Alternative reproductive tactics in fishes. In: Magnhagen, C., Braith-waite, V.A., Fosgren, E., Kapoor, B.G. (Eds.), Fish Behaviour, Science Publishers Inc., Enfield, pp. 411–433.

Lampert, K.P., Schmidt, C., Fischer, P., Volff, J.N., Hoffmann, C., Muck, J., Lohse, M., Ryan, M.J., Schartl, M., 2010. Determination of onset of sexual maturation and mating behavior by mela-nocortin receptor 4 polymorphisms. Curr. Biol. 20, 1–6.

Luo, J., Sanetra, M., Schartl, M., Meyer, A., 2005. Strong reproductive skew among males in the multiply mated swordtail *Xiphophorus multilineatus* (Teleostei). J. Hered. 96, 346–355.

Maddison, W.P., Maddison, D.R., 1992. MacClade, version 3.0. Sinauer, Sunderland.

Mank, J.E., Avise, J.C., 2006. Comparative phylogenetic analysis of male alternative reproductive tactics in ray-finned fishes. Evolution 60, 1311–1316.

Morris, M.R., Gass, L., Ryan, M.J., 1995a. Assessment and individual recognition of opponents in the pygmy swordtails *Xiphophorus nigrensis* and *X. multilineatus*. Behav. Ecol. Sociobiol. 37, 303–310.

Morris, M.R., Mussel, M., Ryan, M.J., 1995b. Vertical bars on male *Xiphophorus multilineatus*: a signal that deters rival males and attracts females. Behav. Ecol. 6, 274–279.

Morris, M.R., Wagner Jr., W.E., Ryan, M.J., 1996. A negative correlation between trait and mate preference in *Xiphophorus pygmaeus*. Anim. Behav. 52, 1193–1203.

Morris, M.R., de Queiroz, K., Morizot, D.C., 2001. Phylogenetic relationships among populations of northern swordtails (*Xiphophorus*) as inferred from allozyme data. Copeia 2001, 65–81.

Morris, M.R., Moretz, J.A., Farley, K., Nicoletto, P., 2005. The role of sexual selection in the loss of sexually selected traits in the swordtail fish *Xiphophorus continens*. Anim. Behav. 69, 1415–1424.

Morris, M.R., Tudor, S.M., Dubois, N.S., 2007. Sexually selected signal attracted females prior to deterring aggression in rival males. Anim. Behav. 74, 1189–1197.

Morris, M.R., Ríos-Cardenas, O., Darrah, A., 2008. Male mating tactics in the Northern Mountain swordtail fish (*Xiphophorus nezahualcoyotl*): coaxing and coercing females to mate. Ethology 114, 977–988.

Morris, M.R., Ríos-Cardenas, O., Brewer, J., 2010. Variation in mating preference within a wild pop-ulation influences the mating success of alternative mating strategies. Anim. Behav. 79, 673–678.

Murphy, C.G., 1998. Interaction-independent sexual selection and the mechanisms of sexual selec-tion. Evolution 52, 8–18.

Neff, B.D., 2008. Alternative mating tactics and mate choice for good genes or good care. In: Oliveira, R.F., Taborsky, M., Brockmann, H.J. (Eds.), Alternative Reproductive Tactics: An Integrative Approach, Cambridge University Press, New York, pp. 421–434.

Oliveira, R.F., Anario, A.V.M., Grober, M.S., 2001. Male sexual polymorphism, alternative repro-ductive tactics, and androgens in combtooth blennies (Pisces: Blenniidae). Horm. Behav. 40, 266–275.

Oliveira, R.F., Taborsky, M., Brockmann, J.H., 2008. Alternative Reproductive Tactics: An Integra-tive Approach. Cambridge University Press, Cambridge.

Parenti, L.R., Rauchenberger, M., 1989. Systematic overview of the poeciliines. In: Meffe, G.K., Snelson, F.F. (Eds.), Ecology and Evolution of Livebearing Fishes (Poeciliidae), Prentice Hall, Englewood Cliffs, pp. 3–12.

Peters, G., 1964. Vergleichende Untersuchungen an drei Subspecies von *Xiphophorus helleri* (Pisces). J. Zool. Syst. Evol. Res. 2, 185–271.

Perrill, S.A., Gerhardt, D., Daniel, R., 1978. Sexual parasitism in the green tree frog (*Hyla cinerea*). Science 200, 1179–1180.

Rauchenberger, M., Kallman, K.D., Morizot, D.C., 1990. Monophyly and geography of the Río Pánuco basin swordtails (Genus *Xiphophorus*) with descriptions of four new species. Am. Mus. Novi. 2975, 1–41.

Reichard, M., Le Comber, S.C., Smith, C., 2007. Sneaking from a female perspective. Anim. Behav. 74, 679–688.

Ríos-Cardenas, O., Tudor, M.S., Morris, M.R., 2007. Female preference variation has implications for the maintenance of an alternative mating strategy in a swordtail fish. Anim. Behav. 74, 633–640.

Ríos-Cardenas, O., Darrah, A., Morris, M.R., 2010. Female mimicry indirectly enhances a male sexually selected trait; what does it take to fool a male? Behaviour 147, 1443–1460.

Rosen, D.E., 1973. Suborder Cyprinodontoidei; Superfamily Cyprinodontoidea; Families Cyprinodontidae, Poeciliidae, and Anablepidae. Mem. Sears Found. Mar. Res. 1, 229–262.

Rosen, D.E., Tucker, A., 1961. Evolution of secondary sexual characters and sexual behavior patterns in a family of viviparous fishes (Cyprinodontiformes: Poeciliidae). Copeia 1961, 201–212.

Ryan, M.J., Causey, B.A., 1989. "Alternative" mating behavior in the swordtails *Xiphophorus nigrensis* and *Xiphophorus pygmaeus* (Pisces: Poeciliidae). Behav. Ecol. Sociobiol. 24, 341–348.

Ryan, M.J., Hews, D.K., Wagner Jr., W.E., 1990. Sexual selection on alleles that determine body size in the swordtail *Xiphophorus nigrensis* (Pisces: Poeciliidae). Behav. Ecol. Sociobiol. 26, 231–237.

Ryan, M.J., Pease, C.M., Morris, M.R., 1992. A genetic polymorphism in the swordtail *Xiphophorus nigrensis*: testing the prediction of equal fitnesses. Am. Nat. 139, 21–31.

Shuster, S.M., Wade, M.H., 2003. Mating Systems and Strategies. Princeton University Press, Princeton.

Taborsky, M., 1994. Sneakers, satellites, and helpers: parasitic and cooperative behavior in fish reproduction. Adv. Stud. Behav. 23, 1–100.

Taborsky, M., 1998. Sperm competition in fish: "bourgeois" males and parasitic spawning. Trends Ecol. Evol. 13, 222–227.

Taborsky, M., 2001. The evolution of bourgeois, parasitic, and cooperative reproductive behaviors in fishes. J. Hered 92, 100–110.

Taborsky, M., 2008. Alternative reproductive tactics in fish. In: Oliveira, R.F., Taborsky, M., Brockmann, H.J. (Eds.), Alternative Reproductive Tactics: An Integrative Approach, Cambridge University Press, New York, pp. 251–299.

Taborsky, M., Oliveira, R.F., Brockmann, H.J., 2008. The evolution of alternative reproductive tactics: concepts and questions. In: Oliveira, R.F., Taborsky, M., Brockmann, J.H. (Eds.), Alternative Reproductive Tactics: An Integrative Approach, Cambridge University Press, Cambridge, pp. 1–21.

Tudor, M.S., Morris, M.R., 2011. Frequencies of alternative mating strategies influence female mate preference in the swordtail *Xiphophorus multilineatus*. Anim. Behav. 82, 1313–1318.

West-Eberhard, M.J., 1986. Alternative adaptations, speciation, and phylogeny (a review). Proc. Natl. Acad. Sci. U. S. A. 83, 1388–1392.

West-Eberhard, M.J., 1989. Phenotypic plasticity and the origins of diversity. Annu. Rev. Ecol. Syst. 20, 249–278.

West-Eberhard, M.J., 2003. Developmental Plasticity and Evolution. Oxford Univ Press, New York p. 794.

Westneat, D.F., Sih, A., 2009. Sexual conflict as a partitioning of selection. Biol. Lett. 5, 675–677.

Zimmerer, E.J., Kallman, K.D., 1988. The inheritance of vertical barring (aggression and appeasement signals) in the pygmy swordtail, *Xiphophorus nigrensis* (Poeciliidae, Teleostei). Copeia 1988, 299–307.

Zimmerer, E.J., Kallman, K.D., 1989. Genetic basis for alternative reproductive tactics in the pygmy swordtail, *Xiphophorus nigrensis*. Evolution 43, 1298–1307.

Chapter 10

Mode of Reproduction, Mate Choice, and Species Richness in Goodeid Fish

Constantino Macías Garcia

Laboratorio de Conducta Animal, Instituto de Ecología, Universidad Nacional Autónoma de México, Mexico

BIOGEOGRAPHY: NORTHERN ORIGIN AND SOUTHERN DIFFERENTIATION

One aim in science is to produce general explanations that may accommodate a collection of related facts. However, the great diversity of biological phenomena means that biologists often sit astride a narrow fence, simultaneously exploring what is it that makes their study systems distinctive while attempting to fit its attributes within some general theoretical framework. This is in part due to the generally imperfect replication of biological – and biogeographical – phenomena. The uneven distribution of biodiversity on earth is a case in point. We may expect that the trends towards greater biodiversity in the tropics than in temperate regions, being evident around the globe, should be the consequence of the same processes acting everywhere (see, for example, MacArthur, 1969, 1972). Yet while in Australasia there is a substantial replacement of biota across the Wallace divide, in the Americas the southern and northern biotas more or less intermingle along a vast latitudinal extension (albeit the southwards invasion is more complete than the expansion of neotropical biota into North America). Thus, while the general pattern is the same, the relative weight of the historical and ecological factors that explain biodiversity patterns may or must differ between regions.

At specific taxonomic levels there is even more variation in the extent to which organisms conform to the proposed explanations of biodiversity patterns (Cox and Moore, 1985). This may relate to differences in the organism's capacity to track environmental change or to disperse into suitable habitats. Thus, the distribution of biodiversity in some taxa, such as birds, may be more likely to respond to variation in ecological factors than that of taxa with limited means

Sexual Selection. http://dx.doi.org/10.1016/B978-0-12-416028-6.00010-4

of dispersion. Freshwater biota such as fish mostly move between and through freshwater habitats, thus hydrography (largely influenced by topography) is a major determinant of the composition of fish communities and is regarded by cladistic biogeographers as the main explanation for fish diversity patterns. This view, however, cannot explain the frequent observation that different clades occupying the same geographic region and for comparable time spans have responded differently to the same hydrographic history. I propose that such mismatches are the consequence of differences in reproductive biology, which determine the facility with which reproductive isolation arises. I use evidence from the Goodeidae, a family of North American cyprinodontid fish, to explore this argument.

The family Goodeidae was originally described as a clade composed exclusively of Mexican viviparous fish (e.g., Hubbs and Turner, 1939), but it was subsequently recognized that members of the few species in the genera *Empetrichthys* and *Crenichthys* (subfamily Empetrichthynae) from the southern USA are their closest relatives and should be grouped in the same family (Parenti, 1981; Doadrio and Domínguez, 2004; Webb *et al.*, 2004). These are oviparous, but otherwise closely resemble the Mexican Goodeinae. The outgroup to the Goodeidae is probably a northern species related to the current *Profundulus* (see Uyeno and Miller, 1962) from which it separated some 23 mya (Webb *et al.*, 2004), thus placing the origin of the Goodeidae in the south of the U.S.A. The subfamily Empetrichthynae is now regarded as being composed of only four species, two in the genus *Crenichthys* (*C. baileyi* and *C. nevadae*) and two in *Empetrichthys* (*E. latos* and *E. merriami*, extinct since the late 1950s). Each species is geographically restricted, with some found in only a handful of spring-fed pools, yet population divergence is substantial and has been regarded as evidence of incipient speciation (Miller, 1948, 1950; see below). The Empetrichthynae are shallow-water cyprinodontids that live in small springs and adjacent marshes in the southwestern Great Basin (USA). Little is known of their ecology, but they appear to be omnivores (Kopec, 1949), a fact consistent with the length of their intestines (Sigler and Sigler, 1987).

Miller (1950) recognized three subspecies of *Empetrichthys latos* based on their morphology. He also noted that populations of *Chrenichtys baileyi* (variously regarded as comprising up to five subspecies) were physiologically adapted to different temperatures in their native desert springs (Sumner and Lanham, 1942), and argued that recurrent fragmentation of habitats during prolonged periods should lead not only to frequent speciation but also to local extinction among desert fish (Miller, 1948, 1950). Indeed, the geological scenario in which the Empetrichthynae evolved has been the subject of iterated fragmentation through block-faulting, desiccation (interspersed with the merging of isolated lakes during the pluvial periods), and a trend towards desertification since the Pleistocene (e.g., Harvey *et al.*, 1999). The process was probably very gradual in the Lahontan Pleistocene Pluvial lake, and perhaps more abrupt, although still prolonged, in the Mohave. Since both genetic drift and local adaptation operate faster in small populations, those locked in small lakes and pools for prolonged periods would have ample opportunity for divergence if they did not become extinct.

Extinction has wiped out one species, as well as several populations of the rest of the Empetrichthynae, which remain endangered. They declined during the first half of the 20th century, at least in some cases as a result of anthropogenic processes (cf. Williams, 1996). Despite the prolonged trend of desertification in their area of origin, it is still unlikely that the subfamily would have persisted for long. It is thus difficult at present to decide whether the Empetrichthynae constitute a clade where speciation is rare, or a rather speciose group where extinction has been common – a question I shall return to later in the chapter.

We know that the sister clade, the Goodeinae, is very diverse, and may have been more so in the past (Grudzien et al., 1992; Doadrio and Domínguez, 2004; Webb et al., 2004). Depending on the author, the Goodeinae comprises from 36 to about 42 species distributed in 17–20 genera – a taxonomic uncertainty discussed below. The Goodeinae diverged from the Empetrichthynae some 16.8 mya ago, during the middle Miocene. A fossil Goodeinae (Tapatia occidentalis) confirms that by the late Miocene to early Pliocene the Goodeinae had already reached the center of their current distribution in Mexico (Álvarez and Arriola-Longoria, 1972). The two subfamilies are now separated by over 1500 km – an area where no fossil records of either subfamily have been found. The fossil E. erdisi lived in southern California, and the accepted distribution of extinct Goodeinae Characodon garmani does not bring the subfamilies any closer. There is thus a frustrating gap in our knowledge of where the main features of the Goodeinae – viviparity and internal fertilization – first appeared; the northernmost (genus Characodon) as well as the oldest (the fossil T. occidentalis) members of the Goodeinae possess all the attributes that characterize their complex reproductive system.

ECOLOGICAL CONSEQUENCES OF INVADING THE NEOTROPICS

The Goodeidae seem to conform to the general pattern of greater diversity in the tropics than in temperate areas. We know that this is not due to the age of the clades that compose the family, since the neotropical Goodeinae are younger than the temperate Empetrichthynae. Two contrasting processes may have generated this pattern: a higher incidence of extinction in the northern Goodeidae, or a higher speciation rate in the southern clade (MacArthur, 1969).

Speciation and Extinction in the Empetrichthynae

Greater extinction rates in the Empetrichthynae cannot be ruled out, and would be consistent with the long-term trend of desertification in southern USA/northern Mexico. Whether the populations that became extinct belonged to a limited number of species distributed across vast territories, or were locally evolved species, is something that we do not know.

Local diversification in extant Empetrichthynae may have resulted from ecological adaptation – for example, to differences in thermal regimes, which, judging by the present conditions, may have been remarkably stable at a local level. This was proposed by Sumner and Lanham (1942) based on differential mortality of fish from different springs when exposed to the typical temperature experienced by other populations. Unfortunately the experiment was conducted in the field, and we cannot rule out the possibility that the observed thermal intolerance was the result of phenotypic plasticity rather than genetic adaptation. Since we have no reason to believe that the attributes involved in the putative thermal adaptation are the same as (or correlated with) those used by ichthyologists to nominate subspecies (mostly morphometrics; Miller, 1948), I am inclined to believe that local diversification in the Empetrichthynae was not the consequence of ecological adaptation.

At least in *C. baileyi*, the local variation in morphology is still detected 60 years after it was originally described (e.g., Jelks *et al.*, 2008), which may suggest, given a modicum of environmental change in six decades of intense anthropogenic influence, that it is not a consequence of phenotypic plasticity. Arguably, if population divergence of Empetrichthynae is not due to ecology, then it could be the consequence of sexual selection/conflict or of chance (which is difficult to demonstrate). We have no evidence of reproductive isolation among the extant subspecies of *C. baileyi*, but this may be due to a lack of studies. There is one uncommonly detailed description of the mating behavior of *C. baileyi* by Kopec (1949). He observed only one male and two females, but found that:

1. There is sexual dichromatism that becomes exaggerated during courtship, when male colors are brighter;
2. Males direct their courtship to ripe females;
3. When ready to spawn, females perform behaviors that may attract males;
4. During courtship, males attack intruding males, but otherwise live in shoals and do not fight;
5. Mating involves a prolonged, agitated embrace during which the male anal fin is folded over the female's enlarged ovipositor and presumably conveys sperm;
6. Females lay one egg at a time and move away from the male;
7. Females lay several, but not a great many, eggs (10–17).

From all of the above, it seems that there is the opportunity for female mate choice, as mating is not collective, and that females can move between males for laying consecutive eggs. There seems to be little opportunity for male monopolization of females, except during the brief period of producing and inseminating one egg. There also seems to be scope for strategic male mating, as well as male mate preferences for large females. It is noteworthy that the mechanics of the mating process is similar to that of the Goodeinae (see below).

In sum, although we may not be able to give a satisfactory answer to the question of whether the Empetrichthynae were likely to have produced many species that subsequently became extinct, informed guesses are possible if we establish (1) whether there is any degree of pre-mating isolation between populations, and (2) whether subspecies differ in locally adaptive, genetically determined attributes. That this research can only be conducted in one species (*C. baileyi*) should not mean that the endeavor is not worthwhile. With the limited information available, I provisionally propose that the Empetrichthynae are unlikely to have diverged into many ecologically distinct species, but that pre-mating isolation between populations would have been frequent given their likelihood of being subject to inter- and intrasexual selection.

Speciation and Extinction in the Goodeinae

The other side of the coin is to ask whether the Goodeinae are more likely to speciate or less likely to become extinct than the Empetrichthynae. They too encountered an unstable habitat that was at times occupied by massive lakes (De Cserna, 1989) repeatedly fragmented into a variety of basins. Thus, habitat loss through desiccation is probably a more recent phenomenon in central Mexico than in the Death Valley, although it has clearly been promoted by human activities. Populations of some species that have been studied are morphologically and genetically distinct (González-Zuarth, 2006; Macías Garcia *et al.*, 2012), and there is some evidence linking this with pre-mating isolation (González-Zuarth, 2006) and restricted gene flow (Ritchie *et al.*, 2007). This would suggest that local differentiation is frequent, although direct comparison with the degree of differentiation in the extant Empetricthynae is complicated by the fact that subspecies are rarely described in the Goodeinae, whose local variants are instead given species status (e.g., *Zoogoneticus purepechus*; Domínguez-Domínguez *et al.*, 2008). This ongoing trend stems from the use of different species concepts, with molecular systematists using differences in the sequences of one gene, or at best a few genes, to decide whether two populations or sets of populations should be regarded as different species. This is inevitably at odds with the use of the biological species concept (Mayr, 1942) among students of microevolution.

Ecology and Species Richness

Ecology can influence biodiversity in many ways. I shall not cover all of them, in part because basic information regarding Goodeidae ecology in general is lacking. Instead, I have identified some ecological factors that could impact Goodeinae diversity and for which there is at least some information available.

Trophic Divergence

Trophic divergence is frequently involved in the origin of morphs/species of freshwater fish. Among the best-studied examples are the benthic–limnetic morphs into which three-spine stickleback (*Gasterosteous aculeatus*) often differentiate following the colonization of freshwater habitats (see review by Wootton, 2009). An extreme example of the role of feeding specialization in speciation – or at least in the consolidation of the locally generated species – is seen in the species swarms of Lake Tanganyika cichlids (Salzburger *et al.*, 2005). Less spectacular examples abound, often comprising species pairs independently diverged within isolated lakes, such as the palearctic *Coregonus* whitefish species pairs (Bernatchez, 2004) or the neotropical characids *Astyanax–Bramocharax* pairs (Ornelas-García *et al.*, 2008).

Within the several isolated lakes currently inhabited by goodeids, the silversides that were formerly grouped under the Mexican endemic genera *Chirostoma* and *Poblana* also evolved as pairs of species differing in size and morphology (Barbour, 1973). Tellingly, about 20 species evolved within the Trans-Mexican Volcanic Belt/ Mexican High Plateau and only diverged to the point of generating trophically different species pairs, and current systematists do not even recognize them as generically different from the widespread North American *Menidia* silversides (Bloom *et al.*, 2009). This is in sharp contrast with the many-genera Goodeinae species, which are seldom ecologically specialized. Most Goodeinae are feeding generalists that live near the surface of ponds and lakes, and only three closely related genera are composed of species adapted to life in relatively fast-flowing rivers (Miller *et al.*, 2005; Fig. 10.1). Species within a genus occupy similar habitats, normally in adjacent watersheds (Webb *et al.*, 2004), have similar morphologies, and are largely of comparable size. It is thus unlikely that local adaptation has played a substantial role in Goodeinae speciation. One possible exception occurs in the genus *Ilyodon*. Current phylogenies recognize two (but sometimes as many as four) geographically separated species. However, a feeding morph (*Ilyiodon xantusi*) of the westernmost species *I. furcidens* was described in the late 1970s (Turner and Grosse, 1980), and although allele frequencies suggest that the surface-feeding and the more pelagic morph interbreed (Grudzien and Turner, 1984), the prevalence of the two within the same rivers, together with differences in sexual dimorphism and habitat use, suggest that the two morphs experience a degree of mating isolation. This example represents an oddity among the Goodeinae, which generally do not seem to have evolved through trophic specialization.

Feeding specialization is not the only way in which ecology can contribute to speciation, as adaptive trait divergence may indirectly generate barriers to gene flow. Attributes that have such side effects have been termed "magic traits" (Kisdi and Priklopil, 2011), and may lead to speciation (reviewed in Butlin *et al.*, 2011; Servedio *et al.*, 2011). For example, habitat preferences of two diverging morphs of three-spined stickleback found in Icelandic lakes have led to a degree of pre-mating isolation (Ólafsdóttir *et al.*, 2007). Similarly, local differences in physiology (e.g., Berdan and Fuller, 2012) and in attributes directly linked to signal transmission (e.g., background/water color; Castillo Cajas *et al.*, 2012; Cooke *et al.*, 2012), have been shown to result in phenotypic divergence associated with mating isolation.

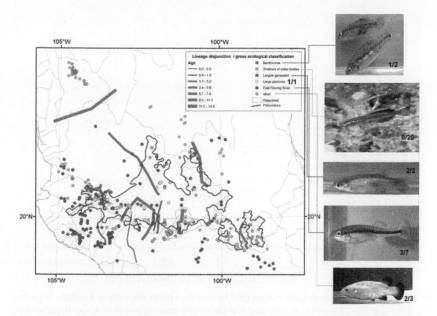

FIGURE 10.1   Biogeography of the Goodeinae. The map shows the distribution of Goodeinae records classified by ecological type, and "lines of disjunction" between sister clade ranges indicating possible allopatric diversification. Lines are equidistant between boundaries of minimum convex polygons (MCPs) built around clade ranges. Lines are depicted for all disjunct sister clades, and for clades with marginally overlapping ranges (<0%). Phylogenetic relationships and node ages are derived from Webb *et al.* (2004). MCPs were generated with Geophylobuilder for ArcGIS v1.2 (Kidd and Lui, 2008), with the disjunction lines drawn by hand. Representative species of the different ecological types are illustrated at the right side of the figure, followed by the number of genera/species belonging to each type. As indicated in the text, trophic diversification is neither particularly frequent, nor leads to extreme diversification among the Goodeinae. See color plate at the back of the book. *Photographs by the author, and courtesy of R. Rodríguez Tejeda and Y. Saldívar Lémus.*

As indicated above (Fig. 10.1), one Goodeinae clade with rather specialized habitat preferences diversified following its dispersal into the fast-flowing rivers of the Pacific-draining watersheds on the western slopes of the western Sierra Madre and the Trans-Mexican Volcanic Belt. Clearly, some degree of ecological divergence took place following their split from the rest of the Goodeinae, yet there is no evidence that their diversification into three genera and seven or so species was also the result of ecological factors.

Geographic Range and Diversification

If local differences in ecology usually promote divergence in some aspect(s) of the species recognition system (Paterson, 1985), it would be reasonable to expect greater phenotypic variation and greater opportunity for the evolution of sexual dimorphism in species that occupy large, environmentally heterogeneous areas than in species with restricted distributions. A proper test of this prediction

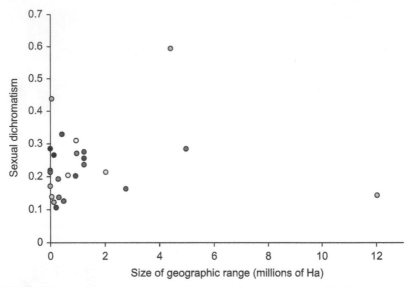

FIGURE 10.2   Large geographic ranges tend to contain a great diversity of habitats, imposing divergent selective pressures on male and female color patterns, and thus it was hypothesized that larger geographic ranges would provide greater opportunities for the evolution of sexual dimorphism. A quantification of sexual dichromatism shows that it evolves independently of (current) size of geographic range. See color plate at the back of the book.

is complicated in the small subfamily Goodeinae because sister clades often diversify in adjacent watersheds, and the prediction requires that the geographic ranges differ only in size, and thus in the probability of encompassing diverse habitats, but not in the likelihood that each particular habitat occurs. Nevertheless, a preliminary comparison using size of geographic range (from Gesundheit and Macías Garcia, 2005) as a proxy for opportunity to encounter ecologically distinct habitats reveals no evidence that sexual dichromatism is greater in more widespread species (Fig. 10.2). Thus, even if it is true that water color and transparency differ between Goodeinae localities (e.g., Moyaho *et al.*, 2005), it seems rather the exception for a species to show a concomitant degree of local variation in dimorphism (e.g., Moyaho *et al.*, 2005), whereas the most widespread species actually show a low level of sexual dichromatism (Fig. 10.2).

The above argument blends two variables: sexual dichromatism and variance in sexual dimorphism. This is due to lack of information on the variance of sexual dimorphism, but it is perhaps not inappropriate given that the default condition should be zero sexual dimorphism, and thus an increase in variance can only be achieved with a corresponding increase in the mean value. Therefore, greater geographic ranges should lead to greater opportunity for local divergence in sexually selected attributes, and thus to greater average and variance in sexual dimorphism. Such a prediction is not supported with our limited information on the Goodeidae, and means that we should look elsewhere for an explanation for their comparatively large degree of diversification.

Multiplicity of Biotic Interactions

Ecology and species diversity have been also conceptually linked in the proposal that the larger number of species in the tropics could be the consequence of an increase in biotic interactions in comparison with temperate areas (Dobshansky, 1950). This seems an unlikely explanation of the greater diversity of the Goodeinae when compared to the Empetrichthynae, since the fish fauna in central Mexico, which is dominated by the Goodeinae, is rather poor (Miller *et al.*, 2005). More generally, the biotic interactions in the freshwater ecosystems that may promote speciation would include predation, trophic competition, and parasitism; can any of these explain the greater diversity of the southern Goodeidae?

Predation

There is little information on the role of predators in shaping the ecology of Goodeid fish. From distribution data, we see that the community of predators consuming the Goodeidae is unlikely to have been very different between the ancient lakes in Nevada and those in central Mexico. The two areas currently share all the important clades of piscivorous birds (mainly herons, egrets, and grebes; see Phillips, 1986) and arthropods (including the Belostomatidae water bugs; see Thorp and Covich, 2001) known to prey on cyprinodontid fish (e.g., Tobler *et al.*, 2008), and this is likely to have been the case through the Pleistocene. Piscivorous snakes do play a role in the evolution of the Mexican Goodeidae (Macías Garcia *et al.*, 1994; see below). However, the species involved (*Thamnophis rufipunctatus, T. melanogaster,* and *T. eques*; Macías Garcia *et al.*, 1998) are also of northern origin (de Queiroz *et al.*, 2002), and their nearest ancestors coexisted with those of the Goodeinae (e.g. Holman, 1995) and are closely related to the snakes that potentially prey or may have preyed on the Empetrichthynae (e.g., *T. elegans*; Kephart, 1982).

Native piscivorous fish are uncommon in central Mexico and in the Nevada springs where the Empetrichthynae evolved. Among the Goodeinae, one species potentially preys on small fish. With a standard length of up to 14 cm, *Alloophorus robustus* is the largest Goodeinae and seems to be the more "highly developed carnivore" (Miller *et al.*, 2005) in the subfamily. Possibly, this predator influenced the evolutionary trajectories of some fish in its clade, yet its geographic range only coincides with that of a handful of Goodeinae species; thus, its presence in central Mexico is also unlikely to explain the evolutionary divergence of the Goodeinae.

Trophic Interactions

As discussed above, trophic differentiation is unlikely to explain the origin of the greater diversity of Goodeinae fish in comparison to their sister clade. Yet competition for food may lead to subtle shifts in the use of habitat or microhabitat among species sharing (primarily or secondarily) the same habitat, even if they are non-sister species. Such a process may not generate new species, but could allow their coexistence. There is increasing evidence that such intraguild interactions are important in shaping community structure and may influence speciation rates (see Polis and Holt, 1992), and this process may have provided the "ratchet" necessary for species interactions to promote speciation. Since Goodeinae species often share segments of their habitat (see Fig. 10.1) and are largely omnivores, it is possible that interspecific competition and predation have led to microhabitat segregation, or to subtle differences in feeding habits that allow species coexistence and multiply their ecological interactions. However, this process requires the previous occurrence of several species interacting in the same habitat, and thus it may help to explain why various Goodeinae species coexist in the same localities, but it would be necessary to invoke a different process to explain why the Empetrichthynae did not generate species, or why they did not evolve the capability to coexist. While the potential effect of intraguild interactions in promoting Goodeinae speciosity cannot be dismissed, this concept is of very limited use to answer the question of why the number of species in the Empetrichthynae and the Goodeinae differ so markedly.

Parasitism

Because of their short life cycles and often significant effects on the fitness of their hosts, parasites (including pathogens) have the potential to accelerate the rate of diversification between populations, and may affect biological diversity. One mechanism relates to the hosts' need to overcome the rapid evolutionary rates of parasites. Since sexual reproduction constantly generates novel genetic combinations, it possibly helps the hosts to remain abreast of their rapidly evolving parasites in their evolutionary race (Van Valen, 1973), thereby constituting a possible reason why costly sexual reproduction is maintained (Maynard Smith, 1976a). Sexual reproduction would arbitrarily generate genetically diverse offspring with a random capacity to overcome parasitism. A more efficient process would ensure that breeding pairs are formed according to their potential to generate greater genetic diversity and/or the genetic combinations that are more likely to cope with parasites. As several studies (see Reusch *et al.*, 2001) have shown, pair formation is not random in relation to the genes involved in fighting infections (major histotocompatibility complex, or MHC), but instead optimizes the number of alleles to be passed to the offspring (too few would expose them to parasites, too many would increase the risk of self-immune disease; Reusch *et al.*, 2001). As there may be local differences in the prevalence/composition of parasites, MHC-based mate choice would promote population differentiation through the selection of local optima. We have no information on the (inter- or intraspecific) diversity of MHC alleles among the Goodeidae, and no studies have looked into the possibility that these genes are involved in mate choice in these fishes. If Goodeinae speciation has been linked to the local divergence in MHC allele optima, I would predict greater interpopulation differences (F_{st}) in MHC alleles among the Goodeinae than among the Empetrichthynae.

Another parasite-related process may be more directly linked to speciation: sexual selection for indirect benefits, often known as the good genes model of sexual selection. When Zahavi (1975) proposed that ornaments are attractive to females because they impose costs ("handicaps") on the bearers, thereby facilitating the selection of fit partners, he was harshly criticized for overlooking the population genetics consequences of his model. As pointed out by Maynard-Smith (1976b), among others, such a handicap process would rapidly be halted by the erosion of genetic diversity, following Fisher's (1958) fundamental theorem of natural selection; the stronger the natural selection pressures bearing on an attribute, the faster the genetic diversity involved in its expression should be eroded. Thus, the handicap principle could not work because any selection for highly fit mates would remove the genetic variance on which that selection depended, leading to uniformly fit males (or females), thereby removing the benefits of mate choice. The obvious genetic flaw in the original version of the handicap principle was overcome by Hamilton and Zuk (1982), who proposed that whenever more than one parasite is involved, the evolutionary arms races between parasites and hosts (involving different alleles/genes and being out of sync) would provide a mechanism to restore genetic diversity in fitness.

One consequence of the Hamilton–Zuk principle is that exposure to a rich and locally diversified parasite fauna should promote rapid population differentiation and act as a speciation engine. Thus, we can ask whether the Goodeinae have been exposed to a more diverse parasite fauna than the Empetrichthynae. This can be inferred from the frequency with which species show a diversity of ornaments (a likely consequence of parasite-led ornamental evolution), and can also be directly evaluated from parasite inventories. Defining what constitutes a diversity of ornaments is complicated by the fact that attributes that appear to us to be separate ornaments may be part of the same display – for example, the modifications of body shape that accompany the development of sexually dimorphic fins (see below). Alternatively, attributes may be amplifiers (Hasson, 1991) of an ornament, as in the black stripe that enhances the terminal yellow band in the tail fin of several Goodeinae males (see below). Nevertheless, the males of many Goodeinae species differ from females in several color marks, in courtship displays, in the size of fins and in body shape, and only rarely exhibit a single ornament (possibly a signature of sexual selection for reliable condition-dependent ornaments; Møller and Pomiankowski, 1993). This is consistent with the idea that interactions with parasites may have promoted Goodeinae speciation via sexual selection for costly ornaments. As secondary sexual dimorphism in the extant Empetrichthynae involves only a few attributes, this points to a possible role of sexual selection in driving the differences in biological diversity between the two clades – a possibility explored below.

Complete parasite inventories are available for some Goodeidae species, but in order to shed any light on whether parasite diversity prompted speciation in these fish, it would be necessary to know the complete inventories of fish parasites in the areas where the two subfamilies evolved, which is not feasible. On the other hand, as parasites experience the same events that generate vicariance in their hosts, we may ask whether there is evidence of comparable speciation rates among the Goodeidae and their parasites. The helminth fauna of the Goodeinae is reasonably well known (e.g., Salgado-Maldonado *et al.*, 2001a, 2001b; Pineda-López *et al.*, 2005), and the phylogeography of some species has been thoroughly investigated (Sánchez-Nava *et al.*, 2004). This fauna includes a few widespread species and some exotics (e.g., *Bothriocephalus acheilognathi*) inadvertently introduced with farmed fish such as carp (*Cyprinus carpio*), but it is a poor fauna, composed of only a few species whose diversity shows little resemblance to that of their hosts. For instance, the Goodeinae gut parasite *Rhabdochona lichtenfelsi*, member of a cosmopolitan genus that initially parasitized silurid catfish, shows very little divergence among what can be recognized as subclades, despite the fact that it parasitizes at least 15 Goodeinae species distributed in 10 genera. It seems that it has accompanied its Goodeinae hosts as they dispersed into distant watersheds during the last million years, yet whereas the latter became different species and then genera, *R. lichtenfelsi* remained a single species (Mejia-Madrid *et al.*, 2007; Mejía-Madrid, 2012). Thus, although host–parasite interactions may have promoted Goodeinae differentiation through handicap-type sexual selection, there is little evidence that interactions with a diverse and fast-evolving parasite fauna has promoted speciation in this clade. Unfortunately, no comparable information is available for the Empetrichthynae. Parasites described for *C. baileyi* (Wilson *et al.*, 1966; Deacon, 1979) are largely introduced species transported by exotic fish, and those found in *Empetrichthys latos* have been described for "refugea" populations where the fish were translocated following the destruction of their original habitat (Heckmann, 2009). Thus, we cannot say anything about the possible role of parasite species diversity in the evolution of Empetrichthynae diversity.

Paleoclimate, Glaciations, and Volcanism

Can differences in climate explain the disparity in species richness between the Goodeinae and the Empetrichthynae? This could happen if the climatic conditions of the last 16 my allowed the Goodeinae to have more reproductive events per year than their northern cousins. The question is therefore not whether the mean temperatures in the two geographic areas were different through the Neogene, when most Goodeinae genera originated (see Webb *et al.*, 2004), but whether the Goodeinae experienced milder/shorter winters during which reproduction did not happen. This is a question about seasonality and its effects on Goodeidae reproduction.

Seasonality

Yearly fluctuations in temperature and rainfall determine the length of the breeding season and generate short-term cycles of population expansion and contraction. Such cycles have been documented in several fish species. For instance, shallow-lake population density of the Goodeinae *Girardinichthys multiradiatus* can change from approximately 10 fish per m^2 in the display areas during the dry season to fewer than 1 in 36 m^2 in the rainy season (Macías García *et al.*, 1998). This is partly due to the fact that reproduction can start in central Mexico in late February, as soon as the temperature is sufficiently high, but as the dry season progresses many shallow ponds and lakes shrink, thereby sometimes fatally increasing the fish density. Since temperatures in the Mexican High Plateau are low in winter, most Goodeinae experience a reduction/cessation of reproductive activity during the cold months, unless they inhabit warm springs (e.g., Mendoza, 1940; Miller *et al.*, 2005). Some Goodeinae species evolved in the warmer coastal basins outside the highlands of central Mexico. They experience milder weather and a less marked seasonality than the rest of the Goodeinae, but because they are mostly members of a river-specialist clade (see Fig. 10.1) it would be impossible to allocate any difference in speciation rate between them and the other Goodeinae to the effect of seasonality (and there seems to be no such difference; see Webb *et al.*, 2004).

Extant populations of Empetrichthynae may also experience protracted breeding seasons, since at least some of them live in thermal springs (Sumner and Lanham, 1942), yet it is unlikely that this was the norm for this fish clade. They must have experienced at least some variation in seasonality over their vast geographic range, and also over time since the split of the two subfamilies. Still, central Mexico and the region of the Great Basin had the same vegetation type during prolonged periods (e.g., mid-Pliocene; Dowsett *et al.*, 1999), which would suggest that they shared gross climatic features. It is also likely that the two areas experienced similar rain patterns; both were influenced every summer by the Atlantic monsoon, and since the Sierra Nevada was already present through most of this period, it produced a rain shadow which stopped the westerly winter rains that would have affected only the region of the Great Basin (Mulch *et al.*, 2008).

Thus, although the climate may have been somehow different, we have no evidence to suggest that longer breeding seasons were the norm in central Mexico, or that these led to faster speciation rates among the Goodeinae than among the Empetrichthynae. Clearly, lack of evidence is no evidence that the effect does not exist, but it seems unlikely since a prolonged reproductive season is not the norm among the Goodeinae.

Geological Instability

As mentioned above, Miller (1948, 1950) proposed that the Empetrichthynae would have experienced recurrent habitat fragmentation, leading to geographic variation such as that observed in *Crenichthys baileyi*, which was at the time regarded as comprising up to five subspecies. The possible association between habitat stability and fish species richness was carefully explored by Smith (1981) in a comparison of the fish faunas of the eastern and western USA through the

Neogene. His main conclusion was that instead of promoting fish diversity, habitat stability in the western USA, coupled with barriers to dispersal, produced a poorer fish fauna than that of the eastern USA. Habitat fragmentation was the result of tectonic activity causing block-faulting and thus desiccation, and of recurrent periods of aridity from the Pleistocene onwards (see, for example, Harvey *et al.*, 1999). This, as Miller (1948, 1950) proposed, would have generated local diversification because small, isolated populations respond rapidly to local adaptation and genetic drift; yet, such a process working over prolonged periods would have also increased the risk of local extinctions. As pointed out by Smith (1981), colonization from outside the region was prevented by barriers to fish dispersal, such as the Rocky Mountains to the west and the Sierra Nevada to the east. A limited colonization rate may be linked to the relative poverty of the fish fauna in the Great Basin, yet it cannot explain the low diversity of the endemic taxa such as the Empetrichthynae. Thus, there is only a handful of species in this subfamily either because they did not produce many species, or because they produced many species that subsequently became extinct, leaving no traces behind.

Geologic stability was also unknown during the Neogene in central Mexico. This is best illustrated by the recent *tour de force* published by Ferrari *et al.* (2011) that shows the extent to which the topography of central Mexico has changed since the Miocene. As a consequence of volcanic activity and uplifting of the southern High Plateau, river basins were formed, fragmented, merged, and emptied through the Miocene and Pliocene, and a dynamic lacustrine landscape appeared in the area in the Pleistocene, of which several lakes remain (see De Cserna, 1989). This provided many opportunities for vicariance and the Goodeinae seem to have responded to them with speciation, given that sister species and even larger clades tend to occupy adjacent basins (Gesundheit and Macías Garcia, 2005; Webb *et al.*, 2004; Domínguez-Domínguez *et al.*, 2006; see also Kidd and Lui, 2008).

Why did the frequent habitat fragmentation not lead to Goodeinae extinctions to the point of preventing an accumulation of clades through time? The Goodeinae may have been more fortunate than the Empetrichthynae in part because the nature of the habitats in which they radiated differed. The Great Basin was essentially a lake district, central Mexico was a patchwork of lakes and rivers through most of the Miocene, and possibly in periods of drought rivers provide more resilient refugia than lakes. This may be one reason why fish faunas in the eastern USA endured periods while the western fish faunas decreased (Smith, 1981), and may similarly explain the survival of many Goodeinae clades through the same periods. Large shallow lakes, which are prone to fragmentation, have been present in central Mexico since the late Pleistocene (see above), but although the region has undergone aridification, this has been less severe than in Nevada and adjacent zones.

Is it then the relative paucity of local extinction that explains the greater species diversity in the Goodeinae than in the Empetrichthynae? It must be a contributing factor. Yet we know that all the major Goodeinae clades (or tribes) were already present some 9 mya, and virtually all genera existed 6 mya (Webb *et al.*, 2004). This pattern of early radiation, perhaps also coupled with a later increase in extinction rate, was also suggested by an analysis using a subsample of the family (Ritchie *et al.*, 2005). Thus, even if the extinction rate was lower than in the Empetrichthynae, we still need to account for the origin of the early diversity of the Goodeinae.

Recent Climatic Instability

There is no evidence of substantial shifts of fish fauna in the Great Basin associated with the Pleistocene glaciations, which contrasts with the post-glacial evolution of some species flocks in the western USA. Possibly, fish species in the western USA were able to track climatic change without much phenotypic change, in accordance with Smith's (1981) proposal that changes in species take longer than climatic or geographic changes. It is also possible that hot springs persisted through the glaciations and acted as refugia. In the south, glaciations contributed to changes in water volume, which dropped towards the east of the Trans-Mexican Volcanic Belt, but the lakes to the west of this region, where most Goodeinae species are found, retained relatively high water levels (Lozano-Garcia *et al.*, 2007; Caballero *et al.*, 2010). Again, species seem to be more stable than the habitat they occupy, as highlighted by Smith (1981). The Goodeinae weathered the cold climate of the glaciations within their ancestral highland lakes, and today several populations are found in montane lakes surrounded by pine and fir forests. It seems that the glacial periods affected neither the Empetrichthynae nor the Goodeinae species' diversity.

VIVIPARITY AND ITS CONSEQUENCES

The convex lineage-through-time plot for the Goodeinae (Ritchie *et al.*, 2005) indicates that they experienced an early radiation, or at least that the ratio of lineage generation/extinction was >1 early in the clade's evolutionary history. This type of adaptive radiation is expected following the occupation of vacant ecological spaces (Schluter, 2000), which is a possible explanation in the case of the Goodeinae as they advanced through the relatively empty freshwater habitats of central Mexico. Adaptive radiations are also expected following the evolution of ecological novelties, which is similar to being new in a vacant ecological space. It has been argued that viviparity constituted such an evolutionary innovation by the Goodeinae (Hubs and Turner, 1939; Ritchie *et al.*, 2005). I now explore some ways in which this reproductive adaptation may be related to Goodeinae species richness.

Viviparity

The Goodeinae differ from the Empetrichthynae in a key reproductive attribute: they are all viviparous. Insemination of ova occurs within the ovarian lumen (Guerrero-Estévez and Moreno-Mendoza, 2012), where the embryos develop, using nutrients drawn from the mother, for about 2 months (Vega-López *et al.*, 2007). The trophotaenia, a hind-gut specialized protrusion of the embryos (Lombardi and Wourms, 1988), is responsible for the embryonic uptake of nutrients in the form of proteins and proteinaceous substances that are scavenged by protein-binding sites (e.g., Schindler, 2003). This constitutes a complex adaptation that is matched by physiological (Schindler and de Vries, 1988; Schindler and Hamlett, 1993), anatomical (the highly vascularized lining of the maternal ovarian cavity; Turner, 1933; Mendoza, 1940; Uribe *et al.*, 2004), and life-history traits in Goodeinae females. Trophotaeniae are found in embryos of all species but one, the monospecific genus *Ataeniobius*. Based on the morphology of the trophotaenia, early workers placed *A. toweri* at the base of the Goodeinae (Hubbs and Turner, 1939), but molecular phylogenies indicate that lack of trophotaenia is a derived condition (Webb *et al.*, 2004). There are no clues about the early stages of the evolution of trophotaenia (and thus of the particular brand of Goodeinae viviparity), and the probability that there is a link between Goodeidae viviparity and speciation cannot be established since the number of contrasts is one. This latter problem can be overcome by comparing species richness between several pairs of viviparous–oviparous sister clades. Fish have evolved viviparity in several independent instances – a circumstance that was used by Mank and Avise (2006) to evaluate the possible link between viviparity and species richness.

Population Size and Dispersal

Mank and Avise (2006) explored the possibility that viviparity and species richness are linked among the Atherinomorpha (to which the Goodeidae belong) using a supertree that allowed sister-clade analyses. They found that invariably the viviparous clade had more species than its oviparous sister clade, although the overall comparison was only marginally significant ($P=0.0625$). Tellingly, they found no evidence of a potential link between viviparity and extinction rates (see below).

Relative to the Goodeidae, there is a potentially large complication with the analysis of Mank and Avise (2006); namely, that in this group the viviparous and oviparous sister clades occupy different geographic areas, which may impose different restrictions on speciation and/or promote different extinction rates. This does not imply that the approach used by Mank and Avise (2006) to explore the association between viviparity and biodiversity is flawed; it is our best tool for the job, and if enough clades are incorporated it should statistically account for the foibles of particular groups. But the Goodeidae is a group full of foibles, and may not correspond to some of the expectations/assumptions of the Mank and Avise (2006) analysis.

The motivations of Mank and Avise (2006) to investigate the association between viviparity and speciation were (1) the observation that viviparity has evolved repeatedly among the Atherinomorpha, and (2) the reasonable expectation that viviparity should imply substantial changes in life history and dispersal capabilities. They expected that viviparity, by reducing the average brood size, may lead to small population size, and thus to extinction, more often than oviparity – an expectation that was not borne by their results ($P = 0.09$). These expectations are only approximately met in the Goodeidae. The Goodeinae, and indeed other viviparous fish, may in fact be more able to colonize new habitats successfully since single pregnant females can originate new populations, as demonstrated by Deacon *et al.* (2011), although stringent Goodeinae female mate choice may counteract this advantage (e.g., Macías Garcia *et al.*, 1998). Also, brood size among the oviparous Empetrichthynae is not larger than that of the Goodeinae (some of which can have broods composed of several dozen fish; Macías Garcia and Saborío, 2004). The mechanics of copulation appear to be very similar in the Empetrichthynae and the Goodeinae, as the latter also embrace and vibrate during copulation and, lacking an intromittent organ, males also fold the anal fin over the female cloacae to conduct sperm (Nelson, 1975). The above, as well as some anatomical similarities in their ovaries (Uribe *et al.*, 2012), suggest that several attributes that may appear to be adaptations to viviparity were already present in the Empetrichthynae.

Apart from the demographic effects proposed by Mank and Avise (2006), there are two potential consequences of viviparity that may influence speciation rate: (1) facilitation of the spread of chromosomal rearrangements, and (2) promotion of sexual asymmetries in potential re-mating rates.

Chromosomal Rearrangements

Chromosomal rearrangements have been associated with rapid speciation in some organisms, such as mammals (Bush *et al.*, 1977; Bengtsson, 1980). Indeed, once it reaches fixation, a population of a chromosomal variant would be unlikely to produce viable hybrids with its ancestral form, hence leading to speciation. However, it is not clear how the variant can initially become established in the population, particularly because of the low fertility of heterozygous carriers, which produce recombinant gametes and zygotes with unbalanced chromosomal genotypes. It has been proposed that a small population size and inbreeding can promote the fixation of chromosomal variants (Bush *et al.*, 1977; White, 1978; Lande, 1979; Hedrick, 1981; see also Huai and Woodruff, 1998). Another factor that may help fixation is the occurrence of low, or "soft", selection against carriers, such as may occur *in utero* (Lande, 1979; Bengtsson, 1980). The Goodeinae reproductive biology would meet the conditions for the operation of this type of soft selection. This was proposed by Turner (1983), who also realized that the Goodeinae exhibit a remarkable variation of karyotypes (see Uyeno *et al.*, 1983). There are cases of sister species with different karyotypes (e.g., *Zoogoneticus tequila*/*Z. quitzeoensis*, $2n = 46/28$; *Allotoca* spp. $2n = 26$, 46, or 48; Uyeno *et al.*, 1983; Webb and Miller, 1998). Such differences are not evidence that karyotypic divergence lies at the origin of the speciation processes (i.e., it is not necessarily the case that those chromosomal rearrangements provoked automatic speciation), but they can and may have contributed to completing the speciation process in a few instances, especially by constituting post-mating barriers to reproduction upon secondary contact of sister or congeneric species (e.g., *Zoogoneticus tequila/Z. quitzeoensis*; *Allotoca dugesi/A. diazi*; Gesundheit and Macías Garcia, 2005). However, about two-thirds of the Goodeinae have karyotypes with $2n = 48$, thus, again, the origin of the bulk of the Goodeinae diversity must lie elsewhere.

Differential Investment and Sexual Conflict

Because viviparity results in a reduction in the number of eggs that the female produces, it can only evolve if it provides an initial advantage to the female in terms of either offspring survival or insemination rate. In aquatic organisms, the cost of sperm waste and the risk of facing sperm competition would impose selection on males to deliver sperm increasingly closer to the female genital pore at the time of egg deposition. Indeed, internal insemination is commonly seen in oviparous fish clades (see details in, for example, Nelson, 1964 [Characidae]; Breder and Rosen, 1966; Munehara *et al.*, 1989 [Cottidae]; Regan, 1913; Parenti, 1989 [Phallostethidae]), and at least in one instance we know that it antecedes viviparity (Meyer and Lydeard, 1993). The possible benefits that males derive from internal fertilization may be met by female benefits in the form of increased fertilization rate. The initial benefits that females get from egg retention are less evident. Because of space limitations, internally fertilized females that retain eggs should produce fewer offspring than internally fertilized females that lay their eggs. Unless egg predation was very high and eggs were not laid until close to hatching (an unlikely situation in the early evolutionary stages of egg retention), this cost would not be compensated by any survival advantage for the offspring (which would accrue from further adaptations on the mother and their embryos). Yet egg retention would be beneficial for the male siring the clutch, since this would further reduce the risk of sperm competition. In other words, if eggs were laid as they were fertilized, males would have to compete for each one, especially if there is a time lag between the laying of successive eggs, as is the case in the Empetrichthynae. Consequently, any adaptation that allows the male to manipulate the probability that females retain the fertilized eggs would spread in the population. Female reproductive physiology can be manipulated directly, for instance through substances in the ejaculate, or through the zygotes via genomic imprinting. This form of sexual conflict can influence the evolutionary trajectories of species, and may also promote speciation.

Genomic imprinting, often the result of chromosomal methylation, is an epigenetic process that can lead to differential expression of the same allele depending on whether the copy is transmitted by one parent or the other (Wilkins, 2005). This process potentially allows the parents to influence the expression of genes in their offspring. Since offspring growth rates can influence postnatal survival, any increase in the expression of genes that influence embryonic growth would be beneficial for parents of the sex that does not invest/carry the unborn offspring. Conversely, excessive investment in current offspring could be detrimental for the members of the other sex, as it may reduce their chances of successfully reproducing in the future. This conflict between sexes should lead to genomic imprinting arms races in sequentially polygamous species where investment in offspring is made mostly by the members of one sex (Haig, 2004); in monogamous species, the optimum level of investment should be similar for both sexes. These conditions describe the reproductive situation of all non-monogamous, iteroparous, viviparous organisms such as the Goodeinae and other viviparous fish.

Genomic conflict can be unmasked by hybridizing distant populations of one species, or closely related species; crosses in one direction should generate offspring that are larger than normal (as female defenses against foreign male manipulation fail), whereas in the other direction the products should be small or non-viable (females over-defend against manipulation by less competitive foreign males; see Tilghman, 1999; Constância *et al.*, 2002). One study used the viviparous poeciliid *Heterandria formosa* as the model, as it is the most extreme case of female provisioning of embryos (or matrotrophy) in that family. The results showed phenotypic evidence consistent with the occurrence of genomic conflict in crosses between two distinct populations (Schrader and Travis, 2009; see also O'Neil *et al.*, 2007).

Conflict over the provisioning of embryos involves intrapopulation genomic arms races. Because of the arbitrary and probably random nature of the mutations involved, the genetic control of DNA methylation is likely to be different across populations, generating a strong post-zygotic barrier upon secondary contact. We do not know yet whether this conflict occurs in the Goodeinae, or whether it may have been involved in the generation of Goodeinae diversity, but it is clearly a good candidate to explain the diversity of this clade of viviparous fish.

Viviparity, an innovation that distinguishes the Goodeinae from the Empetrichthynae, can potentially promote dispersal and colonization by small propagules, thereby favoring random as well as adaptive divergence between small populations. It can facilitate the spread of new chromosomal arrangements that may promote, or at least complete, speciation. Viviparity can also lead to genomic conflict and the subsequent evolution of post-zygotic reproductive barriers. Another consequence of viviparity is that it locks the females in a reproductive mode that, at least initially, increases the difference in the rates at which males and females can re-mate; this in turn determines the intensity with which sexual selection operates (Clutton-Brock and Parker, 1992) and therefore influences speciation rates, although evidence of this last link is elusive (see below, and Ritchie, 2007).

Potential Re-Mating Rates

Judging by the scant information available, Empetrichthynae females seem to be able to mate repeatedly, at least for a few hours/days, but perhaps for longer periods (Kopec, 1949), whereas presumably males are ready to mate frequently every day and throughout the breeding season, which may last most of the year in warm springs. Thus, the potential re-mating rate of Empetrichthynae males must be somewhat higher than that of females. Among the Goodeinae this asymmetry is much more pronounced; gestation periods last about 2 months (e.g., Macías Garcia and Saborío, 2004), and females do not store sperm (Mendoza, 1962; Fitzsimons, 1972), thus adult females are receptive only every 2 months and only for about 1 week. Receptivity periods have been ascertained on only two species (*Girardinichthys multiradiatus* and *G. viviparous*), but laboratory records suggest that these are similar for females from all genera (Macías Garcia and Saborío, 2004; E. Saborío unpublished data; E. Ávila-Luna, unpublished data). Males, on the other hand, constantly seek females during the day, and engage in courtship and copulation attempts at a rate that is plainly above the optimal for females (e.g., Valero *et al.*, 2005). This asymmetry in potential re-mating rates leads to significant male biases in the operational sex ratio (Macías Garcia, 1994), and should favor the expression of male attributes (e.g., ornaments or weaponry) that allow them to capitalize on such potential. Even if the link between intensity of sexual selection and potential re-mating rates is not as straightforward as previously thought (Klug *et al.*, 2010), viviparity sets the scenario for sexual selection to operate more strongly among the Goodeinae than among their northern, oviparous cousins.

Non-Coercive Internal Fertilization

Males can monopolize mating opportunities through physically excluding other males from the areas visited by receptive females or from the females' vicinity, or they can be more effective than their competitors in tracking females, or in attracting and coaxing them to mate. This dichotomy between intrasexual selection (normally male–male competition) and intersexual selection (commonly female mate choice) was initially proposed by Darwin (1859, 1871). It has remained popular because although the function of weaponry can be intuitively understood, the reasons why ornaments evolve and are maintained, and why females pay attention to ornaments, are much less easily understood. But in nature the two forms of sexual selection often merge, as in the Goodeinae. Males are very pugnacious and can victimize smaller rivals (e.g., Macías Garcia and Valero, 2001; Kelley *et al.*, 2006), but that is not sufficient to secure matings. Unlike other viviparous fish (see Constantz, 1984), the Goodeinae lack an intromittent organ, and the relatively non-specialized structure that the males use to direct sperm into the female cloacae may not be properly referred to as a gonopodium (*cf.* Miller and Fitzsimons, 1971). Copulation requires the cooperation of the female; it takes place during a copulatory embrace, and females can avoid insemination from unwanted males indefinitely (Macías Garcia *et al.*, 1998). Such effective control over mating may facilitate sexual selection through female mate choice (Macías Garcia and Valero, 2010), although the relatively primitive insemination method also forces females to mate with males within a narrow range of relative sizes (Macías Garcia, 1994; Bisazza, 1997). Both the highly developed system of viviparity and the nature of their internal insemination process would suggest that sexual selection through female choice is common among the Goodeinae, especially relative to the Empetrichthynae. I shall now briefly go to the evidence for sexual selection in this clade.

Sexual Selection in the Goodeidae

Ecological divergence may not lead to the production of new species because it can be eroded by interbreeding once the populations meet again. Effective barriers, which can be adaptive in origin, need to influence the probability of mating (e.g., the so-called magic traits; Kisdi and Priklopil, 2011). Thus, random population divergence in reproductive attributes (e.g., chromosomal number) or in mating behavior (such as mate preferences) has the potential to generate and maintain the reproductive isolation required for speciation to take place; hence the interest in studying the association between sexual selection and speciation (Panhuis *et al.*, 2001; Ritchie, 2007; Butlin *et al.*, 2011; Kraaijeveld *et al.*, 2011). The intensity of sexual selection has not been quantified in any Goodeidae, but several studies have investigated the effect of male ornaments on female behavior, as well as some of the costs imposed by those ornaments on the bearers. This literature was recently reviewed in Macías Garcia and Valero (2010), and only brief accounts will be given here.

Male Epigamic Characters

Males in all Goodeinae species have larger dorsal, anal, and caudal fins than females, and in some species the differences are very pronounced (see Figs 10.3–10.5). Large fins are obvious candidates for the study of condition dependence, as the cost of bearing them increases exponentially with size. They were found to constitute a handicap that increases the risk of being captured by predators (Macías Garcia *et al.*, 1994, 1998), and their development is incomplete in fish exposed to contaminants (Arellano-Aguilar and Macías Garcia, 2008). Fin size has not been experimentally manipulated, but its effect on female mating preferences has been assessed indirectly from the responses of females to differences in male body shape, which is a function of fin size (e.g., Macías Garcia *et al.*, 1994; González-Zuarth and Macías Garcia, 2006). The magnitude of the sexual dimorphism in fin size and body shape varies between populations (González-Zuarth and Macías Garcia, 2006), but at this level is not clearly linked to genetic differentiation (Macías Garcia *et al.*, 2012). Fin size/body shape dimorphism also

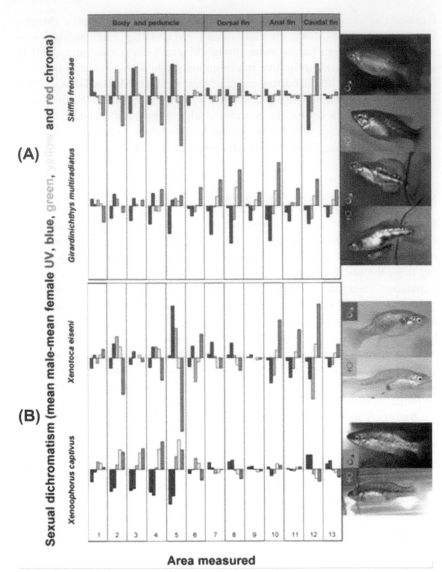

FIGURE 10.3   For each of 25 species, the mean reflectance at 13 points (see Fig. 10.6) on the flanks and fins of 10 males and 10 females was obtained and used to calculate the mean UV (300–400 nm), blue (400–475 nm), green (475–550 nm), yellow (550–625 nm), and red (625–700 nm) chroma at each point for each species and sex, through dividing the sum within the respective range by the total sum across all the spectrum. Female chromas were subtracted from the males'; positive values indicate that at that particular point males have a larger (positive) or smaller (negative) value of chroma than females. Closely related species can differ in the color pattern and in the magnitude of sexual dimorphism (general size of the bars, but formally calculated as the SD of the values for each fish measure). See color plate at the back of the book. *Photographs by the author, and courtesy of R. Rodríguez Tejeda and Y. Saldívar Lémus*

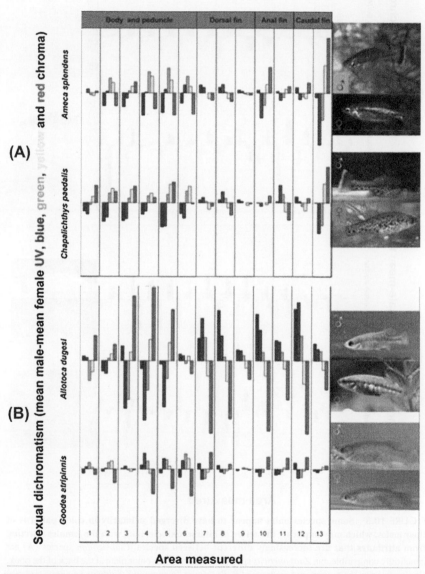

FIGURE 10.4   Sometimes closely related species share both the color pattern of both sexes and the magnitude of the dimorphism (relative size of bars), whereas in one tribe it is possible to find the largest (*A. dugesi*) and the smallest (*Goodea atripinnis*) degree of sexual dichromatism. See color plate at the back of the book. *Photographs by the author, and courtesy of R. Rodríguez Tejeda and Y. Saldívar Lémus*

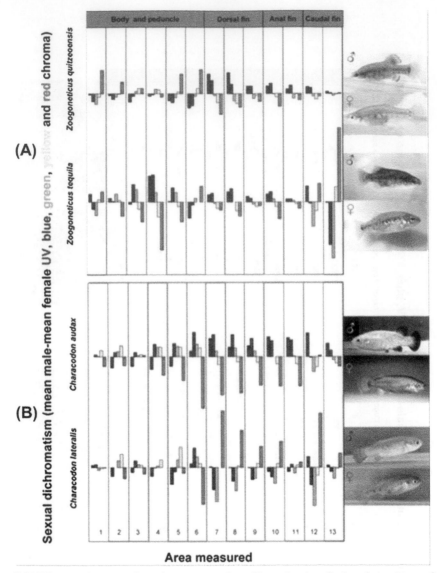

FIGURE 10.5 Some species pairs appear to have diverged primarily in color patterns of their males, which may implicate the action of intersexual selection, with the females selecting from attributes that are increasingly different between species. *Characodon* species (A) are genetically compatible, but *Zoogoneticus* species (B) are not. See color plate at the back of the book. *Photographs by the author, and courtesy of R. Rodríguez Tejeda and Y. Saldívar Lémus.*

varies between species, where it predicts differences in the rate at which geographic distance generates genetic differences (i.e., it predicts genetic divergence; Ritchie *et al.*, 2007), although its role in driving speciation has not been demonstrated (Ritchie *et al.*, 2005).

Male courtship is a target of female preferences that can influence the paternity of offspring in multiply inseminated females (Macías Garcia and Saborío, 2004). Courtship expression reflects the condition of the male (Arellano-Aguilar and Macías Garcia, 2008), and may also increase exposure to parasites, thereby constituting an honest signal of male condition (Ávila *et al.*, 2011). Male courtship behavior may be very different between populations (González-Zuarth and Macías Garcia, 2006), partly at least as a consequence of local differences in female responsiveness (González-Zuarth *et al.*, 2011). The limited information on the Empetrichthynae suggests that some basic courtship displays are common to both subfamilies, but there are many more distinctive courtship displays among the Goodeinae, and state transitions (gains and losses) of specific courtship displays tend to accumulate towards the tips of the phylogeny, suggesting a link (not necessarily causal) with speciation (M. Méndez-Janovitz, A. González-Voyer, M. G. Ritchie, and C. Macías Garcia, unpublished data).

Finally, Goodeinae males are frequently more colorful than females (Figs 10.3–10.5; see below). Color differences can be concentrated on the dimorphic fins, but they may also involve the rest of the body. As in many other biological systems, the carotenoid-based coloration of many species is a likely indicator of health (see Milinski and Bakker, 1990), and in *G. multiradiatus* its expression is reduced if fish are exposed to pesticides early in life (Arellano-Aguilar and Macías Garcia, 2008). Other male ornamental color markings that are attractive to females include shiny scales (speckles) such as those of *Xenotoca variata* (Moyaho *et al.*, 2005), which increase the chance of being detected by predatory snakes (Moyaho *et al.*, 2004), and ultraviolet reflections (Macías Garcia and Burt de Perera, 2002), but there is no evidence that females may be attracted to melanic male coloration (Moyaho *et al.*, 2010).

The design of one color mark in the Goodeinae evolved though the exploitation of pre-existing female feeding biases (Macías Garcia and Ramírez, 2005). A yellow stripe on the distal edge of the undulating tail fin resembles a larva or worm, which evokes female feeding behavior (including bites) to the point of distracting them from chasing real prey (Macías Garcia and Saldívar Lémus, 2012). In more advanced stages this interaction is largely reproductive, and the females no longer pay a foraging cost by responding to the ornament, which becomes a male condition-dependent signal, as it is nevertheless bitten by some fish and has to be regenerated.

Thus, Goodeinae males can enjoy very high mating rates, yet the price they pay to gain access to females is also very high. From the few species studied so far, we know that large fins hamper maneuvers when avoiding incoming predator attacks; and some color markings consume useful antioxidant carotenoids, others attract predators, and yet others are nibbled by other fish and have to be regenerated regularly. Courtship behavior is also costly and exposes males to parasites and, conceivably, also to predators. If males pay such high costs to gain access to females, is it possible that they, in turn, select the females to which they direct their courtship? This is a relevant question here because mutual mate choice can lead to ornamentation in both sexes, adding to the potential for sexual selection to promote divergent local evolution.

Male Mate Choice?

Males of several fish species are attracted to large females because size predicts fecundity (Sargent *et al.*, 1986; Herdman *et al.*, 2004). This may have promoted the evolution of small, highly maneuverable males among the Poeciliidae (Bisazza, 1993), and it would be expected that Goodeinae males showed the same preference for large females. However, any such tendency would be opposed by the mechanical incompatibility that renders small males unable to inseminate large females (Bisazza, 1997) and/or by their tendency to mate size-assortatively (Macías Garcia, 1994). Therefore, males may either mate indiscriminately with any female of the right size or, if there is sufficient variation in female quality unrelated to size, direct more courtship effort to females of better perceived quality. Size-independent male mate choice has been demonstrated in a handful of fish groups, often involving carotenoid-based coloration around the belly (e.g., Amundsen and Forsgren, 2001), which has been shown to be an honest signal of female quality (Massironi *et al.*, 2005).

To test the above prediction that costly male ornamentation has led to male choosiness in the Goodeidae, it would be necessary to demonstrate that (1) there is substantial variation in female fecundity within size classes; (2) this variation is correlated with one or more female attributes; and (3) males react to variation in such attributes in an adaptive manner by preferentially responding to attribute values that correlate with high fecundity. As a first approximation, I examine here the patterns of color dimorphism in a large sample of Goodeinae species and ask (1) whether color dimorphism is due to male ornamental color, female ornamental color, or both; (2) whether there is a correlation between male and female colorfulness, which would suggest that color has been naturally selected; and (3) whether sexual dichromatism is greater in species that occupy larger geographic ranges, and thus a greater diversity of habitats; this was discussed when looking at the possible effects of geographic range on Goodeinae diversification.

I obtained reflectance measures (Minolta CM-2600D spectrophotometer; pulsed xenon lamps; $\lambda = 360\text{--}740$ nm; readings at each 10 nm) at 13 standardized points in the fins and flanks of approximately 10 fish of each sex from 25 species, plus data on males from another species (mean $\pm$ SD $= 9.8 \pm 0.8$, and 9.4 ± 2.2 males and females per species, respectively). The points were selected to encompass all the areas where I have observed sexual dichromatism.

1. *Is color dimorphism due to the ornamentation of males, females, or both?*
   Males and females can differ in their color and color patterns in a great variety of ways (see photographs in Figs 10.3–10.5), thus calculating a single measure of chromatic dimorphism (dichromatism) that is comparable across all the species is difficult. I used the procedure outlined in Figure 10.3 to

calculate differences between males and females in five chromatic values (UV, blue, green, yellow, and red). The variation in these chromatic differences indicates the magnitude of sexual dichromatism at each of the 13 areas measured, and the species' degree of dichromatism is the sum of these 13 values.

The above procedure shows that the pattern of sexual dichromatism is very variable among the Goodeinae. In some species the differences in color are concentrated on the anterior part of the body, and in others on the fins or in the caudal region (Fig. 10.3A). Species also differ in the distribution of colors between sexes: males of one species can have bluer bodies than females, and in another species females may be the ones with blue bodies (Fig. 10.3). There are groups of species with similar color patterns and displaying a similar degree of dichromatism (Fig. 10.4A), but it is also possible to find the most extreme values of sexual dichromatism within the same tribe (Fig. 10.4B). There are also frequent cases of sister species differing in the magnitude and type of sexual dichromatism (Fig. 10.5A, B).

Although females are not normally plain, only in a few cases are they more colorful than the males. Males in the genus *Allotoca* (Fig. 10.4B), the most speciose in the family, are very bright, and females are very colorful (chroma saturation). The design of the female color patterns in this (and other) Goodeinae genera, of vertical bars across the abdomen, gives the female a rounded aspect; thus both the intensity of their colors, and their color patterns, may constitute signals used during reproductive interactions.

2. *Is there a correlation between male and female colorfulness?* If color and color patterns were constrained by natural selection, fish of both sexes would be darker in some places and brighter in others, as the environment allowed. To evaluate this possibility I calculated one value of brightness for the males and one for the females of each species. The result (Fig.10.6) indicates that, in most species, the overall brightness of males and females is similar, regardless of whether they differ in chroma. This suggests that the environment plays a role in driving the evolution of color patterns in the Goodeinae. Tellingly, the correlation between male and female brightness is broken in the members of the genus *Allotoca*, whose males are brighter, or whose females are darker (chroma-saturated), than expected. This again suggests that this may be a case of female ornamentation, and it would be interesting to explore whether it may be linked to species diversity in a genus that includes about 20% of the total Goodeinae species described.

3. *Is size of geographic range associated with increased sexual dichromatism?* This question was addressed above (Fig. 10.2). There is an enormous variation in the size of the Goodeinae geographic ranges, but this is due to the existence of only two or three massively widespread species. Of these, two are rather monochromatic and one is dichromatic; the cluster composed by the remaining species suggests that there is no association between those variables even if the apparent outliers are removed.

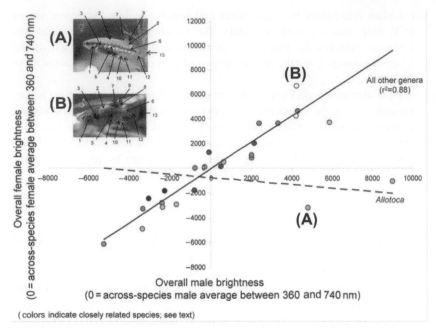

(colors indicate closely related species; see text)

FIGURE 10.6   The species-specific brightness at each point was measured for each sex as the sum of the reflectance values across all lambdas, and the sum of all brightness values across the 13 points was the measure of overall brightness. I calculated one value of overall brightness for the males and one for the females of each species. Then the average brightness across species was calculated for males ($n=26$ species) and for females ($n=25$) separately, and sex-specific deviations from these means were plotted for each species. This provides independent measures of brightness for both sexes, thus the observed correlation indicates that males and females respond to similar selective pressures but does not tell us anything about the degree of dimorphism in brightness. Species share color code if they belong to the same or a closely related genus, according to Webb *et al.* (2004), thus *Hubbsina turneri* shares color with the two species of *Girardinichthys*; *G. atripinnis* has the same color as *Ataeniobius toweri*, etc. See color plate at the back of the book. *Photographs by the author, and courtesy of R. Rodríguez Tejeda and Y. Saldívar Lémus.*

The analysis of color patterns and color dimorphism in the Goodeinae shows that sexual dichromatism is the norm, it can be striking in magnitude, and it varies in a way that is consistent with a degree of independent arbitrary divergence rather than with strict phylogenetic constraints; there is no evidence that habitat diversity promotes sexual dimorphism. Brightness, a component of color, also varies between species but is correlated between sexes, suggesting a degree of environmental control.

A SINGLE EXPLANATION? IF SO, WHICH ONE?

Recent trends from which the Empetrichthynae are exempt may account for a small overestimation of the Goodeinae species richness, and perhaps a larger underestimation of Empetrichthynae diversity. However, given that the distinctive feature of Goodeinae diversity is the high ratio of genera per species, this nomenclatural inconvenience is not really a problem. Since their separation in the Miocene, the Empetrichthynae and the Goodeinae have evolved in areas of comparable geological and ecological instability. It is possible that extinction rates were somewhat higher in the north, since shallow lacustrine environments are less resilient than rivers, but this probably played a minor role in shaping overall Empetrichthynae diversity. Ecological specialization has occurred among the Goodeinae only to a limited degree, but it may nevertheless explain some of the difference in species diversity between the two clades. Still, the greatest contribution to the disparity in species richness must be due to the evolution of viviparity. Its origins are a puzzle, but, once established, it has had the opportunity to influence speciation through chromosomal rearrangement and probably through facilitating the expression of antagonistic genomic imprinting, but mostly because it generates sexual asymmetries in potential mating rates. This, coupled with a sub-efficient mechanism of internal fertilization, is linked to a peculiar style of sexual selection through female mate choice that has been associated with at least some population differentiation, and constitutes the best explanation for the great diversity in ornaments that can be seen among the Goodeinae. Finally, intersexual selection has led to increasingly costly male sexual displays, which in turn may be generating a novel process: differentiation of populations driven by male preferences for female ornaments. Is all this the result of moving into the tropics, though? Apparently not, unless it is found that viviparity evolved as a consequence of living in a more tropical area.

ACKNOWLEDGMENTS

Many students, colleagues, and friends have enriched my research on the Goodeinae, and their names appear in the authors' lists of several papers cited here. Mike Ritchie, Anne Magurran, and Jeff Graves have been my academic life-vests, and Edgar A. Luna has collected, maintained, and measured the color of countless fish. Ruth Rodríguez Tejeda has facilitated the logistics at my laboratory during some critical periods, and together with Marcela Méndez Janovitz provided some editorial help. Gerardo Rodríguez Tapia of the Unidad de Geomática calculated the size of the species' geographic ranges. I conducted all the analyses and writing for this chapter while on a sabbatical leave supported by DGAPA (UNAM), with the forbearance and support of my local host, Diego Gil, of the MNCN-CSIC.

REFERENCES

Álvarez, J., Arriola-Longoria, J., 1972. Primer goodeido fosil procedente del Plioceno jalisciense (Pisces, Teleostomi). Bol. Sociedad Cien. Naturales Jalisco. 6, 6–15.

Amundsen, T., Forsgren, E., 2001. Male mate choice selects for female coloration in a fish. Proc. Natl. Acad. Sci. U. S. A. 98, 13155–13160.

Arellano-Aguilar, O., Macías Garcia, C., 2008. Exposure to pesticides impairs the expression of fish ornaments reducing the availability of attractive males. Proc. R. Soc. B. 275, 1343–1350.

Ávila, E., Valero, A., Macías Garcia, C., 2011. Componentes conductuales del cortejo incrementan la exposición a parásitos en el pez vivíparo *Xenoophorus captivus*. TIP Revista Especializada Ciencias Químico-Biológicas 14, 75–81.

Barbour, C.D., 1973. The systematics and evolution of the genus *Chirostoma* Swainson (Pisces: Atherinidae). Tulane Stud. Zool. Bot. 18, 97–141.

Bengtsson, B.O., 1980. Rates of karyotype evolution in placental mammals. Hereditas 92, 37–47.

Berdan, E.L., Fuller, R.C., 2012. A test for environmental effects on behavioral isolation in two species of killifish. Evolution 66, 3224–3237.

Bernatchez, L., 2004. Ecological theory of adaptive radiation. An empirical assessment from Coregonine fishes (Salmoniformes). In: Hendry APSS (Ed.), Evolution Illuminated, Salmon and their Relatives. Oxford University Press, New York, pp. 175–207.

Bisazza, A., 1993. Male competition, female mate choice and sexual size dimorphism in poeciliid fishes. In: Hungtingford, F., Torricelli, P. (Eds.), The Behavioural Ecology of Fishes, Harwood Academic Press, London, pp. 257–286.

Bisazza, A., 1997. Sexual selection constrained by internal fertilization in the livebearing fish *Xenotoca eiseni*. Anim. Behav. 54, 1347–1355.

Bloom, D.D., Piller, K.R., Lyons, J., Mercado-Silva, N., Medina-Nava, M., 2009. Systematics and biogeography of the silverside tribe Menidiini (Teleostomi: Atherinopsidae) based on the mitochondrial ND2 gene. Copeia 2, 408–417.

Breder, C.M., Rosen, D.E., 1966. Modes of Reproduction in Fishes. The American Museum of Natural History. New York.

Bush, G.L., Case, S.M., Wilson, A.C., Patton, J.L., 1977. Rapid speciation and chromosomal evolution in mammals. Proc. Natl. Acad. Sci. U. S. A. 74, 3942–3946.

Butlin, R., Debelle, A., Kerth, C., Snook, R.R., Beukeboom, L.W., Cajas Castillo, R.F., The Marie Curie SPECIATION Network, 2011. What do we need to know about speciation? Trends Ecol. Evol. 27, 27–39.

Caballero, M., Lozano-García, M.S., Vázquez-Selem, L., Ortega, B., 2010. Evidencias de cambio climático y ambiental en registros glaciales y en cuencas lacustres del centro de México durante el último máximo glacial. Bol. Sociedad Geológica Mexicana 62, 359–377.

Castillo Cajas, R.F., Selz, O.M., Ripmeester, E.A.P., Seehausen, O., Maan, M.E., 2012. Species-specific relationships between water transparency and male coloration within and between two closely related Lake Victoria cichlid species. Int. J. Evol. Biol. (epub ahead of print). ID 161306, doi: 10.1155/2012/161306.

Clutton-Brock, T.H., Parker, G.A., 1992. Potential reproductive rates and the operation of sexual selection. Q. Rev. Biol. 67, 437–456.

Constância, M., Hemberger, M., Hughes, J., Dean, W., Ferguson-Smith, A., Fundele, R., Stewart, F., Kelsey, G., Fowden, A., Sibley, C., Reik, W., 2002. Placental-specific IGF-II is a major modulator of placental and fetal growth. Nature 417, 945–948.

Constantz, G.D., 1984. Sperm competition in poeciliid fishes. In: Smith, R.L. (Ed.), Sperm Competition and the Evolution of Animal Fighting Systems. Academic Press, London, pp. 465–485.

Cooke, G.M., Chao, N.L., Beheregaray, L.B., 2012. Natural selection in the water: freshwater invasion and adaptation by water colour in the Amazonian pufferfish. J. Evol. Biol. 25, 1305–1320.

Cox, C.B., Moore, P.D., 1985. Biogeography. An Ecological and Evolutionary Approach, fourth ed. Blackwell Scientific Publications, London.

Darwin, C., 1859. On the Origin of Species by Means of Natural Selection, or the Preservation of Favoured Races in the Struggle for Life. John Murray, London.

Darwin, C., 1871. The Descent of Man, and Selection in Relation to Sex. John Murray, London.

Deacon, A.E., Ramnarine, I.W., Magurran, A.E., 2011. How reproductive ecology contributes to the spread of a globally invasive fish. PLoS One 6, e24416.

Deacon, J.E., 1979. Endangered and threatened fishes of the West. Great Basin Nat. 3, 41–64.

De Cserna, Z., 1989. An outline of the geology of Mexico. In: Bally, A.W., Palmer, A.R. (Eds.), The Geology of North America – An Overview, Vol. A. Geological Society of North America, Boulder, pp. 233–264.

de Queiroz, A., Lawson, R., Lemos-Espinal, J.A., 2002. Phylogenetic relationships of North American garter snakes (*Thamnophis*) based on four mitochondrial genes: How much DNA sequence is enough? Mol. Phylogenet. Evol. 22, 315–329.

Doadrio, I., Domínguez, O., 2004. Phylogenetic relationships within the fish family Goodeidae based on cytochrome *b* sequence data. Mol. Phylogenet. Evol. 31, 416–430.

Dobzhansky, T., 1950. Evolution in the tropics. Am. Sci. 38, 209–221.

Domínguez-Domínguez, O., Doadrio, I., Pérez-Ponce de León, G., 2006. Historical biogeography of some river basins in Central Mexico evidenced by their goodeine freshwater fishes: a preliminary hypothesis using secondary Brooks parsimony analysis (BPA). J. Biogeogr. 33, 1437–1447.

Domínguez-Domínguez, O., Alda, F., Pérez-Ponce de León, G., García-Garitagoitia, J.L., Doadrio, I., 2008. Evolutionary history of the endangered fish *Zoogoneticus quitzeoensis* (Bean, 1898) (Cyprinodontiformes: Goodeidae) using a sequential approach to phylogeography based on mitochondrial and nuclear DNA data. BMC Evol. Biol. 8, 161.

Dowsett, H.J., Barron, J.A., Poore, R.Z., Thompson, R.S., Cronin, T.M., Ishman, S.E., Willard, D.A., 1999. Middle Pliocene Paleoenvironmental Reconstruction: PRISM2. US Geological Survey Open File Report 99-535. Available at http://pubs.usgs.gov/of/1999/of99-535/#fig4.

Ferrari, L., Orozco-Esquivel, T., Manea, V., Manea, M., 2011. The dynamic history of the Trans-Mexican Volcanic Belt and the Mexico subduction zone. Tectonophysics 522–523, 122–149.

Fisher, R.A., 1958. The Genetical Theory of Natural Selection, second ed. Dover Publications, New York.

Fitzsimons, J.M., 1972. A revision of two genera of goodeid fishes (Cyprinodontiformes, Osteichthyes) from the Mexican Plateau. Copeia 4, 728–756.

Gesundheit, P., Macías Garcia, C., 2005. Biogeografía cladística de la familia Goodeidae. Cap. 19. In: Llorente Bousquets, J., Morrone, J.J. (Eds.), Regionalización Biogeográfica en Iberoamérica y Tópicos Afines: Primeras Jornadas Biogeográficas de la Red Iberoamericana de Biogeografía y Entomología Sistemática (RIBES XII.I-CYTED). Facultad de Ciencias, UNAM. México, pp. 319–338.

González-Zuarth, C., Macías Garcia, C., 2006. Phenotypic differentiation and pre-mating isolation between allopatric populations of *Girardinichthys multiradiatus*. Proc. R. Soc. Lond. B. 273, 301–307.

González-Zuarth, C., Vallarino, A., Macías Garcia, C., 2011. Female responsiveness underlies the evolution of geographic variation in male courtship between allopatric populations of the fish *Girardinichthys multiradiatus*. Evol. Ecol. 25, 831–843.

Grudzien, A., Turner, B.J., 1984. Direct evidence that the *Ilyodon* morphs are a single biological species. Evolution 38, 402–407.

Grudzien, T., White, M., Turner, B.J., 1992. Biochemical systematics of the viviparous fish family Goodeidae. J. Fish Biol. 40, 801–814.

Guerrero-Estévez, S., Moreno-Mendoza, N., 2012. Gonadal morphogenesis and sex differentiation in the viviparous fish *Chapalichthys encaustus* (Teleostei, Cyprinodontiformes, Goodeidae). J. Fish Biol. 80, 572–594.

Haig, D., 2004. Genomic imprinting and kinship: how good is the evidence? Annu. Rev. Genet. 38, 553–585.

Hamilton, W.D., Zuk, M., 1982. Heritable true fitness and bright birds: A role for parasites? Science 218, 384–387.

Harvey, A.M., Wigand, P.E., Wells, S.G., 1999. Response of alluvial fan systems to the late Pleistocene to Holocene climatic transition: contrasts between the margins of pluvial Lakes Lahontan and Mojave, Nevada and California, U.S.A. Catena. 36, 255–281.

Hasson, O., 1991. Sexual displays as amplifiers: practical examples with an emphasis on feather decorations. Behav. Ecol. 2, 189–197.

Heckmann, R.A., 2009. The adaptive characteristics and parasitofauna of the three refugea populations of Pahrump poolfish, *Empetrichthys latos latos* (Miller), Nevada (USA). Proc. Parasitol. J. 47, 1–32.

Hedrick, P.W., 1981. The establishment of chromosomal variants. Evolution 35, 322–332.

Herdman, E.J.E., Kelly, C.D., Godin, J.G.J., 2004. Male mate choice in the guppy (*Poecilia reticulata*): do males prefer larger females as mates? Ethology 110, 97–111.

Holman, J.A., 1955. Pleistocene Amphibians and Reptiles in North America. Oxford University Press, New York Clarendon Press, Oxford.

Huai, H., Woodruff, R.C., 1998. Clusters of new identical mutants and the fate of underdominant mutations. Genetica 102–103, 489–505.

Hubbs, C.L., Turner, C.L., 1939. Studies of fishes of the order Cyprinodontes. XVI. A revision of the Goodeidae. Misc. Publ. Mus. Zool. Univ. Michigan 252, 1–80.

Jelks, H.L., Walsh, S.J., Burkhead, N.M., Contreras-Balderas, S., Díaz-Pardo, E., Hendrickson, D.A., Lyons, J., Mandrak, N.E., McCormick, F., Nelson, J.S., Platania, S.P., Porter, B.A., Renaud, C.B., Schmitter-Soto, J.J., Taylor, E.B., Warren Jr., M.L., 2008. Conservation status of imperiled North American freshwater and diadromous fishes. Fisheries 33, 372–407.

Kelley, J., Magurran, A.E., Macías Garcia, C., 2006. Captive breeding promotes aggression in an endangered Mexican fish. Biol. Cons. 133, 169–177.

Kephart, D.G., 1982. Microgeographic variation in the diets of garter snakes. Oecologia 52, 287–291.

Kidd, D.M., Lui, X., 2008. GEOPHYLOBUILDER 1.0: an ArcGIS extension for creating "geophylogenies". Mol. Ecol. Res. 8, 88–91.

Kisdi, E., Priklopil, T., 2011. Evolutionary branching of a magic trait. J. Math. Biol. 63, 361–397.

Klug, H., Heuschele, J., Jennions, M.D., Kokko, H., 2010. The mismeasurement of sexual selection. J. Evol. Biol. 23, 447–462.

Kopec, J., 1949. Ecology, breeding habits and young stages of *Crenichthys baileyi*, a Cyprinodont fish of Nevada. Copeia 1949, 56–61.

Kraaijeveld, K., Kraaijeveld-Smit, F., Maan, M., 2011. Sexual selection and speciation: the comparative evidence revised. Biol. Rev. Camb. Philos. Soc. 86, 367–377.

Lande, R., 1979. Efective deme size during long term evolution estimated from rates of chromosomal rearrangement. Evolution 33, 234–251.

Lombardi, J., Wourms, J.P., 1988. Embryonic growth and trophotaenial development in Goodeid fishes (Teleostei: Atheriniformes). J. Morphol. 197, 193–208.

Lozano-García, M.S., Caballero, M., Ortega, B., Rodríguez, A., Sosa, S., 2007. Tracing the effects of the Little Ice Age in the tropical lowlands of eastern Mesoamerica. Proc. Natl. Acad. Sci. U. S. A. Vol. 104, 16200–16203.

MacArthur, R.H., 1969. Patterns of communities in the tropics. Biol.. J. Linn. Soc. 1, 19–30.

MacArthur, R.H., 1972. Geographical Ecology. Harper and Row, New York.

Macías Garcia, C., 1994. Social behavior and operational sex ratios in the viviparous fish *Girardinichthys multiradiatus*. Copeia 1994, 919–925.

Macías Garcia, C., Burt de Perera, T., 2002. Ultraviolet-based female preferences in a viviparous fish. Behav. Ecol. Sociobiol. 52, 1–6.

Macías Garcia, C., Ramírez, E., 2005. Evidence that sensory traps can evolve into honest signals. Nature 434, 501–504.

Macías Garcia, C., Saborío, E., 2004. Sperm competition in viviparous fish. Environ. Biol. Fishes 70, 211–217.

Macías Garcia, C., Saldívar Lémus, Y., 2012. Foraging costs drive female resistance to a sensory trap. Proc. R. Soc. B. 279, 2262–2268.

Macías Garcia, C., Valero, A., 2001. Context-dependent sexual mimicry in the viviparous fish *Girardinichthys multiradiatus*. Ethol. Ecol. Evol. 13, 331–339.

Macías Garcia, C., Valero, A., 2010. Sexual conflict and sexual selection in the Goodeidae, a family of viviparous fish with effective female mate choice. In: Macedo, R.H. (Ed.), Advances in the Study of Behaviour, Vol. 42. Academic Press, New York, p. 54.

Macías Garcia, C., Jimenez, G., Contreras, B., 1994. Correlational evidence of a sexually-selected handicap. Behav. Ecol. Sociobiol. 35, 253–259.

Macías Garcia, C., Saborío, E., Berea, C., 1998. Does male biased predation lead to male scarcity in viviparous fish? J. Fish Biol. 53, 104–117.

Macías Garcia, C., Smith, G., González-Zuarth, C., Graves, J.A., Ritchie, G.M., 2012. Variation in sexual dimorphism and assortative mating do not predict genetic divergence in the sexually dimorphic Goodeid fish *Girardinichthys multiradiatus*. Curr. Zool 58, 440–452.

Mank, J., Avise, J., 2006. Supertree analysis of the role of viviparity and habitat in the evolution of atherinomorph fishes. J. Evol. Biol. 19, 734–740.

Massironi, M., Rasotto, M., Mazzoldi, C., 2005. A reliable indicator of female fecundity: the case of the yellow belly in *Knipowitschia panizzae* (Teleostei: Gobiidae). Mar. Biol. 147, 71–76.

Maynard Smith, J., 1976a. A comment on the Red Queen. Am. Nat. 110, 331–338.

Maynard Smith, J., 1976b. Sexual selection and the handicap principle. J. Theor. Biol. 57, 239–242.

Mayr, E., 1942. Systematics and the Origin of Species. Columbia University Press, New York.

Mejía-Madrid, H.H., Vázquez-Domínguez, E., Pérez-Ponce de León, G., 2007. Phylogeography and freshwater basins in central Mexico: recent history as revealed by the fish parasite *Rhabdochona lichtenfelsi* (Nematoda). J. Biogeogr. 34, 787–801.

Mejía-Madrid, H.H., 2012. Biogeographic Hierarchical Levels and Parasite Speciation, Global Advances in Biogeography. In: Stevens, Lawrence (Ed.), InTech, New York, Available from http://www.intechopen.com/books/global-advances-in-biogeography/biogeographic-hierarchical-levels-and-parasite-speciation.

Mendoza, G., 1940. The reproductive cycle of the viviparous teleost *Neotoca bilineata*, a member of the family Goodeidae. II. The cyclic changes in the ovarian stroma during gestation. Biol. Bull. 78, 349–365.

Mendoza, G., 1962. The reproductive cycles of three viviparous teleosts, *Alloophorus robustus, Goodea luitpoldii* and *Neoophorus diazi*. Biol. Bull. 123, 351–365.

Meyer, A., Lydeard, C., 1993. The evolution of copulatory organs, internal fertilization, placentae and viviparity in killifishes (Cyprinodontiformes) inferred from a DNA phylogeny of the tyrosine kinase gene X-src. Proc. R. Soc. B. 254, 153–162.

Milinski, M., Bakker, T.C.M., 1990. Female sticklebacks use male coloration in mate choice and hence avoid parasitized males. Nature 344, 330–333.

Miller, R.R., 1948. The Cyprinodont fishes of the Death Valley system of eastern California and southwestern Nevada. Miscellaneous Publications of the Museum of Zoology. Vol. 68 University of Michigan, Ann Arbor.

Miller, R.R., 1950. Notes on the cutthroat and rainbow trouts with the description of a new species from the Gila River, New Mexico. Occasional Papers of the Museum of Zoology University of Michigan 529: 1–43.

Miller, R.R., Fitzsimons, J.M., 1971. *Ameca splendens*, a new genus and species of goodeid fish from western Mexico, with remarks on the classification of the Goodeidae. Copeia 1971, 1–13.

Miller, R.R., Minkley, W.L., Norris, S.M., 2005. Freshwater Fishes of Mexico, first ed. University of Chicago Press, Chicago.

Møller, A.P., Pomiankowski, A., 1993. Why have birds got multiple sexual ornaments? Behav. Ecol. Sociobiol. 32, 167–176.

Moyaho, A., Macías Garcia, C., Manjarrez, J., 2004. Predation risk is associated with the geographic variation of a sexually selected trait in a viviparous fish, *Xenotoca variata*. J. Zool. 262, 265–270.

Moyaho, A., Macías Garcia, C., Ávila-Luna, E., 2005. Mate choice and visibility in the expression of a sexually dimorphic trait in a goodeid fish, *Xenotoca variatus*. Can. J. Zool. 82, 1917–1922.

Moyaho, A., Guevara-Fiore, P., Beristain-Castillo, E., Macías Garcia, C., 2010. Females of a viviparous fish (*Skiffia multipunctata*) reject males with black colouration. J. Ethol. 8, 165–170.

Mulch, A., Sarna-Wojcicki, A.M., Perkins, M.E., Chamberlain, C.P., 2008. A Miocene to Pleistocene climate and elevation record of the Sierra Nevada (California). Proc. Natl. Acad. Sci. U. S. A. 105, 6819–6824.

Munehara, H., Takano, K., Koya, Y., 1989. Internal gametic association and external fertilization in the elkhorn sculpin, *Alcichthys alcicornis*. Copeia 1989, 673–678.

Nelson, G.G., 1975. Anatomy of the male urogenital organs of *Goodea atripinnis* and *Characodon lateralis* (Atheriniformes: Cyprinodontoidei) and *G. atripinnis* courtship. Copeia 1975, 475–482.

Nelson, K.., 1964. Behavior and morphology in the glandulocaudine fishes (Ostariophysi, Characidae). University of California Publications in Zoology, 75, 59–152.

O'Neil, M.J., Lawton, B.R., Mateos, M., Carone, D.M., Ferreri, G.C., Hrbek, T., Meredith, R.W., Reznick, D.N., O'Neill, R.J., 2007. Ancient and continuing Darwinian selection on insulin-like growth factor II in placental fishes. Proc. Natl Acad. Sci. U. S. A. 104, 12404–12409.

Ólafsdóttir, G.Á., Snorrason, S.S., Ritchie, M.G., 2007. Postglacial intra-lacustrine divergence of Icelandic three-spine stickleback morphs in three neovolcanic lakes. J. Evol. Biol. 20, 1870–1881.

Ornelas-García, C.P., Domínguez-Domínguez, O., Doadrio, I., 2008. Evolutionary history of the fish genus *Astyanax* Baird & Girard (1854) (Actinopterygii, Characidae) in Mesoamerica reveals multiple morphological homoplasies. BMC Evol. Biol. 2008, 340.

Panhuis, T.M., Butlin, R., Zuk, M., Tregenza, T., 2001. Sexual selection and speciation. Trends Ecol. Evolut. 16, 364–370.

Parenti, L., 1981. A phylogenetic and biogeographic analysis of cyprinodontiform fishes (Teleostei, Atherinomorpha). Bull. Am. Mus. Nat. Hist. 168, 335–557.

Parenti, L.R., 1989. A phylogenetic revision of the phallostethid fishes (Atherinomorpha, Phallostethidae). Proc. Calif. Acad. Sci. (Series 4) 46, 243–277.

Paterson, H.E.H., 1985. The recognition concept of species. En. In: Vrba, E.S. (Ed.), Species and Speciation. Transvaal Museum, Pretoria, pp. 21–29.

Phillips, A.R., 1986. The Known Birds of North and Middle America. A. R. Phillips, Denver.

Pineda-López, R., Salgado-Maldonado, G., Soto-Galera, E., Hernández-Camacho, N., Orozco-Zamorano, A., Contreras-Robledo, S., Cabañas-Carranza, G., Aguilar-Aguilar, R., 2005. Helminth parasites of Viviparous Fishes in Mexico. In: Uribe y, M.C., Grier, H. (Eds.), Viviparous Fishes. New Life Publications, Homestead, pp. 437–456.

Polis, G.A., Holt, R.D., 1992. Intraguild predation: the dynamics of complex trophic interactions. Trends Ecol. Evol. 7, 151–154.

Regan, C.T., 1913. *Phallostethus dunckeri*, a remarkable new cyprinodont fish from Johore. Ann. Mag. Nat. Hist. (Series 8) 12, 548–555.

Reusch, T.B.H., Häberli, M.A., Aeschlimann, P.B., Milinski, M., 2001. Female sticklebacks count alleles in a strategy of sexual selection explaining MHC polymorphism. Nature 414, 300–302.

Ritchie, M.G., Hamil, R.M., Graves, J.A., Magurran, A.E., Webb, S.A., Macías Garcia, C., 2007. Sex and differentiation: population genetic divergence and sexual dimorphism in Mexican goodeid fish. J. Evol. Biol. 20, 2048–2055.

Ritchie, M.G., 2007. Sexual selection and speciation. Annu. Rev. Ecol. Syst. 38, 79–102.

Ritchie, M.G., Webb, S.A., Graves, J.A., Magurran, A.E., Macías Garcia, C., 2005. Patterns of speciation in endemic Mexican Goodeid fish: sexual conflict or early radiation? J. Evol. Biol. 18, 922–929.

Salgado-Maldonado, G., Cabañas-Carranza, G., Caspeta-Mandujano, J.M., Soto-Galera, E., Mayén-Peña, E., Brailovsky, D., Báez-Vale, R., 2001a. Helminth parasites of freshwater fishes of the Balsas River drainage, southwestern Mexico. Compar. Parasitol. 68, 196–203.

Salgado-Maldonado, G., Cabañas-Carranza, G., Soto-Galera, E., Caspeta-Mandujano, J.M., Moreno-Navarrete, R.G., Sánchez-Nava, P., Aguilar-Aguilar, R., 2001b. A checklist of helminth parasites of freshwater fishes from the Lerma–Santiago River Basin, Mexico. Compar. Parasitol. 68, 204–218.

Salzburger, W., Mack, T., Verheyen, E., Meyer, A., 2005. Out of Tanganyika: genesis, explosive speciation, key-innovations and phylogeography of the haplochromine cichlid fishes. BMC Evol. Biol. 5, 17.

Sánchez-Nava, P., Salgado Maldonado, G., Soto-Galera, G., Jaimes, B., 2004. Helminth parasites of *Girardinichthys multiradiatus* (Pisces: Goodeidae) in the upper Lerma River sub-basin, Mexico. J. Parasitol. Res. 93, 396–402.

Sargent, R., Gross, M., Van Den Berghe, E., 1986. Male mate choice in fishes. Anim. Behav. 34, 545–550.

Schindler, J.F., 2003. Scavenger receptors facilitate protein transport in the trophotaenial placenta of the Goodeid fish, *Ameca splendens* (Teleostei: Atheriniformes). J. Exp. Zool. 299A, 197–212.

Schindler, J.F., de Vries, U., 1988. Maternal-embryonic relationships in the goodeid teleost, *Xenoophorus captivus*. The vacuolar apparatus in trophotaenial absorptive cells and its role in macromolecular transport. Cell Tissue. Res. 253, 115–128.

Schindler, J.F., Hamlett, W.C., 1993. Maternal-embryonic relations in viviparous teleosts. J. Exp. Zool. 266, 378–393.

Schluter, D., 2000. The Ecology of Adaptive Radiation. Oxford University Press, Oxford.

Schrader, M., Travis, J., 2009. Do embryos influence maternal investment? Evaluating maternal-fetal coadaptation and the potential for parent-offspring conflict in a placental fish. Evolution 63, 2805–2815.

Servedio, M.R., Van Doorn, G.S., Kopp, M., Frame, A.M., Nosil, P., 2011. Magic traits in speciation: "magic" but not rare? Trends Ecol. Evol. 26, 389–397.

Sigler, W.F., Sigler, J.W., 1987. Fishes of the Great Basin. University of Nevada Press, Reno.

Smith, G.R., 1981. Late Cenozoic freshwater fishes of North America. Annu. Rev. Ecol. Syst. 12, 163–193.

Sumner, F.B., Lanham, U.N., 1942. Studies of the respiratory metabolism of warm and cool spring fishes. Biol. Bull. 82, 313–327.

Thorp, J.H., Covich, A.P., 2001. Ecology and Classification of North American Freshwater Invertebrates, Second ed. Academic Press, San Diego.

Tilghman, S.M., 1999. The sins of fathers and mothers: Genomic imprinting in mammal development. Cell 96, 185–193.

Tobler, M., Franssen, C.M., Plath, M., 2008. Male-biased predation of a cave fish by a giant water bug. Naturwissenschaften 95, 775–779.

Turner, B.J., 1983. Does matrotrophy promote chromosomal evolution in viviparous fishes? Am. Nat. 122, 152–154.

Turner, B.J., Grosse, D.J., 1980. Trophic differentiation in *Ilyodon*, a genus of stream-dwelling goodeid fishes: speciation versus ecological polymorphism. Evolution 34, 259–270.

Turner, C.L., 1933. Viviparity superimposed upon ovoviparity in the Goodeidae, a family of cyprinodont teleost fishes of the Mexican Plateau. J. Morphol. 55, 207–251.

Uribe, M.C., De la Rosa-Cruz, G., Guerrero- Estévez, S.M., García-Alarcón, A., Aguilar-Morales, M.E., 2004. Estructura del ovario de teleósteos vivíparos. Gestación intraovárica: Intraluminal en *Ilyodon whitei* (Goodeidae) e intrafolicular en *Poeciliopsis gracilis* (Poeciliidae). In: Lozano Vilano, M.L., Contreras Balderas, A.J. (Eds.), Homenaje al Dr A. Reséndez Medina, Publicaciones, UANL, México: Dir, pp. 31–45.

Uribe, M.C., Grier, H.J., Parenti, L.R., 2012. Ovarian structure and oogenesis of the oviparous Goodeids *Crenichthys baileyi* (Gilbert, 1893) and *Empetrichthys latos* Miller, 1948 (Teleostei, Cyprinodontiformes). J. Morphol. 273, 371–387.

Uyeno, T., Miller, R.R., 1962. Relationships of *Empetrichthys erdisi*, a Pliocene cyprinodontid fish from California, with remarks on the Fundulinae and Cyprinodontinae. Copeia 1962, 520–532.

Uyeno, T., Miller, R.R., Fitzsimons, J.M., 1983. Karyology of the Cyprinodontoid fishes of the Mexican Family Goodeidae. Copeia 1983, 497–510.

Valero, A., Hudson, R., Ávila Luna, E., Macías Garcia, C., 2005. A cost worth paying: energetically expensive interactions with males protect females from intrasexual aggression. Behav. Ecol. Sociobiol. 59, 262–269.

Van Valen, L., 1973. A new evolutionary law. Evol. Theor. 1, 1–30.

Vega-López, A., Ortiz-Ordóñez, E., Uría-Galicia, E., Mendoza-Santana, E.L., Hernández-Cornejo, R., Atondo-Mexia, R., García-Gasca, A., García-Latorre, E., Domínguez-López, M.L., 2007. The role of vitellogenin during gestation of *Girardinichthys viviparus* and *Ameca splendens*; two goodeid fish with matrotrophic viviparity. Comp. Biochem. Physiol. A. 147, 731–742.

Webb, S.A., Miller, R.R., 1998. *Zoogoneticus tequila*, a new goodeid fish (Cyprinodontiformes) from the Ameca drainage of Mexico, and a rediagnosis of the genus. Occasional Papers of the Museum of Zoology, University of Michigan.

Webb, S.A., Graves, J.A., Macías Garcia, C., Magurran, A.E., Foighil, D.O., Ritchie, M.G., 2004. Molecular phylogeny of the livebearing Goodeidae (Cyprinodontiformes). Mol. Phylogenet. Evol. 30, 527–544.

White, M.J., 1978. Chain processes in chromosomal speciation. Syst. Biol. 27, 285–298.

Wilkins, J.F., 2005. Genomic imprinting and methylation: epigenetic canalization and conflict. Trends Genet. 21, 356–365.

Williams, J.E., 1996. Threatened fishes of the world: *Empetrichthys latos* Miller, 1948 (Cyprinodontidae). Environ. Biol. Fishes 45, 272.

Wilson, B.L., Deacon, J.E., Bradley, W.G., 1966. Parasitism in the fishes of the Moapa River, Clark County, Nevada. Trans. Calif. Nev. Sec. Wildl. Soc., 12–23.

Wootton, R.J., 2009. The Darwinian stickleback *Gasterosteus aculeatus*: a history of evolutionary studies. J. Fish Biol. 75, 1919–1942.

Zahavi, A., 1975. Mate selection – a selection for a handicap. J. Theor. Biol. 53, 205–214.

Chapter 11

Parental Care, Sexual Selection, and Mating Systems in Neotropical Poison Frogs

Kyle Summers[1] and James Tumulty[2]

[1]*Department of Biology, East Carolina University, Greenville, North Carolina, USA*, [2]*Department of Ecology, Evolution and Behavior, University of Minnesota, Saint Paul, Minnesota, USA*

INTRODUCTION

The neotropical poison frogs of the family Dendrobatidae are distributed from Nicaragua in Central America to Brazil and Bolivia in South America. These frogs have received considerable attention from scientists and laymen alike, mainly because of the extreme toxicity and bright coloration in many species that has evolved in the context of aposematism (Myers and Daly, 1976, 1983). However, many species in this clade also show elaborate and conspicuous social behaviors, particularly in the contexts of mating strategies and parental care (Summers, 1992a; Summers and McKeon, 2004; Wells, 2007). Whether these conspicuous behaviors have evolved in part because toxicity has permitted the evolution of such behaviors, or because bright coloration provides a visual signal that can then evolve in response to evolutionary forces in the context of sexual selection, is an open question. We note, however, that there are many species of frogs in this family that are not, in fact, toxic or brightly colored, yet do display elaborate and conspicuous patterns of territoriality, courtship, mating, and parental care (some of these species will be discussed below). The presence of these complex yet observable social interactions have made dendrobatid frogs a focal point of interest for researchers interested in the influence of ecological and social factors on the evolution of parental care and mating strategies. In this chapter we review some general theoretical developments focused on the interrelationships of sexual selection and parental care, focusing on the work of Trivers (1972) (see also Chapter 8 in this volume), and then explore how research on neotropical poison frogs has contributed to our understanding of specific issues of interest in this broad area.

Sexual Selection. http://dx.doi.org/10.1016/B978-0-12-416028-6.00011-6

THEORETICAL BACKGROUND

Building on key insights by Bateman (1948) concerning the relationship between the gamete size dimorphism that defines male and female identities and the effect of multiple mating on fecundity in males and females, and Williams (1966) on the effect of parental care on sex roles, Trivers (1972) developed the general argument that sexual selection is controlled by patterns of relative parental investment between the sexes. Trivers (1972) argued that if one sex provides substantially more parental investment per offspring (defined as investment that reduces the parents' ability to produce other offspring), that sex will become a limiting resource for which the opposite sex will compete. In most species that have been studied, females provide more parental investment than males, and males compete intensely for mating opportunities (Andersson, 1994). This argument has dominated the discussion of this topic in the literature ever since, and although various alternative viewpoints have been presented (see, for example, Tang-Martinez and Ryder, 2005; Roughgarden et al., 2006), Trivers' general argument continues to be the dominant paradigm of sexual selection (Kokko and Jennions, 2008).

Trivers (1972) proposed that key empirical tests of the hypothesized relationship between parental investment and mating systems would come from species in which the typical patterns of parental investment are reversed (i.e., males invest substantially more per offspring than females). Under these circumstances, he predicted that females would evolve to be the more competitive sex, and males should be relatively selective about mating. In this review, we highlight work on sexual selection and parental investment in several species of poison frogs that provide an alternative perspective on how parental investment and sexual selection can interact.

Following Trivers' (1972) paper, a number of key contributions highlighted other factors that influence sexual selection. Emlen and Oring (1977) emphasized the effect of ecological factors on the "environmental polygamy potential", and developed the concept of the operational sex ratio (OSR) as a key influence on sexual selection and mating systems. While some authors have advocated the OSR (and related statistics) as a key measure of the strength of sexual selection (e.g., Shuster and Wade, 2003), there are significant problems with this approach (reviewed in Klug et al., 2010). However, the role of ecological factors in the evolution of both parental care and mating systems has continued to be a major theme in behavioral ecology and evolution (e.g., Okuda, 1999; Kokko and Monaghan, 2001; Kokko and Rankin, 2006).

Clutton-Brock and Vincent (1991) developed the concept of the "Potential Reproductive Rate", which was argued to be a bridge between relative parental investment and sexual selection, but more easily measured than parental investment. Several authors (e.g., Clutton-Brock and Parker, 1992; Arnold and Duvall, 1994) developed mathematical models of the influence of relative parental investment on sexual selection, focusing on the idea of "time-out" from the mating pool. The results of these models strongly supported the logic developed by Trivers in his 1972 paper. More recent mathematical models (e.g., Kokko and Jennions, 2008) support many of Trivers' (1972) original arguments, although with some differences concerning the effect of sexual selection on the evolution of parental care (see below). It should be noted that the relationship between parental investment and sexual selection was only one among many important concepts that have emerged from Trivers' (1972) paper. For example, the paper also made major contributions to our understanding of the evolution of parental care (Kokko and Jennions, 2008), and sexual conflict (Lessells, 2012).

With regard to the evolution of parental care (and specifically which sex should provide care when care is provided), Trivers (1972) made three major arguments: first, that high levels of investment in nutrient-rich eggs by females would make them more likely than males to provide care; second, that higher uncertainty of parentage for males would militate against male parental care; and third, that sexual selection acting on males would select against paternal care. Each of these arguments has generated substantial controversy. Dawkins and Carlisle (1976) contended that the first argument involved a "Concorde Fallacy", in that previous investment (in gametes) does not reliably predict future returns on subsequent investment (in parental care). They proposed instead that the sex which has the opportunity to desert first will do so. The Concorde Fallacy argument has been influential, although some authors have pointed out that past investment will predict likely future returns on investment under certain circumstances (e.g., Coleman and Gross, 1991). With regard to the effect of opportunities for desertion, substantial empirical evidence contradicts the predictions of this hypothesis (Clutton-Brock, 1991).

The second argument was criticized extensively by Maynard Smith (1978) and Werren et al. (1980), who argued that, assuming paternity remained constant across broods, uncertainty of paternity should not influence the evolution of male care. This argument was in turn criticized as incomplete (e.g., Westneat and Sherman, 1993). In 1997, Queller developed a simple yet elegant model that demonstrated the general validity of Trivers' (1972) original argument that uncertainty of paternity will select against the evolution of male parental care. More recent modeling efforts have confirmed this general point (e.g., Kokko and Jennions, 2008). Large-scale comparative analyses have produced results consistent with a strong influence of certainty of paternity on the evolution of male versus female parental care (e.g., Møller and Cuervo, 2000; Arnold and Owens, 2002; Ah-King et al., 2005; Mank et al., 2005), and there is substantial evidence from experimental studies that uncertainty of paternity does influence the likelihood and extent of male parental care in some species (e.g., Neff and Gross, 2001; Neff, 2003). However, numerous exceptions to this trend have been identified (e.g., Alonzo and Heckman, 2010; Brennan, 2012; Kamel and Grosberg, 2012; reviewed in Sheldon, 2002; Alonzo, 2010). Recent theoretical models have identified factors that can drive the evolution of male parental care in spite of low paternity, such as female choice for caring males (e.g., Alonzo, 2012). Researchers have also identified various factors that may mitigate the impact of uncertainty of paternity on the evolution of parental care (reviewed in Alonzo and Klug, 2012).

Queller (1997) also developed a model that confirmed Trivers' (1972) arguments concerning the influence of sexual selection on the evolution of male parental care, demonstrating that males that are successful in intrasexual competition for matings should be less inclined to perform parental care, whereas unsuccessful males are not in a position to provide care. This result was confirmed by a more extensive model developed by Kokko and Jennions (2008), although they emphasize that the effect of asymmetric parental investment on the operational sex ratio should actually favor increased parental care by the sex that is present in excess, and this may override the effect of sexual selection to reduce care under some circumstances.

As mentioned above, another major contribution made by Trivers (1972) was to emphasize the importance of sexual conflict. This phenomenon has become the subject of a separate field of investigation in its own right (e.g., Parker, 1979; Arnqvist and Rowe, 2005), and has led to many important insights into the evolution of mating systems and parental care (Davies, 1989; Brown *et al.*, 1997; Lessells, 2012).

Below we review the evolution of parental care in neotropical poison frogs in the context of previous conceptual developments concerning sexual selection and the evolution of parental care.

THE EVOLUTION OF PARENTAL CARE IN TROPICAL FROGS

The reproductive strategies of frogs are highly diverse. While many are familiar with the life cycle of the leopard frog and other common temperate species, which involves short annual bouts of mating in the spring and the deposition of unattended eggs in aquatic habitats, this is only one tiny part of the stunning diversity of reproductive strategies practiced by frogs (Duellman and Trueb, 1986). These strategies have been arrayed into a plethora of reproductive modes (at least 39: Duellman and Trueb, 1986; Haddad and Prado, 2005; Wells, 2007), including variation in egg deposition site, parental care behaviors, and developmental modes, among other features.

Parental care is rare in frogs (occurring in about 10–20% of extant species: McDiarmid, 1978; Lehtinen and Nussbaum, 2003), but it has evolved multiple times across the evolutionary tree of the Anura (Summers *et al.*, 2006, 2007; Gomez-Mestre *et al.*, 2012). Latitude stands out as a key correlate of the evolution of parental care in frogs: most species with parental care occur in tropical latitudes (Duellman and Trueb, 1986; Magnusson and Hero, 1991). Why this is the case has been the subject of considerable speculation, and a number of authors have proposed that it may have been associated with a trend toward terrestrial reproduction in tropical frogs (e.g., Salthe and Duellman, 1973; McDiarmid, 1978; Crump, 1995). Predation in aquatic environments has been frequently proposed as a key factor favoring the evolution of terrestrial reproduction (Lutz, 1947; Goin and Goin, 1962; Duellman and Trueb, 1986). Research on anuran species in the Brazilian Amazon indicated that predation by other anuran larvae was likely a key factor that may have favored the evolution of terrestrial reproduction in neotropical frogs (Magnusson and Hero, 1991). The risk of desiccation is also a key factor that is likely to have affected the evolution of terrestrial reproduction (Touchon and Warkentin, 2008), with wetter, more humid conditions likely favoring terrestrial reproduction. Recently, Touchon (2012) summarized data on aquatic and terrestrial egg mortality in the neotropical hylid frog *Dendrosophus ebbracatus*, a species that shows variation in reproductive mode (aquatic versus terrestrial). This research revealed that predation and desiccation are the major factors impacting the relative fitness of these reproductive modes, providing a rare intraspecific window into the factors that are likely to have influenced the evolution of terrestrial reproduction.

Terrestrial reproduction does not necessarily involve the evolution of parental care, but the two traits are strongly associated in a phylogenetic context (Gomez-Mestre *et al.*, 2012). Hence, terrestrial reproduction has evolved repeatedly in tropical frogs, and in many cases this is associated with the evolution of male, female, or biparental care (Duellman and Trueb, 1986). This trend is seen in the Neotropics as well as the Old World tropics.

In many taxa (e.g., mammals, reptiles, insects), male parental care is rare and female care is the predominant form of care when care is provided (Clutton-Brock, 1991). In contrast, uniparental male care is the most common form of care in teleost fish (Gross, 2005; Mank *et al.*, 2005; Balshine, 2012). Frogs show an intermediate pattern, with similar numbers of species showing male and female parental care (Wells, 2007).

Given the fact that external fertilization is the rule rather than the exception in frogs, a number of authors have argued that certainty of parentage may have played a role in the relatively high frequency of male parental care in frogs (reviewed in Clutton-Brock, 1991). A comparative analysis of parental care and mode of fertilization by Beck (1998) concluded that parentage did not influence the evolution of male versus female care in frogs (assuming certainty of paternity is higher with external fertilization), but there were significant problems and deficiencies in this analysis (reviewed in Wells, 2007) and the issue is not resolved. In fish, the results of comparative analyses do suggest a role for certainty of paternity (Ah-King *et al.*, 2005; Mank *et al.*, 2005), but it is also clear that male parental care exists in many species that have substantial frequencies of sneaker males and associated compromises in paternity (Alonzo and Klug, 2012).

The evolution of parental care in fish has received more theoretical and empirical attention than in any other taxa (reviewed in Balshine, 2012), and provides a valuable starting point to consider the evolution of sex-specific parental care patterns in tropical frogs. A number of researchers focused on the evolution of parental care in fish have argued that male parental care is particularly likely to evolve in circumstances where females lay eggs directly on a territory defended by a male (Williams, 1975; Ridley, 1978; Perrone and Zaret, 1979; Baylis, 1981; Gross and Sargent, 1985). There are likely several interrelated reasons for this association. The first is simple direct association: when the eggs are laid on the male's territory, they are readily available to receive care (Williams 1966). This association also means that some forms of parental care are likely to have a relatively low cost for male fish in terms of lost mating opportunities (Loiselle, 1978; Blumer, 1979; Gross and Sargent, 1985): they can continue to attract and mate with females while simultaneously performing some forms of care (e.g., defending clutches from interspecific and intraspecific predation). In fact, parental care in association with territory defense may actually enhance male mating success in some cases: there is evidence that females in some species prefer to mate with males that have clutches from previous matings (Marconato and Bissaza, 1985; Knapp and Sargent, 1989), and some authors have argued that intersexual selection may have a strong influence on the evolution of male care (e.g., Tallamy, 2000 for arthropods). Recent theory confirms the logic of the argument that intersexual selection via female choice can be a powerful and general selective agent favoring the evolution of male parental care (Alonzo, 2012). Comparative analyses (Ah-King *et al.*, 2005) support the argument that male parental care is likely to evolve following the evolution of male territoriality and pairwise-spawning (which likely increased certainty of paternity relative to group-spawning systems). Other researchers have emphasized the importance of differential costs of parental care to males and females (Gross and Sargent, 1985; Gross, 2005). In species with indeterminate growth, parental care may have higher costs to female lifetime reproductive success, in terms of reduced fecundity, than to males. This is another factor that may have favored the evolution of uniparental male parental care in fish.

In frogs, territoriality may also play a crucial role in the evolution of male versus female parental care (Wells, 1977, 1981, 2007). Just as in fish, the defense of territories that include oviposition sites may result in a low cost of parental care in terms of reduced mating opportunities. However, this hypothesis has not been tested in a comparative phylogenetic framework, as it has been in fish. The costs of parental care to females may be particularly high in frogs (as in fish), owing to the effect of indeterminate growth on fecundity (Gross, 2005). Again, further work is required to test this hypothesis. Wells (2007) hypothesized that male parental care may be especially likely to evolve in species in which males defend elevated sites (e.g., treeholes or leaf axils). This may expose clutches to particularly dry conditions, hence favoring the evolution of specialized oviposition sites that provide a humid microenvironment. Such sites are expected to be in short supply, and hence attractive to females and economically defendable by males (Townsend, 1989). The association of male parental care with economically defendable small territories centered on oviposition sites could also be consistent with a role for certainty of paternity in the evolution of male parental care, but there are few comparative data on paternity between territorial and non-territorial species, or between species with or without paternal care. In fact, there has only been one study of paternity in a species of frog with male parental care outside of poison frogs (see below): *Kurixalus eiffingeri* in Taiwan, a rhacophorid treefrog that breeds in micropools in bamboo (Chen *et al.*, 2011). In this species, mixed paternity was found in spite of the presence of uniparental male care, although there was a positive relationship between parental investment by caring males and levels of paternity in the clutches attended.

PATTERNS OF PARENTAL CARE IN NEOTROPICAL POISON FROGS

Nearly all members of the neotropical poison frogs (families Aromobatidae and Dendrobatidae) have some form of parental care, at a minimum involving the transport of tadpoles from terrestrial egg clutches to bodies of water. It is thought that the ancestors of poison frogs lived and dwelled along streams, as seen in *Aromobates nocturnus*, a basal lineage, which are nocturnal, stream-dwelling frogs (Myers *et al.*, 1991; Summers and McKeon, 2004). Many poison frogs can be classified as terrestrial- or stream-breeders; typically, egg clutches are laid terrestrially and often attended by a parent. When the eggs hatch, one of the parents, usually the male, transports the tadpoles on his back to a terrestrial pool of water or a stream. Most members of the family Aromobatidae display this pattern of reproduction and parental care; however, phytotelm breeding (Bourne *et al.*, 2001) and endotrophic, nidicolous tadpoles that are not transported (Juncá *et al.*, 1994; Caldwell and Lima, 2003) have evolved in this family. Terrestrial breeding with tadpole transport is also characteristic of the genera *Ameerega*, *Epipedobates*, *Colostethus*, *Hyloxalus*, and *Silverstoneia* in the family Dendrobatidae (Figs 11.1 and 11.2).

Phytotelm breeding has evolved at least three times in poison frogs (see Fig. 11.3), and is characterized by the deposition of tadpoles in arboreal pools of water that collect in leaf axils (such as those of bromeliads, *Heliconia*, *Xanthosoma*, and *Dieffenbachia*), treeholes, bamboo stalks, palm fronds, fallen fruit husks, etc. For a nearly exhaustive list of patterns of parental care and tadpole habitat among dendrobatid frogs, see Wells (2007: Ch.11). An ancestral state reconstruction using parsimony suggests that terrestrial breeding and male parental care are ancestral to the clade, with phytotelm breeding, as well as biparental care and uniparental female care, being derived states (Summers and McKeon, 2004; see also Fig. 11.3).

A variety of factors have been posited to influence the transition to phytotelm breeding, including predation by fish and aquatic invertebrates, egg predation by tadpoles, competition, and possibly parasites in larger terrestrial

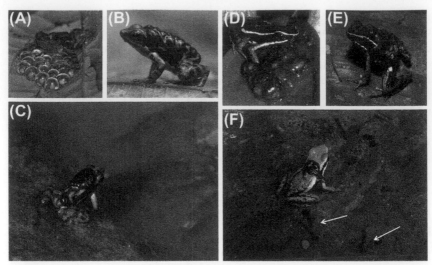

FIGURE 11.1   Parental care in stream- and terrestrial-breeding dendrobatids. (A–C) *Hyloxalus nexipus* is a stream-breeding species with male parental care. (A) A male *H. nexipus* attends an egg clutch of ~16 eggs laid on a leaf on the forest floor, and (B) transports all the tadpoles from one clutch on his back. (C) A male *H. nexipus* with tadpoles next to a flowing stream where the tadpoles will be deposited. (D–F) Many members of the genus *Ameerega* are classified as terrestrial breeders. (D) A male *A. hahneli* attends a clutch of seven eggs, and (E) transports them once they hatch. (F) A male *A. bassleri* deposits tadpoles in a terrestrial pool; arrows indicate tadpoles already in the water. See color plate at the back of the book. *Photographs courtesy of Jason Brown (B, D), Adam Stuckert (A, E), and Evan Twomey (C, F).*

pools and streams, as well as abiotic factors such as pool-drying (Summers and McKeon, 2004; McKeon and Summers, 2013). Recent work on Peruvian populations of *Allobates femoralis*, a species that breeds in terrestrial pools on the ground, highlights the potential importance of predation and indirect effects on the transition to phytotelm breeding (McKeon and Summers, 2013). Experiments with artificial basins placed in the forest investigated the effects of both pool size and presence of a large belostomatid insect predator on tadpole deposition in this species. Large pools were preferred for tadpole deposition, but, surprisingly, pools with the large predators were also preferred. This puzzling result coincided with the observation that these large sit-and-wait predators significantly reduced the presence of a major small, active predator (the dytiscid beetle), which also preys on tadpoles. It appears that the preference of *Allobates* for pools containing belostomatids is caused by the indirect effects of these predators on another, more common predator (a case of "my enemy's enemy is my friend"). In experiments on the colonization by predators of pools of the same size placed at different heights in the forest, it was found that dytiscids (and other predators) rarely colonize pools that are raised above the forest floor. This result is consistent with previous surveys of predator abundance in phytotelmata (Summers, 1990, 1999; Summers and McKeon, 2004).

FIGURE 11.2   Parental care in phytotelm breeders. (A) A male *Ranitomeya variabilis* transports three tadpoles that will be deposited in phytotelmata such as bromeliad axils. Similar to terrestrial- and stream-breeders, no further care is provided after tadpole deposition. (B) A male *R. imitator* transports each tadpole from a clutch, individually, to small, nutrient-poor phytotelmata. (C) After tadpole deposition the male calls to the female, leads her to the pools occupied by their offspring, and continues to call, stimulating her to provision tadpoles with unfertilized trophic eggs. See color plate at the back of the book. *Photographs courtesy of Jason Brown (A), James Tumulty (B), and Adam Stuckert (C).*

Hence these pools are likely to serve as a refuge from predation, and this may have been a critical factor favoring the evolution of phytotelm breeding in dendrobatid frogs. While the most important factors driving the transition to phytotelm breeding are still under investigation, this transition has undoubtedly had profound impacts on parental care, sexual selection, and mating systems in these frogs.

Evolution of Male Care

As stated above, many authors have suggested that male territoriality is an important correlate of paternal care in externally fertilizing taxa (Williams, 1975; Wells, 1977; Ridley, 1978; Baylis, 1981). When males are territorial to attract females, and oviposition occurs within a male's territory, males could be selected to care through association with the offspring, especially if it is beneficial for the male to maintain his territory to attract future mates and thus less costly for the male to remain with the offspring than the female (Ridley, 1978). This scenario has also been developed in detail by Kent Wells, who suggests

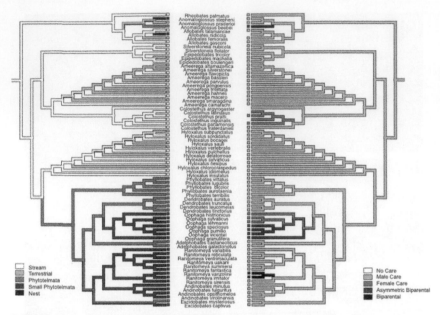

FIGURE 11.3 **Mirror tree illustrating the evolution of parental care and breeding site in dendrobatid frogs.** The topology of the tree is based on Brown *et al.* (2011) (for relationships within and between *Dendrobates*, *Oophaga*, *Ranitomeya*, *Adelphobates*, *Andinobates*, and *Excidobates*) and Pyron and Wiens (2011) (for the rest of the tree). Character states for each species were taken from the literature (see Wells, 2007, for a summary). Ancestral character states were estimated using parsimony in the software package Mesquite (Maddison and Maddison, 2011), and the mirror tree was constructed with that package. Note that in some species that have generally been considered to have male parental care, females have also been observed to transport tadpoles (see Grant *et al.*, 2006). Here we assign the character state generally associated with a particular taxon (summarized in Wells, 2007).

that paternal care in anurans is likely an "outgrowth" of territorial defense of oviposition sites (Wells, 1977, 1981, 2007). Currently we lack the power to investigate this hypothesis from a comparative perspective within poison frogs, as male care is ancestral to the clade (Summers and McKeon, 2004). Nevertheless, some observations appear consistent with this hypothesis and are worth noting.

An important assumption of this hypothesis is that parental care should not preclude mating success – i.e., males should be able to attract additional mates while attending egg clutches within their territory. This assumption is valid for male poison frogs, many of which continue to call while attending eggs, and males have been found guarding multiple clutches (Summers, 1989; Roithmair, 1992; Juncá *et al.*, 1994; Pröhl and Hödl, 1999; Ursprung *et al.*, 2011). However, this is not a general rule, as male *Rheobates* (= *Colostethus*) *palmatus* were observed to sit on eggs constantly and ceased calling throughout the entire period of egg development (Lüddecke, 1999).

Another complication, noted by Wells (1981), is that these predictions about territoriality and egg attendance may not apply to selection for tadpole transport, as tadpoles usually need to be transported well outside a male's territory. Thus, a male will have to leave his territory, and his venue for attracting females, to transport tadpoles. However, confidence of parentage is likely to be an important consideration; females who oviposit in male territories but do not remain with egg clutches may have low confidence of maternity when males guard clutches from multiple females, and selection may still favor male tadpole transport on account of higher confidence of paternity on the part of the male. Furthermore, field observations show that males that transport tadpoles outside of their territory often return and continue territory defense shortly thereafter, thus the cost of tadpole transport to territory defense may not be that great (Summers, 1989; Ringler *et al.*, 2009). This complication also may not apply to male phytotelm breeders defending territories that include tadpole deposition sites (Summers and Amos, 1997; Poelman and Dicke, 2008; Brown *et al.*, 2009). However, costs of tadpole transport in species with particularly strong male–male competition for territories could have driven the evolution of female tadpole transport in *Colostethus panamensis* and closely related congeners (see below).

Along a separate, but not mutually exclusive, line of reasoning, male parental care could correlate with increasing certainty of paternity. Males should only care for offspring for which they have high confidence of paternity (Trivers, 1972), and defending a territory would prevent other males from fertilizing eggs laid in their territory. Relative to other taxonomic groups, multiple paternity has not been widely documented in anurans, but evidence has been accruing (e.g., Laurila and Seppa, 1998; Prado and Haddad, 2003; Lodé and Lesbarrères, 2004; Vieites *et al.*, 2004; Chen *et al.*, 2011). The fact that multiple paternity is not widely documented in anurans is in part due to the lack of genetic analyses of paternity of egg clutches. Given the dense breeding aggregations of many anuran amphibians, it seems likely that multiple paternity is widespread in this order, and the territorial spacing of species with male parental care could be an adaptation to avoid cuckoldry. Consistent with this hypothesis in poison frogs are observations of some males being most vigilant in egg attendance in the first few days after oviposition and subsequently spending less time with the eggs and more time calling as the risk of fertilization decreases (*Ameerega picta*, Weygoldt,1987; *Anomaloglossus stepheni*, Juncá, 1998). Furthermore, genetic analysis revealed no evidence of multiple paternity in *Allobates femoralis*, a species with male care and male territoriality (Ringler *et al.*, 2012). More recently, observations of *Oophaga pumilio* males that adopted a satellite strategy and followed courting pairs to oviposition sites provided evidence that cuckoldry could indeed be a selective force favoring territoriality in species with male care (Meuche and Pröhl, 2011).

Evolution of Female Care

Our ancestral state reconstruction (Fig. 11.3) indicates that female care has evolved once in stream-breeding frogs of the genus *Colostethus*, and is characterized by female tadpole transport (Wells, 1980a, 1981). Female care has evolved three times independently in phytotelm breeders, being characterized by the provisioning of tadpoles with trophic eggs (Brust, 1993; Bourne *et al.*, 2001; Brown *et al.*, 2008a).

Several species in the *Oophaga* group have asymmetric biparental care, in which the male attends the clutches periodically, but the female carries out the far more intensive care involved in carrying the tadpoles to phytotelmata and returning to feed them over the course of tadpole development (Weygoldt, 1980; Brust, 1993). The asymmetric biparental care of the genus *Oophaga* is a derived state within the family Dendrobatidae (Summers *et al.*, 1999). Given the ancestral state and widespread prevalence of male-only care, the evolutionary transitions to biparental and female care are of considerable interest. The *vanzolinii* clade is characterized by biparental care; males typically attend eggs and transport tadpoles, then males call to females and stimulate them to provision tadpoles with unfertilized trophic eggs throughout tadpole development (Caldwell and de Oliveira, 1999; Brown *et al.*, 2008b).

The selective forces favoring the evolution of female tadpole transport in stream breeders *Colostethus panamensis* and its close relatives are unclear, but Wells (1981, 2007) pointed to the costs of transporting tadpoles while defending a territory. Comparisons between *C. panamensis* and *Mannophryne trinitatis* revealed similar breeding ecologies but different sex roles of parental care and territory defense (Wells, 1980a, b). *Mannophryne trinitatis* breed along streams in the mountains of Trinidad and Venezuela; males transport tadpoles, and females are highly territorial and aggressive during the reproductive season (Test, 1954; Sexton, 1960; Wells, 1980b). In contrast, *C. panamensis* also breed along streams, but in this species males defend large reproductive territories and females transport tadpoles (Wells, 1980a). He hypothesized that it is costly to defend a territory while transporting tadpoles due to potential injury to the offspring, and that the sex responsible for tadpole transport will not be the sex that defends territories (Wells, 1980a, b, 1981). This pattern breaks down, however, in many other poison frogs in which males defend territories and transport tadpoles (Summers, 1989, 1992a, b, 1999, 2000; Roithmair, 1992; Summers and Amos, 1997; Summers and McKeon, 2004). Nevertheless, this factor could be most relevant when male–male competition for territories is particularly intense and/or tadpoles are carried for long periods of time (Wells, 1981, 2007).

Future research on the costs and benefits of tadpole transport in relation to territory defense is needed to understand this relationship. One useful test of these costs could be conducted in a species where both sexes transport tadpoles but only one is territorial, such as *Allobates femoralis*, where males are territorial and are predominantly responsible for tadpole transport but females also sometimes transport (Weygoldt, 1987; Ursprung *et al.*, 2011).

Summers and Earn (1999) discussed the evolution of female parental care from male or biparental care, and developed game-theoretic models to investigate the influence of a cost of polygyny on the evolution of female care. They reviewed four different factors that might drive the evolution of female care: the cost of polygyny; resource dispersion; reproductive parasitism by females; and the use of pools of small size with decreased nutrient content.

The Cost of Polygyny

Zimmermann and Zimmermann (1984, 1988) argued that female parental care in the *Oophaga histrionicus* group evolved from shared ancestry with members of the *Ranitomeya variabilis (ventrimaculata)* group, which also shows female care in the form of trophic egg-feeding. Frogs in this group breed in small phytotelmata, and tadpoles are highly cannibalistic (Summers, 1999). Weygoldt (1987) speculated that females in species in this group suffered a high cost of polygyny when males fed the offspring of some mates to the tadpoles of others. He argued that this could have selected for the evolution of female parental care. More recent phylogenetic analyses (e.g., Summers *et al.*, 1999; Grant *et al.*, 2006; Santos *et al.*, 2009; Brown *et al.*, 2011) indicate that female parental care as found in the *histrionicus* clade (with asymmetric biparental and uniparental female care) evolved independently from female parental care in the *variabilis* (*ventrimaculata*) clade (with biparental species). Nevertheless, the question of whether a cost of polygyny could drive the evolution of female care remains valid, and this could have been an important factor in one or more of the clades where female care has evolved. Game-theoretic analyses indicate that a cost of polygyny on its own is unlikely to drive the evolution of a pure uniparental female care strategy from male-only care. However, given specific assumptions (Summers and Earn, 1999), a high cost of polygyny could drive the evolution of biparental care, and it could also interact with a cost of lost mating opportunities in males in a synergistic manner, ultimately resulting in the evolution of a pure female care strategy.

There is evidence for substantial costs of polygyny in several species of dendrobatids that have been studied intensively (Summers, 1989, 1990, 1992a, b; Summers and Amos, 1997), but variation in the cost of polygyny does not appear to be associated with the evolution of parental care in a comparative context (Fig. 11.3). *Ranitomeya imitator*, which shows biparental care with relatively equal male and female parental effort (Brown *et al.*, 2008a), appears to have a low cost of polygyny (Brown *et al.*, 2010a; Tumulty *et al.*, unpublished observations). This does not rule out the possibility that a high cost of polygyny selected for biparental care ancestrally in this lineage, but it does indicate that it is unlikely that a cost of polygyny would drive a transition from biparental to pure female care. In the *histrionicus* lineage, there is both asymmetric biparental care, in which the male performs some care (clutch attendance) but the female invests substantially more time and effort into parental care without the assistance of the male, and pure female care. Ancestral reconstructions of the evolution of parental care in this clade (Fig. 11.3) do not clearly support the prediction from the cost of polygyny hypothesis that female parental care evolved from biparental care rather than the reverse. Further, the nature of asymmetric biparental care in this group is such that a cost of polygyny is unlikely to exist under this pattern of parental care, given that the male has been completely disconnected from tadpole transport and feeding. Hence, it appears that a high cost of polygyny is unlikely to have driven the evolution of female parental care in these groups of poison frogs, although it is likely to have had a profound influence on the mating systems of some species (see below).

Resource Dispersion

It is possible that a change in the dispersion of resources critical to reproduction could have favored male desertion and the evolution of female parental care. However, males do not appear to monopolize reproductive resources (i.e., breeding pools or oviposition sites) in species with female parental care (e.g., *Oophaga pumilio*; Pröhl and Hödl, 1999, and references above). Hence, this hypothesis seems unlikely.

Reproductive Parasitism by Females

Reproductive parasitism by females could also drive the evolution of female care (theoretically) and shift the balance of selection in favor of male desertion, leading to uniparental female parental care by default. Reproductive parasitism, in which males attempt to place tadpoles into pools that contain embryos of other individuals that can be cannibalized, has been demonstrated to occur experimentally in *Ranitomeya variabilis* (Brown *et al.*, 2009), and it is possible that females engage in similar behavior. However, female transport of tadpoles in species with male parental care is rare, and it seems unlikely that females would engage in this behavior frequently enough to drive selection in favor of male desertion.

Use of Pools of Small Size with Decreased Nutrient Content

Female parental care may have evolved in the context of the use of pools of small size with decreased nutrient content (Summers and Earn, 1999; Brown *et al.*, 2010a). Small pools are widely available in many tropical rainforests, yet the ability of frogs to use them is limited because of their extremely low nutrient content. The evolution of trophic egg-feeding may have been a "key innovation" that allowed poison frogs to exploit a new ecological niche (extremely small phytotelmata) (Brown *et al.*, 2010a). This innovation depended on the evolution of female care, because only females can provide trophic eggs. Phylogenetic reconstructions suggest that this innovation has evolved at least twice in the poison frogs (Summers *et al.*, 1999; Summers and McKeon, 2004; Fig. 11.3),

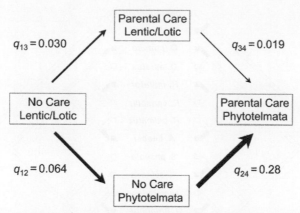

FIGURE 11.4   Path diagram of evolutionary pathways from ancestral terrestrial-breeding strategies (lentic – pool-breeding; lotic – stream-breeding) to phytotelm breeding, and from ancestral breeding strategies without parental care, to breeding with parental care, across the order Anura (frogs). The q values indicate the transition rate for that transition, estimated via maximum likelihood (Pagel, 1994). Statistical analyses using likelihood ratio tests indicate that the most common pathway involved the evolution of phytotelm breeding first, followed by the evolution of parental care. *Figure reprinted from Brown* et al. *(2010a), with permission from the American Society of Naturalists.*

and may have evolved in *Anomaloglossus beebei* as well (Bourne *et al.*, 2001). Comparative analyses across all frogs indicate that the evolution of small-pool use is associated with the evolution of parental care (Brown *et al.*, 2010a). Furthermore, directional analyses using maximum likelihood methods (Pagel, 1994) indicate that transitions to phytotelm breeding (from stream/pond breeding) preceded the evolution of parental care, rather than the reverse, indicating that this key ecological factor drove the evolution of parental care, rather than the reverse (Brown *et al.*, 2010a; Fig. 11.4). These comparative analyses also revealed a significant association between the use of extremely small pools (phytotelm breeding) and the evolution of egg-feeding by females (Brown *et al.*, 2010a; Fig. 11.5). Hence, current research indicates that the transition to the use of extremely small phytotelmata was probably a major factor selecting for the evolution of female parental care in the context of both biparental care and uniparental female care.

Territoriality and Mating Systems

Territoriality and complex courtship seem to be important correlates of terrestrial reproduction and parental care in Dendrobatidae, Aromobatidae, and other anuran families. In contrast to lek-breeding frogs, which usually defend short-term calling sites and oviposit elsewhere, many terrestrial-breeding poison frogs that have been studied defend long-term, multipurpose territories that include calling sites, venues for courtship, oviposition sites, and feeding

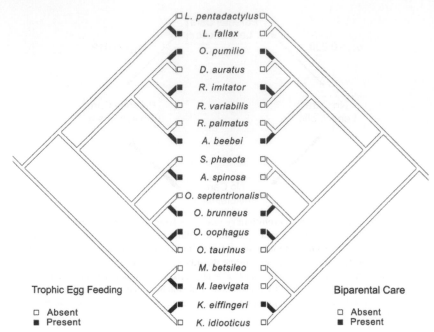

FIGURE 11.5 Mirror tree illustrating the correlation between the evolution of egg-feeding and the evolution of biparental care across all frogs (order Anura). Each taxon with egg-feeding and/or biparental care is paired with one outgroup taxon that does not show the trait. A concentrated changes test (Maddison, 1990) indicates a highly significant correlation between the evolution of egg-feeding and biparental care. *Figure reprinted from Brown* et al. *(2010a), with permission from the American Society of Naturalists.*

grounds (Pröhl, 2005), while some phytotelm breeders defend territories that also include tadpole deposition sites (Bourne *et al.*, 2001; Poelman and Dicke, 2008; Brown *et al.*, 2009).

Despite the general observation that poison frogs often defend multipurpose territories, it is not always clear what exactly is being defended (Pröhl, 2005). Pröhl (2005) identified two categories of territories in dendrobatid frogs: non-reproductive territories that include feeding sites, access to moisture, and retreat sites, as seen in some stream breeders (Wells 1980a, b); and reproductive territories for attracting and courting mates, which are defended specifically against intraspecific competitors (Pröhl, 2005, and citations therein). The most conspicuous and commonly documented territorial behaviors relate primarily to reproduction. Species defending reproductive territories do not seem to defend feeding grounds (but see Meuche *et al.*, 2011) as generally only one sex (usually the male) is territorial (e.g., Crump, 1972; Wells, 1980a; Roithmair, 1992, 1994; Summers, 1992b, 1999; Juncá, 1998); both sexes would be expected to benefit from territorial behavior if a primary function of territoriality was defense of feeding grounds. Furthermore, male territoriality is often only directed at other calling males. For example, male *Allobates femoralis* attacked calling males but ignored non-calling males foraging in their territories (Roithmair, 1992), indicating that the primary role of territory defense relates to intrasexual competition for mates.

Given the reliance of poison frogs on terrestrial resources for reproduction, it is reasonable to speculate that these resources may be limiting and, as such, should be defended. Male territoriality in this case could be classified as resource defense polygyny, where males can monopolize females through defense of resources needed by females for reproduction (Emlen and Oring, 1977). In North American bullfrogs, for example, large males control high-quality oviposition sites needed by females, and experiments show that these sites have lower egg mortality (Howard, 1978). Oviposition sites for poison frogs, however, are usually dead leaves on the forest floor that are generally assumed to be too plentiful to be worth defending (Roithmair, 1992, 1994; Pröhl, 2005). The same is probably true of calling sites themselves, which usually take the form of raised microhabitats such as logs, leaves of understory plants, branches, and raised tree roots (Pröhl, 2005). In contrast, many authors conclude that the main function of territory defense by male poison frogs is the defense of an area in which males can advertise to and court females without interruption from rivals (reviewed in Pröhl, 2005), and male mating success is often correlated with territory size and overall calling activity (Roithmair, 1992, 1994; Pröhl, 2003).

Phytotelmata used for tadpole deposition, however, can be a limiting reproductive resource for phytotelm breeders (Donnelly, 1989a, b; Poelman and Dicke, 2008; Brown *et al.*, 2009). The strawberry poison frog *Oophaga pumilio* is a phytotelm breeder in which females provide the majority of the parental care. Females transport tadpoles to small phytotelmata and provision them with unfertilized eggs throughout development (Limerick, 1980; Weygoldt, 1980; Brust, 1993). Donnelly (1989a, b) examined the effect of reproductive resource limitation on adult density by manipulating the density of oviposition sites (leaf litter) and tadpole-rearing sites (bromeliads) in the field. The density of males and females increased in bromeliad addition plots, but there was no difference in adult densities between leaf litter plots and controls, indicating that tadpole-rearing sites are limiting but oviposition sites are not (Donnelly, 1989b). Male *O. pumilio* exhibit strong intrasexual competition, defending territories vocally as well as through physical aggression (Bunnell, 1973; Weygoldt, 1980; Pröhl, 1997; Bee, 2003). This led Donnelly (1989a, b) to hypothesize that males defend limiting phytotelmata needed by females in order to monopolize access to mates. Others have pointed out, however, that it is unlikely males are defending phytotelmata, as male territories often do not include these tadpole-rearing sites (McVey *et al.*, 1981; Pröhl, 1997; Pröhl and Hödl, 1999). In a more detailed spatial analysis of male and female *O. pumilio*, females were shown to have a clumped distribution around tadpole-rearing sites whereas males did not (Pröhl and Berke, 2001). Furthermore, females were often observed transporting tadpoles to pools outside of the range of the male parent, and even to pools in the ranges of males with whom they had not mated (Pröhl and Berke, 2001). Thus, it appears that resource defense polygyny does not characterize *O. pumilio*, and the space-use pattern seems better explained by females settling around tadpole-rearing sites and males advertising in areas of high densities of females (Pröhl and Hödl, 1999; Pröhl and Berke, 2001; Meuche *et al.*, 2011).

There is some evidence that female *Oophaga pumilio* defend the phyto-telmata in which they are caring for tadpoles (Haase and Pröhl, 2002). These observations are consistent with other egg-feeding species that defend territories closely associated with phytotelmata (Caldwell and de Oliveira, 1999; Brown *et al.*, 2009). The risk of larval and egg cannibalism in small phytotelmata can be extreme (Summers and Amos, 1997; Summers, 1999; Poelman and Dicke, 2008; Brown *et al.*, 2009). In *Ranitomeya variabilis (ventrimaculata)*, multiple embryos and tadpoles often wind up in the same pools and larger tadpoles con-sume eggs deposited at or below the water line as well as newly hatched smaller tadpoles (Summers and Amos, 1997; Summers, 1999). In fact, Brown *et al.* (2009) documented tactical reproductive parasitism in a Peruvian population of this species; males preferentially deposited tadpoles in pools that already con-tained embryos, and newly hatched tadpoles were quickly cannibalized. Given this risk, it is not surprising that species in which parents provision tadpoles with trophic eggs should be especially protective of phytotelmata containing their offspring. Consistent with this is the lack of territorial behavior in species that breed in small phytotelmata but do not provision tadpoles (Brown *et al.*, 2008a; 2009; Werner *et al.*, 2010; see also Poelman and Dicke, 2008). Addition-ally, mate guarding may also be an important function of territorial behavior in species with biparental care and pair bonding (Brown *et al.* 2008a, 2010a).

Parental Investment and Sexual Selection

In the genus *Oophaga*, females put more effort into parental care than do males. Females transport tadpoles when they hatch, and deposit them in small, nutrient-poor phytotelmata, to which they return to provision the tadpoles with unfertil-ized trophic eggs (Weygoldt, 1980; Brust, 1993). Observations of egg-feeding in *Oophaga pumilio* suggest that females respond to cues given by the tadpoles; tadpoles were observed wriggling and nibbling at females when they entered the water, possibly signaling hunger (Brust, 1993). This tactile stimulation appears to be important for initiating trophic egg-feeding, as has been demon-strated in the egg-feeding treefrog *Anotheca spinosa* (Jungfer, 1996). Female removal experiments in *Oophaga pumilio* confirm the need for female trophic egg provisioning of tadpoles in these small pools (Brust, 1993). In contrast, male parental care in *O. pumilio* is limited to egg attendance (Weygoldt, 1980; Pröhl and Hödl, 1999). This pattern of parental investment is similar to mam-mals, where females invest substantially more into offspring post-fertilization than do males, including nutritive provisioning. This pattern is also predicted to result in a similar mating system to many mammals with a highly skewed operational sex ratio (OSR), where males compete intensely for access to females and display a high level of polygyny (Trivers, 1972; Emlen and Oring, 1977).

Pröhl and Hödl (1999) quantified the potential reproductive rates (Clutton-Brock and Vincent, 1991; Clutton-Brock and Parker, 1992) of male and female *Oophaga pumilio* in the field. Calculations of "time out" of the mating pool for each sex – time mating and performing parental care – revealed that the ratio of female to male "time out" was over 73 : 1, mostly due to the 22 days that females spend caring for tadpoles, during which time they do not mate (Pröhl and Hödl, 1999). Males, on the other hand, could mate sequentially with multiple females even on the same day (Pröhl and Hödl, 1999). The mating system was polygamous, with sequential and simultaneous polygyny and sequential polyandry. Strong intrasexual competition (male territoriality) and high variance in male mating success, as well as the observations of females sampling males, indicates strong sexual selection on males (Pröhl and Hödl, 1999). Male mating success correlated most strongly with overall calling activity, a condition-dependent trait (Pröhl, 2003).

The distribution of phytotelmata used for tadpole deposition was found to have an important effect on the mating system of *Oophaga pumilio*. Pröhl (2002) compared the population densities and adult sex ratios of two sites: a primary forest site with relatively few phytotelmata (bromeliads and *Dieffenbachia*), and a secondary forest site with a high density of potential tadpole habitats (mainly *Heliconia* and banana plants). The adult sex ratio (ASR) was more highly skewed towards females in the secondary forest, presumably as a result of females clumping around the available phytotelmata. Male density, on the other hand, may be limited by territorial spacing; males defending small territories in areas of high female density are generally in better condition than those defending larger territories in areas of low female density (Meuche *et al.*, 2011). Taking into account potential reproductive rates, the more even ASR of the primary forest resulted in a more highly male-skewed OSR, and greater opportunity for sexual selection, in the primary forest than the secondary forest (Pröhl, 2002). These results demonstrate the intimate linkage between the limiting ecological trait of phytotelmata and the mating system of these frogs.

Sexual selection on males, in the form of intrasexual competition between males and female mate choice, is also common in many poison frogs with male-only parental care (Summers, 1989, 1992a, b, 1999; Roithmair, 1992, 1994). As shown above, this usually manifests itself as male territory defense of venues for attracting and courting females. When explaining the phenomenon of sexual selection acting more strongly on males despite male paternal care, it is important to realize that male-only care does not necessarily mean that male parental investment is greater than that of females (Wells, 1981). Territorial males often continue to advertise to females while attending egg clutches, and can care for multiple clutches from several females; thus the reproductive rate of males need not be substantially limited by parental care (Summers, 1989; Ursprung *et al.*, 2011).

Strong male–male competition has apparently led to non-choosy females in *Allobates femoralis*, a terrestrial breeder with male egg attendance and tadpole transport. Using molecular markers to conduct parentage analysis of individual frogs over 2 years, Ursprung *et al.* (2011) were able to track the reproductive success of males and females as measured by the number of offspring they produced that reached adulthood in the next generation. More males than females obtained zero matings, due to the failure of these males to hold territories, but among breeders no difference in reproductive success was found between males and females (Ursprung *et al.*, 2011). Their analysis revealed high levels of polygynandry and found that females displayed site fidelity (Ringler *et al.*, 2009), mainly mating with nearby males (Ringler *et al.*, 2012). They concluded that strong intrasexual competition between males for territories, as well as "bet-hedging" benefits of sequential polyandry, have selected against restrictive female choice in this species (Ringler *et al.*, 2012).

Parental Investment, Sex Role Reversal, and Sexual Conflict

Trivers (1972) originally cited *Dendrobates auratus*, the green poison frog, as a possible example of sex role reversal, based on evidence for extensive paternal care and active courtship of males by females cited in the literature. Wells (1977) attempted to test this hypothesis, and discovered that females were more active in courtship than males, and that females competed aggressively for males in captivity. Hence, this species was considered a promising example of sex role reversal. However, Summers (1989, 1990, 1992a, b) tested the sex role reversal hypothesis and concluded that it was not supported in this case. Time budgets derived from long-term observations in the field indicated that males did not spend more time caring for each offspring than the time required for females to produce them. In fact, males could care for multiple clutches simultaneously, and frequently did so. Male territoriality was associated with attracting females to a high-quality habitat and maintaining an area, free of other males, in which courtship and breeding could occur. In contrast, females did not defend specific areas, but rather specific males (see below). Females were highly aggressive, but female–female aggression was not more frequent or intense than male–male aggression. Female–female aggression resulted from mate guarding: large females remained in the territory of a specific male, and attacked any other female that attempted to approach that male (Summers, 1989). This provides a clear example of sexual conflict, because the male (in all cases) would actively court both females when approached by two females simultaneously. Comparative studies of closely related species with male (*Dendrobates leucomelas*) and female (*Oophaga sylvatica*) parental care support this interpretation of female aggression and sexual conflict (Summers, 1992a). The underlying cause of the conflict appears to be costs of polygyny to females, in terms of reduced quality of paternal care.

There are a number of possible sources for such costs, but Summers (1990) focused on the potential costs of multiple tadpole deposition. *Dendrobates auratus* males carry offspring individually to small pools (phytotelmata) that form in treeholes. Males do not have access to an unlimited number of such pools and so they return to pools where they have previously deposited tadpoles, carrying additional ones, especially if they are caring for multiple clutches from multiple females. Experiments on the effect of multiple deposition showed that increasing the number of tadpoles significantly reduced the average growth rate. Furthermore, tadpoles are highly cannibalistic, and pools generally have only a single surviving tadpole when multiple tadpoles overlap in the same pool (Summers, 1990). From the male's perspective, cannibalism may be a case of sacrificing one offspring for the benefit of another, but from the perspective of the female parent that has a tadpole eaten by another (unrelated) female's offspring, such cannibalism imposes a severe fitness cost with no compensating benefit. Note that *D. auratus* tadpoles readily attack and cannibalize both related and unrelated tadpoles (Gray *et al.*, 2009).

One interesting feature noted by Wells (1977) was that females appear to be more active during courtship than males. This was confirmed by Summers (1989), who observed multiple complete courtships in the field. During courtship, males lead females through the leaf litter, searching for an oviposition site. The female follows the male, actively stroking, nudging, and even jumping on him as they proceed through the leaf litter. This process can be lengthy (over 6 hours). However, active courtship by the female does not necessarily indicate that the female is less selective about mating than the male (which is predicted in sex role reversal: Trivers, 1972). In the case of *D. auratus*, active courtship by the female is ultimately not a good indicator of willingness to mate in a particular male–female interaction, and female rejection of males is significantly more common than male rejection of females (Summers, 1989, 1992a). In some cases, females appear to use active courtship as a mechanism to distract males from courting other females that approach them. When a mate-guarding female detects another female approaching the male she is guarding, she will alternate between attacking the second female and actively courting the male. The male actively calls at and courts both females, but ultimately the mate-guarding female prevails and drives away the secondary female and continues to actively court the male for approximately 20–30 minutes. However, guarding females often discontinue courting the male in these interactions while the male pursues and courts the female, but to no avail. It appears that the female is unwilling to mate in such types of interaction, although she will ultimately mate with that male at a later date. This kind of behavior parallels similar tactics seen in some species of birds involving increased solicitation of copulations to prevent males from engaging in extra-pair copulations (e.g., Eens and Pinxten, 1996), although external fertilization prevents the use of complete copulations for this purpose in frogs. Courtship in this context is apparently a form of deception, used by the female to prevent the male from mating with other females that he has attracted to his territory (Summers, 1992b).

The mating system of *D. auratus* appears to be polygynandrous (both males and females may have multiple partners), but is also characterized by high levels of sexual conflict. Females, in particular, suffer costs from male polygyny, and some females (presumably those with high fighting ability) attempt to guard specific males that control high-quality territories. Hence, although sex roles are not reversed, sexual conflict exerts a strong influence on the mating system of this species. In this regard, *D. auratus* shares similarities with various species of birds that exhibit sexual conflict over paternal care, such as dunnocks (Davies, 1985) and starlings (Eens and Pinxten, 1995; Smith and Sandell, 2005). The importance of sexual conflict in structuring mating systems and mating strategies was developed in a seminal paper by Davies (1989) and further discussed by Brown *et al.* (1997). The mating system of *Dendrobates auratus* provides an excellent example of the effects of sexual conflict over paternal care in a non-avian vertebrate (Summers, 1992b).

Biparental Care and Monogamy

Frogs in the *vanzolinii* clade have apparently experienced a decrease in sexual conflict, as both sexes provide substantial investment into offspring (Figs 11.2, 11.3). Biparental care and long-term pair bonding have been documented in both *Ranitomeya imitator* (Brown *et al.*, 2008a) and its sister species *R. vanzolinii* (Caldwell, 1997; Caldwell and de Oliveira, 1999), and further genetic parentage analysis has revealed that *R. imitator* is genetically monogamous (Brown *et al.*, 2010a). Biparental care has long been recognized as an important factor favoring the evolution of monogamy (e.g., Lack, 1968; Kleiman, 1977; Wittenberger and Tilson, 1980), but empirical support has been largely limited to birds (Møller, 2000) and some mammals (e.g., Gubernick and Teferi, 2000). Monogamy is extremely rare among ectothermic vertebrates, and *R. imitator* and *R. vanzolinii* have offered unique examples of biparental care and monogamy in amphibians.

In *Ranitomeya imitator* and *R. vanzolinii*, egg clutches are laid in arboreal oviposition sites, and are usually attached to phytotelmata above the water level (Caldwell and dc Oliveira, 1999; Brown *et al.*, 2008a) or laid in leaf axils that do not necessarily hold water (J. Tumulty, personal observation). Males attend egg clutches and transport tadpoles individually (Fig. 11.2), depositing them in small phytotelmata – water-filled cavities of saplings and vines in *R. vanzolinii*, and *Dieffenbachia*, and *Heliconia* axils in *R. imitator*. Females have also been observed to occasionally attend egg clutches and transport tadpoles in *R. imitator* when males were experimentally removed (Tumulty *et al.*, unpublished observations), but observations of unmanipulated pairs show that males predominantly perform these parental behaviors. Throughout their development, tadpoles are provisioned with unfertilized trophic eggs, but, unlike *Oophaga*, males coordinate provisioning events by leading females to individual phytotelmata while calling and stimulating them to lay trophic eggs in a way that appears similar to courtship (Caldwell and de Oliveira, 1999; Brown *et al.*, 2008a; Fig. 11.2). However, in contrast to egg clutches resulting from courtship, trophic eggs are laid in the water and are not fertilized (Caldwell and de Oliveira, 1999; Brown *et al.*, 2008a). Observations of parent–tadpole interactions show similarities to egg-feeding behavior in *O. pumilio*, in that tadpoles are often observed wriggling and nibbling against parents when they enter the water (Brown *et al.*, 2008a).

Brown *et al.* (2008a, b, 2009, 2010b) compared phytotelm size, parental care strategy, and the mating system of *R. imitator* with that of a sympatric close relative, *R. variabilis*, to demonstrate the critical importance of pool size in the evolutionary transition to biparental care and monogamy in *R. imitator*. In contrast to *R. imitator*, *R. variabilis* has uniparental male care characterized by male tadpole transport, and a promiscuous mating system; there was no evidence of mate fidelity, neither sex defended territories, and males displayed scramble competition for phytotelmata (Brown *et al.*, 2008a, 2009).

This striking difference in parental care strategies and mating systems of these two species is associated with the use of different sized pools for tadpole deposition. *Ranitomeya imitator* typically breed in *Heliconia* and *Dieffenbachia* host plants, which retain an average of 24 mL of water in their axils (Brown *et al.*, 2008a). In contrast, *R. variabilis* deposit tadpoles in bromeliads axils averaging 112 mL in volume (Brown *et al.*, 2008a). Brown *et al.* (2008b) conducted a pool choice experiment using pairs of artificial pools differing in size attached to vegetation throughout a field site in Peru to compare the tadpole deposition preferences of *R. imitator* with that of *R. variabilis*. When given a choice between small and medium sized pools, and between medium and large sized pools, *R. variabilis* males preferentially deposited tadpoles in larger pools. *R. imitator* preferred smaller pools for tadpole deposition, and avoided both medium and large sized pools.

One longstanding hypothesis for the evolution of monogamy posits that when biparental care becomes crucial for offspring survival, males and females can obtain greater reproductive success through exclusive cooperation in the care of mutual offspring than either can from polygamy (Wittenberger and Tilson, 1980). Brown *et al.* (2010a) and Tumulty *et al.* (unpublished observations) tested this hypothesis by examining the adaptive value of male and female care in *R. imitator*. Reciprocal transplants of tadpoles in natural pools revealed that trophic egg provisioning, and hence female parental care, is critical for growth of *R. imitator* tadpoles in the small nutrient-poor pools typically used by this species (Brown *et al.*, 2010a). While tadpoles of both species grew moderately well in the large pools typically used by *R. variabilis*, neither species showed substantial growth when placed in the small pools typically used by *R. imitator* and not provisioned with trophic eggs. Brown *et al.* (2010a) accomplished this by placing a screen over the phytotelmata of *R. imitator* tadpoles so that adults could not access the pool to provision tadpoles, which showed significantly lower growth than control tadpoles that continued to be fed trophic eggs. More recent male removal experiments have revealed the critical importance of male care throughout tadpole development in *R. imitator*. Males were removed 3 weeks after tadpole deposition, and tadpoles in this removal treatment experienced significantly lower growth and survival over the following 3 weeks compared with unmanipulated control pairs (Tumulty *et al.*, unpublished observations). Although a few females were apparently able to provision tadpoles with trophic eggs without male stimulation, as shown by the presence of trophic eggs in the pools of several tadpoles shortly after male removal, the lower growth and survival of tadpoles indicates that females did not maintain provisioning at the level of control parents.

Coupled with the results of a comparative analysis across all frogs, showing the critical importance of pool size in driving the evolution of parental care in anurans (Fig. 11.4), and the association between trophic egg-feeding and biparental care in small pools (Fig. 11.5), these results make a strong case that the transition to breeding in small phytotelmata drove the evolution of biparental care and monogamy in *R. imitator* (Brown *et al.*, 2010a). The extended high levels of biparental investment necessary to rear tadpoles in small nutrient-poor pools are apparently enough to make polygamy unprofitable for *R. imitator*. Pairs were not observed to produce additional fertilized egg clutches while caring for tadpoles, indicating that rates of egg production limit reproductive rate as well as the number of tadpoles that can be cared for simultaneously (Tumulty *et al.*, unpublished observations). This limitation likely prevents females from practicing simultaneous polyandry. The importance of male care, as revealed by the male removal experiments, also shows that males may be limited in the number of offspring they can rear. It could also be difficult for males to monopolize more than one female, given their role in defending territories, surveying pools, and coordinating feeding events. Based on genetic analyses, 1 out of 12 males in monitored pairs was polygynous (Brown *et al.*, 2010a), so it is possible, although apparently very difficult, for a male to monopolize two females. It would indeed be interesting to compare the reproductive success of polygynous and monogamous male *R. imitator*, and the potential cost of polygyny to females, but the rarity of polygynous males in this species makes this sort of study unfeasible.

While the territorial nature of male *R. imitator*, and occasional observations of female–female aggression observed in captivity (Brown *et al.*, 2008a), indicate that mate guarding could be an important factor in the mating system of this species, it is unlikely to be a sufficient explanation for maintenance of monogamy. As shown in the previous section, male territoriality and intense female mate guarding characterizes several species with uniparental male care, yet these frogs still maintain a polygynandrous mating system (Summers, 1989, 1990, 1992a). Instead, the salient feature that uniquely characterizes monogamy in *R. imitator* is cooperative biparental care of eggs and tadpoles, with similar high levels of parental investment by males and females.

CONCLUSIONS AND FUTURE RESEARCH

Much remains to be learned about parental care and sexual selection in dendrobatids. To date, only a handful of species have been the subject of long-term studies of behavior in the field. Many surprises may await us in terms of the interaction of parental care and mating strategies in these frogs. The costs of parental care to males and females have been difficult to investigate, and it remains an important challenge for the future to quantify these costs in the field. There are intriguing hints of links between parental care and territoriality in frogs (similar to those in fish), and comparative analyses would be useful to illuminate the interaction between these two categories of behavior. As mentioned above, few studies have quantified uncertainty of parentage in anuran mating systems, yet this factor may have an important impact on both patterns of parental care and mating strategies in dendrobatid frogs and other species of anurans. Detailed research on particular species (e.g., Pröhl and Berke, 2001) indicate strong influences of ecological factors on sexual selection and mating systems, but research on a wider range of species is needed before general conclusions can be reached. Sexual conflict is known to have an important impact on mating strategies in some species (Summers, 1992b), but the nature and extent of sexual conflict is not known for most species.

One area that has been receiving increasing attention in the past few years is the interaction of sexual selection and population divergence in the poison frogs. For example, the strawberry poison frog *Oophaga* (*Dendrobates*) *pumilio* shows extreme color pattern divergence among populations in the Bocas del Toro region of Panama, and on the nearby mainland (Daly and Myers, 1967). Summers *et al.* (1997) demonstrated that this diversity arose very rapidly in the recent past, and posited that sexual selection in the form of female mate choice for different colors in different populations could have driven such rapid divergence. Subsequent research has confirmed the plausibility of sexual selection as a diversifying force acting on these populations (Summers *et al.*, 1999; Reynolds and Fitzpatrick, 2007; Rudh *et al.*, 2007, 2010; Maan and Cummings, 2008, 2009; Brown *et al.*, 2010b; Tazzyman and Iwasa, 2010; Richards-Zawacki and Cummings, 2011; Richards-Zawacki *et al.*, 2012; Gehara *et al.*, 2013). Whether sexual selection acts to increase color pattern divergence between populations in other species of poison frogs is currently unknown, but there are a number of species with high levels of color pattern variation among populations in which sexual selection may play a role in diversification (e.g., Roberts *et al.*, 2007; Yeager *et al.*, 2012).

REFERENCES

Ah-King, M., Kvarnemo, C., Tullberg, B.S., 2005. The influence of territoriality and mating system on the evolution of male care: a phylogenetic study on fish. J. Evol. Biol. 18, 371–382.

Alonzo, S.H., 2010. Social and coevolutionary feedbacks between mating and parental investment. Trends Ecol. Evol. 25, 99–108.

Alonzo, S.H., 2012. Sexual selection favours male parental care, when females can choose. Proc. R. Soc. B 279, 1784–1790.

Alonzo, S.H., Heckman, K.L., 2010. The unexpected but understandable dynamics of mating, paternity and paternal care in the ocellated wrasse. Proc. R. Soc. B 277, 115–122.

Alonzo, S.H., Klug, H., 2012. Paternity, maternity, and parental care. In: Royle, N., Smiseth, P., Kölliker, M. (Eds.), The Evolution of Parental Care, Oxford University Press, Oxford, pp. 189–205.

Andersson, M., 1994. Sexual Selection. Princeton University Press, Princeton.

Arnold, K.E., Owens, I.P.F., 2002. Extra-pair paternity and egg dumping in birds: life history, parental care and the risk of retaliation. Proc. R. Soc. Lond. B 269, 1263–1269.

Arnold, S.J., Duvall, D., 1994. Animal mating systems: a synthesis based on selection theory. Am. Nat. 143, 317–348.

Arnqvist, G., Rowe, L., 2005. Sexual Conflict. Princeton University Press, Princeton.

Balshine, S., 2012. Patterns of parental care in vertebrates. In: Royle, N., Smiseth, P., Kölliker, M. (Eds.), The Evolution of Parental Care, Oxford University Press, Oxford, pp. 62–80.

Bateman, A.J., 1948. Intra-sexual selection in *Drosophila*. Heredity 2, 349–368.

Baylis, J.R., 1981. The evolution of parental care in fishes, with reference to Darwin's rule of male sexual selection. Environ. Biol. Fish. 6, 223–251.

Beck, C.W., 1998. Mode of fertilization and parental care in anurans. Anim. Behav. 55, 439–449.

Bee, M.A., 2003. A test of the "dear enemy effect" in the strawberry dart-poison frog (*Dendrobates pumilio*). Behav. Ecol. Sociobiol. 54, 601–610.

Blumer, L., 1979. Male parental care in bony fishes. Q. Rev. Biol. 54, 149–161.

Bourne, G., Collins, A., Holder, A., McCarthy, C., 2001. Vocal communication and reproductive behavior of the frog *Colostethus beebei* in Guyana. J. Herpetol. 35, 272–281.

Brennan, P.L.R., 2012. Mixed paternity despite high male parental care in great tinamous and other Palaeognathes. Anim. Behav. 84, 693–699.

Brown, J.L., Twomey, E., Morales, V., Summers, K., 2008a. Phytotelm size in relation to parental care and mating strategies in two species of Peruvian poison frogs. Behaviour 145, 1139–1165.

Brown, J.L., Morales, V., Summers, K., 2008b. Divergence in parental care, habitat selection and larval life history between two species of Peruvian poison frogs: an experimental analysis. J. Evol. Biol. 21, 1534–1543.

Brown, J.L., Morales, V., Summers, K., 2009. Home range size and location in relation to reproductive resources in poison frogs (Dendrobatidae): a Monte Carlo approach using GIS data. Anim. Behav. 77, 547–554.

Brown, J.L., Morales, V., Summers, K., 2010a. A key ecological trait drove the evolution of biparental care and monogamy in an amphibian. Am. Nat. 175, 436–446.

Brown, J.L., Maan, M.E., Cummings, M.E., Summers, K., 2010b. Evidence for selection on coloration in a Panamanian poison frog: a coalescent-based approach. J. Biogeogr. 37, 891–901.

Brown, J.L., Twomey, E., Amézquita, A., De Souza, M.B., Caldwell, J.P., Lötters, S., Von May, R., Melo-Sampaio, P.R., Mejía-Vargas, D., Perez-Peña, P., Pepper, M., Poelman, E.H., Sanchez-Rodriguez, M., Summers, K., 2011. A taxonomic revision of the neotropical poison frog genus *Ranitomeya* (Amphibia: Dendrobatidae). Zootaxa 3083, 1–120.

Brown, W.D., Crespi, B.J., Choe, J.C., 1997. Sexual conflict and the evolution of mating systems. In: Choe, J.C., Crespi, B.J. (Eds.), The Evolution of Mating Systems in Insects and Arachnids, Cambridge University Press, Cambridge, pp. 352–377.

Brust, D.G., 1993. Maternal brood care by *Dendrobates pumilio*: A frog that feeds its young. J. Herpetol. 27, 96–98.

Bunnell, P., 1973. Vocalizations in the territorial behavior of the frog *Dendrobates pumilio*. Copeia 1973, 277–284.

Caldwell, J.P., 1997. Pair bonding in spotted poison frogs. Nature 385, 211.

Caldwell, J.P., De Oliveira, V.R.L., 1999. Determinants of biparental care in the spotted poison frog, *Dendrobates vanzolinii* (Anura: Dendrobatidae). Copeia 1999, 565.

Caldwell, J.P., Lima, A.P., 2003. A new Amazonian species of *Colostethus* (Anura: Dendrobatidae) with a nidicolous tadpole. Herpetologica 59, 219–234.

Chen, Y.H., Cheng, W.C., Yu, H.T., Kam, Y.C., 2011. Genetic relationship between offspring and guardian adults of a rhacophorid frog and its care effort in response to paternal share. Behav. Ecol. Sociobiol. 65, 2329–2339.

Clutton-Brock, T.H., 1991. The Evolution of Parental Care. Princeton University Press, Princeton.

Clutton-Brock, T.H., Parker, G.A., 1992. Potential reproductive rates and the operation of sexual selection. Q. Rev. Biol. 67, 437–456.

Clutton-Brock, T.H., Vincent, A.C.J., 1991. Sexual selection and the potential reproductive rates of males and females. Nature 351, 58–60.

Coleman, R.M., Gross, M.R., 1991. Parental investment theory: the role of past investment. Trends Ecol. Evol. 6, 404–406.

Crump, M.L., 1972. Territoriality and mating behavior in *Dendrobates granuliferus* (Anura: Dendrobatidae). Herpetologica 28, 195–198.

Crump, M.L., 1995. Parental care. In: Heatwole, H., Sullivan, B.K. (Eds.), Amphibian Biology, Social Behaviour, Vol. 2. Surrey Beatty & Sons, Chipping Norton, NSW, Australia, pp. 518–567.

Daly, J.W., Myers, C.W., 1967. Toxicity of Panamanian poison frogs (*Dendrobates*): some biological and chemical aspects. Science 156, 970–973.

Davies, N.B., 1985. Cooperation and conflict among dunnocks, *Prunella modularis*, in a variable mating system. Anim. Behav. 33, 628–648.

Davies, N.B., 1989. Sexual conflict and the polygamy threshold. Anim. Behav. 38, 226–234.

Dawkins, R., Carlisle, T.R., 1976. Parental investment, mate desertion and a fallacy. Nature 262, 131–132.

Donnelly, M.A., 1989a. Effects of reproductive resource supplementation on space-use patterns in *Dendrobates pumilio*. Oecologia 81, 212–218.

Donnelly, M.A., 1989b. Demographic effects of reproductive resource supplementation in a territorial frog, *Dendrobates pumilio*. Ecol. Monogr. 59, 207–221.

Duellman, W.E., Trueb, L., 1986. The Biology of Amphibians. McGraw-Hill, New York.

Eens, M., Pinxten, R., 1995. Inter-sexual conflicts over copulations in the European starling: evidence for the female mate-guarding hypothesis. Behav. Ecol. Sociobiol. 36, 71–81.

Eens, M., Pinxten, R., 1996. Female European starlings increase their copulation solicitation rate when faced with the risk of polygyny. Anim. Behav. 51, 1141–1147.

Emlen, S.T., Oring, L.W., 1977. Ecology, sexual selection, and the evolution of mating systems. Science 197, 215–223.

Gehara, M., Summers, K., Vences, M., Brown, J.L., 2013. Population expansion, isolation and selection: novel insights into the evolution of coloration in the strawberry poison frog. Evol. Ecol. (in press).

Goin, O.B., Goin, C.J., 1962. Amphibian eggs and the montane environment. Evolution 16, 364–371.

Gomez-Mestre, I., Pyron, R.A., Wiens, J.J., 2012. Phylogenetic analyses reveal unexpected patterns in the evolution of reproductive modes in frogs. Evolution 66, 3687–3700.

Grant, T., Frost, D.R., Caldwell, J.P., Gagliardo, R., Haddad, C.F.B., Kok, P.J.R., Means, D.B., Noonan, B.P., Schargel, W.E., Wheeler, W.C., 2006. Phylogenetic systematics of dart-poison frogs and their relatives (Amphibia: Athesphatanura: Dendro-batidae). Bull. Am. Museum Nat. Hist. 299, 1–262.

Gray, H.M., Summers, K., Ibáñez, D.R., 2009. Kin discrimination in cannibalistic tadpoles of the green poison frog, *Dendrobates auratus* (Anura, Dendrobatidae). Phyllomedusa 8, 41–50.

Gross, M.R., 2005. The evolution of parental care. Q. Rev. Biol. 80, 37–45.

Gross, M.R., Sargent, R.C., 1985. The evolution of male and female parental care in fishes. Am. Zool. 25, 807–822.

Gubernick, D.J., Teferi, T., 2000. Adaptive significance of male parental care in a monogamous mammal. Proc. R. Soc. Lond. B 267, 147–150.

Haase, A., Pröhl, H., 2002. Female activity patterns and aggressiveness in the strawberry poison frog *Dendrobates pumilio* (Anura: Dendrobatidae). Amphibia-Reptilia 23, 129–140.

Haddad, C.F.B., Prado, C.P.A., 2005. Reproductive modes in frogs and their unexpected diversity in the Atlantic forest of Brazil. Bioscience 55, 207–218.

Howard, R.D., 1978. The influence of male-defended oviposition sites on early embryo mortality in bullfrogs. Ecology 59, 789–798.

Juncá, F.A., 1998. Reproductive biology of *Colostethus stepheni* and *Colostethus marchesianus* (Dendrobatidae), with a description of a new anuran mating behavior. Herpelogica 54, 377–387.

Juncá, F.A., Altig, R., Gascon, C., 1994. Breeding biology of *Colostethus stepheni*, a dendrobatid frog with a nontransported nidicolous tadpole. Copeia 1994, 747–750.

Jungfer, K.H., 1996. Reproduction and parental care of the coronated treefrog, *Anotheca spinosa* (Steindachner, 1864) (Anura: Hylidae). Herpetologica 52, 25–32.

Kamel, S.J., Grosberg, R.K., 2012. Exclusive male care despite extreme female promiscuity and low paternity in a marine snail. Ecol. Lett. 15, 1167–1173.

Kleiman, D.G., 1977. Monogamy in mammals. Q. Rev. Biol. 52, 39–69.

Klug, H., Heuschele, J., Jennions, M.D., Kokko, H., 2010. The mismeasurement of sexual selection. J. Evol. Biol. 23, 447–462.

Knapp, R.A., Sargent, R.C., 1989. Egg-mimicry as a mating strategy in the fantail darter, *Etheostoma flabellare*: females prefer males with eggs. Behav. Ecol. Sociobiol. 25, 321–326.

Kokko, H., Jennions, M.D., 2008. Parental investment, sexual selection and sex ratios. J. Evol. Biol. 21, 919–948.

Kokko, H., Monaghan, P., 2001. Predicting the direction of sexual selection. Ecol. Lett. 4, 159–165.

Kokko, H., Rankin, D.J., 2006. Lonely hearts or sex in the city? Density-dependent effects in mating systems. Phil. Trans. R. Soc. B 361, 319–334.

Lack, D., 1968. Ecological Adaptations for Breeding in Birds. Methuen, London.

Laurila, A., Seppä, P., 1998. Multiple paternity in the common frog (*Rana temporaria*): genetic evidence from tadpole kin groups. Biol. J. Linn. Soc. 63, 221–232.

Lehtinen, R.M., Nussbaum, R.A., 2003. Parental care: a phylogenetic perspective. In: Jamieson, B.G.M. (Ed.), Reproductive Biology and Phylogeny of Anura, Science Publishers Inc., Enfield, NH, USA, pp. 343–386.

Lessells, C.M., 2012. Sexual conflict. In: Royle, N., Smiseth, P.T., Kölliker, M. (Eds.), The Evolution of Parental Care, Oxford University Press, Oxford.

Limerick, S., 1980. Courtship behavior and oviposition of the poison-arrow frog *Dendrobates pumilio*. Herpetologica 36, 69–71.

Lodé, T., Lesbarrères, D., 2004. Multiple paternity in *Rana dalmatina*, a monogamous territorial breeding anuran. Naturwissenschaften 91, 44–47.

Loiselle, P.V., 1978. Prevalence of male brood care in teleosts. Nature 276, 98.

Lüddecke, H., 1999. Behavioral aspects of the reproductive biology of the Andean frog *Colostethus palmatus*. Rev. Acad. Colomb. Cienc. 23 (Suppl.), 303–316.

Lutz, B., 1947. Trends towards non-aquatic and direct development in frogs. Copeia 1947, 242–252.

Maan, M.E., Cummings, M.E., 2008. Female preferences for aposematic signal components in a polymorphic poison frog. Evolution 62, 2334–2345.

Maan, M.E., Cummings, M.E., 2009. Sexual dimorphism and directional sexual selection on aposematic signals in a poison frog. Proc. Natl. Acad. Sci. U. S. A. 106, 19072–19077.

Maddison, W.P., 1990. A method for testing the correlated evolution of two binary characters: are gains or losses concentrated on certain branches of a phylogenetic tree? Evolution 44, 539–557.

Maddison, W.P., Maddison, D.R., 2011. Mesquite: a modular system for evolutionary analysis. Version 2.75 http://mesquiteproject.org.

Magnusson, W.E., Hero, J.M., 1991. Predation and the evolution of complex oviposition behaviour in Amazon rainforest frogs. Oecologia 86, 310–318.

Mank, J.E., Promislow, D.E.L., Avise, J.C., 2005. Phylogenetic perspectives in the evolution of parental care in ray-finned fishes. Evolution 59, 1570–1578.

Marconato, A., Bisazza, A., 1985. Males whose nests contain eggs are preferred by female *Cottus gobio* L. (Pisces, Cottidae). Anim. Behav. 34, 1580–1582.

Maynard Smith, J., 1978. The Evolution of Sex. Cambridge University Press, Cambridge.

McDiarmid, R.W., 1978. Evolution of parental care in frogs. In: Burghardt, G., Bekoff, M. (Eds.), The Development of Behavior: Comparative and Evolutionary Aspects, Garland STPM Press, New York, pp. 127–147.

McKeon, C.S., Summers, K., 2013. Predator driven reproductive behavior in a tropical frog. Evol. Ecol. 27, 725–737.

McVey, M.E., Zahary, R.G., Perry, D., MacDougal, J., 1981. Territoriality and homing behavior in the poison dart frog (*Dendrobates pumilio*). Copeia 1981, 1–8.

Meuche, I., Pröhl, H., 2011. Alternative mating tactics in the strawberry poison frog (*Oophaga pumilio*). J. Herpetol. 21, 275–277.

Meuche, I., Linsenmair, K.E., Pröhl, H., 2011. Female territoriality in the strawberry poison frog (*Oophaga pumilio*). Copeia 2011, 351–356.

Møller, A.P., 2000. Male parental care, female reproductive success, and extrapair paternity. Behav. Ecol. 11, 161–168.

Møller, A.P., Cuervo, J.J., 2000. The evolution of paternity and paternal care in birds. Behav. Ecol. 11, 472–485.

Myers, C.W., Daly, J.W., 1976. Preliminary evaluation of skin toxins and vocalizations in taxonomic and evolutionary studies of poison-dart frogs (Dendrobatidae). Bull. Am. Mus. Nat. Hist. 157, 157–173.

Myers, C.W., Daly, J.W., 1983. Dart-poison frogs. Sci. Am. 248, 120–133.

Myers, C.W., Paolillo, A.O., Daly, J.W., 1991. Discovery of a defensively malodorous and nocturnal frog in the family Dendrobatidae: phylogenetic significance of a new genus and species from the Venezuelan Andes. Am. Mus. Novit. 3002, 1–33.

Neff, B.D., 2003. Decisions about parental care in response to perceived paternity. Nature 422, 716–719.

Neff, B.D., Gross, M.R., 2001. Dynamic adjustment of parental care in response to perceived paternity. Proc. R. Soc. B 268, 1559–1565.

Okuda, N., 1999. Sex roles are not always reversed when the potential reproductive rate is higher in females. Am. Nat. 153, 540–548.

Pagel, M., 1994. Detecting correlated evolution on phylogenies: a general method for the comparative analysis of discrete characters. Proc. R. Soc. Lond. B 255, 37–45.

Parker, G.A., 1979. Sexual selection and sexual conflict. In: Blum, M.S., Blum, N.A. (Eds.), Sexual Selection and Reproductive Competition in Insects, Academic Press, New York, pp. 123–166.

Perrone, M.J., Zaret, T.M., 1979. Parental care patterns of fishes. Am. Nat. 113, 351–361.

Poelman, E.H., Dicke, M., 2008. Space use of Amazonian poison frogs: testing the reproductive resource defense hypothesis. J. Herpetol. 42, 270–278.

Prado, C.P., de, A., Haddad, C.F.B., 2003. Testes size in leptodactylid frogs and occurrence of multimale spawning in the genus *Leptodactylus* in Brazil. J. Herpetol. 37, 354–362.

Pröhl, H., 1997. Territorial behavior of the strawberry poison-dart frog, *Dendrobates pumilio*. Amphibia-Reptilia 18, 437–442.

Pröhl, H., 2002. Population differences in female resource abundance, adult sex ratio, and male mating success in *Dendrobates pumilio*. Behav. Ecol. 13, 175–181.

Pröhl, H., 2003. Variation in male calling behaviour and relation to male mating success in the strawberry poison frog (*Dendrobates pumilio*). Ethology 109, 273–290.

Pröhl, H., 2005. Territorial behavior in dendrobatid frogs. J. Herpetol. 39, 354–365.

Pröhl, H., Berke, O., 2001. Spatial distributions of male and female strawberry poison frogs and their relation to female reproductive resources. Oecologia 129, 534–542.

Pröhl, H., Hödl, W., 1999. Parental investment, potential reproductive rates, and mating system in the strawberry dart-poison frog, *Dendrobates pumilio*. Behav. Ecol. Sociobiol. 46, 215–220.

Pyron, R.A., Wiens, J.J., 2011. A large-scale phylogeny of Amphibia with over 2,800 species, and a revised classification of extant frogs, salamanders, and caecilians. Mol. Phylogenet. Evol. 61, 543–583.

Queller, D.C., 1997. Why do females care more than males? Proc. R. Soc. B 264, 1555–1557.

Reynolds, R.G., Fitzpatrick, B.M., 2007. Assortative mating in poison-dart frogs based on an ecologically important trait. Evolution 61, 2253–2259.

Richards-Zawacki, C.L., Cummings, M.E., 2011. Intraspecific reproductive character displacement in a polymorphic poison dart frog, *Dendrobates pumilio*. Evolution 65, 259–267.

Richards-Zawacki, C.L., Wang, I.J., Summers, K., 2012. Mate choice and the genetic basis for colour variation in a polymorphic dart frog: inferences from a wild pedigree. Mol. Ecol. 21, 3879–3892.

Ridley, M., 1978. Paternal care. Anim. Behav. 26, 904–932.

Ringler, E., Ringler, M., Jehle, R., Hödl, W., 2012. The female perspective of mating in *A. femoralis*, a territorial frog with paternal care – a spatial and genetic analysis. PLoS ONE 7, e40237.

Ringler, M., Ursprung, E., Hödl, W., 2009. Site fidelity and patterns of short- and long-term movement in the brilliant-thighed poison frog *Allobates femoralis* (Aromobatidae). Behav. Ecol. Sociobiol. 63, 1281–1293.

Roberts, J.L., Brown, J.L., Schulte, R., Arizabal, W., Summers, K., 2007. Rapid diversification of colouration among populations of a poison frog isolated on sky peninsulas in the central cordilleras of Peru. J. Biogeogr. 34, 417–426.

Roithmair, M.E., 1992. Territoriality and male mating success in the dart-poison frog, *Epipedobates femoralis* (Dendrobatidae, Anura). Ethology 92, 331–343.

Roithmair, M.E., 1994. Male territoriality and female mate selection in the dart-poison frog *Epipedobates trivittatus* (Dendrobatidae, Anura). Copeia 1, 107–115.

Roughgarden, J., Oishi, M., Akçay, E., 2006. Reproductive social behavior: cooperative games to replace sexual selection. Science 311, 965–969.

Rudh, A., Rogell, B., Höglund, J., 2007. Non-gradual variation in colour morphs of the strawberry poison frog *Dendrobates pumilio*: genetic and geographical isolation suggest a role for selection in maintaining polymorphism. Mol. Ecol. 16, 4284–4294.

Salthe, S., Duellman, W.E., 1973. Quantitative constraints associated with reproductive mode in anurans. In: Vial, J. (Ed.), Evolutionary Biology of the Anurans: Contemporary Research on Major Problems, University of Missouri Press, Columbia, pp. 229–249.

Santos, J.C., Coloma, L.A., Summers, K., Caldwell, J.P., Ree, R., Cannatella, D.C., 2009. Amazonian amphibian diversity is primarily derived from late Miocene Andean lineages. PLoS Biol. 7, e1000056.

Sexton, O.J., 1960. Some aspects of the behavior and of the territory of a dendrobatid frog, *Prostherapis trinitatis*. Ecology 41, 107–115.

Sheldon, B.C., 2002. Relating paternity to paternal care. Phil. Trans. R. Soc. Lond. B 357, 341–350.

Shuster, S.M., Wade, M.J., 2003. Mating Systems and Strategies. Princeton University Press, Princeton.

Smith, H.G., Sandell, M.I., 2005. The starling mating system as an outcome of the sexual conflict. Evol. Ecol. 19, 151–165.

Summers, K., 1989. Sexual selection and intra-female competition in the green poison-dart frog, *Dendrobates auratus*. Anim. Behav. 37, 797–805.

Summers, K., 1990. Paternal care and the cost of polygyny in the green dart-poison frog. Behav. Ecol. Sociobiol. 27, 307–313.

Summers, K., 1992a. Mating strategies in two species of dart-poison frogs: a comparative study. Anim. Behav. 43, 907–919.

Summers, K., 1992b. Dart-poison frogs and the control of sexual selection. Ethology 91, 89–107.

Summers, K., 1999. The effects of cannibalism on Amazonian poison frog egg and tadpole deposition and survivorship in *Heliconia* axil pools. Oecologia 119, 557–564.

Summers, K., 2000. Mating and aggressive behaviour in dendrobatid frogs from Corcovado National Park, Costa Rica: a comparative study. Behaviour 137, 7–24.

Summers, K., Amos, W., 1997. Behavioral, ecological, and molecular genetic analyses of reproductive strategies in the Amazonian dart-poison frog, *Dendrobates ventrimaculatus*. Behav. Ecol. 8, 260–267.

Summers, K., Earn, D.J.D., 1999. The cost of polygyny and the evolution of female care in poison frogs. Biol. J. Linn. Soc. 66, 515–538.

Summers, K., McKeon, C.S., 2004. The evolutionary ecology of phytotelmata use in neotropical poison frogs. Misc. Pub. Mus. Zool. Univ. Mich. 193, 55–73.

Summers, K., Bermingham, E., Weigt, L., McCafferty, S., Dahlstrom, L., 1997. Phenotypic and genetic divergence in three species of dart-poison frogs with contrasting parental behavior. J. Hered. 88, 8–13.

Summers, K., Weight, L.A., Boag, P., Bermingham, E., 1999. The evolution of female parental care in poison frogs of the genus *Dendrobates*: evidence from mitochondrial DNA sequences. Herpetologica 55, 254–270.

Summers, K., McKeon, C.S., Heying, H., 2006. The evolution of parental care and egg size: a comparative analysis in frogs. Proc. R. Soc. B 273, 687–692.

Summers, K., McKeon, C.S., Heying, H., Hall, J., Patrick, W., 2007. Social and environmental influences on egg size evolution in frogs. J. Zool. 271, 225–232.

Tallamy, D.W., 2000. Sexual selection and the evolution of exclusive paternal care in arthropods. Anim. Behav. 60, 559–567.

Tang-Martinez, Z., Ryder, T.B., 2005. The problem with paradigms: Bateman's worldview as a case study. Integr. Comp. Biol. 45, 821–830.

Tazzyman, S.J., Iwasa, Y., 2010. Sexual selection can increase the effect of random genetic drift – a quantitative genetic model of polymorphism in *Oophaga pumilio*, the strawberry poison-dart frog. Evolution 64, 1719–1728.

Test, F.H., 1954. Social aggressiveness in an amphibian. Science 120, 140–141.

Touchon, J.C., 2012. A treefrog with reproductive mode plasticity reveals a changing balance of selection for nonaquatic egg laying. Am. Nat. 180, 733–743.

Touchon, J.C., Warkentin, K.M., 2008. Reproductive mode plasticity: aquatic and terrestrial oviposition in a treefrog. Proc. Natl. Acad. Sci. U. S. A. 105, 7495–7499.

Townsend, D.S., 1989. Sexual selection, natural selection, and a fitness trade-off in a tropical frog with male parental care. Am. Nat. 133, 266–272.

Trivers, R.L., 1972. Parental investment and sexual selection. In: Campbell, B. (Ed.), Sexual Selection and the Descent of Man 1871–1971, Aldine Press, Chicago, pp. 136–179.

Ursprung, E., Ringler, M., Jehle, R., Hödl, W., 2011. Strong male/male competition allows for nonchoosy females: high levels of polygynandry in a territorial frog with paternal care. Mol. Ecol. 20, 1759–1771.

Vieites, D.R., Nieto-Román, S., Barluenga, M., Palanca, A., Vences, M., Meyer, A., 2004. Post-mating clutch piracy in an amphibian. Nature 431, 305–308.

Wells, K.D., 1977. The social behaviour of anuran amphibians. Anim. Behav. 25, 666–693.

Wells, K.D., 1980a. Behavioral ecology and social organization of a dendrobatid frog (*Colostethus inguinalis*). Behav. Ecol. Sociobiol. 6, 199–209.

Wells, K.D., 1980b. Social behavior and communication of a dendrobatid frog (*Colostethus trinitatis*). Herpetologica 36, 189–199.

Wells, K.D., 1981. Parental behavior of male and female frogs. In: Alexander, R.D., Tinkle, D.W. (Eds.), Natural Selection and Social Behavior, Chiron Press, New York, pp. 184–197.

Wells, K.D., 2007. The Ecology and Behavior of Amphibians. University of Chicago Press, Chicago.

Werner, P., Elle, O., Schulte, L.M., Lötters, S., 2010. Home range behaviour in male and female poison frogs in Amazonian Peru (Dendrobatidae: *Ranitomeya reticulata*). J. Nat. Hist. 45, 15–27.

Werren, J.H., Gross, M.R., Shine, R., 1980. Paternity and the evolution of male parental care. J. Theor. Biol. 82, 619–631.

Westneat, D.F., Sherman, P.W., 1993. Parentage and the evolution of parental behavior. Behav. Ecol. 4, 66–77.

Weygoldt, P., 1980. Complex brood care and reproductive behavior in captive poison-arrow frogs, *Dendrobates pumilio* O. Schmidt. Behav. Ecol. Sociobiol. 7, 329–332.

Weygoldt, P., 1987. Evolution of parental care in dart poison frogs (Amphibia: Anura: Dendrobatidae). J. Zool. Syst. Evol. 25, 51–67.

Williams, G.C., 1966. Adaptation and Natural Selection: A Critique of Some Current Evolutionary Thought. Princeton University Press, Princeton.

Williams, G.C., 1975. Sex and Evolution. Princeton University Press, Princeton.

Wittenberger, J.F., Tilson, R.L., 1980. The evolution of monogamy: hypotheses and evidence. Annu. Rev. Ecol. Syst. 11, 197–232.

Yeager, J., Brown, J.L., Morales, V., Cummings, M., Summers, K., 2012. Testing for selection on color and pattern in a mimetic radiation. Curr. Zool. 58, 668–677.

Zimmermann, E., Zimmermann, H., 1984. Durch nachtzucht erhalten: baumsteigerfrosche *Dendrobates quinquevittatus* und *D. reticulatus*. Aquarien Magatin 18, 35–41.

Zimmermann, H., Zimmermann, E., 1988. Ethotaxonomie und zoographische artengruppenbildung bei pfeilgiftfroschen (Anura: Dendrobatidae). Salamandra 24, 125–146.

Chapter 12

Testosterone, Territoriality, and Social Interactions in Neotropical Birds

John C. Wingfield,[1] Rodrigo A. Vasquez[2] and Ignacio T. Moore[3]

[1]*Department of Neurobiology, Physiology and Behavior, University of California, Davis, California, USA,* [2]*Institute of Ecology and Biodiversity, Departamento de Ciencias Ecologicas, Facultad de Ciencias, Universidad de Chile, Santiago, Chile,* [3]*Department of Biological Sciences, Virginia Tech. University, Blacksburg, Virginia, USA*

INTRODUCTION

Inspection of a map of the Earth quickly reveals that the northern and southern hemispheres differ markedly (Chapter 1, Fig. 1.1). The north includes large land masses surrounding an ocean at the pole. In contrast, the south is made up of a vast southern ocean surrounding an ice-bound land mass. Southern hemisphere tips of continents and Sub-Antarctic islands have the same range of vegetation types as in the northern hemisphere, although diversity at high latitudes in the south tends to be lower (French and Smith, 1985). These authors suggest that Sub-Antarctic habitats combine elements from Sub-Arctic habitats to form qualitatively different ecosystems. Furthermore, large land masses in the north show great variation in annual range of temperature compared with similar latitudes in the south that are dominated by oceans with much less annual variation (Ghalambor *et al.*, 2006). Additionally, it appears that reduced seasonal variation in temperature in the southern hemisphere is a result of a greater range of minimum temperature rather than variation in maximum temperature (Ghalambor *et al.*, 2006), although snow events can occur in every month of a year in the southernmost area of the New World, namely southern Patagonia (Rozzi *et al.*, 2006). This suggests that greater variation in temperature and lows in the north may result in more migration and greater seasonality than at similar latitudes in the south. Although seasonality in day length in both hemispheres is identical, other abiotic components of seasons, such as temperature and rainfall, among others, are very different. Does this mean that avian species breeding at mid- to high latitudes in the north and south have solved problems of regulating life cycles and seasonality in the same way, or differently? Are altitudinal

Sexual Selection. http://dx.doi.org/10.1016/B978-0-12-416028-6.00012-8

gradients of seasonality of migration and breeding the same in the north and south? What can we predict about such life-history stages in the lowlands and highlands in the Neotropics that separate New World Arctic and Sub-Antarctic regions? Here, the focus will be on changes in the hypothalamic–pituitary–gonadal axis (HPG), particularly in relation to patterns of testosterone secretion and control mechanisms.

LIFE CYCLES AND ENDOCRINE REGULATORY SYSTEMS: A NORTHERN PERSPECTIVE

Regulatory pathways of the HPG system in relation to pubertal, seasonal, and social effects have been studied almost exclusively in northern hemisphere avian species. Although the hormones involved and many of the mechanisms by which their secretions are controlled – their transport, metabolism, and actions – likely have been conserved throughout vertebrates (Adkins-Regan, 2005; Wingfield 2006, 2012a; Norris 2007), the ways by which environmental and social factors affect the HPG system and the actions of testosterone are much less well known (see, for example, Wingfield and Farner, 1993). Furthermore, recent research has also highlighted that within species there is much stable variability in behavior and physiology, possibly including feedbacks between behavior and endocrine regulatory systems (Korte *et al.*, 2005; McGlothlin *et al.*, 2007; see also Hau and Wingfield, 2011). This may be a result of diverse evolutionary origins, including equatorial and more septentrional and/or meridional species, or due to the influence of diverse types of ecological variability in time and space. Again, much work has been done on northern hemisphere species, but recently several studies have focused on tropical and southern hemisphere species and these now warrant a summary of where the current knowledge stands and where there are gaps for future research.

Control mechanisms for the secretion of testosterone and territorial aggression in birds have three major components (Fig. 12.1): first, the regulation of hormone secretion from the HPG axis (left-hand part of Fig. 12.1); second, the transport of testosterone in the blood (broken lines); and third, the mechanisms associated with action of the hormone in the target cell – in this case, a neuron in the brain (central part of Fig. 12.1). The net result is regulation of territorial aggression by neural networks (right-hand part of Fig. 12.1). This three-part system of control of hormone secretion, transport, and effects on target organs is an important concept (Wingfield 2012a, b) because it outlines many points of potential regulatory mechanisms. The secretion cascade on the left of Fig. 12.1 summarizes how sensory information such as social interactions is transduced through neurotransmitters, such as glutamate (NMDA), and neuroendocrine secretions such as gonadotropin-releasing hormone (GnRH) and gonadotropin-inhibiting hormone (GnIH), into the release of the gonadotropins luteinizing hormone (LH) and follicle-stimulating hormone (FSH – not shown) from the anterior pituitary into the blood. LH circulates to the gonad, where it acts on

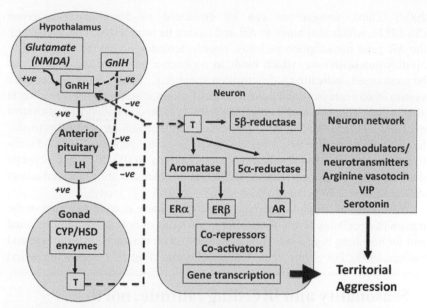

FIGURE 12.1   Control mechanisms for territorial aggression in birds in spring. There are three major components to control mechanisms: the regulation of hormone secretion from the hypothalamic–pituitary–gonadal (HPG) axis (left-hand part of the figure); transport of hormones such as testosterone in the blood (broken lines); and the mechanisms associated with action of the hormone in the target cell, in this case a neuron in the brain (central part of the figure). The net result is regulation of territorial aggression by neural networks (right-hand part of the figure). This three-part system of control of hormone secretion, transport, and effects on target organs is an important concept because it proffers many points of potential regulatory mechanisms. *Reproduced from Wingfield (2012b), courtesy of Springer-Verlag, Berlin.*

cells that express steroidogenic enzymes to stimulate secretion of the sex steroid hormone testosterone that is in turn released into the blood. Local actions of testosterone in the testis include regulation of spermatogenesis, but it is also released into the blood in many avian species when breeding. Among many actions of testosterone are effects on territorial aggression, and negative feedback on neuroendocrine and pituitary secretions (broken lines in Fig. 12.1). In birds, testosterone circulates bound weakly to corticosterone-binding globulin (CBG) before entering target neurons involved in the expression of territorial aggression (center of Fig. 12.1). Once testosterone has entered a target neuron, it has four potential fates. First, it can bind directly to the androgen receptor (AR), a member of the type 1 genomic receptors that become gene transcription factors once they are bound to testosterone. Second, testosterone can be converted to estradiol (E2) by the enzyme aromatase. E2 can then bind to either estrogen receptor alpha (ERα) or estrogen receptor beta (ERβ), both of which are genomic receptors that regulate gene transcription, but different genes from those regulated by AR (e.g., Schlinger, 1994; Schlinger and Brenowitz,

2002). Third, testosterone can be converted to 5α-dihydrotestosterone (5α-DHT), which also binds to AR and cannot be aromatized, thus enhancing the AR gene transcription pathway. Fourth, testosterone can be converted to 5β-dihydrotestosterone, which binds to no known receptors and also cannot be aromatized, indicating a deactivation shunt (e.g., Soma, 2006). A complex system of co-repressors and co-activators of genomic steroid receptor action is also known. The end result is regulatory action on neural networks that control expression of territorial aggression. Several neurotransmitters and neuromodulators, such as arginine vasotocin, vasoactive intestinal peptide (VIP), and serotonin, are also involved at this level (e.g., Goodson, 2005). Evidence suggests that the basic secretory, transport, and action mechanisms are conserved across vertebrates (e.g., Wingfield, 2012a, b).

Field and laboratory endocrine studies show there is wide variation in the pattern of circulating levels of testosterone in terms of both the amount secreted and for how long high levels of testosterone can be maintained (Wingfield and Farmer, 1993). For example, a schematic diagram (Fig. 12.2) shows the period

Seasonality and breeding latitude, north

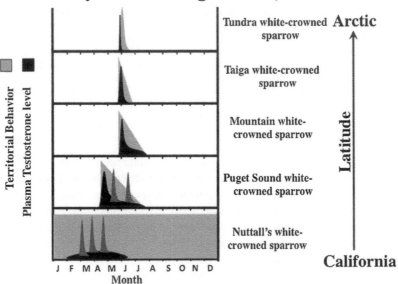

FIGURE 12.2 Schematic diagram showing the period of territoriality (grey areas) and plasma concentrations of testosterone (black shapes) in male white-crowned sparrows, *Zonotrichia leucophrys*. The most northerly tundra (tundra *Z.l. gambelii*) and taiga populations of *Z.l. gambelii* have the shortest breeding season and briefest period of high testosterone levels. More southerly, multiple-brooded Puget Sound white-crowned sparrows, *Z.l. pugetensis*, have longer breeding seasons. Only *Z.l. pugetensis* populations show social modulation of testosterone secretion (bottom panel: two peaks with checkered shading indicates how testosterone levels can peak after a social challenge. *Reproduced from Wingfield et al. (2007) (after Wingfield and Farner 1978, 1979; Wingfield and Hahn, 1994; Meddle et al., 2002), courtesy of Springer-Verlag, Berlin.*

of territoriality (grey areas) and plasma concentrations of testosterone (black shapes) within populations of male white-crowned sparrows, *Zonotrichia leucophrys*, with several distinct subspecies and migration/breeding strategies. The most northerly population (tundra *Z.l. gambelii*) and the taiga populations of *Z.l. gambelii* have the shortest breeding season and briefest period of high testosterone levels, at just a few days. More southerly populations, such as the mid-latitude breeding and multiple-brooded Puget Sound white-crowned sparrows, *Z.l. pugetensis*, have longer breeding seasons (Fig. 12.2). Furthermore, males of *Z.l. pugetensis* populations show social modulation of testosterone secretion: in the bottom panel of Figure 12.2, two peaks with checkered shading indicate how testosterone levels can peak after a social challenge (Wingfield and Farner, 1978, 1979; Wingfield and Hahn, 1994; Meddle *et al.*, 2002). More recently, field investigations of the non-migratory white-crowned sparrow of coastal California revealed that plasma testosterone levels remain low throughout the prolonged breeding season, but surges of plasma testosterone levels can be induced by social challenges such as simulated territorial intrusions (STIs) (Fig. 12.2; Wingfield *et al.*, 2007; J. C. Wingfield unpublished).

Determining the regulatory mechanisms underlying these responses of testosterone secretion to seasonal and social environmental cues in changing environments is important if we are to understand the evolution of hormonally mediated traits, including those resulting from sexual selection (Hau, 2007; Hau and Wingfield, 2011). A framework of how diverse environmental factors might interact to regulate the life cycles of birds, indeed all vertebrates, is based largely on investigations from the northern hemisphere (e.g., Wingfield, 2008). This includes effects of social interactions that synchronize individuals of a group or a breeding pair and integrate behavioral changes such as the establishment of a breeding territory and attracting a mate, then the transition from sexual to parental behavior (Wingfield *et al.*, 1999; Wingfield, 2006). However, neotropical birds offer many model systems, and closely related taxa have ranges extending into the northern and southern hemispheres, providing opportunities to determine whether the hormonal mechanisms are similar (evolutionary constraints hypothesis; Reed *et al.*, 2006; Hau, 2007), or whether multiple unique mechanisms have evolved in response to similar perturbations (evolutionary potential hypothesis, Hau, 2007). Reproductive aggression, including territoriality, is controlled in part by the steroid hormone testosterone (see Fig. 12.1), but the relationship of plasma levels of testosterone to aggression is inconsistent (Wingfield *et al.*, 1990, 2005; Hirschenhauser *et al.*, 2003; Adkins-Regan, 2005; Hirschenhauser and Oliveira, 2006; Wingfield, 2006; Hau, 2007). Nonetheless, the interrelationships of testosterone and aggression have been the focus of theoretical approaches, from behavioral ecology, evolution, and endocrinology down to cell and molecular biology (Wingfield, 2012a, b), although the potentially strong influences of latitude and altitude remain much less well known but are likely important determinants of mechanistic pathways (Goymann *et al.*, 2004), particularly considering recent studies showing latitudinal intraspecific variation in behavioral and other physiological traits (e.g., Goymann *et al.*, 2007; Goymann and Landys, 2011; Maldonado *et al.*, 2012). Next, it is important to reflect on the actions of testosterone, which may be highly conserved across vertebrates.

What Does Testosterone Do?

Circulating hormones rarely have one simple action. Many can affect a complex suite of physiological, behavioral, and morphological parameters simultaneously. Suites of effects may occur by season or in social contexts, and secretion of those hormones at the wrong time of year could be deleterious and could incur costs that reduce fitness (e.g., Wingfield *et al.*, 1990, 1999; Ketterson and Nolan, 1999; Reed *et al.*, 2006; Hau, 2007; Ketterson *et al.*, 2009). The steroid hormone testosterone is a classical example, having a range of specific actions to regulate spermatogenesis, male reproductive associated structures, development of some secondary sex characters, muscle hypertrophy, and activation of reproductive behaviors (see Wingfield, 2006). On the other hand, a large number of experimental studies in males show that prolonged high levels of testosterone, even within the breeding season, result in decreased expression of paternal care, increased likelihood of injury and susceptibility to predation, decreased fat stores and over-winter survival, and, in many cases, a compromised immune system (Wingfield *et al.*, 1990, 1999; Reed *et al.*, 2006; Hau, 2007; Ketterson *et al.*, 2009). High circulating levels of testosterone may also interfere with the temporal progression of life-history stages such as development of molt (e.g., Runfeldt and Wingfield, 1985; Schleussner *et al.*, 1995; Nolan *et al.*, 1992). Clearly, the patterns, and thus regulation, of testosterone titers circulating in blood must be a reflection of trade-offs of benefits and potential deleterious effects at different times of the year and in social contexts (e.g., Ketterson and Nolan, 1999; Hau, 2007; Hau and Wingfield, 2011). In the long term, they are also a reflection of strong ecological pressures producing evolutionary tendencies and/or plasticities (Piggliucci, 2001; Ricklefs and Wikelski, 2002), which are the result of evolutionary processes, where endocrine physiology has become adapted to ecological conditions, producing certain more or less fixed or flexible (= plastic) trade-offs (see Ricklefs and Wikelski, 2002; Korte *et al.*, 2005). These in turn might have resulted in diverse suites of control mechanisms, as outlined in Fig. 12.1 (see also Wingfield, 2012a, b).

An aspect of testosterone investigation that is becoming important is distinguishing between differences in baseline and elevated levels. In terms of social responsiveness, it is important to know if a bird is responsive or not to a social cue, or if it is even physiologically capable of responding. Goymann *et al.* (2007) described a series of experiments that investigators should perform to develop a more complete understanding of the HPG axis in male birds. By adhering to a similar protocol, future comparisons among species will not be confounded by differences in protocols, and thus true environmental and species differences or similarities will become more apparent and easier to quantify.

Social Modulation of Testosterone and the Challenge Hypothesis

Despite the evidence that testosterone activates aggression associated with male–male competition over territories and mates, correlations of plasma levels of testosterone with expression of territorial aggression when breeding are highly variable (e.g., Adkins-Regan, 2005). Numerous explanations for this variation cannot avoid the conclusion that circulating testosterone concentrations do not necessarily correlate with actual expression of territorial aggression on a day-to-day or even hour-to-hour basis (Wingfield et al., 1990). However, on a seasonal basis, testosterone titers in blood do correlate roughly with expression of territorial aggression, especially in species that are not territorial at other times of the year (Wingfield et al., 2007). Further, there is now evidence that neural activity of the androgens can explain some of the variation that plasma levels of the hormone do not explain (Rosval et al., 2012). These neural activities include androgen metabolizing enzymes and receptors for sex steroids (see Fig. 12.1).

An early survey of the literature (Wingfield and Ramenofsky, 1985) on testosterone and aggression in birds and mammals revealed that those studies showing a positive correlation were split equally with those studies that found no correlation. Moreover, it appears that those showing a positive correlation of testosterone levels and aggression involved varying degrees of social instability, whereas those finding no correlation tended to have been conducted with animals in socially stable groups. Since then many field studies of vertebrates, now numbering over 160 species (Hirschenhauser and Oliveira, 2006), have revealed that circulating testosterone levels show transient surges leading to much higher concentrations at usually brief periods often occurring at times of heightened male–male competition. For example, plasma levels of testosterone are particularly high when establishing a territory, when challenged by another male, and when mate guarding (Wingfield et al., 1990, 1999; Fig. 12.2). Earlier analyses suggested that, at least in birds, variation in social modulation of testosterone secretion involves the mating system of the species. For example, simple comparisons of testosterone patterns in males of polygynous species tended to show higher levels of testosterone for longer periods than males of socially monogamous and polyandrous species (Wingfield et al., 1990). Further neural correlates of androgen action (Rosvall et al., 2012; Fig. 12.1) will likely provide further insight.

As mentioned above, if there were "costs" associated with high levels of testosterone at certain times or for long periods, then there may have evolved many patterns of testosterone secretion to suit particular contexts. Experimental manipulations of the pattern of testosterone secretion confirmed that high circulating levels of testosterone for long periods may indeed be deleterious, resulting in, for example, reduced male paternal care and lower reproductive success (e.g., Wingfield *et al.*, 1990; Ketterson and Nolan, 1999; Reed *et al.*, 2006; Hau, 2007). The earlier studies led to the "challenge hypothesis", which states that high plasma levels of testosterone occur during periods of social instability in the breeding season (resulting from male–male competition for territories and mates), but are at a lower breeding baseline in stable social conditions, thus allowing paternal care to be expressed (Wingfield *et al.*, 1990, 1999). This hypothesis has been tested widely over recent years, but a major confound when comparing patterns of testosterone to aggression among many species is phylogeny. In an extensive meta-analysis of all avian investigations in relation to patterns of testosterone secretion and behavior, Hirschenhauser *et al.* (2003) tested predictions of the "trade-off" scenarios involving male–male interactions resulting in an increase of testosterone secretion and male parental care requiring a decrease in testosterone. The analysis revealed that, after adjustment for phylogeny, the overall prediction of an effect of paternal care disappeared, but the effects of mating system, male–male interactions, and, possibly, male participation in incubation persisted. Testosterone patterns may vary according to mating success and testis size as well (Garamszegi *et al.*, 2005).

Hirschenhauser and Oliveira (2006) extended the meta-analyses to all vertebrates and found that tremendous variation in presence and types of parental care obscured any relationship of testosterone pattern with paternal behavior, but a general relationship with mating systems and male–male competition persisted, consistent with some of the earlier predictions of the challenge hypothesis. This is significant, because integration of mechanistic approaches such as endocrinology with ecology and evolutionary biology are not only important, but may also be critical in providing significant new insight into the biodiversity of mechanisms as well as species. For example, recent insight into why there is variation in mechanisms of hormone action among populations comes from the evolutionary constraints hypothesis, which suggests that if the mechanisms by which testosterone acts are highly conserved across all species, then the pattern of secretion should be very important for adjustments to environmental and social changes (Reed *et al.*, 2006; Hau, 2007). Alternatively, the evolutionary potential hypothesis states that there may be diversity in the mechanisms by which testosterone acts at the organismal, cellular, and molecular levels, and multiple ways by which a suite of mechanisms may evolve in response to environmental and social change (Hau, 2007; Rosval *et al.*, 2012). These influential papers set the stage for many experimental tests to explore the biodiversity of mechanisms. Birds provide a particularly tractable model system because nowadays there is an abundance of experimental information studies in a broad range of species, avian phylogenies for different groups are often well known, and closely related taxa have often invaded higher latitudes and altitudes independently – such as in the northern and southern hemispheres. A question now arises: have northern and southern populations of birds solved problems of seasonality and control of aggression patterns in reproduction by the same or different mechanisms?

Goymann *et al.* (2004), in a phylogenetically controlled analysis of patterns of testosterone in birds in relation to latitude and altitude, found that variable patterns in the tropics were related to environmental factors such as short breeding seasons rather than to phylogeny *per se*. The occurrence of short breeding seasons (seasonality) tends to increase with latitude and altitude. Garamszegi *et al.* (2005) generally confirmed these findings. As the numbers of species and populations studied under natural conditions increases, analyses of this sort will be critical to tease apart phylogeny and ecological constraints, leading to insights into how hormone–behavior interrelationships evolved.

LIFE CYCLES AND TESTOSTERONE, A NEOTROPICAL PERSPECTIVE

As reviewed by Goymann *et al.* (2004) and Garamszegi *et al.* (2005), comparisons of the patterns of testosterone secretion show increasing amplitude in relation to latitude, altitude and seasonality. Species breeding in lowland tropical regions, including neotropical species, tend to have extended breeding seasons with low circulating testosterone concentrations (Figs 12.3, 12.4). The lowland Neotropics are some of the most biodiverse areas on Earth, with birds displaying an immense range of life-history traits. As such, one would predict an equally diverse variety of mechanisms mediating the life-history variation, but one theme in the lowland neotropical birds that has been investigated to date is low plasma testosterone levels (Figs 12.3, 12.4). Tropical birds tend to have lower plasma testosterone levels than temperate zone species, but there are many variables that influence this (Goymann *et al.*, 2004), including effects related to elevation, with higher altitude species displaying higher levels (Goymann *et al.*, 2004). Also, social classes may differ in testosterone levels (DuVal and Gorman, 2011; Ryder *et al.*, 2011). Further, migratory species tend to have higher levels than non-migratory species (Garamszegi *et al.*, 2005).

Patterns of Testosterone Secretion and Action in Neotropical Birds

Only in the past 20 years or so have investigations begun on tropical species, despite the fact that approximately 60% or more of all avian species spend at least part of their life cycles in this region. Initially, studies on tropical species showed that males of most species had very low levels of testosterone, including some whose levels were below the detection limits of the assay systems (Figs 12.3, 12.4; Wikelski *et al.*, 2003; Busch *et al.*, 2008; see also Levin and Wingfield, 1992; Goymann *et al.*, 2004 for reviews). Comparisons of the pattern of north temperate zone species (e.g., Emberizines) versus lowland tropical species indicate dramatic differences (Wikelski *et al.*, 2000; Fig. 12.5). Goymann *et al.* (2004) analyzed all field studies of tropical species up to that time, and

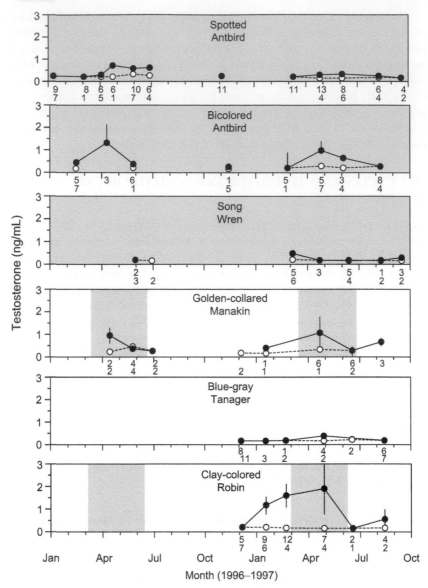

FIGURE 12.3   Seasonal changes (means ± 95% CI) in testosterone concentrations of males (filled circles) and females (unfilled circles) of six neotropical passerines. Shading indicates the extent of territorial behavior. Bicolored antbirds are territorial, but also commute into territories of conspecifics to forage around army-ant swarms. Sample sizes are indicated below each graph, males over females. *Reproduced from Wikelski* et al. *(2003), courtesy of the Cooper Ornithological Society.*

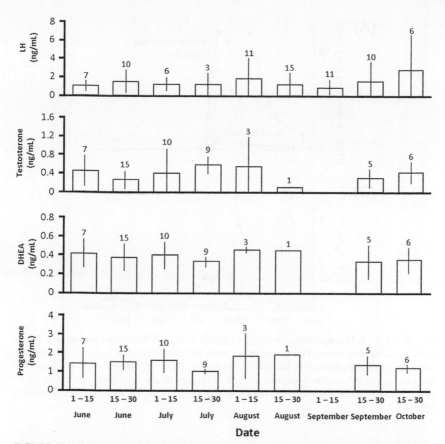

FIGURE 12.4   **Mean ± SD hormone titers (ng/mL) in male song wrens in Panama over the course of the 2002 reproductive season.** The sample size for each group is shown above its bar. LH, luteinizing hormone; DHEA, dihydroepiandosterone. *Reproduced from Busch* et al. *(2008), courtesy of the Cooper Ornithological Society.*

showed that species that are clearly seasonal in reproductive function tend to show higher magnitude of changes in plasma testosterone levels than species that breed and/or are territorial year round or are colonial (Fig. 12.6). However, testicular concentrations of testosterone of tropical species appear to be more similar to those of north temperate species (Fig. 12.7), suggesting that low levels of plasma testosterone in tropical species are not the result of low production of testosterone, but that it is somehow sequestered in the testis, probably by androgen binding proteins (Levin and Wingfield, 1992). This is consistent with the action of testosterone on spermatogenesis – i.e., a local paracrine action that does not require release of testosterone into the blood (e.g., Hillgarth and Wingfield, 1997). An interesting aspect of testosterone action that is the focus of many current studies is differences in hormone action independent of plasma

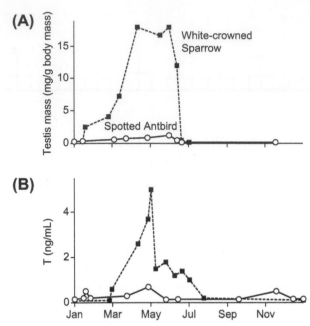

FIGURE 12.5 Schematic comparison of **(A)** gonad sizes and **(B)** plasma testosterone levels of white-crowned sparrows (north temperate zone, 50°N latitude; solid squares) and spotted antbirds (tropical Panama, 9°N latitude; open circles). Note that testosterone levels in spotted antbirds can stay at base-line throughout the year or increase at any time of the year during times of social instability. *Reproduced from Wikelski et al. (2000), courtesy of the Ecological Society of America.*

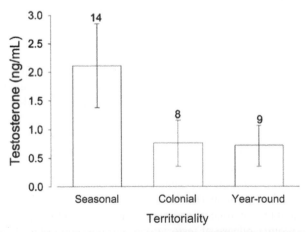

FIGURE 12.6 Mean testosterone concentrations (95% confidence intervals) in seasonal territorial, colonial, and year-round territorial tropical bird species (numbers above bars refer to sample size). *Reproduced from Goymann et al. (2004), courtesy of University of Chicago Press.*

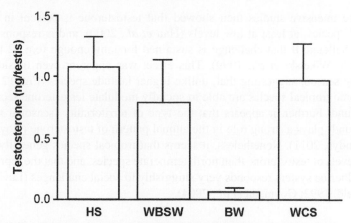

FIGURE 12.7   Testicular content of testosterone in temperate zone house sparrows, _Passer domesticus_ (HS), white-crowned sparrows (WCS), tropical white-browed sparrow weavers, _Plocepasser mahali_ (WBSW), and bay wrens, _Thryophorus nigricapillus_ (BW). Histograms represent means ± standard errors. _Reproduced from Levin and Wingfield (1992), courtesy of Wiley Blackwell._

levels of the hormone. For example, Canoine _et al._ (2007) found that low testosterone levels in the spotted antbird, _Hylophylax naevioides_, are compensated for by elevated expression on androgen receptors in the brain. Moreover, expression of the enzymes that metabolize testosterone to affect their activity (see above; also Rosvall _et al._, 2012; Wingfield, 2012a, b) could further play a role in modulating activity of relatively low plasma hormone levels to high levels of activity in the brain.

Hormones, Behavior, and Sociality in the Neotropics

The effects of social interactions on patterns of testosterone secretion have been investigated primarily in laboratory animals, and field work has focused largely on north temperate species (e.g., Hirschenhauser _et al._, 2003; Hirschenhauser and Oliveira, 2006). Field experimental studies on tropical and southern hemisphere birds are beginning to explore the potential roles of social interactions and regulation of reproductive function. Lekking behavior in manakins (Pipridae) has been a recent focus of neuroendocrine studies. Manakins provide an interesting system for such investigations because of their intricate social behaviors. In the wire-tailed manakin, _Pipra filicauda_, males dance together to attract mates, and territorial males with definitive plumage have higher testosterone levels than do floaters with either definitive or pre-definitive plumage (Ryder _et al._, 2011). Interestingly, this elevated testosterone does not appear to make the territorial males more aggressive, as they need to cooperate and dance with other males. A similar situation appears to be occurring in the lance-tailed manakin, _Chiroxiphia lanceolata_ (DuVal and Goymann, 2011).

More intensive studies then showed that testosterone is present in some tropical species, at least at low levels (Hau *et al.*, 2000), and is responsive to social challenge if that challenge is sustained for long enough (e.g., 2 hours, Fig. 12.8; Wikelski *et al.*, 1999). This effect was apparent even outside the breeding season, suggesting that, unlike higher latitude species (Fig. 12.2), at least some tropical species are able to socially modulate testosterone secretion year round. Further, it appears that the type of territoriality (seasonal versus year-round) plays a strong role in the annual pattern of testosterone (Goymann and Landys, 2011). Nonetheless, it seems that tropical species generally have lower levels of testosterone than north temperate species, and that the reproductive endocrine system responds very sluggishly to social challenges (Levin and Wingfield, 1992; Goymann *et al.*, 2004).

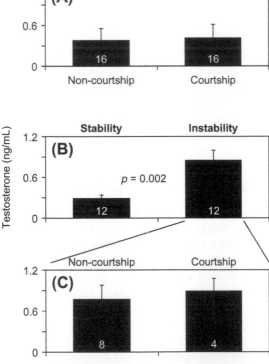

FIGURE 12.8 In spotted antbirds, the maximum testosterone levels for individual males were similar during the courtship season and the non-courtship season (A). However, the maximal testosterone levels were significantly lower when individual males experienced social stability compared to times when those individuals experienced social instability (B). This was not due to the fact that testosterone levels for socially challenged males were higher during the breeding season (C). *Reproduced from Wikelski* et al. *(1999), courtesy of the Royal Society, London.*

Tropical species that show more marked seasonality in breeding may also show greater magnitude of testosterone secretion synchronized with reproductive activity (Goymann *et al.*, 2004). Field investigations of equatorial populations of the rufous-collared sparrow, *Z. capensis*, at around 3300 m in elevation in Ecuador gave further insight. Males of these populations are territorial and sing mostly during the breeding season (up to 6 months), and have high circulating levels of testosterone similar to the northern white-crowned sparrow (Fig. 12.2, Moore *et al.*, 2004a). However, the pattern of testosterone levels in plasma was shown to be different from those of northern, socially monogamous songbirds, remaining elevated throughout the breeding season, and showed no clear decrease during the parental phase (Moore *et al.*, 2004a). Additionally, males from this population do not modulate testosterone levels in response to male–male interactions (Moore *et al*, 2004a). Moreover, treatment of free-living male rufous-collared sparrows with anti-androgen and an aromatase inhibitor (to block the pathway of action of testosterone through estrogen receptors; Fig. 12.1) had no effect on territorial aggression as measured by responsiveness of males to STIs (Moore *et al.*, 2004b). However, volume of song control nuclei in the brain showed seasonal changes similar to those of northern congeners (Moore *et al.*, 2004c).

Field studies of a Costa Rican population of rufous-collared sparrows have documented similar relationships between testosterone, reproduction, and behavior to those shown in the Ecuadorian studies. Specifically, testosterone levels of breeding and non-breeding males sorted by season in a Costa Rican population showed significant variation with season and breeding condition, with highest testosterone levels during the dry breeding season and lowest levels during the rainy breeding season (Addis *et al.*, 2010). There is similar variation in the amplitude of testosterone secretion, with a peak during the major breeding season (dry season) and lower levels in the secondary breeding season (rainy season) (Addis *et al.*, 2010). Simulated territorial intrusions (STIs) failed to indicate any effects of male–male interactions on testosterone pattern. These studies are important because they suggest that even within closely related groups of populations (high- and low-latitude breeding *Zonotrichia*), mechanisms by which testosterone affects reproductive behaviors may be similar within the tropics but differ markedly from northern congeners.

Research on under-studied families from the Neotropics, such as the Furnariidae (= ovenbirds) might also contribute to a deeper understanding of life cycles and the influence of testosterone in southern hemisphere avifauna. For example, the southern Patagonian furnariid *Aphrastura spinicauda* (thorn-tailed rayadito) has breeding traits similar to tropical species, such as low seasonal fecundity, large eggs, and prolonged dependence periods, reflecting a "slow" life history similar to that of tropical passerines (Moreno *et al.*, 2005), but it reproduces probably up to 55°S of latitude, where it is the most abundant forest bird species (see also Ippi *et al.*, 2011).

LIFE CYCLES AND TESTOSTERONE, A SOUTHERN PERSPECTIVE

An intriguing question with many implications for evolution of mechanisms underlying social interactions, and for reproduction, is whether processes determined from investigations of northern hemisphere species are the mirror image of those in the southern hemisphere. Field investigations will be essential to explore this issue because of trade-offs of benefits and costs in variable environments with latitude and altitude (e.g., Hau, 2007; Hau and Wingfield, 2011). The Neotropics is in many ways ideal for such studies because the southern hemisphere land mass extends further south than anywhere else.

The rufous-collared sparrow, *Zonotrichia capensis*, presents an interesting model because the species ranges widely from the Neotropics to 56°S latitude, with migratory and non-migratory populations occupying similar seasonal habitats to *Z. leucophrys* in the northern hemisphere (Wingfield *et al.*, 2007). Field studies on a mid-latitude breeding and non-migratory population of rufous-collared sparrows, *Z.c. chilensis*, in the central valley of Chile also revealed a high-magnitude pattern of circulating testosterone that is similar early and later in the breeding season (Addis *et al.*, 2011). This is very unlike northern populations of *Z. leucophrys* (Fig. 12.3). The Chilean *Z. c. chilensis* central valley population breeds at similar latitudes and is non-migratory as compared with the northern *Z.l. nuttallii* of coastal California, but testosterone patterns are very different. Furthermore, the Chilean population shows no effects of STIs on testosterone. In contrast, early and mid-breeding androgen levels of highland *Z.c. chilensis* show a peak of testosterone early in the breeding season and a decline as the parental phase of reproductive function begins. This pattern is very consistent with those of northern *Zonotrichia* (Fig. 12.2) including at high altitude (Lynn *et al.*, 2007; Addis *et al.*, 2011). Moreover, STIs during the parental phase of highland *Z.c. chilensis* did have a significant effect upon testosterone levels (Addis *et al.*, 2011). However, this was unlike northern high altitude breeding *Z.l. oriantha* in California that showed no changes in plasma testosterone in response to STI (Lynn *et al.*, 2007). Much more work is needed to confirm these differences and what their ecological bases may be.

High-latitude southern, and migratory, populations of *Z.c. australis* at 55°S on Isla Navarino, southern Chile, provide a direct comparison with the northern high-latitude and migratory *Z.l. gambelii*. Baseline testosterone levels were lower in the mid-breeding season than in early breeding (Addis *et al.*, 2011), as in *Z.l. gambelii* (Fig. 12.2, Wingfield and Farner, 1978). Similarly, these high-latitude breeding birds showed no response of testosterone secretion to STIs (Meddle *et al.*, 2002; Addis *et al.*, 2011).

Taken together, these comparisons of congeners from northern and southern hemispheres and tropical regions in the lowlands and highlands show some different patterns and some similar traits. This is a very useful model system of populations under different environmental constraints but shared phylogenetic histories. However, much more work is now needed to build on this foundation and determine why patterns of testosterone secretion vary in these ways, and what their significance is for life-history evolution.

CONCLUSIONS

The finding that neotropical populations of the rufous-collared sparrow have very different patterns of testosterone secretion during the breeding season compared to breeding northern songbirds is highly note-worthy. Furthermore, preliminary studies with chemical blockers of testosterone action suggest that expression of territorial aggression and singing behavior may not be dependent upon circulating testosterone levels. Whether or not males of populations of rufous-collared sparrows at southern mid- to high latitudes are more like populations of white-crowned sparrows in similar habitats in the north is currently under investigation. A phylogeny of the *Zonotrichia* (Zink and Blackwell, 1996) suggests that *Z. capensis* and *Z. leucophrys* are derived from a common ancestor, but the two species colonized high-latitude and high-altitude habitats independently. Studies of this type will further our general understanding of the role of environment versus phylogeny in defining these hormone–behavior relationships. Given that the territorial behavior of the two species is very similar, this is an ideal opportunity to test whether these species regulate seasonal territorial aggression by the same mechanisms (evolutionary constraints hypothesis, Reed *et al.*, 2006; Hau, 2007) or whether they have evolved different mechanisms (evolutionary potential hypothesis; Hau, 2007).

ACKNOWLEDGMENTS

Much of the work summarized in this paper was supported by grant number IOS 0750540 from the National Science Foundation and the Endowment in Physiology at the University of California, Davis to JCW. I.T.M. acknowledges the National Science Foundation for grant number IOB-0545735. RAV acknowledges support from the Institute of Ecology and Biodiversity (ICM-P05-002, PFB-23-CONICYT-Chile), and FONDECYT 1090794.

REFERENCES

Addis, E.A., Busch, D.S., Clark, A.D., Wingfield, J.C., 2010. Seasonal and social modulation of testosterone in Costa Rican rufous-collared sparrows (*Zonotrichia capensis costaricensis*). Gen. Comp. Endocrinol. 166, 581–589.

Addis, E.A., Clark, A.D., Wingfield, J.C., 2011. Modulation of androgens in southern hemisphere temperate breeding sparrows (*Zonotrichia capensis*): an altitudinal comparison. Horm. Behav. 60, 195–201.

Adkins-Regan, E., 2005. Hormones and Animal Social Behavior. Princeton University Press, Princeton.

Busch, D.S., Robinson, T.R., Hahn, T.P., Wingfield, J.C., 2008. Sex hormones in the song wren: variation with time of year, molt, gonadotropin releasing hormone, and social challenge. Condor 110, 125–133.

Canoine, V., Fusani, L., Schlinger, B.A., Hau, M., 2007. Low sex steroids, high steroid receptors: increasing sensitivity of the non-reproductive brain. J. Neurobiol. 67, 57–67.

DuVal, E.H., Goymann, W., 2011. Hormonal correlates of social status and courtship display in the cooperatively lekking lance-tailed manakin. Horm. Behav. 59, 44–50.

French, D.D., Smith, V.R., 1985. A comparison between northern and southern hemisphere tundras and related ecosystems. Polar Biol. 5, 515–521.

Garamszegi, L.Z., Eens, M., Hurtrez-Boussás, S., Møller, A.P., 2005. Testosterone, testes size, and mating success in birds: a comparative study. Horm. Behav. 47, 389–409.

Ghalambor, C.K., Huey, R.B., Martin, P.R., Tewksbury, J.J., Wang, G., 2006. Are mountain passes higher in the tropics? Janzen's hypothesis revisited. Integr. Comp. Biol. 46, 5–17.

Goodson, J.L., 2005. The vertebrate social behavior network: evolutionary themes and variations. Horm. Behav. 48, 11–22.

Goymann, W., Landys, M.M., 2011. Testosterone and year round territoriality in tropical and non-tropical songbirds. J. Avian Biol. 42, 485–489.

Goymann, W., Moore, I.T., Scheuerlein, A., Hirschenhauser, K., Grafen, A., Wingfield, J.C., 2004. Testosterone in tropical birds: effects of environmental and social factors. Am. Nat. 164, 327–334.

Goymann, W., Landys, M.M., Wingfield, J.C., 2007. Distinguishing seasonal androgen responses from male-male androgen responsiveness – revisiting the challenge hypothesis. Horm. Behav. 51, 463–476.

Hau, M., 2007. Regulation of male traits by testosterone: implications for the evolution of vertebrate life histories. BioEssays 29, 133–144.

Hau, M., Wingfeld, J.C., 2011. Hormonally-regulated trade-offs: evolutionary variability and phenotypic plasticity in testosterone signaling pathways. In: Flatt, T., Heyland, A. (Eds.), History Mechanisms of Life Evolution, Oxford University Press, Oxford, pp. 349–361.

Hau, M., Wikelski, M., Soma, K.K., Wingfield, J.C., 2000. Testosterone and year-round territorial aggression in a tropical bird. Gen. Comp. Endocrinol. 117, 20–33.

Hillgarth, N., Wingfield, J.C., 1997. Testosterone and immunosuppression in vertebrates: implications for parasite-mediated sexual selection. In: Beckage, N.E. (Ed.), Parasites and Pathogens: Effects on Host Hormones and Behavior, Chapman and Hall, New York, pp. 143–155.

Hirschenhauser, K., Oliveira, R., 2006. Social modulation of androgens in male vertebrates: meta-analyses of the challenge hypothesis. Anim. Behav. 71, 265–277.

Hirschenhauser, K., Winkler, H., Oliveira, R.F., 2003. Comparative analysis of male androgen responsiveness to social environment in birds: the effects of mating system and paternal incubation. Horm. Behav. 43, 508–519.

Ippi, S., Vasquez, R.A., van Dongen, W.F.D., Lazzoni, I., 2011. Geographical variation in the vocalizations of the suboscine thorn tailed rayadito *Aphrastura spinicauda*. Ibis 153, 789–805.

Ketterson, E.D., Nolan Jr, V., 1999. Adaptation, exaptation and constraint: a hormonal perspective. Am. Nat. 154, S4–S25.

Ketterson, E.D., Atwell, J.W., McGlothlin, J.W., 2009. Phenotypic integration and independence: hormones, performance and response to environmental change. Intgr. Comp. Biol. 49, 365–379.

Korte, S.M., Koolhaas, J.M., Wingfield, J.C., McEwen, B.S., 2005. The Darwinian concept of stress: benefits of allostasis and costs of allostatic load and the trade-offs in health and disease. Neurosci. Biobehav. Rev. 29, 3–38.

Levin, R.N., Wingfield, J.C., 1992. The hormonal control of territorial aggression in tropical birds. Ornis Scand. 23, 284–291.

Lynn, S.E., Hahn, T.P., Breuner, C.W., 2007. Free-living male mountain white-crowned sparrows exhibit territorial aggression without modulating total or free plasma testosterone. Condor 109, 173–180.

Maldonado, K., van Dongen, W.F.D., Vasquez, R.A., Sabat, P., 2012. Geographic variation in the association between exploratory behavior and physiology in rufous-collared sparrows. Physiol. Biochem. Zool. 85, 618–624.

McGlothlin, J.W., Jawor, J.M., Ketterson, E.D., 2007. Natural variation in a testosterone-mediated trade-off between mating effort and parental effort. Am. Nat. 170, 864–875.

Meddle, S.L., Romero, M., Astheimer, L.B., Buttemer, W.A., Wingfield, J.C., 2002. Steroid hormone interrelationships with territorial aggression in an arctic-breeding songbird, Gambel's white-crowned sparrow, *Zonotrichia leucophrys gambelii*. Horm. Behav. 42, 212–221.

Moore, I.T., Wada, H., Perfito, N., Busch, D.S., Hahn, T.P., Wingfield, J.C., 2004a. Territoriality and testosterone in an equatorial population of rufous-collared sparrows, *Zonotrichia capensis*. Anim. Behav. 67, 411–420.

Moore, I.T., Walker, B.G., Wingfield, J.C., 2004b. The effects of aromatase inhibitor and anti-androgen on male territorial aggression in a tropical population of rufous-collared sparrows, *Zonotrichia capensis*. Gen. Comp. Endocrinol. 135, 223–229.

Moore, I.T., Wingfield, J.C., Brenowitz, E.A., 2004c. Plasticity of the avian song control system in response to localized environmental cues in an equatorial songbird. J. Neurosci. 24, 10182–10185.

Moreno, J., Merino, S., Vásquez, R.A., Armesto, J.J., 2005. Breeding biology of the thorn-tailed rayadito (Furnariidae) in south-temperate rainforests of Chile. Condor 107, 69–77.

Nolan Jr., V., Ketterson, E.D., Ziegenfus, C., Cullen, D.P., Chandler, C.R., 1992. Testosterone and avian life histories: effects of experimentally elevated testosterone on prebasic moult and survival in male dark-eyed juncos. Condor 94, 364–370.

Norris, D.O., 2007. Vertebrate Endocrinology. Academic Press, San Diego, CA.

Piggliuicci, M., 2001. Phenotypic Plasticity: Beyond Nature and Nurture. John Hopkins University Press.

Reed, W.L., Clark, M.E., Parker, P.G., Raouf, S.A., Arguedas, N., Monk, D.S., Snajdr, E., Nolan Jr., V., Ketterson, E.D., 2006. Physiological effects on demography: a long term experimental study of testosterone's effects on fitness. Am. Nat. 167, 667–683.

Ricklefs, R.E., Wikelski, M., 2002. The physiology/life-history nexus. Trends Ecol. Evol. 17, 462–468.

Rosvall, K.A., Bergeon Burns, C.M., Barske, J., Goodson, J.L., Sengelaub, D., Schlinger, B.A., Ketterson, E.D., 2012. Neural sensitivity to sex steroids predicts individual differences in aggression: implications for behavioral evolution. Proc. R. Soc. B. 279, 3547–3555.

Rozzi, R., Massardo, F., Anderson, C., Berghoefer, A., Mansilla, A., Mansilla, M., Plana, J., Berghoefer, U., Barros, E., Araya, P., 2006. The Cape Horn Biosphere Reserve: A proposal for Conservation and Tourism to Achieve Sustainable Development at the Southern End of the Americas. Ediciones Universidad de Magallanes, Punta Arenas, Chile.

Runfeldt, S., Wingfield, J.C., 1985. Experimentally prolonged sexual activity in female sparrows delays termination of reproductive activity in their untreated mates. Anim. Behav. 33, 403–410.

Ryder, T.P., Horton, B.M., Moore, I.T., 2011. Understanding testosterone variation in a tropical lek-breeding bird. Biol. Lett. 7, 506–509.

Schleussner, G., Dittami, J.P., Gwinner, E., 1995. Testosterone implants affect molt in male European starlings., *Sturnus vulgaris*. Physiol. Zool. 58, 597–604.

Schlinger, B.A., 1994. Estrogens to song: picograms to sonograms. Horm. Behav. 28, 191–198.

Schlinger, B.A., Brenowitz, E.A., 2002. Neural and hormonal control of birdsong. In: Pfaff, D. (Ed.), Hormonal Brain Behavior, Vol. 2. Elsevier, Amsterdam, pp. 799–839.

Soma, K., 2006. Testosterone and aggression: Berthold, birds and beyond. J. Neuroendocrinol. 18, 543–551.

Wikelski, M., Hau, M., Wingfield, J.C., 1999. Social instability increases plasma testosterone in a year round territorial neotropical bird. Proc. R. Soc. B 266, 551–556.

Wikelski, M., Hau, M., Wingfield, J.C., 2000. Seasonality of reproduction in a neotropical rainforest bird. Ecology 81, 2458–2472.

Wikelski, M., Hau, M., Robinson, W.D., Wingfield, J.C., 2003. Reproductive seasonality of seven neotropical passerine species. Condor 105, 683–695.

Wingfield, J.C., 2006. Communicative behaviors, hormone-behavior interactions, and reproduction in vertebrates. In: Neill, J.D. (Ed.), Physiology of Reproduction, Academic Press, New York, pp. 1995–2040.

Wingfield, J.C., 2008. Organization of vertebrate annual cycles: implications for control mechanisms. Phil. Trans. R. Soc. B. 363, 425–441.

Wingfield, J.C., 2012a. Regulatory mechanisms that underlie phenology, behavior, and coping with environmental perturbations: an alternate look at biodiversity. Auk 129, 1–7.

Wingfield, J.C., 2012b. The challenge hypothesis: behavioral ecology to neurogenomics. J. Ornithol. http://dx.doi.org/10.1007/s10336-012-0857-8, (epub before print).

Wingfield, J.C., Farner, D.S., 1978. The annual cycle in plasma irLH and steroid hormones in feral populations of the white-crowned sparrow, *Zonotrichia leucophrys gambelii*. Biol. Reprod. 19, 1046–1056.

Wingfield, J.C., Farner, D.S., 1979. Some endocrine correlates of renesting after loss of clutch or brood in the white-crowned sparrow, *Zonotrichia leucophrys gambelii*. Gen. Comp. Endocrinol. 38, 322–331.

Wingfield, J.C., Farner, D.S., 1993. Endocrinology of reproduction in wild species. In: Farner, D.S., King, J.R., Parkes, K.C. (Eds.), Avian Biology, Vol. IX. Academic Press, New York, pp. 164–327.

Wingfield, J.C., Hahn, T.P., 1994. Testosterone and territorial behavior in sedentary and migratory sparrows. Anim. Behav. 47, 77–89.

Wingfield, J.C., Hegner, R.E., Dufty Jr., A.M., Ball, G.F., 1990. The "challenge hypothesis": theoretical implications for patterns of testosterone secretion, mating systems, and breeding strategies. Am. Nat. 136, 829–846.

Wingfield, J.C., Ramenofsky, M., 1985. Hormonal and environmental control of aggression in birds. In: Gilles, R., Balthazart, J. (Eds.), Neurobiology, Springer-Verlag, Berlin, pp. 92–104.

Wingfield, J.C., Jacobs, J.D., Tramontin, A.D., Perfito, N., Meddle, S., Maney, D.L., Soma, K., 1999. Toward an ecological basis of hormone–behavior interactions in reproduction of birds. In: Wallen, K., Schneider, J. (Eds.), Reproduction in Context, MIT Press, Cambridge, pp. 85–128.

Wingfield, J.C., Moore, I.T., Goymann, W., Wacker, D.W., Sperry, T., 2005. Contexts and ethology of vertebrate aggression: implications for the evolution of hormone-behavior interactions. In: Nelson, R.J. (Ed.), Biology of Aggression, Oxford University Press, New York, pp. 179–210.

Wingfield, J.C., Meddle, S.L., Moore, I.T., Busch, S., Wacker, D., Lynn, S., Clark, A., Vasquez, R.A., Addis, E., 2007. Endocrine responsiveness to social challenges in northern and southern hemisphere populations of *Zonotrichia*. J. Ornithol. 148 (Suppl.), S435–S441.

Zink, R.M., Blackwell, R.C., 1996. Patterns of allozyme, mitochondrial DNA, and morphometric variation in four sparrow genera. Auk 113, 59–67.

Chapter 13

Sexual Selection and the Evolution of Vocal Mating Signals: Lessons from Neotropical Songbirds

Jeffrey Podos

Department of Biology, University of Massachusetts, Amherst, Massachussetts, USA

INTRODUCTION

Darwin's (1871) theory of sexual selection aimed to explain the evolution of elaborate animal displays and ornaments, traits that on the surface seem counterintuitive in the context of natural selection and the struggle for survival. The solution Darwin proposed is that elaborate traits are favored because of the advantages they confer in mating success – advantages that must somehow outweigh their survival costs. Darwin supported his thesis by presenting wide-ranging examples of interactions both within males and between the sexes, considering "how it is that the males which conquer other males, or those which prove the most attractive to the females, leave a greater number of offspring to inherit their superiority than the beaten and less attractive males" (1871: 260–261).

Since Darwin, and especially in the past half-century, we have amassed detailed knowledge about how sexual selection proceeds, and have also learned that sexual selection is influenced not just by inter-rival conflict and mate attraction, but also by diverse ecological and behavioral factors such as parental investment, habitat heterogeneity, life-history allocation, and mating systems (Trivers, 1972; Emlen and Oring, 1977; Andersson, 1994; Dunn *et al.*, 2001; Kokko and Monaghan, 2001; Pribil and Searcy, 2001; Clutton-Brock, 2007). These ecological and behavioral factors often vary systematically by taxon and geographical locality, and it follows that patterns of sexual selection might vary in accordance. To take one well-known example, Trinidadian guppies *Poecilia reticulata* inhabit streams with diverse predator regimes, and this heterogeneity is seen to correspond with both the expression of visual signals in males, and

Sexual Selection. http://dx.doi.org/10.1016/B978-0-12-416028-6.00013-X

341

preferences for different signal variants in females (Endler, 1983; Endler and Houde, 1995).

The present volume addresses the impact of taxonomic and geographical heterogeneity on sexual selection by asking whether sexual selection operates differently in neotropical environments as compared to temperate zone environments. If so, are the differences qualitative or rather a matter of degree? Does understanding sexual selection in neotropical species require distinct conceptual frameworks, or merely the extension of frameworks already developed for temperate zone taxa? These are timely questions, as the scientific community accumulates increasingly precise information about broad-scale regional and phylogenetic effects on ecology and organismal evolution (Ricklefs and Schluter, 1993; Gaston, 2000; Webb *et al.*, 2002). One of the main challenges in drawing comparisons across temperate and tropical zones is that the natural history of a vast majority of tropical taxa remains poorly characterized. While Darwin's (1871) original argument about sexual selection made copious reference to neotropical animals, the functions of exaggerated male traits in many of his examples were inferred from second-hand, anecdotal observations.

Direct study of tropical taxa can advance our understanding of sexual selection in at least two important ways. First, it can provide fertile ground for testing hypotheses about sexual selection and its interface with ecology and behavior. For instance, a hypothesis presently under consideration for socially monogamous birds – that sexual selection is strengthened by high breeding synchrony and density (Griffith *et al.*, 2002) – is tested most profitably by examining not just temperate zone species, many of which breed synchronously and in high density, but also tropical species, many of which show limited breeding synchrony and density (Stutchbury and Morton, 1995; Macedo *et al.*, 2008). The broader range of variation achieved by including tropical species enables the focal hypothesis to be tested with greater rigor. Second, studies of neotropical taxa can reveal novel contexts in which sexual selection operates, not just in the specifics of the sexual selection/ecology/behavior interface but also in more basic elements such as the diversity of traits on which sexual selection acts. A prominent example that comes to mind is the discovery of electrocommunication in neotropical gymnotiform fishes (Lissmann, 1958; Hopkins, 1988). This communication modality, which has no analog in the temperate zone, appears to be molded by selection not just for mating but also for other signal functions, including navigation, orientation, and predator evasion (Hagedorn and Heiligenberg, 1985; Stoddard, 1999).

The present chapter addresses a topic that has provided a focal point of discussion about sexual selection since Darwin (1871): avian vocal mating signals. The wealth of available evidence suggests that bird songs are a product of sexual selection, serving two main functions: to attract mates, and to mediate aggressive interactions among rivals (Searcy and Andersson, 1986; Catchpole and Slater, 2008). Our knowledge about song – not just its function but also its diversity, mechanics, and development – was initially established through the study of temperate zone birds (reviewed by Slater, 2003; Marler, 2004). Descriptive analyses of song variation, previously conducted by ear, were powered by the invention of the sound spectrograph in the 1940s, and initially applied to temperate zone species such as North American sparrows, thrushes, and wrens (e.g., Borror and Reese, 1953; Borror, 1956). Temperate zone species, notably chaffinches (*Fringilla coelebs*) and white-crowned sparrows (*Zonotrichia leucophrys*), were also the initial focus of experimental studies of song function and learning (Thorpe, 1961; Marler and Tamura, 1964), and temperate zone birds, or tropical species subsequently domesticated, have served as the primary study species for research on neurobiology, mechanics, and the hormonal control of singing behavior (Nottebohm *et al.*, 1976; Konishi, 1985; Bolhuis and Gahr, 2006).

This is not to say that opportunities in the tropics have gone unrecognized. On the contrary, researchers over the decades have generated a steady output of influential studies on vocal communication in neotropical birds, with particular emphasis on vocal variation within and across species (see, for example, Nottebohm, 1969; Wiley, 1971; Lougheed and Handford, 1992; Ridgley and Tudor, 1994; Isler *et al.*, 2007), and habitat-specific effects on song propagation (e.g., Marten *et al.*, 1977; Ryan and Brenowitz, 1985; Seddon, 2005). Nevertheless, taking the long view, tropical birds have been relatively overlooked in studies of vocal communication. In 1996, Kroodsma *et al.* (1996) challenged song researchers to redouble their focus on neotropical species. They noted that

for the vast majority of [neotropical] species, essentially nothing is known about
their ontogeny, vocal repertoires, geographic variation, how vocalizations are used
in communication, or even the barest essentials of their life histories. Almost every
evolutionary experiment imaginable seems to have been played out among Neotropical
species, and countless opportunities exist for studying every aspect of acoustic
communication.

(Kroodsma *et al.*, 1996: 269)

Examples of topics they found especially promising for studies on vocal behavior in neotropical birds included sexual selection, speciation, geographic variation, functions of duetting, origins of mimicry, and strategies of vocal development. Consistent with the exhortations of Kroodsma *et al.* (1996), the study of song in neotropical bird taxa appears to have been accelerating, especially in terms of contributions from researchers associated with neotropical institutions.

I begin this chapter by first considering the fundamental question of how bird song is shaped by sexual selection, and then examine, in turn, three additional factors that together shape song expression and evolution: selection for signaling efficacy, evolution of vocal morphology, and drift. For each factor I summarize insights established first through studies of temperate zone birds, and then assess how these insights are being enriched through work on neotropical species. My ultimate goal with this chapter is to ask whether sexual selection operates differently in the Neotropics than in temperate zones, as applied to the specific case of bird song and its evolution.

SONG AND SEXUAL SELECTION

Bird songs are typically produced by males competing for access to mates, and are closely attended by females prospecting for mates (Kroodsma and Byers, 1991). Songs are thus thought to evolve and diverge mainly as a consequence of sexual selection (Catchpole and Slater, 2008). In this context we presume that males achieve greater mating success by singing versus not singing (Peek, 1972; Eriksson and Wallin, 1986), by singing often versus rarely (Alatalo et al., 1990), and also by singing certain types of vocal patterns. Traditionally, studies on vocal variation and its fitness consequences focused on note or song-type repertoires, in line with other research in sexual selection attempting to explain the evolution of signal complexity (Andersson, 1994). In some but not all species studied, repertoire size provides a reliable indicator of aspects of male quality, and males who sing larger note or song repertoires sometimes enjoy greater mating success, measured either directly or in its correlates (e.g., Catchpole, 1980; Hasselquist et al., 1996; Reid et al., 2004; Kipper et al., 2006; reviewed by Searcy and Andersson, 1986, and Searcy and Yasukawa, 1996; but see Byers and Kroodsma, 2009).

More recent work has focused on variation among males in vocal performance, which can be characterized in terms of both vigor and skill (Darwin, 1871; Byers et al., 2010a). Vocal vigor, measured most readily as singing rate, appears to predict male breeding success in some species (e.g., Alatalo et al., 1990). However, it is often difficult to conclude whether singing rate itself is responsible for enhanced breeding success, given that singing rate tends to covary with other factors that influence breeding success, such as territory quality and the timing of arrival on breeding grounds (Nystrom, 1997). Research on vocal skill has focused on the consistency of song or note structure across renditions, the production of song phrases that are challenging to produce, and the accuracy with which particular song elements are imitated (e.g., Podos, 1996;, 1997; Nelson, 2000; Draganoiu et al., 2002; Ballentine et al., 2004; Byers, 2007; Grava et al., 2012; reviewed by Podos et al., 2009, and Sakata and Vehrencamp, 2012). Ultimately, sexual selection on song within any given lineage is best viewed as a multidimensional process, favoring a range of outcomes, including (but not restricted to) enhanced vocal complexity, performance, learning accuracy, and precision of repertoire use (Gil and Gahr, 2002; Podos et al., 2004a; Beecher and Brenowitz, 2005; Nowicki and Searcy, 2005), with selection for some of these factors potentially constraining the evolution of others (Gil and Gahr, 2002; Logue and Forstmeier, 2008; Cardoso and Hu, 2011; Lahti et al., 2011).

The Neotropics provides fertile testing grounds for diverse facets of sexual selection theory, as it applies to vocal function and evolution in birds. Three recent directions seem particularly productive, for which I now offer a brief synopsis. First, recent research on neotropical birds is expanding our knowledge about the relationship between song function and vocal performance. In two neotropical species, the banded wren *Thryothorus pleurostictus* and the tropical mockingbird *Mimus gilvus*, older birds have been found to sing with greater inter-note consistency (Botero *et al.*, 2009; de Kort *et al.*, 2009a), and in tropical mockingbirds syllable consistency predicts male reproductive success, as measured through genetic analyses of paternity (Botero *et al.*, 2009). Territorial male banded wrens are seen to respond differently to playbacks of high-performance versus low-performance trills, in ways that suggest the former signals an intruder of greater threat (Illes *et al.*, 2006; de Kort *et al.*, 2009b). These results nicely parallel other recent findings from temperate zone birds (e.g., Cramer and Price, 2007; Ballentine, 2009).

Second, research on neotropical birds is revealing a remarkably broad diversity of song phenotypes and functions, including fascinating basic discoveries about vocal natural history (e.g., Morton, 1996; Rios-Chelen and Macías Garcia, 2004, 2007; Fitzsimmons *et al.*, 2008; Brumm *et al.*, 2010; Ippi *et al.*, 2011; Kirschel *et al.*, 2011). An area that has received particularly close attention is the production and function of song by females. In many tropical songbirds, both sexes sing and defend breeding territories. In some of these species, males and females interweave their songs in stereophonic duets, some of which achieve remarkable levels of intricacy and precision (e.g., Mennill and Vehrencamp, 2005, 2008). Possible functions for vocal duets have been explored through playback experiments, and it seems that duets contribute to varied functions, including joint resource defense, mate guarding, and reproductive synchrony (e.g., Seddon and Tobias, 2006; Mennill, 2006; reviewed by Hall, 2009). Illes and Yunes-Jimenez (2009) recently offered a fascinating account of a species, the stripe-headed sparrow *Aimophila ruficauda*, in which females exceed males in their song output and aggressive responses to song playback. The authors attribute this "role-reversed" behavior to a polyandrous mating system that rewards females who compete more vigorously with each other for access to males and breeding territories (Illes and Yunes-Jimenez, 2009).

Third, patterns of sexual selection and song evolution in neotropical birds are being described with renewed rigor through comparative analyses across species, as exemplified by studies of *Thryothorus* wrens and New World blackbirds. *Thryothorus* wrens exhibit a wide diversity of vocal patterns, including tight duets, loosely coordinated duets, chorusing, or a virtual lack of singing by females (Mann *et al.*, 2009). A comparative study of the 27 *Thryothorus* species reveals that signing styles have been relatively conserved within closely related groups – i.e., that songs retain a strong phylogenetic signal (Mann *et al.*, 2009). This result suggests that ecological and behavioral factors associated with deep divergence events may have driven patterns of song evolution in this genus. New World blackbirds also exhibit wide-ranging variation in the occurrence of female song, and a phylogenetic analysis indicates that the tendency of females to sing correlates closely with latitude, females tending to sing in tropical but not temperate zone taxa (Price *et al.*, 2009). The phylogenetic analysis also suggests that New World blackbirds originated in the tropics and subsequently diverged multiple times into temperate zone taxa, each time accompanied by losses of female song (Price *et al.*, 2009). Moreover, within the oropendola and cacique lineages, sexual selection appears to have targeted different structural elements of song, illustrating further the diversity of outcomes resulting from sexual selection in the New World blackbirds (Price and Lanyon, 2004).

While sexual selection is without question the primary driving force behind bird song evolution, other factors may shape how song evolves. Prominent among these are selection for signal efficacy, evolutionary divergence of vocal morphology, and evolutionary drift. I now consider each of these factors in turn, as first established for temperate zone birds and then as presently being explored in studies of bird species from the Neotropics.

SONG AND SIGNAL EFFICACY

For signals to function properly they must be transmitted successfully from sender to receiver. In many circumstances, however, signal transmission is impeded by environmental factors that degrade or mask signal features, thus rendering signal information more difficult for receivers to detect and decode. To take a recent example, visual communication in many cichlid fish species hinges on color pattern variation, the detection of which can be degraded by water eutrophication and turbidity, and resulting interference with color transmission en route from sender to receiver (Seehausen *et al.*, 1997). Across all communication modalities, signals (and signalers) have evolved structural (and behavioral) adaptations that help minimize signal degradation (reviewed by Bradbury and Vehrencamp, 2011).

In the acoustic modality, transmission-related signal degradation occurs in two main ways. First, acoustic signals tend to attenuate and reverberate as they propagate from a sound source (Wiley and Richards, 1978; Slabbekoorn, 2004). Attenuation occurs as signal energy dissipates over distance, and reverberation occurs as signal waves are reflected off objects in the landscape (e.g., trees, rocks) or encounter turbulence in the medium (e.g., wind, waves). The second major source of signal degradation is interference from other sound sources, both abiotic (wind, water, anthropogenic noise) and biotic (other vocalizing species, reviewed by Brumm and Slabbekoorn, 2005).

In response to these challenges to the efficacy of acoustic signaling, birds have evolved a wide range of adaptations in song structure and patterns of usage (reviewed by Slabbekoorn, 2004; Brumm and Naguib, 2009). Of particular interest has been the discovery of habitat-specific influences on sound attenuation and reverberation (Morton, 1975; Wiley and Richards, 1978). Sounds transmitting through forests are subject to high attenuation and scatter, given the high density of leaves and other reflecting surfaces. Species inhabiting forests have thus evolved songs that emphasize low vocal frequencies, helping to maintain a sufficient signaling "active space" (Brumm and Naguib, 2009). Sounds in forests are also subject to high reverberation, and as a result forest species tend to repeat notes at slow to moderate rates, thus minimizing the masking effect of note echoes. Species inhabiting grasslands, by contrast, tend to sing with relatively high rates of note repetition, presumably to ensure the successful transmission of song information in the face of wind turbulence. These and other predictions of "acoustic adaptation hypotheses", as applied to bird groups, have received consistent but not universal support (e.g., Wiley, 1991; Boncoraglio and Saino, 2007).

When faced with the second major class of signal degradation, interference from other sound sources, bird songs show evidence of divergence in numerous parameters (amplitude, frequency structure, or timing) that enhance their distinctiveness compared to songs of other bird species (e.g., Becker, 1982; Nelson and Marler, 1990), and/or their detectability against background noise (Lohr et al., 2003). The evolution of song distinctiveness and detectability against biotic and abiotic interference is presumably driven by pressures to mate within – rather than across – species and by pressures to successfully detect and capitalize on potential mating opportunities within species (i.e., by pressures to minimize, respectively, Type I and Type II errors; Wiley, 1994). Much recent study in the evolutionary divergence of vocal structure concerns birds' responses to low-frequency noise in urban populations. Populations of some species in urban settings are seen to circumvent this noise by singing at higher frequencies (e.g., Slabbekoorn and Peet, 2003; Halfwerk and Slabbekoorn, 2009; Ripmeester et al., 2010; Luther and Derryberry, 2012), by singing at higher amplitudes (Nemeth and Brumm, 2010), or by occupying quieter locations (Goodwin and Shriver, 2011). Populations in noisy environments also show numerous adaptations in song perception (reviewed by Brumm and Naguib, 2009).

While much research on song and signaling efficacy has focused on temperate zone birds, studies in the Neotropics have been ongoing and influential, describing in fundamental ways the relationship between song evolution and environmental acoustics (e.g., Morton, 1975; Nottebohm, 1975; Marten et al., 1977; Bowman, 1979; Handford, 1981; Ryan and Brenowitz, 1985; Handford and Lougheed, 1991). Part of the reason why the Neotropics has proven so useful for such studies is that it supports an unusually broad natural diversity, both in the sheer number of bird taxa and in the acoustic properties of habitats – for example, in the abundance and diversity of insect species competing for a limited acoustic space. The study of acoustic adaptation in neotropical birds proceeds apace and, as I illustrate here with several select examples, continues to offer compelling insights.

A primary recent insight is that acoustic adaptation can occur at extremely fine spatial, ecological, and taxonomic scales. In a study of gray-breasted woodwrens (*Henicorhina leucophrys*), to illustrate, Dingle et al. (2008) described song variations across a pair of subspecies with parapatric distributions along an Andean altitudinal gradient. These subspecies generally occupy acoustically distinctive habitats, and are found to sing songs that match those habitats in both spectral and temporal structure. More specifically, *H. l. hilaris*, which lives at lower elevations, encounters high-frequency noise from insects (probably cicadas) and accordingly sings songs with relatively low maximum frequencies, whereas *H.l. leucophrys*, which lives at higher elevations, encounters denser understory vegetation and accordingly sings songs with lower rates of note repetition. These acoustic differences are maintained in spite of the subspecies' presumed recent divergence and their current spatial proximity, which allows for continued meme and gene flow. A study by Tobias *et al.* (2010) assessed song variations across bird species occupying two adjacent and superficially similar forest types (bamboo forest, and terra firma forest) at their study site in the Peruvian Amazon. In spite of these habitats' similarities, bamboo forest was found to have a somewhat denser understory and, accordingly, a greater propensity to attenuate high-frequency sounds; by contrast, terra firma forest was found to have a greater density of tall trees, and a correspondingly higher propensity for reverberation-related distortion of songs with rapid rates of note repetition. The research team focused their analyses on the songs of suboscine species, which do not learn to sing and for which interspecific vocal divergence can thus be attributed to genetic factors. The songs of closely related species pairs were compared, and, in accord with acoustic adaptation hypotheses, species living in bamboo forests were found consistently to emphasize low frequencies and high rates of note repetition, whereas relatives from terra firma forests tended to sing songs at higher frequencies and with lower rates of note repetition (Tobias *et al.*, 2010). In another study of suboscines, this time within the Thamnophilid antbirds, Seddon (2005) assessed song variations between species occupying different forest strata, and found that species occupying more densely vegetated strata (canopy and understory) emphasized lower frequencies than did species occupying the less dense midstory. Results from all three of these studies offer novel illustrations of the power and precision by which selection for optimal sound transmission can shape vocal evolution.

Other recent studies on neotropical birds have focused on selection for distinctiveness against background interference. Luther (2009) recorded and analyzed, at multiple spatial and temporal scales, the songs of 82 bird species in a southern Amazonian rain forest. These species showed considerable divergence across a range of acoustic features, i.e., they occupied a broad "acoustic space". This result – which nicely parallels Nelson and Marler's (1990) description of song variation in a North American bird community – implies a history of selection for acoustic distinctiveness. Further supporting this interpretation, Luther (2009) also found that species that sing at the same time or in the same forest strata (i.e., with high temporal or spatial overlap) tend to diverge most in their song features, compared to random sets of species in the sample. This pattern suggests an evolutionary history of selection favoring distinct songs, although the pattern might also occur if birds actively choose the times, locations, and types of songs they sing so as to minimize interference (Luther, 2008, 2009).

Results from these and similar studies presume a history of acoustic adaptation; better yet would be to observe the actual trajectory of acoustic adaptation. This has been achieved in a recent analysis of song evolution in Darwin's finches of the Galápagos Islands, Ecuador. Grant and Grant have documented the founding and expansion of a population of large ground finches, *Geospiza magnirostris*, on Daphne Major, a small island which previously supported only two other breeding species, *G. scandens* and *G. fortis* (Grant and Grant, 2006). Acoustic analyses reveal that songs of the two original resident species have become, in very short order, increasingly distinct from those of *G. magnirostris* (Grant and Grant, 2010). For instance, the songs of *G. magnirostris* are characterized by slow trill rates, just below 2 Hz, and the songs of *G. fortis* have subsequently increased from average values of 4.6 Hz to 6 Hz since the arrival of *G. magnirostris*. Trill rates of *G. scandens* have also increased during this time, from average values of 8.3 Hz to 10.6 Hz, perhaps as a result of pressures to maintain distinctiveness from the newly-shifted songs of *G. fortis*. Trajectories of vocal evolution might be mediated by a learning mechanism that renders highly distinctive songs preferable as learning models (Grant and Grant, 2010).

The mechanisms of song evolution discussed above – selection for mating signal function and transmission efficacy – are fundamentally adaptive in nature. It is also possible, as outlined in the next section, for songs to evolve as an indirect byproduct of evolutionary changes in the morphological, physiological, or neural mechanisms that underlie vocal behavior. Incorporating data about all of these mechanisms in analyses of song evolution has become a growing trend in studies of neotropical birds, and provides critical contextual information when interpreting the effects of sexual selection on song evolution.

SONG AND MORPHOLOGICAL EVOLUTION

A body of work conducted mainly on temperate zone birds has revealed song production to be an impressively complex behavioral phenotype. Vocalizations are generated by the syrinx, an organ unique to birds that produces sound when activated by respiratory airflow (e.g., Suthers, 1990; Suthers *et al.*, 1994; Goller and Larsen, 1997). Birds modulate source frequencies by adjusting the tension of vibrating source tissues, and tightly coordinate patterns of syrinx modulation and respiration (e.g., Zollinger and Suthers, 2004). Moreover, singing birds continuously adjust the configuration of their vocal tracts, i.e., the trachea, larynx, and beak, and the airsac through which the trachea passes (Westneat *et al.*, 1993; Riede *et al.*, 2006). Reconfigurations of the vocal tract help birds to retain its function as a resonance filter, thus maintaining the pure tonal structure of song across varying source frequencies (Nowicki, 1987; Hoese *et al.*, 2000; Beckers *et al.*, 2003; Riede *et al.*, 2006). Finally, the outputs of the syrinx, respiratory, and vocal tract systems are coordinated in their activity by a specialized, hierarchical neural control system, which in oscine songbirds also mediates the process of vocal imitation (Brainard and Doupe, 2002). From all of this it follows that limits on birds' capacities to activate and coordinate the multiple components of the vocal apparatus (i.e., constraints on vocal performance) should set limits on the acoustic structure of songs (reviewed by Podos and Nowicki, 2004; Suthers, 2004; Podos *et al.*, 2009).

The most direct evidence for performance limits on song production has come from studies of temperate zone songbird species. Performance limits arise as one or more elements of the vocal production mechanism constrain the types and quantitative values of song features that are possible (Podos *et al.*, 2004a, especially their Figure 2). In three studies, young male swamp sparrows (*Melospiza georgiana*, Podos, 1996; Podos *et al.*, 2004b) and northern mockingbirds (*Mimus polyglottis*, Zollinger and Suthers, 2004) were hand-reared and trained with song models in which trill rates, i.e., rates of note repetition, were artificially elevated above normal values. In these studies, young birds proved unable to produce accurate copies of learning models, instead reproducing song models with reduced trill rates, note omissions, or extended pauses between syllables. These outcomes are consistent with a hypothesis of motor constraints on vocal production and development.

Performance constraints on vocal production and evolution have also been inferred through descriptive studies of trade-offs among vocal features, again primarily for temperate zone birds. One such trade-off is between trill rate and the consistency of note frequency, as demonstrated for European great tits *Parus major* (Lambrechts, 1997). Presumably this trade-off occurs because birds singing fast trills with broad frequency changes have limited time and capacity to achieve target frequencies, especially in reaching frequency maxima and minima (Lambrechts, 1997). A second, related trade-off is between trill rate and frequency bandwidth, as has now been demonstrated within and across species-rich North American bird groups, especially sparrows and warblers (Podos, 1997; Cardoso and Hu, 2011; reviewed by Podos *et al.*, 2009). Male songs that achieve high levels of performance in trill rate and frequency bandwidth seem to gain a functional advantage in both intra- and intersexual functions, suggesting that vocal performance is itself a target of sexual selection (e.g., Vallet and Kreutzer, 1995; Draganoiu *et al.*, 2002; Ballentine *et al.*, 2004; Schmidt *et al.*, 2008; reviewed by Podos *et al.*, 2009).

Studies on neotropical birds have been particularly useful in defining and exploring possible constraints on vocal performance imposed by beak morphology. As mentioned above, the beak is part of the vocal tract, and songbirds typically modify beak gape while singing, in their efforts to maintain the vocal tract's resonance-filtering function. It follows that limits on birds' abilities to modulate beak gape could set indirect constraints on song structure (Nowicki *et al.*, 1992). In particular, in seed-eating birds that have evolved beaks specialized for force production, biomechanical trade-offs between force production and kinematic versatility might set limits on song features shaped by performance limits, including trill rate and frequency bandwidth (Podos and Nowicki, 2004). This prediction was first borne out in a study of song and morphology in Darwin's finches, in which beak size was shown to correlate negatively with song parameters associated with vocal performance (Podos, 2001). The negative relationship between beaks and song was shown across eight species using independent contrasts analysis (Podos, 2001), as well as within a species of ground finch, *Geospiza fortis*, that exhibits unusually broad diversity in both morphology and vocal structure (Podos, 2001; Huber and Podos, 2006; see also Christensen *et al.*, 2006). Analyses of beak movements during song production in Darwin's finches offer further support for the vocal tract constraint hypothesis. These analyses illustrate tight correlations between vocal frequency and beak gape (Podos *et al.*, 2004c), and show that birds that can modulate beak gapes more rapidly also have smaller beaks and weaker bite forces (Herrel *et al.*, 2009). Correlations between beak morphology and vocal performance have since been reported for two other species-rich neotropical bird groups, both suboscines: the Thamnophilid antbirds (Seddon, 2005), and the Dendrocolaptid woodcreepers (Derryberry *et al.*, 2012). The latter study offers a particularly striking parallel to data from North American sparrows (Podos, 1997) in a clear triangular relationship between trill rate and frequency bandwidth; in upper-bound regressions for trill rate × frequency bandwidth plots with statistically indistinguishable slopes; and in tight relationships between beak size and vocal performance, as measured via deviations of songs from upper-bound regressions. The parallel nature of these patterns across such distinct and distant bird groups suggests the patterns may apply with even broader taxonomic generality than previously supposed.

Song evolution may also be circumscribed by variation in body size. The majority of research on this topic has focused on vocal frequencies, guided by the expectation that larger-bodied animals should sing at lower frequencies. This expectation is based on the observation that body size typically scales positively with the mass of the tissues that generate sound, and by the presumption that larger, more massive tissues should vibrate more slowly and thus generate lower sound frequencies. Indeed, vocal frequency has been shown to scale negatively with body size across a broad array of taxa (e.g., Wallschlager, 1980; Ophir *et al.*, 2010; reviewed by Searcy and Nowicki, 2005), including neotropical birds (e.g., Ryan and Brenowitz, 1985; Bertelli and Tubaro, 2002). A particularly strong body size × vocal frequency relationship has been established in Darwin's finches by Cutler (1970), who demonstrated that in these birds body mass and syrinx size co-vary with near perfect isometry. Thus body mass is a precise proxy for syrinx size, which explains why variation in body size translates closely into variation in emphasized vocal frequencies (Bowman, 1983). In the Dendrocolaptid woodcreepers, which show unusually marked variation in beak morphology, vocal frequency correlates not just with body size but also with beak length, in accordance with the fact that long beaks augment vocal tract volume, and thus provide a more effective resonance chamber for sounds at low frequencies (Palacios and Tubaro, 2000; see also Podos and Nowicki, 2004).

Accumulating information about the influence on song structure of both beak or body size bears on the evolution not just of songs but also of their singers. Both beak morphology and body size are common targets of natural selection, and adaptive evolution in either trait might drive incidental, correlated changes in song structure (Podos, 2001; Ballentine, 2006). Moreover, given that song often acts as a mating signal, incidental changes in song among adaptively diverging lineages might facilitate reproductive isolation and thus perhaps speciation (Podos and Nowicki, 2004; Podos and Hendry, 2006; Herrel et al., 2009; Derryberry et al., 2012). One example of how this might work is illustrated in a series of recent studies of the Darwin's medium ground finch (*Geospiza fortis*) at El Garrapatero, Santa Cruz Island. This *G. fortis* population is bimodal, with a clear separation between large birds with large beaks and small birds with small beaks (Hendry et al., 2006). These large and small "morphs" are favored by disruptive selection (Hendry et al., 2009), and have evolved songs that are acoustically distinct, with large-morphed birds singing songs of lower performance, as predicted by the vocal constraint hypothesis (Huber and Podos, 2006). Females are seen to mate preferentially with males of their own morph (Huber et al., 2007) – decisions likely guided by female preferences for same-morph songs (see Grant and Grant, 1997). Differences in song structure across morphs are now also known to be meaningful to males. Territorial males presented with song playback simulating an intruder respond much more quickly and strongly to playback of songs of the same morph than of the other morph (Podos, 2010). Males probably perceive a greater threat from males of the same morph, because they are more likely to compete with the focal male for both food and mates (Podos, 2010). Continued study of beaks, song, and ecological speciation holds great promise for neotropical species, given the broad morphological diversity that often characterizes these groups. By contrast, temperate zone species rarely show equivalent levels of diversity in beak size, so that beak by song relationships in temperate zone taxa are not as likely to be strong (e.g., Irwin et al., 2008; Mahler and Gil, 2009; but see Ballentine, 2006; Badyaev et al., 2008).

SONG AND EVOLUTIONARY DRIFT

A final factor that I now address, in brief, is that of evolutionary changes to song in the absence of selection – i.e., drift. To illustrate the principle of drift, consider two populations of a species that are separated by a geographic barrier. Even if these populations' ecological and acoustic environments are identical (and thus show parallel selection pressures), and even if both populations initially show coincident value ranges for mating signals and associated preferences, songs can still diverge over time via the accumulation in one or both populations of changes generated through random or stochastic processes. Most documented cases of drift in nature have a genetic basis, which for bird songs might drive inter-population divergence if vocal features (e.g., song frequencies) are tied to randomly evolving morphology (e.g., body size; see Podos *et al.*, 2004a for a more detailed discussion). For species that learn to sing by imitation, drift in song features can also have a cultural basis, occurring via the accumulation across generations of inaccuracies in song learning (Slater, 1986; Lachlan and Servedio, 2004). Cultural drift in bird species can result in song variations across both temporal and spatial scales (e.g., as dialects) – patterns that have been studied most rigorously in temperate zone birds (e.g., Thorpe, 1961; Marler and Tamura, 1964; Jenkins, 1978; Ince *et al.*, 1980; Payne, 1996; Podos and Warren, 2007; Irwin *et al.*, 2008; Byers *et al.*, 2010b; see also Koetz *et al.*, 2007 for an example from the Old World tropics).

By contrast, studies of drift and song evolution in neotropical birds have been few and far between. Drift is most likely to shape song divergence when contact and interchange among populations is restricted (Irwin *et al.*, 2008), and vocal differences among isolated populations have been noted for numerous neotropical suboscine bird species (e.g., Isler *et al.*, 2007; Ippi *et al.*, 2011; see also Ridgely and Tudor, 1994). Geographic variation is also prevalent in neotropical oscines, although such variation is typically attributed to selective factors rather than drift (e.g., Handford and Lougheed, 1991). In oscines, one context in which random drift might best account for song variations is among populations that are very recently split and which inhabit very similar habitats. This point is well-illustrated in recent studies of geographic variation in song in *Geospiza fortis* on Santa Cruz Island, Galápagos (11 km separation; Podos, 2007) and in the rufous-collared sparrow, *Zonotrichia capensis*, in the Ecuadorean Andes (25 km separation; Danner *et al.*, 2011). Variants among these populations presumably derived from random processes, and birds in both studies were found to respond more strongly to playback of local songs than the more distant, unfamiliar songs (males only: Podos, 2007; both males and females: Danner *et al.*, 2011). The influence of cultural drift on song evolution can also be studied through longitudinal studies of focal populations (e.g., Jenkins, 1978; Derryberry, 2009), an approach that was recently taken in an analysis of song evolution in the Academy Bay population of *G. fortis* (Goodale and Podos, 2010). Recordings made by R. Bowman at the Academy Bay site in 1961 enabled comparisons with more recent songs, and some song types were found to persist with high fidelity over the intervening decades, thus suggesting, for some song types, a limited influence of drift on song evolution (Goodale and Podos, 2010).

CONCLUSION

Our understanding of sexual selection and song evolution has been greatly enriched by studies on neotropical birds. Through such studies we have gained a more comprehensive appreciation of the factors, adaptive or otherwise, that push songs to diversify, as well as the factors that circumscribe patterns of song divergence. The advantages of conducting research on neotropical birds stem fundamentally from the Neotropics' extended diversity of natural history contexts, within which assumptions that underlie sexual selection theory may be tested. As illustrated in this chapter, tests of hypotheses about song divergence have been facilitated by specific features of the Neotropics, including its diverse and morphologically-variable bird radiations (e.g., Podos, 2001; Seddon, 2005; Derryberry *et al.*, 2012), high species density in bird communities (e.g., Luther, 2009), wide range of mating systems and sex roles (e.g., Illes and Yunes-Jimenez, 2009), and diversity of acoustic habitats (Tobias *et al.*, 2010). This list is not intended to be comprehensive; to be sure there are other advantages to be gleaned through studies of neotropical birds, concerning song evolution in the context of sexual selection, performance, function, ecology, and species divergence (e.g., Tobias and Seddon, 2009; Weir and Wheatcroft, 2011).

Returning, then, to the motivating question of this chapter, as it applies to bird songs: does sexual selection operate differently in neotropical environments as compared to temperate zone environments? My answer to this question would be no; I would argue instead that sexual selection follows the same principles in the Neotropics as in temperate zones, but that it achieves a greater range of outcomes given the broader intrinsic organismal and ecological diversity of neotropical habitats. In other words, outcomes of research on sexual selection and bird song are helping to reinforce and extend conceptual frameworks initially developed for temperate zone taxa, rather than requiring the construction of new frameworks with distinct conceptual bases.

To conclude, I would like to suggest three particularly promising trends and topics for continued research on sexual selection and song evolution in neotropical birds. First, there is a clear growing trend for research papers on song to consider concurrently multiple factors that can shape song evolution (e.g., Seddon, 2005; Irwin *et al.*, 2008; Mahler and Gil, 2009; Tobias *et al.*, 2010; Ippi *et al.*, 2011), rather than the more traditional approach of testing single factors individually. Emerging evidence points to a previously unappreciated diversity of scenarios by which songs evolve, and it is becoming increasingly clear that song evolution is driven not by unitary factors but rather by blends of factors that can interact in manners complex and subtle. Second, there is a trend towards incorporating more information about mechanisms of song production and expression, as a complement to more traditional field-oriented studies that are more descriptive in nature. To illustrate, there is much information to be gained by augmenting descriptive analyses of song structure with data about the organismal processes responsible for song generation (e.g., Podos *et al.*, 2004c; see Bostwick and Prum, 2005 for a parallel illustration). Ongoing progress in this area will ultimately benefit from laboratory or controlled field studies on vocal mechanics, performance, and learning in neotropical species. Last but not least, studies of neotropical birds offer a promising natural laboratory for exploring how song evolution and divergence might translate into patterns of speciation. This is especially relevant given the central role that song divergence plays in reproductive isolation across birds. Might there be a connection, in the Neotropics, between the broad diversity of factors and contexts that can drive song divergence, and the high numerical diversity of bird species the Neotropics supports? This is a focused restatement of a longstanding question (e.g., Raikow, 1986; Baptista and Trail, 1992), and the potential connections between signal evolution and speciation in neotropical taxa have not gone unnoticed (e,g, Podos, 2001, 2010; Slabbekoorn and Smith, 2002; Seddon and Tobias, 2007; Huber *et al.*, 2007; Seddon *et al.*, 2008; Dingle *et al.*, 2010, Danner *et al.*, 2011). Prospects for continued new insights from neotropical bird systems seem better than ever.

ACKNOWLEDGMENTS

I thank Regina Macedo and Glauco Machado for inviting me to contribute to this volume, and Elizabeth Derryberry and Sarah Goodwin for helpful comments on a prior draft. J. P. is supported by the University of Massachusetts, Amherst, and National Science Foundation grant IOS-1028964.

REFERENCES

Alatalo, R.V., Glynn, C., Lundberg, A., 1990. Singing rate and female attraction in the pied flycatcher: an experiment. Anim. Behav. 39, 601–603.

Andersson, M.B., 1994. Sexual Selection. Princeton University Press, Princeton.

Badyaev, A.V., Young, R.L., Oh, K.P., Addison, C., 2008. Evolution on a local scale: Developmental, functional, and genetic bases of divergence in bill form and associated changes in song structure between adjacent habitats. Evolution 62, 1951–1964.

Ballentine, B., 2006. Morphological adaptation influences the evolution of a mating signal. Evolution 60, 1936–1944.

Ballentine, B., 2009. The ability to perform physically challenging songs predicts age and size in male swamp sparrows, *Melospiza georgiana*. Anim. Behav. 77, 973–978.

Ballentine, B., Hyman, J., Nowicki, S., 2004. Vocal performance influences female response to male bird song: An experimental test. Behav. Ecol. 15, 163–168.

Baptista, L.F., Trail, P.W., 1992. The role of song in the evolution of passerine diversity. Syst. Biol. 41, 242–247.

Becker, P.H., 1982. The coding of species-specific characteristics in bird sounds. In: Kroodsma, D.E., Miller, E.H. (Eds.), Acoustic Communication in Birds, Academic Press, New York, pp. 213–252.

Beckers, G.J.L., Suthers, R.A., ten Cate, C., 2003. Pure-tone birdsong by resonance filtering of harmonic overtones. Proc. Natl. Acad. Sci. U. S. A. 100, 7372–7376.

Beecher, M.D., Brenowitz, E.A., 2005. Functional aspects of song learning in songbirds. Trends Ecol. Evol. 20, 143–149.

Bertelli, S., Tubaro, P.L., 2002. Body mass and habitat correlates of song structure in a primitive group of birds. Biol. J. Linn. Soc. 77, 423–430.

Bolhuis, J.J., Gahr, M., 2006. Neural mechanisms of birdsong memory. Nat. Rev. Neurosci. 7, 347–357.

Boncoraglio, G., Saino, N., 2007. Habitat structure and the evolution of bird song: A meta-analysis of the evidence for the acoustic adaptation hypothesis. Funct. Ecol. 21, 134–142.

Borror, D.J., 1956. Variation in Carolina wren songs. Auk 73, 211–229.

Borror, D.J., Reese, C.R., 1953. The analysis of bird songs by means of a vibralyser. Wilson Bull. 65, 271–276.

Bostwick, K.S., Prum, R.O., 2005. Courting bird sings with stridulating wing feathers. Science 309, 736.

Botero, C.A., Rossman, R.J., Caro, L.M., Stenzler, L.M., Lovette, I.J., de Kort, S.R., Vehrencamp, S.L., 2009. Syllable type consistency is related to age, social status and reproductive success in the tropical mockingbird. Anim. Behav. 77, 701–706.

Bowman, R.I., 1979. Adaptive morphology of song dialects in Darwin's finches. J. Ornithol. 120, 353–389.

Bowman, R.I., 1983. The evolution of song in Darwin's finches. In: Bowman, R.I., Berson, M., Leviton, A.E. (Eds.), Patterns of Evolution in Galápagos Organisms, American Association for the Advancement of Science, Pacific Division, San Francisco, CA, pp. 237–537.

Bradbury, J.W., Vehrencamp, S.L., 2011. Principles of Animal Communication. Sinauer Associates, Sunderland, MA.

Brainard, M.S., Doupe, A.J., 2002. What songbirds teach us about learning. Nature 417, 351–358.

Brumm, H., Naguib, M., 2009. Environmental acoustics and the evolution of bird song. Adv. Stud. Behav. 40, 1–33.

Brumm, H., Slabbekoorn, H., 2005. Acoustic communication in noise. Adv. Stud. Behav. 35, 151–209.

Brumm, H., Farrington, H., Petren, K., Fessl, B., 2010. Evolutionary dead end in the Galapagos: divergence of sexual signals in the rarest of Darwin's finches. PLoS One 5(6), e11191.

Byers, B.E., 2007. Extrapair paternity in chestnut-sided warblers is correlated with consistent vocal performance. Behav. Ecol. 18, 130–136.

Byers, B.E., Kroodsma, D.E., 2009. Female mate choice and songbird song repertoires. Anim. Behav. 77, 13–22.

Byers, J., Hebets, E., Podos, J., 2010a. Female mate choice based upon male motor performance. Anim. Behav. 79, 771–778.

Byers, B.E., Belinsky, K.L., Bentley, R.A., 2010b. Independent cultural evolution of two song traditions in the chestnut-sided warbler. Am. Nat. 176, 476–489.

Cardoso, G.C., Hu, Y., 2011. Birdsong performance and the evolution of simple (rather than elaborate) sexual signals. Am. Nat. 178, 679–686.

Catchpole, C.K., 1980. Sexual selection and the evolution of complex songs among European warblers of the genus *Acrocephalus*. Behavior 74, 149–166.

Catchpole, C.K., Slater, P.J.B., 2008. Bird Song: Biological Themes and Variations, second ed. Cambridge University Press, Cambridge.

Christensen, R., Kleindorfer, S., Robertson, J., 2006. Song is a reliable signal of bill morphology in Darwin's small tree finch *Camarhynchus parvulus*, and vocal performance predicts male pairing success. J. Avian Biol. 37, 617–624.

Clutton-Brock, T., 2007. Sexual selection in males and females. Science 318, 1882–1885.

Cramer, E.R.A., Price, J.J., 2007. Red-winged blackbirds *Agelaius phoeniceus* respond differently to song types with different performance levels. J. Avian Biol. 38, 122–127.

Cutler, B., 1970. Anatomical Studies of the Syrinx of Darwin's Finches. San Francisco State University, San Francisco.

Danner, J.E., Danner, R.M., Bonier, F., Martin, P.R., Small, T.W., Moore, I.T., 2011. Female, but not male, tropical sparrows respond more strongly to the local song dialect: Implications for population divergence. Am. Nat. 178, 53–63.

Darwin, C., 1871. The Descent of Man, and Selection in Relation to Sex. John Murray, London.

de Kort, S.R., Eldermire, E.R.B., Valderrama, S., Botero, C.A., Vehrencamp, S.L., 2009a. Trill consistency is an age-related assessment signal in banded wrens. Proc. R. Soc. B 276, 2315–2321.

de Kort, S.R., Eldermire, E.R.B., Cramer, E.R.A., Vehrencamp, S.L., 2009b. The deterrent effect of bird song in territory defense. Behav. Ecol. 20, 200–206.

Derryberry, E.P., 2009. Ecology shapes birdsong evolution: variation in morphology and habitat explains variation in white-crowned sparrow song. Am. Nat. 174, 24–33.

Derryberry, E.P., Seddon, N., Claramunt, S., Tobias, J.A., Baker, A., Aleixo, A., Brumfield, R.T., 2012. Correlated evolution of beak morphology and song in the neotropical woodcreeper radiation. Evolution 66, 2784–2797.

Dingle, C., Halfwerk, W., Slabbekoorn, H., 2008. Habitat-dependent song divergence at subspecies level in the grey-breasted wood-wren. J. Evol. Biol. 21, 1079–1089.

Dingle, C., Poelstra, J.W., Halfwerk, W., Brinkhuizen, D.M., Slabbekoorn, H., 2010. Asymmetric response patterns to subspecies-specific song differences in allopatry and parapatry in the gray-breasted wood-wren. Evolution 64, 3537–3548.

Draganoiu, T.I., Nagle, L., Kreutzer, M., 2002. Directional female preference for an exaggerated male trait in canary (*Serinus canaria*) song. Proc. R. Soc. B 269, 2525–2531.

Dunn, P.O., Whittingham, L.A., Pitcher, T.E., 2001. Mating systems, sperm competition, and the evolution of sexual dimorphism in birds. Evolution 55, 161–175.

Emlen, S.T., Oring, L.W., 1977. Ecology, sexual selection, and evolution of mating systems. Science 197, 215–223.

Endler, J.A., 1983. Natural and sexual selection on color patterns in poeciliid fishes. Environ. Biol. Fishes 9, 173–190.

Endler, J.A., Houde, A.E., 1995. Geographic variation in female preferences for male traits in *Poecilia reticulata*. Evolution 49, 456–468.

Eriksson, D., Wallin, L., 1986. Male bird song attracts females: A field experiment. Behav. Ecol. Sociobiol. 19, 297–299.

Fitzsimmons, L.P., Barker, N.K., Mennill, D.J., 2008. Individual variation and lek-based vocal distinctiveness in songs of the screaming piha (*Lipaugus vociferans*), a suboscine songbird. Auk 125, 908–914.

Gaston, K.J., 2000. Global patterns in biodiversity. Nature 405, 220–227.

Gil, D., Gahr, M., 2002. The honesty of bird song: Multiple constraints for multiple traits. Trends Ecol. Evol. 17, 133–141.

Goller, F., Larsen, O.N., 1997. A new mechanism of sound generation in songbirds. Proc. Natl. Acad. Sci. U. S. A. 94, 14787–14791.

Goodale, E., Podos, J., 2010. Persistence of song types in Darwin's finches, *Geospiza fortis*, over four decades. Biol. Lett. 6, 589–592.

Goodwin, S.E., Shriver, W.G., 2011. Effects of traffic noise on occupancy patterns of forest birds. Conserv. Biol. 25, 406–411.

Grant, B.R., Grant, P.R., 2010. Songs of Darwin's finches diverge when a new species enters the community. Proc. Natl. Acad. Sci. U. S. A. 107, 20156–20163.

Grant, P.R., Grant, B.R., 1997. Hybridization, sexual imprinting, and mate choice. Am. Nat. 149, 1–28.

Grant, P.R., Grant, B.R., 2006. Evolution of character displacement in Darwin's finches. Science 313, 224–226.

Grava, T., Grava, A., Otter, K.A., 2012. Vocal performance varies with habitat quality in black-capped chickadees (*Poecile atricapillus*). Behaviour 149, 35–50.

Griffith, S.C., Owens, I.P.F., Thuman, K.A., 2002. Extra pair paternity in birds: A review of inter-specific variation and adaptive function. Mol. Ecol. 11, 2195–2212.

Hagedorn, M., Heiligenberg, W., 1985. Court and spark: Electric signals in the courtship and mating of gymnotoid fish. Anim. Behav. 33, 254–265.

Halfwerk, W., Slabbekoorn, H., 2009. A behavioural mechanism explaining noise-dependent frequency use in urban birdsong. Anim. Behav. 78, 1301–1307.

Hall, M.L., 2009. A review of vocal duetting in birds. Adv. Stud. Behav. 40, 67–121.

Handford, P., 1981. Vegetational correlates of variation in the song of *Zonotrichia capensis*. Behav. Ecol. Sociobiol. 8, 203–206.

Handford, P., Lougheed, S.C., 1991. Variation in duration and frequency characters in the song of the rufous-collared sparrow, *Zonotrichia capensis*, with respect to habitat, trill dialects and body size. Condor 93, 644–658.

Hasselquist, D., Bensch, S., von Schantz, T., 1996. Correlation between male song repertoire, extra-pair paternity and offspring survival in the great reed warbler. Nature 381, 229–232.

Hendry, A.P., Grant, P.R., Grant, B.R., Ford, H.A., Brewer, M.J., Podos, J., 2006. Possible human impacts on adaptive radiation: beak size bimodality in Darwin's finches. Proc. R. Soc. B 273, 1887–1894.

Hendry, A.P., Huber, S.K., De Leon, L.F., Herrel, A., Podos, J., 2009. Disruptive selection in a bimodal population of Darwin's finches. Proc. R. Soc. B 276, 753–759.

Herrel, A., Podos, J., Vanhooydonck, B., Hendry, A.P., 2009. Force-velocity trade-off in Darwin's finch jaw function: A biochemical basis for ecological speciation? Funct. Ecol. 23, 119–125.

Hoese, W.J., Podos, J., Boetticher, N.C., Nowicki, S., 2000. Vocal tract function in birdsong production: experimental manipulation of beak movements. J. Exp. Biol. 203, 1845–1855.

Hopkins, C.D., 1988. Neuroethology of electric communication. Annu. Rev. Neurosci. 11, 497–535.

Huber, S.K., Podos, J., 2006. Beak morphology and song features covary in a population of Darwin's finches *(Geospiza fortis)*. Biol. J. Linn. Soc. 88, 489–498.

Huber, S.K., De Leon, L.F., Hendry, A.P., Bermingham, E., Podos, J., 2007. Reproductive isolation of sympatric morphs in a population of Darwin's finches. Proc. R. Soc. B 274, 1709–1714.

Illes, A.E., Yunes-Jimenez, L., 2009. A female songbird out-sings male conspecifics during simulated territorial intrusions. Proc. R. Soc. B 276, 981–986.

Illes, A.E., Hall, M.L., Vehrencamp, S.L., 2006. Vocal performance influences male receiver response in the banded wren. Proc. R. Soc. B 273, 1907–1912.

Ince, S.A., Slater, P.J.B., Weismann, C., 1980. Changes with time in the songs of a population of chaffinches. Condor 82, 285–290.

Ippi, S., Vasquez, R.A., van Dongen, W.F.D., Lazzoni, I., 2011. Geographical variation in the vocalizations of the suboscine thorn-tailed rayadito *Aphrastura spinicauda*. Ibis 153, 789–805.

Irwin, D.E., Thimgan, M.P., Irwin, J.H., 2008. Call divergence is correlated with geographic and genetic distance in greenish warblers *(Phylloscopus trochiloides)*: A strong role for stochasticity in signal evolution? J. Evol. Biol. 21, 435–448.

Isler, M.L., Isler, P.R., Whitney, B.M., 2007. Species limits in antbirds (Thamnophilidae): The warbling antbird *(Hypocnemis cantator)* complex. Auk 124, 11–28.

Jenkins, P.F., 1978. Cultural transmission of song patterns and dialect development in a free-living bird population. Anim. Behav. 26, 50–78.

Kipper, S., Mundry, R., Sommer, C., Hultsch, H., Todt, D., 2006. Song repertoire size is correlated with body measures and arrival date in common nightingales, *Luscinia megarhynchos*. Anim. Behav. 71, 211–217.

Kirschel, A.N.G., Cody, M.L., Harlow, Z.T., Promponas, V.J., Vallejo, E.E., Taylor, C.E., 2011. Territorial dynamics of mexican ant-thrushes *Formicarius moniliger* revealed by individual recognition of their songs. Ibis 153, 255–268.

Koetz, A.H., Westcott, D.A., Congdon, B.C., 2007. Geographical variation in song frequency and structure: the effects of vicariant isolation, habitat type and body size. Anim. Behav. 74, 1573–1583.

Kokko, H., Monaghan, P., 2001. Predicting the direction of sexual selection. Ecol. Lett. 4, 159–165.

Konishi, M., 1985. Birdsong: From behavior to neuron. Annu. Rev. Neurosci. 8, 125–170.

Kroodsma, D.E., Byers, B.E., 1991. The function(s) of bird song. Am. Zoologist 31, 318–328.

Kroodsma, D.E., Vielliard, J.M.E., Stiles, F.G., 1996. Study of bird sounds in the Neotropics: urgency and opportunity. In: Kroodsma, D.E., Miller, E.H. (Eds.), Ecology and Evolution of Acoustic Communication in Birds, Cornell University Press, Ithaca, pp. 269–281.

Lachlan, R.F., Servedio, M.R., 2004. Song learning accelerates allopatric speciation. Evolution 58, 2049–2063.

Lahti, D.C., Moseley, D.L., Podos, J., 2011. A tradeoff between performance and accuracy in bird song learning. Ethology 117, 802–811.

Lambrechts, M.M., 1997. Song frequency plasticity and composition of phrase versions in great tits. *Parus major*. Ardea 85, 99–109.

Lissmann, H.W., 1958. On the function and evolution of electric organs in fish. J. Exp. Biol. 35, 156–191.

Logue, D.M., Forstmeier, W., 2008. Constrained performance in a communication network: Implications for the function of song-type matching and for the evolution of multiple ornaments. Am. Nat. 172, 34–41.

Lohr, B., Wright, T.F., Dooling, R.J., 2003. Detection and discrimination of natural calls in masking noise by birds: Estimating the active space of a signal. Anim. Behav. 65, 763–777.

Lougheed, S.C., Handford, P., 1992. Vocal dialects and the structure of geographic variation in morphological and allozymic characters in the rufous-collared sparrow, *Zonotrichia capensis*. Evolution 46, 1443–1456.

Luther, D., 2009. The influence of the acoustic community on songs of birds in a neotropical rain forest. Behav. Ecol. 20, 864–871.

Luther, D.A., 2008. Signaller: receiver coordination and the timing of communication in Amazonian birds. Biol. Lett. 4, 651–654.

Luther, D.A., Derryberry, E.P., 2012. Birdsongs keep pace with city life: Changes in song over time in an urban songbird affects communication. Anim. Behav. 83, 1059–1066.

Macedo, R.H., Karubian, J., Webster, M.S., 2008. Extrapair paternity and sexual selection in socially monogamous birds: Are tropical birds different? Auk 125, 769–777.

Mahler, B., Gil, D., 2009. The evolution of song in the *Phylloscopus* leaf warblers (Aves: Sylviidae): a tale of sexual selection, habitat adaptation, and morphological constraints. Adv. Stud. Behav. 40, 35–66.

Mann, N.I., Dingess, K.A., Barker, F.K., Graves, J.A., Slater, P.J.B., 2009. A comparative study of song form and duetting in neotropical *Thryothorus* wrens. Behaviour 146, 1–43.

Marler, P., 2004. Science and birdsong: the good old days. In: Marler, P., Slabbekoorn, H. (Eds.), Nature's Music: The Science of Birdsong, Elsevier Academic Press, Amsterdam, pp. 1–38.

Marler, P., Tamura, M., 1964. Culturally transmitted patterns of vocal behavior in sparrows. Science 146, 1483–1486.

Marten, K., Quine, D., Marler, P., 1977. Sound transmission and its significance for animal vocalization 2. Tropical forest habitats. Behav. Ecol. Sociobiol. 2, 291–302.

Mennill, D.J., 2006. Aggressive responses of male and female rufous-and-white wrens to stereo duet playback. Anim. Behav. 71, 219–226.

Mennill, D.J., Vehrencamp, S.L., 2005. Sex differences in singing and duetting behavior of neotropical rufous-and-white wrens. (*Thryothorus rufalbus*). Auk 122, 175–186.

Mennill, D.J., Vehrencamp, S.L., 2008. Context-dependent functions of avian duets revealed by microphone-array recordings and multispeaker playback. Curr. Biol. 18, 1314–1319.

Morton, E.S., 1975. Ecological sources of selection on avian sounds. Am. Nat. 109, 17–34.

Morton, E.S., 1996. A comparison of vocal behavior among tropical and temperate passerine birds. In: Kroodsma, D.E., Miller, E.H. (Eds.), Ecology and Evolution of Acoustic Communication in Birds, Cornell University Press, Ithaca, pp. 258–268.

Nelson, D.A., 2000. A preference for own-subspecies' song guides vocal learning in a song bird. Proc. Nat. Acad. Sci. U. S. A. 97, 13348–13353.

Nelson, D.A., Marler, P., 1990. The perception of birdsong and an ecological concept of signal space. In: Stebbins, W.C., Berkley, M.A. (Eds.), Comparative Perception: Complex Signals, John Wiley and Sons, New York, pp. 443–478.

Nemeth, E., Brumm, H., 2010. Birds and anthropogenic noise: Are urban songs adaptive? Am. Nat. 176, 465–475.

Nottebohm, F., 1969. Song of chingolo, *Zonotrichia capensis*, in Argentina – description and evaluation of a system of dialects. Condor 71, 299–315.

Nottebohm, F., 1975. Continental patterns of song variability in *Zonotrichia capensis*: Some possible ecological correlates. Am. Nat. 109, 605–624.

Nottebohm, F., Stokes, T.M., Leonard, C.M., 1976. Central control of song in canary, *Serinus canarius*. J. Comp. Neurol. 165, 457–486.

Nowicki, S., 1987. Vocal tract resonances in oscine bird sound production: evidence from birdsongs in a helium atmosphere. Nature 325, 53–55.

Nowicki, S., Searcy, W.A., 2005. Song and mate choice in birds: How the development of behavior helps us understand function. Auk 122, 1–14.

Nowicki, S., Westneat, M.W., Hoese, W., 1992. Birdsong: motor function and the evolution of communication. Semin. Neurosci. 4, 385–390.

Nystrom, K.G.K., 1997. Food density, song rate, and body condition in territory-establishing willow warblers (*Phylloscopus trochilus*). Can. J. Zool. 75, 47–58.

Ophir, A.G., Schrader, S.B., Gillooly, J.F., 2010. Energetic cost of calling: General constraints and species-specific differences. J. Evol. Biol. 23, 1564–1569.

Palacios, M.G., Tubaro, P.L., 2000. Does beak size affect acoustic frequencies in woodcreepers? Condor 102, 553–560.

Payne, R.B., 1996. Song traditions in indigo buntings: Origin, improvaisation, dispersal, and extinction in cultural evolution. In: Kroodsma, D.E., Miller, E.H. (Eds.), Ecology and Evolution of Acoustic Communication in Birds, Cornell University Press, Ithaca, pp. 198–220.

Peek, F.W., 1972. An experimental study of the territorial function of vocal and visual display in the male red-winged blackbird (*Agelaius phoeniceus*). Anim. Behav. 20, 112–118.

Podos, J., 1996. Motor constraints on vocal development in a songbird. Anim. Behav. 51, 1061–1070.

Podos, J., 1997. A performance constraint on the evolution of trilled vocalizations in a songbird family (Passeriformes: Emberizidae). Evolution 51, 537–551.

Podos, J., 2001. Correlated evolution of morphology and vocal signal structure in Darwin's finches. Nature 409, 185–188.

Podos, J., 2007. Discrimination of geographical song variants by Darwin's finches. Anim. Behav. 73, 833–844.

Podos, J., 2010. Acoustic discrimination of sympatric morphs in Darwin's finches: A behavioural mechanism for assortative mating? Philos. Trans. R. Soc. B 365, 1031–1039.

Podos, J., Hendry, A.P., 2006. The biomechanics of ecological speciation. In: Herrel, A., Speck, T., Rowe, N.P. (Eds.), Ecology and Biomechanics, Edinburgh, pp. 301–321.

Podos, J., Nowicki, S., 2004. Performance limits on birdsong. In: Marler, P., Slabbekoorn, H. (Eds.), Nature's Music: The Science of Birdsong, Academic Press, New York, pp. 318–342.

Podos, J., Warren, P.S., 2007. The evolution of geographic variation in birdsong. Adv. Stud. Behav. 37, 403–458.

Podos, J., Huber, S.K., Taft, B., 2004a. Bird song: The interface of evolution and mechanism. Annu. Rev. Ecol. Evol. Syst. 35, 55–87.

Podos, J., Peters, S., Nowicki, S., 2004b. Calibration of song learning targets during vocal ontogeny in swamp sparrows, *Melospiza georgiana*. Anim. Behav. 68, 929–940.

Podos, J., Southall, J.A., Rossi-Santos, M.R., 2004c. Vocal mechanics in Darwin's finches: Correlation of beak gape and song frequency. J. Exp. Biol. 207, 607–619.

Podos, J., Lahti, D.C., Moseley, D.L., 2009. Vocal performance and sensorimotor learning in songbirds. Adv. Stud. Behav. 40, 159–195.

Pribil, S., Searcy, W.A., 2001. Experimental confirmation of the polygyny threshold model for red-winged blackbirds. Proc. R. Soc. B 268, 1643–1646.

Price, J.J., Lanyon, S.M., 2004. Patterns of song evolution and sexual selection in the oropendolas and caciques. Behav. Ecol. 15, 485–497.

Price, J.J., Lanyon, S.M., Omland, K.E., 2009. Losses of female song with changes from tropical to temperate breeding in the New World blackbirds. Proc. R. Soc. B 276, 1971–1980.

Raikow, R.J., 1986. Why are there so many kinds of passerine birds? Syst. Zool. 35, 255–259.

Reid, J.M., Arcese, P., Cassidy, A.L.E.V., Hiebert, S.M., Smith, J.N.M., Stoddard, P.K., Marr, A.B., Keller, L.F., 2004. Song repertoire size predicts initial mating success in male song sparrows, *Melospiza melodia*. Anim. Behav. 68, 1055–1063.

Ricklefs, R.E., Schluter, D., 1993. Species Diversity in Ecological Communities: Historical and Geographical Perspectives. University of Chicago Press, Chicago IL.

Ridgely, R.S., Tudor, G., 1994. The Birds of South America, *Volume 2*: The Suboscine Passerines. University of Texas Press, Austin TX.

Riede, T., Suthers, R.A., Fletcher, N.H., Blevins, W.E., 2006. Songbirds tune their vocal tract to the fundamental frequency of their song. Proc. Natl. Acad. Sci. U. S. A. 103, 5543–5548.

Rios-Chelen, A.A., Macías Garcia, C., 2004. Flight display song of the vermilion flycatcher. Wilson Bull. 116, 360–362.

Rios-Chelen, A.A., Macías Garcia, C., 2007. Responses of a sub-oscine bird during playback: Effects of different song variants and breeding period. Behav. Proc. 74, 319–325.

Ripmeester, E.A.P., Kok, J.S., van Rijssel, J.C., Slabbekoorn, H., 2010. Habitat-related birdsong divergence: A multi-level study on the influence of territory density and ambient noise in European blackbirds. Behav. Ecol. Sociobiol. 64, 409–418.

Ryan, M.J., Brenowitz, E.A., 1985. The role of body size, phylogeny, and ambient noise in the evolution of bird song. Am. Nat. 126, 87–100.

Sakata, J.T., Vehrencamp, S.L., 2012. Integrating perspectives on vocal performance and consistency. J. Exp. Biol. 215, 201–209.

Schmidt, R., Kunc, H.P., Amrhein, V., Naguib, M., 2008. Aggressive responses to broadband trills are related to subsequent pairing success in nightingales. Behav. Ecol. 19, 635–641.

Searcy, W.A., Andersson, M., 1986. Sexual selection and the evolution of song. Annu. Rev. Ecol. Syst. 17, 507–533.

Searcy, W.A., Nowicki, S., 2005. The Evolution of Animal Communication: Reliability and Deception in Signaling Systems. Princeton University Press, Princeton.

Searcy, W.A., Yasukawa, K., 1996. Song and female choice. In: Kroodsma, D.E., Miller, E.H. (Eds.), Ecology and Evolution of Acoustic Communication in Birds, Cornell University Press, Ithaca, pp. 454–473.

Seddon, N., 2005. Ecological adaptation and species recognition drives vocal evolution in neotropical suboscine birds. Evolution 59, 200–215.

Seddon, N., Tobias, J.A., 2006. Duets defend mates in a suboscine passerine, the warbling antbird (*Hypocnemis cantator*). Behav. Ecol. 17, 73–83.

Seddon, N., Tobias, J.A., 2007. Song divergence at the edge of Amazonia: An empirical test of the peripatric speciation model. Biol. J. Linn. Soc. 90, 173–188.

Seddon, N., Merrill, R.M., Tobias, J.A., 2008. Sexually selected traits predict patterns of species richness in a diverse clade of suboscine birds. Am. Nat. 171, 620–631.

Seehausen, O., van Alphen, J.J.M., Witte, F., 1997. Cichlid fish diversity threatened by eutrophication that curbs sexual selection. Science 277, 1808–1811.

Slabbekoorn, H., 2004. Singing in the wild: The ecology of birdsong. In: Marler, P., Slabbekoorn, H. (Eds.), Nature's Music: The Science of Birdsong, Elsevier Academic Press, Amsterdam, pp. 178–205.

Slabbekoorn, H., Peet, M., 2003. Ecology: Birds sing at a higher pitch in urban noise – Great tits hit the high notes to ensure that their mating calls are heard above the city's din. Nature 424, 267.

Slabbekoorn, H., Smith, T.B., 2002. Bird song, ecology and speciation. Philos. Trans. R. Soc. B 357, 493–503.

Slater, P.J.B., 1986. The cultural transmission of birdsong. Trends Ecol. Evol. 1, 94–97.

Slater, P.J.B., 2003. Fifty years of bird song research: a case study in animal behaviour. Anim. Behav. 65, 633–639.

Stoddard, P.K., 1999. Predation enhances complexity in the evolution of electric fish signals. Nature 400, 254–256.

Stutchbury, B.J., Morton, E.S., 1995. The effect of breeding synchrony on extra-pair mating systems in songbirds. Behaviour 132, 675–690.

Suthers, R.A., 1990. Contributions to birdsong from the left and right sides of the intact syrinx. Nature 347, 473–477.

Suthers, R.A., 2004. How birds sing and why it matters. In: Marler, P., Slabbekoorn, H. (Eds.), Nature's Music: The Science of Birdsong, Elsevier Academic Press, Amsterdam, pp. 272–295.

Suthers, R.A., Goller, F., Hartley, R.S., 1994. Motor dynamics of song production by mimic thrushes. J. Neurobiol. 25, 917–936.

Thorpe, W.H., 1961. Bird Song. Cambridge University Press, Cambridge, UK.

Tobias, J.A., Seddon, N., 2009. Sexual selection and ecological generalism are correlated in antbirds. J. Evol. Biol. 22, 623–636.

Tobias, J.A., Aben, J., Brumfield, R.T., Derryberry, E.P., Halfwerk, W., Slabbekoorn, H., Seddon, N., 2010. Song divergence by sensory drive in amazonian birds. Evolution 64, 2820–2839.

Trivers, R.L., 1972. Parental investment and sexual selection. In: Campbell, B. (Ed.), Sexual Selection and the Descent of Man, Aldine, Chicago, IL, pp. 136–179.

Vallet, E., Kreutzer, M., 1995. Female canaries are sexually responsive to special song phrases. Anim. Behav. 49, 1603–1610.

Wallschlager, D., 1980. Correlation of song frequency and body weight in passerine birds. Experientia 36, 412.

Webb, C.O., Ackerly, D.D., McPeek, M.A., Donoghue, M.J., 2002. Phylogenies and community ecology. Annu. Rev. Ecol. Syst. 33, 475–505.

Weir, J.T., Wheatcroft, D., 2011. A latitudinal gradient in rates of evolution of avian syllable diversity and song length. Proc. R. Soc. B 278, 1713–1720.

Westneat, M.W., Long, J.H., Hoese, W., Nowicki, S., 1993. Kinematics of birdsong: Functional correlation of cranial movements and acoustic features in sparrows. J. Exp. Biol. 182, 147–171.

Wiley, R.H., 1971. Song groups in a singing assembly of little hermits. Condor 73, 28–35.

Wiley, R.H., 1991. Associations of song properties with habitats for territorial oscine birds of eastern North America. Am. Nat. 138, 973–993.

Wiley, R.H., 1994. Errors, exaggeration, and deception in animal communication. In: Real, L.A. (Ed.), Behavioral Mechanisms in Evolutionary Ecology, University of Chicago Press, Chicago, pp. 157–189.

Wiley, R.H., Richards, D.G., 1978. Physical constraints on acoustic communication in the atmosphere: Implications for the evolution of animal vocalizations. Behav. Ecol. Sociobiol. 3, 69–94.

Zollinger, S.A., Suthers, R.A., 2004. Motor mechanisms of a vocal mimic: Implications for birdsong production. Proc R. Soc. B 271, 483–491.

Chapter 14

Impacts of Mating Behavior on Plant–Animal Seed Dispersal Mutualisms: A Case Study from a Neotropical Lek-Breeding Bird

Jordan Karubian and Renata Durães

Department of Ecology and Evolutionary Biology, Tulane University, New Orleans, Louisiana, USA

INTRODUCTION

Sexual selection is a profound evolutionary force that impacts animal phenotype in myriad ways. A core question in the field of sexual selection is how intraspecific variation in phenotypic traits associated with mate choice (i.e., secondary sexual characteristics) affects access to mates and, ultimately, fitness. Examples of secondary sexual characteristics may include visual, acoustic, and olfactory signals, as well as behaviors, body size, and weapons (Andersson, 1994). A rich body of literature on these and other traits robustly supports Darwin's (1871) contention that, within a species, more extreme values for secondary sexual characteristics should be associated with increased attractiveness to mates and enhanced reproductive success (Andersson, 1994). This focus on mate choice has shaped our basic understanding of how sexual selection operates, and is a cornerstone of behavioral ecology and evolutionary biology.

Lek mating systems have proved particularly useful in furthering our understanding of mate choice processes (Wiley, 1991; Andersson, 1994). Lekking is characterized by spatially and temporally clustered aggregations of males in sites where display, mate choice, and copulation take place (i.e., leks). Lekking males invest heavily in attracting mates and provide no parental care; females are exclusively responsible for provisioning young. Reproductive skew is typically pronounced among males of lek-breeding species, such that a small number of males monopolizes the vast majority of matings, and most males have low levels of reproductive success (Payne, 1984; Mackenzie *et al.*, 1995). These conditions are thought to promote heavy investment in mate attraction by males,

Sexual Selection. http://dx.doi.org/10.1016/B978-0-12-416028-6.00014-1

and to underlie the extreme levels of secondary sexual characteristics for which lekking species are known (Andersson, 1994).

Darwin (1871) also recognized that sexual selection does not operate in a vacuum, and that there may be complex, and sometimes counteracting, relationships between sexual and natural selection. Because most animals routinely interact directly or indirectly with other species, natural selection pressures often take the form of interspecific, ecological interactions. The relationship between sexual selection and ecological interactions is likely to work in both directions: ecological interactions may determine the intensity of sexual selection in a given species or population, but at the same time sexual selection may shape the nature and outcome of ecological interactions. For example, natural selection pressure from parasites and/or predators may restrict the intensity of sexual selection and limit expression of secondary sexual characteristics at the individual, population, or species level (Zuk, 1992), while degree of investment in secondary sexual characteristics is also likely to affect parasite or predator populations and evolutionary trajectories (Kirkpatrick, 1986; Knell, 1999). As such, sexually selected traits are frequently thought to represent an optimum balance between mate choice (which may select for exaggerated values), and predation risk and parasite infection, which limit trait expression.

Along with predation and parasitism, sexually selected traits are also likely to be shaped by, and to shape, mutualistic ecological interactions. Consider, for example, frugivorous animals that serve as dispersal vectors for the seeds of plants. In such plant–animal mutualisms, plants provide resources (fruits) for the animals, which in turn disperse seeds, which assist with recruitment and gene movement in plants. In such cases, the spatial distribution of resources will affect the ability to monopolize mates, and hence the opportunity for sexual selection and mating system (the "polygyny threshold"; Emlen and Oring, 1977). The strength of sexual selection and the mating system exhibited by a species are, in turn, likely to affect secondary sexual characters, movement biology, and foraging ecology. When the animal involved is a dispersal vector, these factors may influence its dispersal services, which in turn will contribute to plant distributions and resource availability. Thus, in plant–animal mutualisms, factors such as distribution of resources, mating system, sexual selection, and foraging ecology are likely to interact to affect seed dispersal outcomes in an iterative, and potentially quite complex, manner.

Terborgh (1990) estimated that 85% of all tree species in one study area in the Peruvian Amazon were animal-dispersed, and similar values have been reported for other tropical forest sites across the globe (Foster, 1982; Howe and Smallwood, 1982; Ganesh and Davidar, 2001). Though the intensity and importance of plant–animal seed dispersal mutualisms reaches its acme in tropical rainforests, animal-mediated seed dispersal is a critical ecological and genetic process for plant species across most terrestrial habitats. When viewed through this lens, sexual selection among frugivores may be considered a potentially significant, albeit indirect, factor in determining demographic and genetic characteristics of rainforest plant species. At the same time, natural selection forces associated with frugivory may shape signaling systems and other traits associated with mate choice (see, for example, Schaefer *et al.*, 2004). For instance, sexual selection may drive changes in morphology (e.g., mouthparts or body size), sensory systems (e.g., vision or olfaction), social organization (e.g., degree of territoriality and sociality), display behavior (e.g., use of traditional display areas), diet (e.g., preferential consumption of certain fruits for compounds used in secondary sexual signals), and use of space (e.g., movement patterns and foraging ecology), all of which in turn may impact what a frugivorous animal eats and where it disperses seeds. As such, a better understanding of how sexual selection interfaces with mutualistic ecological interactions in tropical rainforest would expand our appreciation for the forces that shape and are shaped by sexual selection in the tropics and beyond.

In this chapter, we address how sexual selection among frugivorous animals may affect the seed dispersal services that associated plants receive. Our particular focus is on how mating system and associated display behaviors impact foraging ecology and seed dispersal by frugivorous, lek-breeding birds. We first provide an overview of key concepts and predictions, and then go on to illustrate these concepts with our own work on the long-wattled umbrellabird (Cotingidae: *Cephalopterus penduliger*), a lek-breeding bird from the Chocó rainforests of northwest South America. We then compare long-wattled umbrellabirds with other lekking and non-lekking tropical and temperate species whose seed dispersal services may be impacted by sexual selection and/or mating system to varying degrees. Our broad objective in this chapter is to demonstrate the indirect, but biologically significant, effect that sexual selection in general, and mating behavior in particular, may have for the demographic and genetic structure of animal-dispersed plants in tropical rainforest and other habitats.

OVERVIEW AND PREDICTIONS

Lekking in birds is broadly distributed both taxonomically and geographically; this behavior has been recorded for approximately 100 species representing 15 avian families distributed across temperate and tropical regions of the globe (Höglund and Alatalo, 1995). This broad taxonomic range makes it difficult to make broad generalizations about the behavioral ecology of temperate versus tropical lek breeders. However, there does exist one striking difference among tropical versus temperate lekking bird species: diet. Whereas nearly all tropical lek-breeding species of bird are primarily frugivorous or nectarivorous, these food types do not constitue an important part of the diet for lek-breeding species from the temperate zone (Höglund and Alatalo, 1995). Lek-breeding birds from the temperate zone eat primarily grains and seeds (e.g., Ploecidae wydahs and widowbirds, $n=6$ species classified as lek breeders in Höglund and Alatalo, 1995), invertebrates (e.g., Scolopacidae ruff and sandpipers, $n=3$ species), or some combination thereof (e.g., Tetraoninae grouse, $n=9$ species; Phasianinae pheasant and peafowl, $n=2$ species; Meleagrinae turkey, $n=1$ species; and Otidae bustard, $n=1$ species). In contrast, tropical lek breeders are dominated by frugivorous families such as Cotingidae (cotingas, $n=18$ species), Pipridae (manakins, $n=18$ species), and Paradisaeidae (birds of paradise, $n=13$ species), as well as the nectarivorous family Trochilidae (hummingbirds, $n=14$ species).

What factor or factors can explain this marked difference in diet between temperate versus tropical lek breeders? Male emancipation from parental care, and the associated ability for females alone to provide adequate parental care for successful reproduction, is a first requisite for lekking to evolve and be maintained. Clutch sizes typically drop with decreasing latitude, perhaps because of more intense predation pressure (Skutch, 1949) or increased adult survival near the tropics (Martin et al., 2000), but this is unlikely to be related to differences in diet between temperate and tropical lekking species. Although there is strong evidence for the role of phylogenetic inertia in determining the occurrence of lekking, as seen by the fact that this phenomenon is concentrated in just a few families, this also cannot fully explain the high incidence of frugivory among tropical species because of the independent origin of lekking behavior among many of the major families.

We consider it likely that differences in the temporal and spatial distribution of fruit may play a fundamental role in determining why most lekking species are frugivores in the tropics, but not the temperate zone. First and foremost, because fruit is abundant in tropical rainforest relative to most other habitat types, and is available year-round, males from species relying on this resource may be emancipated from parental care and thus able to engage in lekking behavior. Second, as pointed out by Emlen and Oring's (1977) aforementioned paper on the environmental potential for polygyny, the distribution of resources in time and space will likely determine a species' mating system. When resources essential for females are clumped and easily defended, we might expect resource defense polygyny to evolve. When resources are dispersed but female groups can be easily followed, we might expect female defense polygyny to evolve. It is only when resources or females are not defensible that "male dominance" polygyny (e.g., lekking) may evolve (Oring, 1982). In the tropics, fruit may be distributed in such a way that it is sufficiently abundant to emancipate males from parental care, yet diffuse enough that neither fruit nor groups of females may be economically defended by males. In these circumstances, we might expect lekking to evolve. In the temperate zone, invertebrates and grains may exhibit attributes similar to those of fruit in tropical rainforest that promote lekking behavior. As such, on an evolutionary timescale, the underlying distribution of resources can shape the mating system of a given species, which in turn shapes the intensity of sexual selection on that species.

The distribution of resources is also likely to impact animal-mediated seed dispersal services on an ecological timescale. Frugivorous vertebrates vary widely in their seed dispersal services (Dennis and Westcott, 2006; Jordano *et al.*, 2007), both quantitatively (e.g., the number of fruits removed) and qualitatively (e.g., what and where they forage, how seeds are treated during ingestion and digestion, and how they are moved around and finally deposited; Schupp, 1993; Schupp *et al.*, 2010). This variation in dispersal services is directly related to foraging ecology (Murray, 1988; Westcott *et al.*, 2005; Jordano *et al.*, 2007), which on an ecological timescale is driven by the location of fruits in combination with constraints placed on movement by social organization, mating system, or other factors (Karubian and Durães, 2009). Foraging ecology in turn underlies observed patterns of seed movement and deposition (e.g., Karubian *et al.*, 2012a).

The variation in seed dispersal outcomes driven by these forces is likely to have significant demographic and genetic consequences for plant species and communities. Seed dispersal is a fundamental demographic process in plants, in that it determines seedling establishment and the distribution of species within and between populations (Howe and Smallwood, 1982; Levey *et al.*, 2002; Wang and Smith, 2002; Dennis *et al.*, 2007). Dispersal patterns shape the spatial distribution of dispersed seeds, including distance seeds are moved (Clark *et al.*, 2005; Jordano *et al.*, 2007), probability of deposition into microsites that may be particularly advantageous for germination or recruitment (Davidson and Morton, 1981; Reid, 1989; Wenny, 2001), and aggregation patterns (clustered vs scattered) of deposited seeds (Howe, 1989; Vander Wall and Beck, 2012).

The degree to which seeds are dispersed in a clumped manner is of particular interest because clustering can reduce survival of seeds and seedlings due to density-dependent effects (Janzen, 1970; Connell, 1979; Kwit *et al.*, 2004; Jansen *et al.*, 2008). Clustering of non-dispersed seeds falling directly underneath the parent tree is ubiquitous in nature and often leads to near-complete mortality, but frugivores often yield clumped distributions of seeds both underneath fruiting trees and away from them (Jordano and Godoy, 2002). This "spatially contagious" (Schupp *et al.*, 2002) or "destination-based" (Karubian *et al.*, 2010) pattern of seed dispersal can result in few sites receiving many seeds and most sites receiving few to none – a pattern that can have profound demographic consequences for plant populations due to the limited dissemination of propagules (Jordano and Godoy, 2002; Schupp *et al.*, 2002). Based on the microsite characteristics, it may also enhance or diminish probability of seed survival and recruitment (Davidson and Morton, 1981; Reid, 1989; Wenny, 2001; Holland *et al.*, 2009). Long-distance dispersal is thought to be particularly advantageous for plants because it reduces density-dependent mechanisms of competition, predation and disease, and increases arrival into favorable new sites (Nathan and Muller-Landau, 2000; Nathan *et al.*, 2008). Scaling up, these demographic processes also have important consequences for community structure of plant populations, metapopulation dynamics, long-term species persistence, and range expansion (Cain *et al.*, 2000; Pakeman, 2001; Laurance *et al.*, 2006).

Seed dispersal is also a fundamental genetic process in plants, in that it shapes the distribution of genotypes within a population, and gene movement within and among populations (Nathan and Muller-Landau, 2000; Sork and Smouse, 2006; Dennis *et al.*, 2007). While pollen represents the first phase of gametic dispersal, dispersal of seeds may have a larger impact on genetic processes because it determines the location of an individual with all its inherent risks of mortality, and it moves both maternal and paternal gametic genomes (Crawford, 1984; Hamilton, 1999; Grivet *et al.*, 2009). In this sense, the movement of seeds provided by frugivores directly shapes the fine-scale genetic structure of plant populations, as well as connectivity between populations and colonization of new patches (Sork and Smouse, 2006).

The use of genotypes from maternally inherited tissue in seed or fruit (Godoy and Jordano, 2001) provides a powerful tool to assess the genetic consequences of animal-mediated dispersal. Genetic methods have provided novel insights into the distances that seeds are moved by frugivores, and the genetic homogeneity of the seed pools that frugivores generate. Applying recently developed analytical methods to genotypic data derived from natural systems, researchers have found that animal-mediated seed dispersal often appears to result in a non-random and highly structured distribution of maternal genotypes (i.e., seeds) away from the maternal seed source (Grivet *et al.*, 2005; García *et al.*, 2009). However, although many studies consistently point to structured seed populations among vertebrate-dispersed plants, it would be premature to conclude that animal-mediated seed dispersal universally results in genetic bottlenecks because a relatively narrow range of (mostly temperate zone) species has been studied to date.

At present, three factors would significantly aid efforts to understand the demographic and genetic consequences of animal-mediated seed dispersal in general, as well as the effects of sexual selection and mating system on dispersal outcomes in particular. First, it would be useful to establish a direct linkage between a single dispersal agent or behavior and subsequent dispersal outcomes. This is because studies often lack a detailed knowledge of the frugivore species responsible for observed patterns of dispersal, meaning that genetic patterns observed among dispersed seeds could be due to a variety of dispersal agents exhibiting a potentially wide range of behaviors and activities. There has been a recent focus on this linkage (e.g., Grivet *et al.*, 2005; Jordano *et al.*, 2007; García *et al.*, 2009), but data are still scarce.

Second, the vast majority of relevant genetic studies focus on temperate systems in which dispersal agents exhibit similar forms of social organization consisting of social pairs or cooperative family groups that defend and forage within a fixed territory. As such, characterizing dispersal processes for a broader diversity of species, with a range of mating systems including lek breeding, would move us closer to a more robust understanding of the relationship between behavior and seed dispersal outcomes.

Third, a better integration of traditional ecological and more recent molecular methods is needed. Currently, molecular and observational estimates of seed dispersal are rarely combined in a single study system, despite the fact that both approaches have potentially significant limitations when employed in isolation. Direct observational studies of dispersal are likely to miss rare but important longer-distance dispersal events, and can pose significant challenges for tracking movement of individual seeds (Koenig *et al.*, 1996; Nathan, 2006; Scofield *et al.*, 2011). Molecular studies, in turn, may fail to capture the proximate factors driving observed genetic patterns because they rarely identify the seed dispersal vector and/or behavior responsible for the seed arriving at its final location (but see Jordano *et al.*, 2007; Scofield *et al.*, 2010). For these reasons, integrating molecular results with a mechanistic understanding of underlying seed dispersal and deposition processes (e.g., frugivore behavior and movement) is desirable when possible.

The remainder of this chapter focuses on how lek-breeding may influence seed dispersal outcomes. As we have seen, among lek-breeding birds, frugivorous species are clustered in tropical, and especially neotropical, rainforests, and are largely absent from other habitat types. For this reason, the impacts of lek breeding on seed dispersal will be most pronounced in the tropics. Also, the relatively long duration of fruit availability in tropical rainforests allows extended, and in some cases year-round, activity at leks, thereby amplifying the effect of this mating system on seed dispersal outcomes in these habitats. This focus on lek mating systems in the tropics is meant to be illustrative of a broader point: the mating system and mating behavior exhibited by frugivorous animals will shape seed dispersal outcomes, regardless of what form that mating system takes.

In the following sections, we will use the long-wattled umbrellabird as a case study to illustrate how lekking behavior impacts foraging ecology, seed movement and deposition, and, ultimately, patterns of recruitment and genetic structure among plant populations. In doing so, we will test the following expectations: (1) males and females in lek-breeding species will exhibit marked differences in their movement patterns and foraging ecology that are directly attributable to their distinctive mating and reproductive strategies; (2) these differences in movement and foraging ecology will lead to differences in seed movement and deposition, which in the case of long-wattled umbrellabirds will lead to males yielding longer dispersal distances and a higher density of dispersed seeds at leks relative to areas outside the lek; (3) the high density of seeds at leks will be associated with reduced survival relative to areas outside leks unless some other factor, such as leks being particularly favorable microsites for seed recruitment, is relevant; and (4) dispersed seeds in leks will exhibit high degrees of genetic heterogeneity relative to areas outside leks because, over time, displaying males will bring seeds from a variety of seed sources surrounding leks.

FOCAL STUDY SPECIES

Long-wattled umbrellabirds (hereafter "umbrellabirds") are large frugivorous birds endemic to the humid Chocó rain forests of northwestern Ecuador and western Colombia (Snow, 1982, 2004; Fig. 14.1). The species is considered

FIGURE 14.1   A male long-wattled umbrellabird in northwest Ecuador. This male in perched on his display territory on a lek, and is in the act of regurgitating a palm seed. *Photograph courtesy of Murray Cooper.*

"Vulnerable" to extinction, due primarily to widespread deforestation in this area (BirdLife International, 2000; IUCN, 2011). Umbrellabirds belong to the neotropical family Cotingidae, a group known for lekking behavior and exuberant secondary sexual characteristics (Snow, 1982). As is typical for lek-breeding species, males and females exhibit morphological and behavioral attributes that are likely related to their distinctive mating and reproductive strategies. Males are approximately 1.5 times larger than females and have large crests and long wattles, both of which are present but much reduced in females (Tori *et al.*, 2008). Groups of 5–15 males congregate in leks of ~ 1 ha in area, with a peak in sexual display activity in early mornings and late afternoons from August to February, and lower levels of activity at other times of the year (Tori *et al.*, 2008). Most males, which we refer to as "territorial" males, hold small (*ca.* 25 m²), long-term display territories on a single lek, which in turn allows us to link seeds dispersed into these display territories to male umbrellabird mating behavior (see below). Unusually for a lek-breeding bird, males from the same lek often forage together away from the lek in a relatively large, cohesive group (Tori *et al.*, 2008).

A subset of males, refered to as "floater" males, exhibits a qualitatively different strategy. Instead of holding a territory at a single lek, these males move between multiple leks without holding a fixed display territory. This behavior may be relatively common among lek-breeding species (Théry, 1992; Westcott and Smith, 1994; Tello, 2001), especially among younger males, but is poorly understood – in part because of challenges involved with tracking these individuals over large spatial areas. We have confirmed floating behavior in three of the 30 total umbrellabird males (10%) we have tracked with telemetry (below) for at least two radio-tracking sessions. One floater was not fully grown in either body size or sexual ornamentation, but the other two floater males were morphologically indistinguishable from territorial males. As such, we are uncertain at present what the true incidence of this behavior is, and whether this represents a fixed alternative mating strategy or a flexible strategy that may only be employed for a portion of a male's life.

Female umbrellabirds are largely solitary. They appear to visit the lek only for purposes of mate choice and copulation (or occasionally to forage if there happens to be a fruiting tree in the lek), but spend the majority of the time alone. Females provide all parental care (Karubian et al., 2003), and nesting is concentrated from January to May, which, curiously, only overlaps partially with peak male display activity at the leks (Tori et al., 2008). They typically lay a single egg, and our observations suggest that post-fledging parental care lasts for several months (J. Karubian, unpublished data). Most females appear to visit only a single lek each mating season, though in rare instances we have recorded a female visiting two leks in a single season (J. Karubian, unpublished data). Although both sexes are highly frugivorous, females appear to consume a higher proportion of insects and small vertebrates than do males, especially when nesting (Karubian et al., 2003; J. Karubian, unpublished data).

As one of the few large avian frugivores in the Chocó, umbrellabirds are important dispersers of large-seeded fruits typical of mature rainforest. The species consumes fruits of at least 35 plant species in our study area in northwest Ecuador, but exhibits a preference for species of the palm, avocado, and nutmeg families (Arecaceae, Lauraceae, and Myristicaceae, respectively). Fruits of these species present a single, large seed surrounded by a thin, lipid-rich aril, and umbrellabirds ingest fruit at the source tree before regurgitating the seed at some later point, usually away from the source tree. In contrast, umbrellabirds defecate smaller seeds associated with smaller fruits, such as the strangler fig *Ficus crassiuscula* (Moraceae).

In our study area in northwest Ecuador, umbrellabirds have a particularly tight ecological relationship with the canopy palm species *Oenocarpus bataua* (hereafter *Oenocarpus*; see Karubian et al., 2010). *Oenocarpus* produces large-seeded (35×22 mm; Karubian et al., 2012b), lipid-rich fruits in single infructescences of up to 2000 fruits (Goulding and Smith, 2007). Individual trees have ripe fruits for 3–4 weeks, and fruits are an important food source for both humans and large frugivores across the species' range, which extends from Panama to Bolivia on both sides of the Andes (Henderson et al., 1995; Goulding and Smith, 2007). In our study area, umbrellabirds are the primary seed dispersal agents for *Oenocarpus*, with a lesser contribution by toucans (Ramphastidae); primate seed dispersal agents for *Oenocarpus* are absent from the site. We focus specifically on the mutualism between *Oenocarpus* and umbrellabirds for many of the analyses presented below.

GENERAL METHODS

We have been studying umbrellabird lekking behavior and seed dispersal since October 2002 in the Bilsa Biological Station (hereafter BBS; 79°45′W, 0°22′N,

330–730 m asl), a 3500-ha reserve of humid pre-montane Chocó rainforest in northwest Ecuador (see Karubian *et al.*, 2007 for more informaton on BBS). We have located a total of nine lek sites in BBS, and we monitor four of them for activity year round. Leks are typically located on forested ridges, and are separated by 1.7 ± 0.2 km (J. Karubian, unpublished data).

One important advantage of the umbrellabird system is that it allows us to link individual seeds dispersed into display territories on leks to male umbrellabird mating behavior. This is because males spend large amounts of time on these traditional display areas within the lek from which they effectively exclude other large avian frugivores, thereby dominating seed dispersal into these areas (see Karubian *et al.*, 2010, 2012b). It is more challenging to link specific seeds encountered in other areas in the forest to a specific dispersal agent. For example, for the majority of seeds encountered outside the lek we are unable to determine which of the several potential vectors is responsible for the arrival of the seed at that point. This list of potential dispersal agents of *Oenocarpus* outside umbrellabird leks includes female umbrellabirds, male umbrellabirds in transit between foraging sites and the lek, toucans, terrestrial rodents, water, and gravity.

To test our first prediction regarding movement patterns of males vs females (Prediction One), we used two complementary tracking approaches. Umbrellabirds we captured at leks in mist nets placed in the canopy, measured, color-banded, and equipped with either tail-mounted radio-transmitters (Holohil Systems Ltd, Ontario, Canada) or backpack-style GPS tracking devices (e-obs, Hamburg Germany; see Holland *et al.*, 2009). Radio-tracking was conducted on foot using hand-held GPS devices (Garmin LTD) to record locations at 30-minute intervals; GPS tracking devices record locations at 15-minute intervals and download data to a handheld base station. Movement data was visualized in ArcView GIS 3.2 (ESRI™, Redlands, California) and home ranges estimated using the Animal Movement ArcView Extension v. 2.04 (Hooge and Eichenlaub, 2001).

To estimate patterns of seed movement and deposition by umbrellabirds (Prediction Two), we integrated information on movement from radio-tracking with gut retention time, i.e., the time from ingestion to regurgitation (for four large-seeded species: two Arecaceae, *Oenocarpus* and *Bactris setulosa*; and two Myristicaceae, *Virola dixonni* and *Otoba gordonifolia*) or defecation (*F. crassiuscula*) to calculate seed dispersal distributions, following Murray (1988). As described in Karubian *et al.* (2012b), we also extended this method to produce "spatially explicit" seed dispersal distributions which calculate deposition patterns relative to a fixed location, such as the lek. We used these empirically based distributional models to generate predictions about the density of dispersed seeds in lek sites versus "control" sites outside the lek. To test these predictions, on a monthly basis we quantified seed rain into 1-m² seed traps placed in leks. These seed traps consisted of a square of PVC tubing around canvas material supported by strings approximately 1 m above the ground (Karubian *et al.*, 2012b).

We used two approaches to assess demographic consequences of umbrellabird dispersal for the five focal plant species (Prediction Three). First, we compared initial seed rain to density of established seedlings and adults in both leks and control sites to assess the degree to which seed rain density was associated with the probability of transition from one age class to the next (Karubian *et al.*, 2012b). Second, we experimentally planted germinated seedlings (which were raised from seed in common nursery conditions as part of a pollen flow study; see Ottewell *et al.*, 2012) in leks and control sites to assess survival at this life stage. We have also tested possible effects of habitat structure among leks vs control sites in this experiment, as well as degree of relatedness (i.e., siblings or not) among experimental seedlings.

To quantify the genetic consequences of umbrellabird dispersal (Prediction Four), we focused exclusively on the palm *Oenocarpus* and made use of the fact that the pericarp tissue of dispersed seeds is of maternal origin, meaning that the genetic profile of the pericarp of a dispersed seed exactly matches that of the source tree from which it came (Godoy and Jordano, 2001). We used a direct genotype matching procedure based on results from microsatellite markers to link dispersed seeds to their maternal (source) tree (Karubian *et al.*, 2010). Samples were gathered from 10-m diameter patches, which we refer to as "seed pools". Seed pools were randomly situated in a single lek or in control areas within a 30-ha study parcel surrounding the lek in which all adult *Oenocarpus* individuals were mapped and genotyped (Karubian *et al.*, 2010). Each seed pool in the lek is likely to represent seed dispersal by a single male, because males hold and defend stable territories within the lek. Outside the lek, "control" seed pools are likely to be generated by seeds dispersed by a broad range of vectors.

Because we can link dispersed seeds arriving in lek sites to umbrellabird display behavior, we can use Godoy and Jordano's (2001) genetic approach to work backward from the dispersed seed to the source tree to gain insights into umbrellabird foraging ecology and dispersal patterns. We used a seed pool structure approach to assess the degree of overlap among seed sources (i.e., maternal trees) contributing to a single patch of dispersed seeds, as well as the overlap between different seed patches. The probability of maternal identity (PMI) approach (Grivet *et al.*, 2005) estimates the number of seed sources per seed pool, which can tell us, for example, whether clumped distributions represent genetic bottlenecks or areas of unusually high genetic mixing (Scofield *et al.*, 2010, 2011). PMI can also utilize the degree of genetic overlap in seed sources between different seed pools to provide insights into the distance seeds are being moved. An advantage of PMI is that one can test the genetic consequences of seed dispersal without locating and mapping the genotypes of all adults. At the same time, if one has a site where all adult trees are mapped and genotyped, traditional maternity analysis can be used to document exact dispersal distances (e.g., García *et al.*, 2007, 2009) as well as the frequency of immigrant seeds, which do not match any maternal genotype in the study area and are therefore likely to represent long-distance dispersal events (e.g., Jordano *et al.*, 2007).

HOW DOES LEKKING BEHAVIOR AFFECT SEED DISPERSAL OUTCOMES IN LONG-WATTLED UMBRELLABIRDS?

Prediction One: Umbrellabird Movement

In umbrellabirds, as in other lek-breeding species, the vast majority of male display behavior, female inspection and mate choice, and actual matings all take place at the lek. As such, males are predicted to spend as much time as possible at the lek in order to maximize their fitness (Fiske *et al.*, 1998). One might therefore expect that "territorial" male umbrellabirds would forage in close proximity to leks when possible in order to minimize reproductive "opportunity costs" associated with foraging away from the lek. However, the relatively high density of males present at a given lek may exhaust nearby resources, requiring these individuals to travel relatively long distances to find sufficient fruit when local supplies dwindle. Thus, uncertainty exists concerning the degree to which males from a single lek will exhibit highly overlapping, relatively small home ranges around the lek (indicative of foraging at nearby trees, presumably leading to low seed dispersal distances and high overlap in the seed pools generated by different males at the lek) versus larger home ranges with less overlap (indicative of foraging at more distant trees, presumably leading to longer seed dispersal distances, and less overlap between seed pools generated by different males at the lek). Females, in contrast, have no focal location equivalent to the lek except when they are nesting, and we have had no direct evidence that any of the females we studied were nesting during relevant data collection periods. As such, we expected females to maintain largely non-overlapping home ranges and to move evenly across these home ranges. "Floater" males were expected to exhibit a third distinctive pattern of movement, characterized by traveling relatively long distances between multiple leks.

We tested these predictions by tracking the movement of radio-equipped umbrellabirds in 2003–2004, as described in Karubian *et al.* (2012b). As expected, radio-tracking demonstrated that territorial males do indeed spend the vast majority of their time at the lek: these individuals spent 95% of their time in an area of only 7.3 ± 2.4 ha centered on the lek, and 50% in a 1.0 ± 0.2 ha area that corresponds to the lek itself (as demonstrated by 95% and 50% kernels, respectively). However, males also made occasional relatively long foraging trips, such that the overall foraging range (as measured by minimum convex polygons, or MCPs; Mohr, 1947) for radio-tracked territorial males was 37.8 ± 7.4 ha. Within the course of a day, males were present at the lek during peak display periods in the early morning and late afternoon, but traveled between the lek and fruiting trees on foraging trips of varying length and duration during the rest of the day (Fig. 14.2).

A single floater male that we radio-tracked exhibited a movement pattern distinctive from those of territorial males. The floater home range included

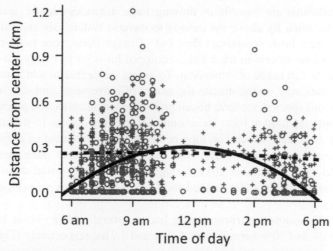

FIGURE 14.2 **Distance moved from home range center for male (open circles) and female (crosses) long-wattled umbrellabirds as a function of time of day.** For males, home range center corresponds to the lek. Lines represent the fit of separate quadratic functions to the distribution of points for males (solid line) and females (dashed line). Males exhibit a strong tendency to be present at the lek during early morning and late afternoon hours, but depart during the middle of the day to forage away from the lek ($R^2 = 0.13$, $F_{2,535} = 39.61$, $P < 0.0001$) whereas females exhibit no discernable movement pattern in relation to time of day ($R^2 = 0.01$, $F_{2,465} = 1.57$, $P = 0.21$).

three leks, and this male remained in the vicinity of a single lek for periods of 3–10 days before moving to another lek. The floater male overall (MCP) home range (596.3 ha), 95% kernel (250.1 ha), and 50% kernel (24.9 ha) were approximately an order of magnitude larger than corresponding values for territorial males (Karubian *et al.*, 2012b).

Females maintained an overall (MCP) home range of 49.2 ± 8.5 ha, which was slightly larger, but not significantly different, than that of territorial males. However, females used their home ranges much more evenly than did males. They spent 95% of their time in an area of 37.3 ± 6.2 ha, and 50% of their time in a core area of 4.7 ± 0.6 ha. These values are approximately five times greater than equivalent measures for territorial males. Unlike territorial males, females showed no discernable daily pattern of movement relative to the center of their home ranges (Fig. 14.2). These findings reflect the fact that females did not focus their movements on a specific location such as the lek, but instead moved across their home-range area without a strong bias towards any particular point. In this sense, the movement patterns of female umbrellabirds may be qualitatively similar to those of many non-lekking species of tropical frugivorous bird (e.g., toucans, Ramphastidae) that maintain regular home ranges and do not have the need to return regularly to a central display site (i.e., the lek). As such, female umbrellabird dispersal patterns may be useful to gain a qualitative sense for dispersal patterns generated by non-lekking species.

Umbrellabirds are capable of moving large distances rapidly, and males in particular often fly above the canopy to traverse watersheds and ridges. As a consequence, birds sometimes flew out of range during our radio-tracking sessions and our efforts to track radio-equipped birds on foot may have failed to capture the full range of movement. To assess the degree to which our radio telemetry data may underestimate the extent of movements and size of home ranges for this species, we have begun preliminary data collection placing GPS tags on birds captured at leks. These units log location data at 15-minute intervals every third day, which can then be remotely downloaded to a hand-held base station. We tracked one territorial male and one floater male captured at a single lek from December 2011 through April 2012 (a similar duration and time period as that of radio-tracked individuals in previous years) using this technology.

The GPS-equipped territorial male had an overall (MCP) home range of 97.4 ha; 95% and 50% kernels were 6.9 ha and 1.7 ha, respectively (Fig. 14.3).

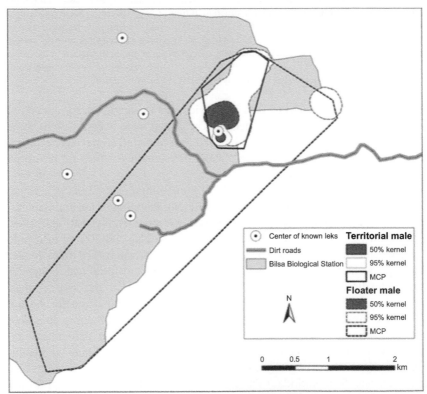

FIGURE 14.3   Home range usage patterns by two male long-wattled umbrellabirds in northwest Ecuador. Data were obtained with GPS tracking devices during a three month period in a single breeding season (2011–2012). Unweighted minimum convex polygon (MCP) estimates and 95% and 50% kernels are shown for a "floater" male (dashed borders) and a territorial male.

Thus, although the overall home range for this individual was two- to three-fold larger than that recorded for other territorial males via conventional radio-tracking, its core usage areas (i.e., 95% and 50% kernels) were virtually identical. The GPS-equipped floater male's overall home range was 871 ha; 95% and 50% kernels were 118 ha and 17 ha, respectively. This GPS-equipped floater male visited two areas, at least one of which is likely to represent a new lek that we were previously unaware of. In this case, the GPS-equipped individual had an overall home range 1.5 times larger than that of the radio-tracked floater (above), and its core usage areas were smaller.

We consider it notable that the core usage areas remained relatively concentrated around leks regardless of the method used, suggesting that the longer foraging trips that underlie the relatively large MCP home ranges for both territorial and floater males may contribute relatively little to overall seed movement in terms of numbers of seeds moved. Instead, the majority of seed movement is likely to occur within the core usage areas, around the leks. On the other hand, these foraging trips away from the lek may be critical to the occurrence of rare but biologically important long-distance seed dispersal events (Nathan *et al.*, 2008).

In sum, animal tracking demonstrates that territorial males, floater males, and females each exhibit dramatically different movement patterns: territorial males concentrate their time in and around their home lek but occasionally forage further afield, females move evenly over moderate-sized home ranges, and floater males travel at a landscape scale that greatly exceeds that of territorial males and females. These differences in foraging ecology can be directly related to the respective mating and reproductive strategies of these three classes of bird. In the following section, we explore how foraging ecology of territorial lek-breeding males impacts seed movement and deposition.

Prediction Two: Seed Movement and Deposition

When, where, and how often a frugivore engages in display behavior should impact the source trees it visits, and the locations where it deposits seeds from those source trees (Wenny and Levey, 1998). We calculated seed dispersal distributions to assess the extent to which distinctive reproductive strategies of territorial male, floater male, and female umbrellabirds would drive biologically meaningful differences in seed transport and deposition (Karubian *et al.*, 2012b). These analyses are restricted to radio-tracked individuals (i.e., they do not include GPS-equipped individuals). Territorial males and females produced similarly shaped distributions of seed dispersal distances from the source tree for both large-seeded and small-seeded tree species, with probability of deposition decreasing sharply as distance from the source tree increased. However, mean dispersal distance from the source tree to deposition site by territorial males was longer than that of females for both large-seeded fruits (257 vs 218 m, respectively) and the small-seeded *F. crassiuscula* (326 vs 244 m). For both fruit types, maximum dispersal distance was also greater for territorial males than for females (1338 vs 1108 m). The single floater male tracked for this study, in contrast, yielded a "flatter" seed dispersal distribution and much longer mean and maximum seed movement distances, reflecting the relatively high frequency with which it undertook long flights, such as those between leks. Mean dispersal distance from the source tree was 542 m for large-seeded fruit and 723 m for *Ficus*, approximately twice the corresponding values for territorial males and females; maximum dispersal distance was 2650 m (Karubian *et al.*, 2012b).

We also predicted that, as a consequence of male display behavior, seed deposition should be spatially aggregated at lek sites relative to areas outside the lek. In keeping with this prediction, "spatially explicit" seed dispersal distributions indicated that territorial males deposit a high proportion of the ingested seeds (>50%) into lek sites. This finding was consistent for both large, regurgitated seeds and smaller, defecated seeds favored by umbrellabirds. Floater males were estimated to deposit 20% of the large seeds they ingest into lek sites. Females, in contrast, were estimated to deposit most seeds >200 m from the center of their home ranges. We confirmed the results of the seed dispersal distribution model empirically in the field; relative to control seed traps located outside the lek, seed traps in leks received more than four times as many dispersed seeds from our five focal tree species (Karubian *et al.*, 2012b). Thus, both the seed movement and deposition patterns we recorded appear to follow directly from the mating behavior and foraging ecology differences our movement analyses uncovered. In the following section, we explore the demographic consequences of clumped seed deposition patterns in leks.

Prediction Three: Demographic Consequences

The high density of seeds deposited beneath display perches by umbrellabirds might be expected to reduce seed and seedling survival via density-dependent mortality and competition processes (Janzen, 1970; Connell, 1971; Kwit *et al.*, 2004; Jansen *et al.*, 2008). Interestingly, we found no evidence for the expected decrease in germination rates among seeds in leks vs outside leks (Karubian *et al.*, 2012b). To further explore this issue, we have implemented a field-based experiment in which seedlings of known provenance that were raised in a common nursery were planted in a standardized design in either leks or control areas outside leks. This experiment, currently in its third year, is still in progress, but preliminary results strongly indicate that seedling survival is significantly higher in leks relative to control areas outside leks (J. Karubian, unpublished data).

These departures from predicted density-dependent mortality suggest that some other factor, such as the lek being a particularly favorable microsite for seed recruitment, may be relevant in this system. However, our measures at leks versus non-leks have uncovered no obvious differences in basic forest structure parameters such as ambient light or canopy height (J. Karubian, unpublished data). We are currently testing the potential impact of soil quality, and in particular the idea that displaying male umbrellabirds may effectively fertilize seeds in leks via repeated defecation from display perches – a pattern reported beneath roosts of other frugivorous birds (T. Carlo, personal communication).

Another possible explanation is that, despite the constraints imposed by lek attendance on foraging behavior, territorial males still forage from multiple trees and move seeds relatively long distances. A relatively high incidence of long-distance dispersal and foraging from multiple trees are both expected to contribute to higher genetic diversity among seed pools at leks, and this genetic variety might in turn enhance survival at leks via a "rare allele" effect (Levin, 1975). "Spatially explicit" seed dispersal distributions used in Karubian *et al.* (2012b) reveal that the average distance territorial males move seeds from source trees to their home range center (i.e., the lek) is shorter than equivalent measures for females (259 vs 325 m for large-seeded fruits and 330 vs 286 m for *Ficus*, respectively). However, maximum dispersal distance from source tree to territory center is greater for territorial males than females (1129 vs 853 m). Also, both ecological observation and genetic analyses (see below) suggest males may bring seeds from a large number of source trees back with them to the lek. However, in the seedling survival experiment described above, we have not been able to detect an effect of relatedness on survival probability.

Regardless of the mechanisms driving this pattern of higher than expected survival in leks versus outside leks, these findings suggest that umbrellabird dispersal may provide an important survival advantage to dispersed seeds. Interestingly, however, this effect appeared not to carry through to adults of the five focal tree species, in that there was no difference in density of adults in leks vs outside leks. This is not necessarily surprising, given the highly stochastic nature of the transition from seedling to adult. Also, leks move over time; we have recorded one lek abandonment and one lek formation over 71 lek-years of monitoring. This uncertainty over how long a lek has been located in a particular location complicates tests of the long-term demographic or genetic consequences of umbrellabird lekking behavior.

Prediction Four: Genetic Consequences

Radio-tracking and opportunistic observations in the field led us to believe that umbrellabird males forage from multiple trees within a day. Over time, we expected that this foraging behavior would lead to accumulation of seeds from multiple source trees beneath each male's perch. We therefore predicted that lekking behavior would lead to high levels of seed-source diversity at fine spatial scales relative to control plots situated outside leks. This in turn may have important consequences for the distribution of genotypes and fine-scale genetic structure exhibited by plant species dispersed by umbrellabirds.

Genetic analyses on *Oenocarpus* seeds collected in the field reveal that the genetic composition of seed pools in a single umbrellabird lek is far more heterogeneous than in areas outside the lek (Karubian *et al.*, 2010). Using the PMI approach to estimate N_{em}, the effective number of seed sources per seed pool (Grivet *et al.*, 2005), we found that seed pools within the focal lek represented over five times more source trees than equivalent seed pools outside the focal lek ($N_{em}=27$ in leks vs 5.2 outside leks). This value fits well with the prediction that males are bringing seeds from multiple source trees back to their territories on the lek. Thus, it does appear that the display behavior of individual umbrellabirds contributes to significant mixing of propagules from different seed sources beneath display perches at the lek, although corroboration from multiple leks is desirable.

Because males often forage in flocks (J. Karubian, unpublished data) and have highly overlapping home ranges that include the same sets of source trees surrounding the lek, one might predict relatively high overlap among the source trees of seeds encountered beneath display perches of different males in the same lek. When conducting between-plot comparisons of overlap, we found that, relative to control plots, lek plots exhibited a more gradual decline in inter-plot maternal overlap (r_{ij}) as inter-plot distance increased (Karubian *et al.*, 2010). This indicates that males from the same lek may be foraging at overlapping but diverse sets of trees, concordant with the ecological observations of overlap in foraging range. This finding is also consistent with the idea that the spatial scale of seed dispersal outside the lek is limited, leading to high levels of overlap among seed plots separated by short distances, a rapid decline of overlap as pairwise distance between seed plots increases, and very low levels of overlap at longer distances between seed plots (see Figure 2 in Karubian *et al.*, 2010).

We also used genetic analyses to directly estimate the spatial scale of seed movement. In keeping with the predictions of our seed dispersal distributions, males are expected to move seeds longer distances than those frugivores contributing to control plots. The proportion of "immigrant seeds" from outside the boundary of our study parcel was higher in the lek than in control plots, suggesting a higher rate of long-distance dispersal from outside the parcel's boundaries into the lek than into the control plots, despite the fact that, on average, control plots were located closer to the parcel's boundaries (Karubian *et al.*, 2010). Genetic analyses of seed movement, using direct matching of genotypes between dispersed seeds and maternal source trees, are currently in progress.

These dispersal differences led to significantly lower spatial genetic autocorrelation at all distance classes analyzed in the lek than outside the lek (Karubian *et al.*, 2010). Thus, destination-based dispersal driven by male umbrellabird display behavior promotes gene movement and homogenizes local genetic structure of *Oenocarpus* seedlings.

COMPARISONS WITH OTHER LEKKING AND NON-LEKKING SPECIES

To what extent can the relationships and patterns we have documented for umbrellabirds be extended to other species of lek-breeding frugivores? We consider it likely that, like umbrellabirds, most if not all lek-breeding species exhibit qualitative differences between the sexes in foraging ecology and movement patterns. This is because sex-specific movement patterns arise directly from the distinctive male and female mating strategies which define the lek mating system. That is, females of lek-breeding species will typically move more evenly across the landscape (when not nesting) relative to males (Westcott and Graham, 2000), which will in turn have smaller core usage areas centered on the lek (Théry, 1992). High attendance at the lek by males should lead to high clustering of seeds at these display sites for all lek-breeding species. Available empirical evidence, though sparse, corroborates this expectation. Lek sites of both manakins and cotingas have higher densities of favored food plants than do control areas outside the leks (Théry and Larpin, 1993; Ryder et al., 2006). There is also a higher density of seeds in the seed bank of manakin leks relative to control areas (Krijger et al., 1997). The degree to which seeds are clustered at leks of different species (or even between different leks of the same species) will depend on the intensity of display behavior, the number of males at a given lek, and the spatial aggregation of displaying males (i.e., classical vs "exploded" leks).

Floater males also appear to be a common element of lek mating systems in many tropical lek-breeding frugivores (e.g., Théry, 1992; Westcott and Smith, 1994; Tello, 2001). However, because this behavior is still so poorly understood, few generalizations can be made for how it should affect seed movement and deposition patterns across species. It does seem likely that floating behavior will lead to longer dispersal distances (e.g., Théry, 1992) and flatter dispersal distributions as birds move between leks, but this issue requires more data. More broadly, the existence of two or even three distinctive reproductive strategies within a single lek-breeding species provides a convenient venue in which to examine the effects of social behavior on seed dispersal outcomes.

The distances that seeds are moved between maternal source tree and lek are likely to vary across, and even within, lek-breeding species. This is because the spatial and temporal distribution of fruiting trees will interact with mating system to shape foraging patterns and space use, which will in turn determine the distances that seeds are moved. Species of bird with a high concentration of fruiting trees in or near the lek (e.g., Ryder et al., 2006) will be expected to transport seeds shorter distances than species forced to forage further afield. Unfortunately, very few data are currently available on seed movement distances attributable to lek-breeding species.

Demographic and genetic consequences of lek breeding are also likely to be system-dependent. Survival probability for seeds dispersed into lek sites is likely to be a function of the microhabitat associated with the lek or display site, though few case studies beyond Wenny and Levey's (1998) aforementioned bellbird study exist. Data are even more sparse for the genetic consequences of dispersal by lek-breeding tropical birds. In the case of umbrellabirds, high levels of genetic mixing among seed pools in the lek can be traced to a relatively high density of maternal source trees around the lek from which males forage. Over time, males bring the seeds from multiple trees back with them to the lek and thereby generate diverse seed pools representing multiple source trees. There are currently few other data available with which to compare these results. In theory, the degree to which this scenario applies to other lekking (or non-lekking) species will depend upon the spatial and temporal distribution of fruiting trees. In contrast to our results from umbrellabirds, for example, if favored food sources are abundant on or adjacent to a male's display territory, as appears to be the case for some manakin species (Ryder *et al.*, 2006), we might expect extremely low dispersal distances and seed source diversity among seed pools generated by these species of bird.

In summary, there currently exist very few data from tropical birds, be they lek-breeding or not, with which to compare umbrellabirds. A priority for future research is to assess animal movement, seed movement, and genetic and demographic consequences in the context of mating system for a broader range of lek-breeding species (as well as other species). A related goal is to account for the distribution of fruiting trees in shaping foraging ecology and dispersal outcomes (e.g., Carlo and Morales, 2008). These are challenging tasks, but are necessary to fully understand the complex dynamics described here.

Comparisons between umbrellabirds and temperate zone species benefit from the fact that much more work has been done in the temperate zone than in the tropics. However, because lek-breeding among frugivorous bird species is essentially absent from the temperate zone, we must limit our comparisons to species that exhibit other mating systems. One particularly well-studied system is that of acorn dispersal of the oak *Quercus lobata* by acorn woodpeckers *Melanerpes formicivorous*. Unlike umbrellabirds, acorn woodpeckers live in highly territorial social groups and tend to forage in the trees proximal to their storage sites, with each territorial group gathering acorns from non-overlapping territories and hence from different trees (Grivet *et al.*, 2005; Scofield *et al.*, 2010, 2011). This mating system and social organization results in pronounced structuring of seed pools in granaries where the acorns are stored (Grivet *et al.*, 2005; Scofield *et al.*, 2010, 2011). In a separate study of dispersed *Q. lobata* seedlings (Grivet *et al.*, 2009), the combined effect of several vertebrate dispersal agents also produced pronounced structuring of seedling populations. Relative to umbrellabird dispersal, then, acorn woodpeckers generate seed patches (i.e., granaries) characterized by very low maternal seed-source diversity and very limited overlap between seed patches on different territories. These differences can be directly attributed to the mating systems of the two species, with the strong territoriality of acorn woodpeckers essentially constraining potential source trees to those located within the territorial boundary.

Similarly, a community of frugivorous birds in Spain (García *et al.*, 2009) dispersed *Prunus mahaleb* seeds long distances from source trees but still yielded strong clustering of maternal genotypes in seed traps. This pattern, which is likely explained by preferences for certain microhabitat types as foraging, resting or roosting sites, and/or by the fact that several fruits may be consumed in a single foraging bout and deposited together, shows that genetic bottlenecks can arise despite long-distance movement of seeds. Although the majority of avian species in this study form socially monogamous pair bonds, it is not clear to what extent the breeding system *per se* contributes to observed seed dispersal outcomes.

CONCLUSIONS

In sum, variation in seed dispersal associated with sexual selection and mating system is likely to be an indirect but nonetheless important factor in determining demography, gene flow, and genetic structure for animal-dispersed plant species. Among umbrellabirds, we observe patterns of seed movement and deposition that can be directly traced back to mating system, and which favorably impact seed survival and fine-scale genetic structure for the plants these birds disperse. This phenomenon is likely to apply across lek-breeding birds, which constitute an important part of the tropical avifauna, but data are lacking and the scale and even the direction of the impacts will vary across lekking species. We expect reduced distances of seed movement, less clustering of seeds, and less genotypic heterogeneity within seed pools for territorial species in both tropical and temperate regions, but again data are sparse. In all cases, the distribution of resources will be an important factor that interacts with mating system to shape foraging ecology and determine seed dispersal outcomes. Also in all cases, but particularly in the tropics, more data are needed to advance our understanding of the relationship between mating system and seed dispersal outcomes.

Our understanding of the relationships between mating system and seed dispersal outcomes is still in its nascent stages. A better understanding of these relationships would illuminate the interaction between sexual selection and mutualistic ecological interactions, which have been relatively neglected in this context relative to antagonistic ecological interactions such as predation and parasitism. A more refined understanding of this phenomenon would also enhance conservation efforts. The breakdown of dispersal syndromes is expected to alter forest dynamics (Terborgh *et al.*, 2008), and a better understanding of how the mating system shapes dispersal services would improve the ability to assess and predict consequences of perturbation to these systems, such as extirpation of dispersal agents or fragmentation of habitats (Karubian and Durães, 2009).

Clearly, animal behaviors directly impacting seed dispersal outcomes can vary dramatically between species and even within a species (e.g., between the sexes or across a geographic range), with potentially important ecological and evolutionary implications for plant species and communities. As such, we hope this chapter highlights the need to expand the breadth and depth of studies that explicitly consider the social behavior of animals when investigating seed dispersal dynamics, and motivates additional research in the field, particularly in the tropics.

ACKNOWLEDGMENTS

We thank R. Macedo and G. Machado for inviting us to contribute to this volume and for helpful edits on earlier drafts; many field researchers, but especially D. Cabrera and J. Olivo, for data collection; and collaborators K. Ottewell, T. B. Smith, and V. L. Sork for feedback while developing these ideas. We also thank the people of northwest Ecuador for leaving their forests intact and contributing to the conservation of functioning ecosystems in our study area. Umbrellabird research was supported by the National Science Foundation (OISE-0402137), National Geographic Society, Wildlife Conservation Society, and Disney Worldwide Conservation Fund.

REFERENCES

Andersson, M.D., 1994. Sexual Selection. Princeton University Press, Princeton.

BirdLife International, 2000. Threatened Birds of the World. BirdLife International and Lynx Ediciones, Barcelona.

Cain, M.L., Milligan, B.G., Strand, A.E., 2000. Long-distance seed dispersal in plant populations. Am. J. Bot. 87, 1217–1227.

Carlo, T.A., Morales, J.M., 2008. Inequalities in fruit-removal and seed dispersal: Consequences of bird behaviour, neighbourhood density and landscape aggregation. J. Ecol. 96, 609–618.

Clark, C.J., Poulsen, J.R., Bolker, B.M., Connor, E.F., Parker, V.T., 2005. Comparative seed shadows of bird-, monkey-, and wind-dispersed trees. Ecology 86, 2684–2694.

Connell, J.H., 1971. On the role of natural enemies in preventing competitive exclusion in some marine animals and in rain forest trees. In: den Boer, P.J., Gradwell, G.R. (Eds.), Dynamics of Populations, Centre for Agricultural Publishing and Documentation, Wageningen, pp. 298–313.

Crawford, T., 1984. The estimation of neighborhood parameters for plant-populations. Heredity 52, 273–283.

Darwin, C., 1871. The Descent of Man and Selection in Relation to Sex. John Murray, London.

Davidson, D.W., Morton, S.R., 1981. Competition for dispersal in ant-dispersed plants. Science 213, 1259–1261.

Dennis, A.J., Westcott, D.A., 2006. Reducing complexity when studying seed dispersal at community scales: A functional classification of vertebrate seed dispersers in tropical forests. Oecologia 149, 620–634.

Dennis, A.J., Green, R.J., Schupp, E.W., Westcott, D.A. (Eds.), 2007. Seed Dispersal: Theory and its Application in a Changing World, CABI Publishing, New York.

Emlen, S.T., Oring, L.W., 1977. Ecology, sexual selection, and the evolution of mating systems. Science 197, 215–223.

Fiske, P., Rintamaki, P.T., Karvonen, E., 1998. Mating success in lekking species: A meta-analysis. Behav. Ecol. 9, 328–338.

Foster, R.B., 1982. The seasonal rhythm of fruitfall on Barro Colorado Island. In: Leigh, E.G., Rand, A.S., Windsor, D.M. (Eds.), The Ecology of a Neotropical Forest: Seasonal Rhythms and Long-term Changes, Smithsonian Institution Press, Washington DC, pp. 151–172.

Ganesh, T., Davidar, P., 2001. Dispersal modes of tree species in the wet forests of southern Western Ghats. Curr. Sci. 80, 394–399.

García, C., Jordano, P., Godoy, J.A., 2007. Contemporary pollen and seed dispersal in a *Prunus mahaleb* population: Patterns in distance and direction. Mol. Ecol. 16, 1947–1955.

García, C., Jordano, P., Arroyo, J.M., Godoy, J.A., 2009. Maternal genetic correlations in the seed rain: effects of frugivore activity in heterogeneous landscapes. J. Ecol. 97, 1424–1435.

Godoy, J.A., Jordano, P., 2001. Seed dispersal by animals: Exact identification of source trees with endocarp DNA microsatellites. Mol. Ecol. 10, 2275–2283.

Goulding, M., Smith, N., 2007. Palms: Sentinels for Amazon Conservation. Amazon Conservation Association. Missouri Botanical Garden Press, Saint Louis.

Grivet, D., Smouse, P.E., Sork, V.L., 2005. A new approach to the study of seed dispersal: A novel approach to an old problem. Mol. Ecol. 14, 3585–3595.

Grivet, D., Robledo-Arnuncio, J.J., Smouse, P.E., Sork, V.L., 2009. Relative contribution of contemporary pollen and seed dispersal to the effective parental size of seedling population of California valley oak (*Quercus lobata*, Née). Mol. Ecol. 18, 3967–3979.

Hamilton, M.B., 1999. Tropical tree gene flow and seed dispersal. Nature 401, 129–130.

Henderson, A., Galeano, G., Bernal, R., 1995. Field Guide to the Palms of Americas. Princeton University Press, Princeton.

Höglund, J., Alatalo, R.V., 1995. Leks. Princeton University Press, Princeton.

Holland, R.A., Wikelski, M., Kümmeth, F., Bosque, C., 2009. The secret life of oilbirds: New insights into the movement ecology of a unique avian frugivore. PLoS One 4, e8264.

Hooge, P.N., Eichenlaub, B., 2001. Animal Movement Extension for Arcview v. 2.04. Alaska Science Center – Biological Sciences Office. US Geological Survey, Anchorage.

Howe, H.F., 1989. Scatter- and clump-dispersal and seedling demography: Hypothesis and implications. Oecologia 79, 417–426.

Howe, H.F., Smallwood, J., 1982. Ecology of seed dispersal. Annu. Rev. Ecol. Syst. 13, 201–228.

IUCN, IUCN Red list of threatened species. Version 2011.1. Available at www.iucnredlist.org2011 (downloaded 6 October 2011).

Jansen, P.A., Bongers, F., van der Meer, P.J., 2008. Is farther seed dispersal better? Spatial patterns of offspring mortality in three rainforest tree species with different dispersal abilities. Ecography 31, 43–52.

Janzen, D.H., 1970. Herbivores and the number of tree species in tropical forests. Am. Nat. 104, 501–528.

Jordano, P., Godoy, J.A., 2002. Frugivore-generated seed shadows: A landscape view of demographic and genetic effects. In: Levey, D.J., Silva, W.R., Galetti, M. (Eds.), Seed Dispersal and Frugivory: Ecology, Evolution and Conservation. CABI Publishing, New York, pp. 305–321.

Jordano, P., García, C., Godoy, J.A., García-Castaño, J.L., 2007. Differential contribution of frugivores to complex seed dispersal patterns. Proc. Natl. Acad. Sci. U. S. A. 104, 3278–3282.

Karubian, J., Durães, R., 2009. Effects of seed disperser social behavior on patterns of seed movement and deposition. Oecologia Brasiliensis 13, 45–57.

Karubian, J., Casteneda, G., Freile, J.F., Santander, T., Smith, T.B., 2003. Breeding biology and nesting behavior of the long-wattled umbrellabird *Cephalopterus penduliger* in northwestern Ecuador. Bird Conserv. Int. 13, 351–360.

Karubian, J., Carrasco, L., Cabrera, D., Cook, A., Olivo, J., 2007. Nesting biology of the banded-ground cuckoo. Wilson J. Ornithol. 119, 222–228.

Karubian, J., Sork, V.L., Roorda, T., Durães, R., Smith, T.B., 2010. Destination-based seed dispersal homogenizes genetic structure of a tropical palm. Mol. Ecol. 19, 1745–1753.

Karubian, J., Browne, L., Bosque, C., Carlo, T., Galetti, M., Loiselle, B.A., Blake, J.G., Cabrera, D., Durães, R., Labecca, F.M., Holbrook, K.M., Holland, R., Jetz, W., Kummeth, F., Olivo, J., Ottewell, K., Papadakis, G., Rivas, G., Steiger, S., Voirin, B., Wikelski, M., 2012a. Seed dispersal by neotropical birds: Emerging patterns and underlying processes. Ornitol. Neotr. 23, 9–24.

Karubian, J., Durães, R., Storey, J.L., Smith, T.B., 2012b. Mating behavior drives seed dispersal by the long-wattled umbrellabird Cephalopterus penduliger. Biotropica 44, 689–698.

Kirkpatrick, M., 1986. Sexual selection and cycling parasites: A simulation study of Hamilton's hypothesis. J. Theoret. Biol. 119, 263–271.

Knell, R.J., 1999. Sexually transmitted disease and parasite-mediated sexual selection. Evolution 53, 957–961.

Koenig, W.D., Van Vuren, D., Hooge, P.N., 1996. Detectability, philopatry, and the distribution of dispersal distances in vertebrates. Trends Ecol. Evol. 11, 514–517.

Krijger, C.L., Opdam, M., Théry, M., Bongers, F., 1997. Courtship behavior of manakins and seed bank composition in a French Guiana rain forest. J. Trop. Ecol. 13, 631–636.

Kwit, C., Levey, D.J., Greenberg, C.H., 2004. Contagious seed dispersal beneath heterospecific fruiting trees and its consequences. Oikos 107, 303–308.

Laurance, W.F., Nascimento, H.E.M., Laurance, S.G., Andrade, A., Ribeiro, J.E.L.S., Giraldo, J.P., Lovejoy, T.E., Condit, R., Chave, J., Harms, K.E., D'Angelo, S., 2006. Rapid decay of tree-community composition in Amazonian forest fragments. Proc. Natl. Acad. Sci. U. S. A. 103, 19010–19014.

Levey, D.J., Silva, W.R., Galetti, M. (Eds.), 2002. Seed Dispersal and Frugivory: Ecology, Evolution, and Conservation. CABI Publishing, New York.

Levin, D.A., 1975. Pest pressure and recombination systems in plants. Am. Nat. 109, 437–457.

Mackenzie, A., Reynolds, J.D., Brown, V.J., Sutherland, W.J., 1995. Variation in male mating success on leks. Am. Nat. 145, 633–652.

Martin, T.E., Martin, P.R., Olson, C.R., Heidinger, B.J., Fontaine, J.J., 2000. Parental care and clutch size in North and South American birds. Science 287, 1482–1485.

Mohr, C.O., 1947. Table of equivalent populations of North American small mammals. Am. Midl. Nat. 37, 223–249.

Murray, K.G., 1988. Avian seed dispersal of three neotropical gap-dependent plants. Ecol. Monogr. 58, 271–298.

Nathan, R., 2006. Long-distance dispersal of plants. Science 313, 786–788.

Nathan, R., Muller-Landau, H.C., 2000. Spatial patterns of seed dispersal, their determinants and consequences for recruitment. Trends Ecol. Evol. 15, 278–285.

Nathan, R., Schurr, F.M., Spiegel, O., Steinitz, O., Trakhtenbrot, A., Tsoar, A., 2008. Mechanisms of long-distance seed dispersal. Trends Ecol. Evol. 23, 638–647.

Oring, L.W., 1982. Avian mating systems. In: Farner, D.S., King, J.R., Parkes, K.C. (Eds.), Avian Biology, Vol. 6. Academic Press, Orlando, pp. 1–92.

Ottewell, K., Gray, E., Castillo, F., Karubian, J., 2012. The pollen dispersal kernel and mating system of an insect-pollinated tropical palm, Oenocarpus bataua. Heredity 109 (6), 332–339.

Pakeman, R.J., 2001. Plant migration rates and seed dispersal mechanisms. J. Biogeogr. 28, 795–800.

Payne, R.B., 1984. Sexual Selection, Lek and Arena Behavior, and Sexual Size Dimorphism in Birds. Ornithological Monographs 33. American Ornithologists's Union, Washington DC.

Reid, N., 1989. Dispersal of mistletoes by honeyeaters and flowerpeckers: Components of seed dispersal quality. Ecology 70, 137–145.

Ryder, T.B., Blake, J.G., Loiselle, B.A., 2006. A test of the environmental hotspot hypothesis for lek placement in three species of manakins (Pipridae) in Ecuador. Auk 123, 247–258.

Schaefer, H.M., Schaefer, V., Levey, D.J., 2004. How plant–animal interactions signal new insights in communication. Trends Ecol. Evol. 19, 577–584.

Schupp, E.W., 1993. Quantity, quality and the effectiveness of seed dispersal by animals. Vegetatio, 107/108. 15–29.

Schupp, E.W., Milleron, T., Russo, S.E., 2002. Dissemination limitation and the origin and main-tenance of species-rich tropical forests. In: Levey, D.J., Silva, W.R., Galetti, M. (Eds.), Seed Dispersal and Frugivory: Ecology, Evolution and Conservation. CABI Publishing, New York, pp. 19–33.

Schupp, E.W., Jordano, P., Gómez, J.M., 2010. Seed dispersal effectiveness revisited: A Conceptual review. New Phytol. 188, 333–353.

Scofield, D.G., Sork, V.L., Smouse, P.E., 2010. Influence of acorn woodpecker social behaviour on transport of coastal live oak (Quercus agrifolia) acorns in a southern California oak savanna. J. Ecol. 98, 561–572.

Scofield, D.G., Alfaro, V.R., Sork, V.L., Grivet, D., Martinez, E., Papp, J., Pluess, A., Koenig, W.D., Smouse, P.E., 2011. Foraging patterns of acorn woodpeckers (Melanerpes formicivorus) on valley oak (Quercus lobata Née) in two California oak savannah-woodlands. Oecologia 166, 187–196.

Skutch, A.F., 1949. Do tropical birds rear as many young as they can nourish? Ibis 91, 430–455.

Snow, D.W., 1982. The Cotingas: Bellbirds, Umbrellabirds and Other Species. Cornell University Press, Ithaca.

Snow, D.W., 2004. Family Cotingidae (Cotingas). In: del Hoyo, J., Elliott, A., Christie, D.A. (Eds.), Handbook of the Birds of the World, Vol. 9, Cotingas to Pipits and Wagtails, Lynx Editions, Barcelona, pp. 32–109.

Sork, V.L., Smouse, P.E., 2006. Genetic analysis of landscape connectivity in tree populations. Land. Ecol. 21, 821–836.

Tello, J.G., 2001. Lekking behavior of the round-tailed manakin. Condor 103, 298–321.

Terborgh, J., 1990. Seed and fruit dispersal – commentary. In: Bawa, K.S., Handley, M. (Eds.), Reproductive Ecology of Tropical Forest Plants. The Parthenon Publishing Group, Paris, pp. 181–190.

Terborgh, J., Nuñez-Iturri, G., Pitman, N.C.A., Valverde, F.H.C., Alvarez, P., Swamy, V., Pringle, E.G., Paine, C.E.T., 2008. Tree recruitment in an empty forest. Ecology 89, 1757–1768.

Théry, M., 1992. The evolution of leks through female choice: Differential clustering and space utilization in six sympatric manakins. Behav. Ecol. Sociobiol. 30, 227–237.

Théry, M., Larpin, D., 1993. Seed dispersal and vegetation dynamics at a cock-of-the-rock's lek in the tropical forest of French Guiana. J. Trop. Ecol. 91, 109–116.

Tori, W.P., Durães, R., Ryder, T.B., Anciães, M., Karubian, J., Macedo, R.H., Uy, J.A.C., Parker, P.G., Smith, T.B., Stein, A.C., Webster, M.S., Blake, J.G., Loiselle, B.A., 2008. Advances in sexual selection theory: Insights from tropical avifauna. Ornitol. Neotrop. 19 (Suppl.), 151–163.

Vander Wall, S.B., Beck, M.J., 2012. A comparison of frugivory and scatter-hoarding seed-dispersal syndromes. Bot. Rev. 78, 10–31.

Wang, B.C., Smith, T.B., 2002. Closing the seed dispersal loop. Trends Ecol. Evol. 17, 379–385.

Wenny, D.G., 2001. Advantages of seed dispersal: A re-evaluation of directed dispersal. Evol. Ecol. Res. 3, 51–74.

Wenny, D.G., Levey, D.J., 1998. Directed seed dispersal by bellbirds in a tropical cloud forest. Proc. Natl. Acad. Sci. U. S. A. 95, 6204–6207.

Westcott, D.A., Graham, D.L., 2000. Patterns of movement and seed dispersal of a tropical frugivore. Oecologia 122, 249–257.

Westcott, D.A., Smith, J.N.M., 1994. Behavior and social organization during the breeding season in *Mionectes oleagineus*, a lekking flycatcher. Condor 96, 672–683.

Westcott, D.A., Bentrupperbaumer, J., Bradford, M.G., McKeown, A., 2005. Incorporating patterns of disperser behaviour into models of seed dispersal and its effects on estimated dispersal curves. Oecologia 146, 57–67.

Wiley, R.H., 1991. Lekking in birds and mammals: behavioral and evolutionary issues. Adv. Stud. Behav. 20, 201–291.

Zuk, M., 1992. The role of parasites in sexual selection – current evidence and future directions. Adv. Stud. Behav. 21, 39–68.

Chapter 15

Flights of Fancy: Mating Behavior, Displays and Ornamentation in a Neotropical Bird

Lilian T. Manica,[1] Jeffrey Podos,[2] Jefferson Graves[3] and Regina H. Macedo[1]

[1]*Departamento de Zoologia, Universidade de Brasília, Brasília, Brazil,* [2]*Department of Biology, University of Massachusetts, Amherst, Massachussetts, USA,* [3]*School of Biology, University of St Andrews, St Andrews, Fife, UK*

INTRODUCTION

Secondary sexual characters such as ornaments and elaborate displays occur in numerous species, and are more common in males than in females. In *The Descent of Man and Selection in Relation to Sex* (1871), Darwin proposed that these traits evolved to increase the reproductive success of individuals bearing such traits through a process he named "sexual selection". Such a process, he argued, would explain both the persistence and the exaggerated nature of elaborate displays and ornamentation, which appeared to be intrinsically costly and thus subject to opposing pressures from natural selection (Darwin, 1871; Andersson, 1994). Presently, sexual selection is one of the most prevalent topics of study in the field of animal behavior and evolutionary biology. Yet, despite the many studies that support Darwin's initial theory (Andersson, 1994), the genetic, behavioral, and physiological mechanisms that underlie and sustain sexual selection are still unclear and occasionally controversial (Kokko *et al.*, 2003, 2006).

By definition, males and females differ in the quality and quantity of the gametes they produce: male gametes are small and numerous, while female gametes are larger but fewer in quantity (Trivers, 1972). This phenomenon, commonly known as "anisogamy", results in different potential reproductive rates for males and females, with more males than females typically being sexually active at any moment in time – that is, the operational sex ratio (OSR) in natural populations tends to be male-biased (Emlen and Oring, 1977; Clutton-Brock and Parker, 1992). Consequently, males compete intensively for

Sexual Selection. http://dx.doi.org/10.1016/B978-0-12-416028-6.00015-3

access to females, and females tend to be much more selective in their choice of sexual partners (Andersson, 1994). Despite these basic premises, the relations between parental investment, potential reproductive rates, and OSR tend to be complex and to vary greatly from species to species. In some cases the standard competitive versus choosy sex roles can be inverted (e.g., Kentish plover, Székely and Cuthill, 2000; jacanas, Emlen and Wrege, 2004; seahorses, Kvarnemo *et al.*, 2007), and factors such as primary sex ratio and sex-specific mortality rates can override the standard effects of anisogamy (Clutton-Brock and Parker, 1992; Clutton-Brock, 2007).

Competition for sexual partners (i.e., intrasexual selection typically by males) can take several forms, including scramble competition where individuals attempt to find a mate before rivals do, maintenance of reproductive capacity for longer periods than competitors, direct contests between competitors either over a resource the female needs or over the females, coercion of females, sperm competition, and infanticide (Andersson and Iwasa, 1996). Such intense competition results in selection that favors the evolution of attributes that enhance male persistence or vigor. Two main examples of such attributes include large body size, and weaponry such as deer antlers and beetle horns (Andersson, 1994). Evidence for female choice for partners (or intersexual selection) also abounds in the literature; female preference in many taxa has favored the evolution of elaborate male displays and ornaments (Andersson, 1994). The benefits of female mate choice for costly traits are still a topic of debate (Kokko *et al.*, 2003; Andersson and Simmons, 2006; Kotiaho and Puurtinen, 2007). Current hypotheses focus on direct adaptive advantages to females (Price *et al.*, 1993; Møller and Jennions, 2001), on indirect advantages in genetic benefits to offspring (Zahavi, 1975; Eshel *et al.*, 2000; Kokko *et al.*, 2002), and on increased genetic compatibility between mates (Zeh and Zeh, 1996).

Female mate choice for males with extravagant traits can confer direct benefits if these traits indicate a male's health status, ability to provision material benefits, or capacity to help with parental care (Kirkpatrick and Ryan, 1991; Andersson, 1994). Alternatively, female preference may evolve through the positive effects generated in terms of her offspring's genetic quality. The choice for more specific types of male can favor offspring inheritance of genes that code for robust body condition and health ("good genes") or attractiveness ("sexy sons") which, respectively, increase offspring survival or reproductive success (Fisher-Zahavi's hypothesis; Zahavi, 1975; Eshel *et al.*, 2000; Kokko *et al.*, 2002). Male signals perceived and favored by females in non-sexual contexts may also promote the selection of sexual traits (sensory drive hypothesis, Endler and Basolo, 1998; Ryan, 1998). According to this hypothesis, female preference for male traits evolves primarily under natural selection and in association with activities such as foraging or detecting predators (Enquist and Arak, 1993; Ryan, 1998), and male traits subsequently evolve as a response to the behavior of females. For example, male water mites *Neumania papillator* court females via vibration of their first and second pairs of legs mimicking the vibrations produced by copepod prey. This apparently stimulates the predatory perceptual systems of females, who approach and clutch the male, allowing for spermatophore transfer by the male (Proctor, 1991).

Sexual selection through female choice may impose costs on females, despite the eventual occurrence of direct or indirect benefits. Sexual conflicts can arise due to different optimal expression of a trait between the sexes, or conflicts over the outcome of male–female interactions (sexually antagonistic coevolution, reviewed in Arnqvist and Nilsson, 2000; Chapman *et al.*, 2003). For instance, females could benefit through enhanced breeding success of sons with extravagant attributes, which presumably increase their attractiveness, at the expense of reduced fitness of daughters if such attributes impose production and maintenance costs for them (Kokko *et al.*, 2003).

SEXUAL SELECTION AND BIRD MATING SYSTEMS

For decades, almost 90% of bird species were considered monogamous (Lack, 1968). Thus, the behavioral consequences of anisogamy (competitive males and selective females) were expected to be mild, since the social bond was seen as a constraint to multiple matings by the most competitive sex (Emlen and Oring, 1977). However, advancing molecular techniques that allow for parent-age assignments have revealed that genetic polygamy occurs in many socially monogamous birds (review in Griffith *et al.*, 2002). Of the socially monoga-mous bird species studied, almost 86% show extra-pair fertilizations, averaging 11% of nestlings and 19% of broods (Griffith *et al.*, 2002). This discovery led to a revolution in the understanding of bird mating systems and, consequently, in studies of the evolutionary mechanisms of sexual selection in species with social monogamy.

Genetic studies have revealed a significant inter- and intraspecific variation in frequencies of extra-pair fertilizations, and several hypotheses have emerged to explain this pattern (Petrie and Kempenaers, 1998; Griffith *et al.*, 2002). Apparently, ecological, genetic and social factors have different impacts on extra-pair copulation behavior at different taxonomic levels (Griffith *et al.*, 2002). Life-history traits, such as mortality rate and the form of parental care, tend to explain variation between families and orders (Bennet and Owens, 2002), while eco-logical factors, such as breeding density (Westneat and Sherman, 1997; Møller and Ninni, 1998), breeding synchrony (Stutchbury and Morton, 1995), and population genetic variability (Petrie and Kempenaers, 1998), tend to explain differences between phylogenetically related species, and among populations or individuals within species. Notably, a single robust explanation for the vari-able patterns of multiple mating found in socially monogamous birds is lacking (Griffith *et al.*, 2002; Arnqvist and Kirkpatrick, 2005; Akçay and Roughgarden, 2007; Griffith, 2007).

Variation in extra-pair fertilization rates among species and populations may also reflect different individual costs and benefits (Petrie and Kempenaers, 1998;

Westneat and Stewart, 2003). For example, females may gain fitness benefits by copulating with males outside the social bond (extra-pair copulations). Such benefits may include females increasing their foraging territory or gaining assistance from extra-pair males to defend the nest or care for offspring (Burke *et al.*, 1989), increasing the attractiveness or quality of offspring by their inheritance of good or attractive genes (Fisher–Zahavi process, Eshel *et al.*, 2000), and increasing the chance of finding more genetically compatible males than their social partners (Zeh and Zeh, 1996; Neff and Pitcher, 2005). Multiple mating may also generate negative consequences for females, such as a higher risk of acquiring ectoparasites and sexually transmitted diseases (Sheldon, 1993), or desertion/retaliation by social partners (Cezilly and Nager, 1995; reviewed in Westneat and Stewart, 2003; Arnqvist and Kirkpatrick, 2005). Nevertheless, females in numerous species actively seek extra-pair copulations (Kempenaers *et al.*, 1992; Graves *et al.*, 1993; Gray, 1996; Double and Cockburn, 2000) or accept mating attempts passively from multiple males (Akçay *et al.*, 2011), suggesting that in these species the benefits to females of polyandry outweigh the costs. This view is supported by the fact that female birds, in general, can control whether or not copulations result in sperm transfer (postcopulatory control of paternity, Birkhead and Møller, 1993; Birkhead, 1998; Petrie and Kempenaers, 1998), which also indicates that cryptic female choice or sperm competition may play important roles in avian reproduction. Polyandry in socially monogamous systems can be an alternative reproductive strategy relative to a previous poor choice when mating options are constrained (Gowaty, 1996). In monogamous reproductive systems female choice does not necessarily reflect female preference, because the availability of preferred potential mates is expected to decline rapidly over the breeding period as high-quality males are more likely to form social pairs first.

EXTRA-PAIR PATERNITY IN TROPICAL BIRDS

Most of what we know about evolutionary mechanisms driving polygamy in socially monogamous species is based on studies conducted in the temperate region; by contrast, little is known about social systems in birds from the tropics (Stutchbury and Morton, 2001; Macedo *et al.*, 2008; Tori *et al.*, 2008). It is generally presumed that tropical species will show lower levels of extra-pair fertilizations as compared to temperate region species, mainly due to the differences in climatic seasonality and life-history characteristics of species in different regions (Macedo *et al.*, 2008). However, it is still far too early for such generalizations, given the scarcity of empirical studies on tropical species coupled with the extraordinary biological diversity and variable climatic conditions that characterize the tropics (Macedo *et al.*, 2008). There are many tropical species that inhabit highly seasonal regions, such as savanna and high-altitude habitats, in which precipitation and temperature may vary largely across the year and constrain the period suitable for breeding. For example, in the Brazilian

Cerrado biome, a typical tropical savanna, the breeding season of several passerines occurs within a short time comprising only 3–4 months (Marini *et al.*, 2012). Similarly, intratropical migrants breed in the Cerrado area during the rainy season, when resources are probably more abundant and conditions more favorable for rearing the offspring (e.g., *Volatinia jacarina*, Almeida and Macedo, 2001; Carvalho *et al.*, 2007; *Elaenia chiriquensis*, Paiva and Marini, 2013).

It is thus clear that conditions in some regions within the tropics can resemble the highly seasonal patterns seen in temperate regions. Yet many predictions concerning mating systems and sexual selection in tropical species have been generated based on the assumption that such species inhabit aseasonal habitats, thus underestimating the potential diversity of mating patterns and behaviors. One example includes the assumption that breeding synchrony in songbirds is rare in the tropics compared with that for the temperate region, resulting in reduced opportunity for simultaneous comparison of multiple sexual partners by females (leading to the prediction of lower extra-pair fertilization rates for the tropical species; Stutchbury and Morton, 1995). Similarly, the general assumption that tropical birds breed in lower densities than temperate region species also leads to the prediction of lower rates of extra-pair fertilizations because of fewer male–female interactions (Møller and Birkhead, 1993). Yet, as pointed out above, seasonality and short breeding seasons are common in many tropical regions, and species in these regions may also show high extra-pair fertilization rates (Macedo *et al.*, 2008). Studies focusing on socially monogamous tropical species are urgently needed to better understand the relationship between ecological and social factors influenced by sexual selection.

In the remainder of this chapter, we review published data on breeding, mating systems, and communication in the blue-black grassquit (*Volatinia jacarina*), a common neotropical passerine, and explore possible mechanisms by which sexual characteristics evolve and are maintained in this species. The blue-black grassquit is an excellent model for investigating predictions of evolutionary models of sexual selection for several reasons. First, this species shows a striking sexual dimorphism: males have a blue-black, iridescent nuptial plumage and engage in aerial courtship displays, whereas females are brownish and do not perform any apparent mating display (Fig.15.1; Sick, 2001). These characteristics suggest a history of male–male competition and female choice selecting the most extravagant attributes, favoring their transmission across generations. Second, the blue-black grassquit is socially monogamous, but extra-pair fertilizations are common (Almeida and Macedo, 2001; Carvalho *et al.*, 2006). The occurrence of a cryptic polyandrous sexual system indicates that female choice for multiple partners may also play an important role in sexual selection in this species. Finally, in specific regions of the Neotropics, particularly in central Brazil, the blue-black grassquit breeds within a constrained period of approximately 5 months (Almeida and Macedo, 2001; Carvalho *et al.*, 2007). Thus, this

FIGURE 15.1 (A) Female (left) and male (right) blue-black grassquits; (B) Image composition sequentially depicting a single male blue-black grassquit leaping courtship display. See color plate at the back of the book.

species contradicts what would be expected for a typical tropical songbird and represents a good study model to explore predictions of extra-pair fertilizations in a seasonal tropical region.

THE BLUE-BLACK GRASSQUIT

The blue-black grassquit is a seedeater in the family Emberizidae that inhabits natural or altered grasslands from Mexico to northern Argentina, Chile, and Brazil (Sick, 2001). From October to April blue-black grassquits migrate to central Brazil for breeding, although their migration routes are unknown. During this period in central Brazil, where our study area is located, males establish relatively small territories (13–72 m²) distributed within aggregations, where males exhibit multimodal courtship displays that vary in complexity (Almeida and Macedo, 2001; Sick, 2001). The "complete display" involves motor, ornamental, and acoustic signals: it starts with a vertical flight, similar to a leap, with a forward and/or reverse rotation of the body at the peak of the leap. During the leap, the males' white wing patches become visible. The motor display is also coupled with a short song produced during the descending part of the leap and usually continuing briefly after the male lands on the perch (Fig. 15.1B). "Incomplete displays" are comprised only of songs, structurally similar to those produced during complete displays, but conducted while the male remains perched. Both types of displays are produced repeatedly during continuous bouts, with brief pauses of a few seconds or intercalating complete and incomplete displays. We have recorded bouts of complete and incomplete displays from dozens of males. Bouts can be very brief, lasting just a minute or two, but sometimes they continue much longer, lasting up to 27 minutes (370 leaps) and 33 minutes (169 songs), respectively (L. Manica and R. Macedo, unpublished data). During the breeding season males also molt into a nuptial blue-black plumage, which enhances male conspicuousness during leap flights and contrasts sharply with their white underwing patches (Doucet, 2002; Maia et al., 2009).

Structural Plumage

The nuptial plumage of male blue-black grassquits is iridescent and reflects mostly in blue, violet, and ultraviolet wavelengths (Doucet, 2002; Maia and Macedo, 2010). This plumage coloration results from structural organization and refraction properties of feather nanostructures (Maia et al., 2009). More specifically, feather barbules are composed of a thin keratin layer positioned over a single layer of melanin granules, as revealed by spectrometric measurements, transmission electron microscopy, and thin-film modeling (Maia et al., 2009). Individual males show considerable variation in plumage spectral properties and molting patterns. While some individuals maintain a brownish coloration during the breeding season, others display a more ultraviolet or bluer plumage (Doucet, 2002; Maia and Macedo, 2010). Similarly, color saturation can vary widely, as a function of the speed of nuptial plumage acquisition, with males molting earlier or faster showing more saturated and ultraviolet-shifted color plumage (Maia and Macedo, 2010). Distinctions in plumage among individuals may have important impacts on female perception and could elicit different responses during mate choice or in competitive contexts among males, especially if this trait provides a reliable index of signalers' physiological, behavioral or genetic quality. While the condition-dependency of melanin plumage coloration has been widely debated (Jawor and Breitwisch, 2003; Griffith et al., 2006), some studies have shown support for this relationship (Veiga and Puerta, 1996; Doucet and Montgomerie, 2003; Hill et al., 2005).

For blue-black grassquits, the relation between plumage color and male condition was tested by assessing correlation between plumage attributes and coccidian parasite infestation estimated from fecal samples (Costa and Macedo, 2005). Results showed that the percent coverage of blue-black feathers was negatively correlated with the coccidian oocyst count. Similarly, a study on a wild population in Mexico showed that brightness of wing covert feathers was negatively correlated with body condition, measured by feather growth rate (Doucet, 2002). Both studies provide evidence that the structural plumage of male blue-black grassquits is a good indicator of body quality, since these two measurements (presence of endoparasites and feather growth rates) are closely linked with individual nutritional condition in birds (Hadley, 1917; Grubb, 1989).

Social context can also influence the expression of structural plumage, probably mediated by variations in hormonal levels such as testosterone (reviewed in McGraw, 2008). Two experiments with captive blue-black grassquits have identified associations between intra- and intersexual interactions and plumage ornamentation. In the first experiment, males were submitted to agonistic encounters over a limited food resource and their plumage condition was evaluated (Santos *et al.*, 2009). The number of aggressive acts and the dispute outcome (wins versus defeats) were registered during trials, and a dominance score was calculated for each male based on the overall outcome. Results showed that spectral properties of the plumage were not related to dominance status (winner or loser), suggesting this trait was not important in predicting fighting ability. However, males losing disputes were more aggressive when their opponents had a higher percentage of nuptial plumage coverage, indicating that darker males are probably seen as a threat in competitions over resources. In the second experiment, spectral properties and molting speed of the nuptial plumage were compared among males in different social environments: (1) paired with a female, (2) in mixed-sex groups, and (3) in all-male groups (Maia *et al.*, 2012). Over 9 months, males in mixed-sex and all-male groups acquired peak nuptial plumage coverage earlier, and maintained the dark plumage for a longer period of time. Males in mixed-sex and all-male groups also developed more violet–UV-shifted plumage coloration, whereas males in the paired treatment were bluer. These patterns were also reflected in differences recorded in levels of testosterone among groups; males interacting with other males had higher testosterone peaks that were achieved faster relative to males paired with females only (Lacava *et al.*, 2011). Together, these studies support the conclusion that plumage and social interaction (at least at the intrasexual level) are correlated, suggesting that nuptial plumage serves social functions and thus evolves via sexual selection.

Another intriguing characteristic of structural plumage in male blue-black grassquits is their white underwing patches, which are displayed during the leap flight in flashes in synchrony with repeated wing beats. Interestingly, to date no study has found evidence that this achromatic plumage is a quality indicator or is important in competition or mate choice contexts. Different studies tested whether spectral color properties and patch size were related to body condition (e.g., ratio of body mass/tarsus length, hematocrit and plasma protein levels), presence of parasites or molting speed, but found no significant results (Costa and Macedo, 2005; Aguilar *et al.*, 2008; Maia and Macedo, 2010). Nevertheless, we expect that patch size is an important signal given that patches increase in size during the breeding period in captive birds (Maia *et al.*, 2012), although the exact information content of this plumage ornament as it is used in mating or social contexts remains unknown.

Acoustic Signals

Male blue-black grassquits produce three types of acoustic signals in association with their complete displays: snapping sounds resulting from the wing flaps, chipping notes, and a short song that may be executed during leap displays or by itself. The mechanics and ultimate effects of the wing-snapping remain unexplored, although these snaps are often clearly audible and can be verified in sonograms of audio recordings of the display (e.g., Webber, 1985). The song lasts approximately half a second, and is comprised of a single note that can span from 2 to 13 kHz (Fandiño-Mariño and Vielliard, 2004). Despite its short duration, the song may include complex acoustic elements such as blocks of repeated frequency modulations (similar to a vibration), blocks of short-range frequency modulations (~1 kHz), or isolated modulations (Fandiño-Mariño and Vielliard, 2004). Songs of different individuals can vary largely in acoustic elements, number of block repetitions, frequency bandwidth, and duration. However, within-individual stereotypy is typically high, suggesting that males could be distinguished and recognized on the basis of vocal features (Fandiño-Mariño and Vielliard, 2004; Dias et al., unpublished data). According to spectrograph cross-correlation analysis, intra-individual song similarity is approximately 70% while inter-individual similarity is much lower at approximately 24% (Dias et al., unpublished data).

Songs can indicate individual quality in songbirds in general by revealing attributes related to developmental history, condition and the ability to produce challenging signals (Nowicki et al., 1998; Gil and Gahr, 2002; Podos et al., 2009). In the blue-black grassquit, available evidence does not suggest that song similarity (i.e., ability to produce consistent songs across renditions) indicates the quality of the signaler, since similarity values were only weakly associated with body condition (body mass/tarsus length index) and plumage attributes (molting speed and presence of feather parasites; Dias et al., unpublished data). A question currently being assessed is whether song output (i.e., the rate at which males sing) correlates with the resource-holding ability of males (Dias et al., unpublished data).

Leap Flight

As part of their complete displays, males execute leaps repeatedly and uninterruptedly for bouts of several minutes, and these bouts are conducted intermittently throughout the day and across several months. Males are apparently able to sustain complete displays even in harsh climatic conditions. An observational study monitoring 21 displaying males showed that the execution of both complete (leap flights and songs) and incomplete displays (only songs) is not influenced by air temperature and relative air humidity (Sicsú *et al.*, 2013). In addition, the same study showed that males increased their rates of both display types, and especially complete displays, when bathed in direct sunlight. This is consistent with the presumption that direct sunlight may enhance the conspicuousness of visual (plumage) signals.

Available evidence suggests that leaping displays may be good indicators of male quality. Two studies, one correlational and another experimental, showed that parasitized males had lower rates of leaping than non-parasitized males (Costa and Macedo, 2005; Aguilar *et al.*, 2008). Moreover, healthier males treated with a coccidiostatic drug, vermifuge, and insecticide talcum in the experimental manipulation had higher leaps and spent more time leaping when compared with non-treated males (Aguilar *et al.*, 2008). These findings concur with the broadly supported hypothesis that infestation by parasites tends to directly affect traits that are under female evaluation (Hamilton and Zuk, 1982).

Multimodality Aspect of the Leap Display

One of the most prominent features of mating behavior in blue-black grassquits concerns the evolution of displays that involve multiple sensory modalities. Multimodal signals that span acoustic, visual, or chemical modalities have been described for many taxa (e.g. ants, Hölldobler, 1999; spiders, Taylor *et al.*, 2006; birds, Patricelli and Krakauer, 2010). The most favored hypotheses explaining the evolution of multimodal signals are that: (1) multiple modalities reveal distinct messages about the signaler's quality (multiple message hypothesis); or (2) taken together, signals reveal a common aspect of the signaler's quality (redundant signal or "backup signals" hypothesis, Møller and Pomiankowski, 1993; Johnstone, 1996; Candolin, 2003; Hebets and Papaj, 2004). As described here, male blue-black grassquits produce a sexual display that integrates both visual (iridescent plumage, white underwing patches, and leap flight) and acoustic (vocalizations and wing-snapping) components. Studies conducted on this species have generally shown that specific aspects of the plumage and motor traits of the display reflect males' physical condition (e.g., Doucet, 2002; Costa and Macedo, 2005; Aguilar *et al.*, 2008), whereas the information content of song remains unclear (Dias *et al.*, unpublished data). It is thus premature to infer such functions for display parameters of the blue-black grassquit, as few experiments have been conducted to test the function of the song and wing-snapping in a natural social context – i.e., to determine how receivers respond to these signals.

A further question concerning multimodal signals in the blue-black grassquit relates to possible constraints and trade-offs in the synchronous expression of these attributes and the consequences this may generate in signal transmission. Trade-offs among multiple signals or signal attributes have been widely documented (Bertram and Warren, 2005; Cardoso *et al.*, 2007; Ornelas *et al.*, 2009; Lahti *et al.*, 2011), and explanations for observed trade-offs typically rely on restrictions in the simultaneous production or development of energetically costly or mechanically demanding signals. For example, emberizid birds are limited in the production of fast trills and wide frequency bandwidth (Podos, 1997; Ballentine *et al.*, 2004) due to a limitation in coordinating syringeal activity and vocal tract movement (Nowicki *et al.*, 1992; Podos, 1996).

Because blue-black grassquit males synchronize their acoustic and motor displays, and because these can vary widely in the quality of expression, their simultaneous production could be challenging and possibly result in trade-offs that depend upon male capacity or condition. Motor and acoustic display components include production of song, wing-snapping, leap height, and body rotation. In contrast, nuptial plumage is produced immediately prior to breeding, when males are not yet engaged in complete displays, and it thus seems less likely that they would experience trade-offs relative to the motor or acoustic signals. These possibilities are currently under investigation.

Sexual Signals and Breeding System

Ornament complexity is known to influence female choice, as it might indicate superior quality or attractivity of a potential partner (Zahavi, 1975; Hamilton and Zuk, 1982; Andersson, 1994). Specific ornament attributes in blue-black grassquits indicate that male multimodal displays may be important in mate choice. First, as mentioned earlier, these traits usually reflect male condition (e.g., Costa and Macedo, 2005; Aguilar *et al.*, 2008) and might be under female evaluation. Second, the cryptic polyandrous sexual mating system suggests alternative scenarios for female choice in addition to the social pairing context, thus there may be strong selection pressure for the evolution of male ornamental or display traits (Almeida and Macedo, 2001; Carvalho *et al.*, 2006). Moreover, the aggregated pattern of breeding territories is intriguing and strengthens the possibility for multiple effects of sexual selection in this species (Alderton, 1963; Webber, 1985; Almeida and Macedo, 2001). In socially monogamous species with bi-parental care, females may prefer dense aggregations of males to seek extra-pair copulations from those of superior quality ("hidden-lek" hypothesis; Wagner, 1998). Except for the presence of individual breeding territories, this situation is similar to traditional lek systems, wherein males of polygamous species aggregate in arenas to display and attract females for copulations (Höglund and Alatalo, 1995). A hidden-lek scenario could explain the aggregated pattern of territorial clustering in blue-black grassquits and the high rates of extra-pair copulations (Almeida and Macedo, 2001; Dias *et al.*, 2009), analogous to that which has been found for the least flycatcher (*Empidonax minimus*; Tarof *et al.*, 2004).

To date, few studies in the blue-black grassquit have shown overall evidence for female choice. The first study associating male traits and breeding success in the field showed that leap height, complete display rates, and total time spent in complete displays or defending the territory predicted whether males were successfully mated with a female (Carvalho *et al.*, 2006). This study was also the first to register extra-pair fertilizations in the blue-black grassquit, including high rates of extra-pair paternity and a few events of female alternative breeding strategies (intraspecific brood parasitism and quasi-parasitism). In another study, males and females were treated to reduce infestation by coccidian parasites, and female preference for male quality was tested experimentally (see above, and Aguilar *et al.*, 2008). However, females appeared to show no preference for males based on their parasitism condition. Furthermore, female health condition also did not affect their choice regarding the two groups of males.

However, none of these studies considered the influence of male traits in both social and sexual mate choice. Recently, we concluded a study based upon 3 years of field breeding data, assessing the genetic parentage of nestlings using 15 pairs of microsatellite markers. Preliminary data appear to be consistent with previous findings of high extra-pair paternity rate and low intraspecific brood parasitism and quasi-parasitism (Carvalho *et al.*, 2006). Our preliminary analyses also suggest some variation in extra-pair fertilization rates across years, suggesting that environmental fluctuations, such as temperature ranges and precipitation levels, or resource distribution or availability, could influence attractiveness and sexual mating choice (Cornwallis and Uller, 2010; Botero and Rubenstein, 2012). We are currently assessing possible correlations between mating success and display attributes. We predict that male offspring that inherit genes from males with higher leaps should be healthier (for example, by being more resistant to diseases that affect the production of sexual signals, Hamilton and Zuk, 1982), or should express the same phenotype as adults and increase their reproductive success, but such predictions remain to be tested.

CONCLUDING REMARKS

This chapter has reviewed studies addressing the evolution of ornaments in male blue-black grassquit through sexual selection. Several aspects have been extensively investigated over the past decade, including the expression of sexual dimorphism, the elaborate courtship behavior, with males performing motor displays and females invariably being inconspicuous. Overall, studies have suggested that some components of male displays reliably reveal individual quality and possibly are under conspecific evaluation in different social contexts. Characteristics such as the blue-black nuptial plumage and motor display are evidently influenced by male health condition, while the investment in displays also indicates resource-holding ability. However, the function of other characteristics remains unclear, including that of the white underwing patches, which do not appear to be condition dependent yet increase in conspicuousness during the breeding period.

Understanding the process of intersexual selection in the blue-black grassquit will continue to be challenging, because in this species female choice for a social partner does not necessarily reflect her sexual choice. As in other songbird species, female blue-black grassquits engage in extra-pair copulations, therefore any robust estimate of male breeding success must assess the genetic paternity of the offspring. We have recently incorporated this component in studies of the social and sexual mating system in this species, and preliminary analyses suggest that motor and acoustic components of male displays may indeed influence the process of female choice. We consider that benefits of being choosy are probably related to offspring quality, through the inheritance of good or attractive genes from fathers. Future research should test the importance of the nuptial plumage and the conspicuousness of the white underwing patches in determining reproductive success of males, and aim to understand female preference for male traits. We are also interested in investigating the relationship between the spatial and temporal distribution of individuals across territories, and the influence this has on breeding patterns. With such information in hand, we will be able to test predictions of the hidden lek model (Wagner, 1998) and resolve whether extra-pair copulation should produce or facilitate aggregation patterns of territories. Finally, we emphasize the importance of studies on sexual selection in the tropical region, and reinforce that strong theoretical patterns in animal behavior will remain incomplete until they are evaluated in the systems and animals found in this region of immense biodiversity.

REFERENCES

Aguilar, T.M., Maia, R., Santos, E.S.A., Macedo, R.H., 2008. Parasite levels in blue-black grassquits correlate with male displays but not female mate preference. Behav. Ecol. 19, 292–301.

Akçay, E., Roughgarden, J., 2007. Extra-pair paternity in birds: review of the genetic benefits. Evol. Ecol. Res. 9, 855–868.

Akçay, C., Searcy, W.A., Campbell, S.E., Reed, V.A., Templeton, C.N., Hardwick, K.M., Beecher, M.D., 2011. Who initiates extra-pair mating in song sparrows? Behav. Ecol. 23, 44–50.

Alderton, C.C., 1963. The breeding behavior of the blue-black grassquit. Condor 65, 154–162.

Almeida, J.B., Macedo, R.H., 2001. Lek-like mating system of the monogamous blue-black grassquit. Auk 118, 404–411.

Andersson, M., 1994. Sexual Selection. Princeton University Press, Princeton.

Andersson, M., Iwasa, Y., 1996. Sexual selection. Trends Ecol. Evol. 11, 53–58.

Andersson, M., Simmons, L.W., 2006. Sexual selection and mate choice. Trends Ecol. Evol. 21, 296–302.

Arnqvist, G., Kirkpatrick, M., 2005. The evolution of infidelity in socially monogamous passerines: The strength of direct and indirect selection on extra-pair copulation behavior in females. Am. Nat. 165, S26–S37.

Arnqvist, G., Nilsson, T., 2000. The evolution of polyandry: Multiple mating and female fitness in insects. Anim. Behav. 60, 145–164.

Ballentine, B., Hyman, J., Nowicki, S., 2004. Vocal performance influences female response to male bird song: An experimental test. Behav. Ecol. 15, 163–168.

Bennet, P.M., Owens, I.P.F., 2002. Evolutionary Ecology of Birds: Life Histories, Mating Systems, and Extinction. Oxford University Press, Oxford.

Bertram, S.M., Warren, P.S., 2005. Trade-offs in signalling components differ with signalling effort. Anim. Behav. 70, 477–484.

Birkhead, T.R., 1998. Sperm competition in birds: Mechanisms and function. In: Birkhead, T.R., Møller, A.P. (Eds.), Sperm Competition and Sexual Selection. Academic Press, London, UK, pp. 579–622.

Birkhead, T.R., Møller, A.P., 1993. Female control of paternity. Trends Ecol. Evol. 8, 100–104.

Botero, C.A., Rubenstein, D.R., 2012. Fluctuating environments, sexual selection and the evolution of flexible mate choice in birds. PloS One 7, e32311.

Burke, T., Davies, N.B., Bruford, M.W., Hatchwell, B.J., 1989. Parental care and mating behaviour of polyandrous dunnocks *Prunella modularis* related to paternity by DNA fingerprinting. Nature 338, 249–251.

Candolin, U., 2003. The use of multiple cues in mate choice. Biol. Rev. 78, 575–595.

Cardoso, G.C., Atwell, J.W., Ketterson, E.D., Price, T.D., 2007. Inferring performance in the songs of dark-eyed juncos (*Junco hyemalis*). Behav. Ecol. 18, 1051–1057.

Carvalho, C.B.V., Macedo, R.H., Graves, J.A., 2006. Breeding strategies of a socially monogamous neotropical passerine: Extra-pair fertilizations, behavior, and morphology. Condor 108, 579–590.

Carvalho, C.B.V., Macedo, R.H.F., Graves, J.A., 2007. Reproduction of blue-black grassquits in central Brazil. Braz. J. Biol. 67, 275–281.

Cezilly, F., Nager, R.G., 1995. Comparative evidence for a positive association between divorce and extra-pair paternity in birds. Proc. R. Soc. B Biol. Sci. 262, 7–12.

Chapman, T., Arnqvist, G., Bangham, J., Rowe, L., 2003. Sexual conflict. Trends Ecol. Evol. 18, 41–47.

Clutton-Brock, T.H., 2007. Sexual selection in males and females. Science 318, 1882–1885.

Clutton-Brock, T., Parker, G.A., 1992. Potential reproductive rates and the operation of sexual selection. Q. Rev. Biol. 67, 437–456.

Cornwallis, C.K., Uller, T., 2010. Towards an evolutionary ecology of sexual traits. Trends Ecol. Evol. 25, 145–152.

Costa, F.J.V., Macedo, R.H., 2005. Coccidian oocyst parasitism in the blue-black grassquit: influence on secondary sex ornaments and body condition. Anim. Behav. 70, 1401–1409.

Darwin, C., 1871. The Descent of Man, and Selection in Relation to Sex. John Murray, London, UK.

Dias, R.I., Kuhlmann, M., Lourenço, L.R., Macedo, R.H., 2009. Territorial clustering in the blue-black grassquit: reproductive strategy in response to habitat and food requirements? Condor 111, 706–714.

Double, M., Cockburn, A., 2000. Pre-dawn infidelity: Females control extra-pair mating in superb fairy-wrens. Proc. R. Soc. B 267, 465–470.

Doucet, S.M., 2002. Structural plumage coloration, male body size, and condition in the blue-black grassquit. Condor 104, 30–38.

Doucet, S.M., Montgomerie, R., 2003. Multiple sexual ornaments in satin bowerbirds: Ultraviolet plumage and bowers signal different aspects of male quality. Behav. Ecol. 14, 503–509.

Emlen, S.T., Oring, L.W., 1977. Ecology, sexual selection, and the evolution of mating systems. Science 197, 214–223.

Emlen, S.T., Wrege, P.H., 2004. Division of labour in parental care behaviour of a sex-role-reversed shorebird, the wattled jacana. Anim. Behav. 68, 847–855.

Endler, J.A., Basolo, A.L., 1998. Sensory ecology, receiver biases and sexual selection. Trends Ecol. Evol. 13, 415–420.

Enquist, M., Arak, A., 1993. Selection of exaggerated male traits by female aesthetic senses. Nature 361, 446–448.

Eshel, I., Volovik, I., Sansone, E., 2000. On Fisher–Zahavi's handicapped sexy son. Evol. Ecol. Res. 2, 509–523.

Fandiño-Mariño, H., Vielliard, J.M.E., 2004. Complex communication signals: the case of the blue-black grassquit *Volatinia jacarina* (Aves, Emberizidae) song. Part I – A structural analysis. Anais da Academia Brasileira de Ciências 76, 325–334.

Gil, D., Gahr, M., 2002. The honesty of bird song: multiple constraints for multiple traits. Trends Ecol. Evol. 17, 133–141.

Gowaty, P.A., 1996. Battles of the sexes and origins of monogamy. In: Black, J.M. (Ed.), Partnerships in Birds. The Study of Monogamy. Oxford University Press, pp. 21–52.

Graves, J., Ortega Ruano, J., Slater, P.J.B., 1993. Extra-pair copulations and paternity in shags – Do females choose better males? Proc. R. Soc. B 253, 3–7.

Gray, E.M., 1996. Female control of offspring paternity in a western population of red-winged blackbirds (*Agelaius phoeniceus*). Behav. Ecol. Sociobiol. 38, 267–278.

Griffith, S.C., 2007. The evolution of infidelity in socially monogamous passerines: Neglected components of direct and indirect selection. Am. Nat. 169, 274–281.

Griffith, S.C., Owens, I.P.F., Thuman, K.A., 2002. Extra-pair paternity in birds: A review of interspecific variation and adaptive function. Mol. Ecol. 11, 2195–2212.

Griffith, S.C., Parker, T.H., Olson, V.A., 2006. Melanin- versus carotenoid-based sexual signals: Is the difference really so black and red? Anim. Behav. 71, 749–763.

Grubb, T.C., 1989. Ptilochronology: Feather growth bars as indicators of nutritional status. Auk 106, 314–320.

Hadley, P.B., 1917. Coccidia in subepithelial infections of the intestines of birds. J. Bacteriol. 2, 73–78.

Hamilton, W.D., Zuk, M., 1982. Heritable true fitness and bright birds: A role for parasites? Science 218, 384–387.

Hebets, E.A., Papaj, D.R., 2004. Complex signal function: Developing a framework of testable hypotheses. Behav. Ecol. Sociobiol. 57, 197–214.

Hill, G.E., Doucet, S.M., Buchholz, R., 2005. The effect of coccidial infection on iridescent plumage coloration in wild turkeys. Anim. Behav. 69, 387–394.

Höglund, J., Alatalo, R.V., 1995. Leks. Princeton University Press, Princeton.

Hölldobler, B., 1999. Multimodal signals in ant communication. J. Comp. Physiol. A. 184, 129–141.

Jawor, J.M., Breitwisch, R., 2003. Melanin ornaments, honesty, and sexual selection. Auk 120, 249–265.

Johnstone, R.A., 1996. Multiple displays in animal communication: "Backup signals" and "multiple messages". Philos. Trans. R. Soc. Lond. B 351, 329–338.

Kempenaers, B., Verheyen, G.R., Van den Broeck M., Burke, T., Van Broeckhoven, C., Dhondt, A.A., 1992. Extra-pair paternity results from female preference for high-quality males in the blue tit. Nature 357, 494–496.

Kirkpatrick, M., Ryan, M.J., 1991. The evolution of mating preferences and the paradox of the lek. Nature 350, 33–38.

Kokko, H., Brooks, R., McNamara, J.M., Houston, A.I., 2002. The sexual selection continuum. Proc. R. Soc. B 269, 1331–1340.

Kokko, H., Brooks, R., Jennions, M.D., Morley, J., 2003. The evolution of mate choice and mating biases. Proc. R. Soc. B 270, 653–664.

Kokko, H., Jennions, M.D., Brooks, R., 2006. Unifying and testing models of sexual selection. Annu. Rev. Ecol. Evol. Syst. 37, 43–66.

Kotiaho, J.S., Puurtinen, M., 2007. Mate choice for indirect genetic benefits: Scrutiny of the current paradigm. Funct. Ecol. 21, 638–644.

Kvarnemo, C., Moore, G.I., Jones, A.G., 2007. Sexually selected females in the monogamous western Australian seahorse. Proc. R. Soc. B. Biol. Sci. 274, 521–525.

Lacava, R.V., Brasileiro, L., Maia, R., Oliveira, R.F., Macedo, R.H., 2011. Social environment affects testosterone level in captive male blue-black grassquits. Horm. Behav. 59, 51–55.

Lack, D., 1968. Ecological Adaptations for Breeding in Birds. Methuen, London, UK.

Lahti, D.C., Moseley, D.L., Podos, J., 2011. A tradeoff between performance and accuracy in bird song learning. Ethology 117, 802–811.

Macedo, R.H., Karubian, J., Webster, M.S., 2008. Extra-pair paternity and sexual selection in socially monogamous birds: Are tropical birds different? Auk 125, 769–777.

Maia, R., Macedo, R.H., 2010. Achieving luster: Prenuptial molt pattern predicts iridescent structural coloration in blue-black grassquits. J. Ornithol. 152, 243–252.

Maia, R., Caetano, J.V.O., Báo, S.N., Macedo, R.H., 2009. Iridescent structural colour production in male blue-black grassquit feather barbules: The role of keratin and melanin. J. R. Soc. Interface 6, S203–S211.

Maia, R., Brasileiro, L., Lacava, R.V., Macedo, R.H., 2012. Social environment affects acquisition and color of structural nuptial plumage in a sexually dimorphic tropical passerine. PloS One 7, e47501.

Marini, M.Â., Borges, F.J.A., Lopes, L.E., Sousa, N.O.M., Gressler, D.T., Santos, L.R., Paiva, L.V., Duca, C., Manica, L.T., Rodrigues, S.S., França, L.F., Costa, P.M., França, L.C., Heming, N.M., Silveira, M.B., Pereira, Z.P., Lobo, Y., Medeiros, R., Roper, J.J., 2012. Breeding biology of birds in the cerrado of central Brazil. Ornitol. Neotrop. 23, 385–405.

McGraw, K.J., 2008. An update on the honesty of melanin-based color signals in birds. Pigment Cell Melanoma Res. 21, 133–138.

Møller, A.P., Birkhead, T.R., 1993. Cuckoldry and sociality: A comparative study of birds. Am. Nat. 142, 118–140.

Møller, A.P., Jennions, M.D., 2001. How important are direct fitness benefits of sexual selection? Naturwissenschaften 88, 401–415.

Møller, A.P., Ninni, P., 1998. Sperm competition and sexual selection: A meta-analysis of paternity studies of birds. Behav. Ecol. Sociobiol. 43, 345–358.

Møller, A.P., Pomiankowski, A., 1993. Why have birds got multiple sexual ornaments? Behav. Ecol. Sociobiol. 32, 167–176.

Neff, B.D., Pitcher, T.E., 2005. Genetic quality and sexual selection: An integrated framework for good genes and compatible genes. Mol. Ecol. 14, 19–38.

Nowicki, S., Westneat, M., Hoese, W., 1992. Birdsong: Motor function and the evolution of communication. Semin. Neurosci. 4, 385–390.

Nowicki, S., Peters, S., Podos, J., 1998. Song learning, early nutrition and sexual selection in songbirds. Am. Zoolog. 38, 179–190.

Ornelas, J.F., González, C., Espinosa de los Monteros, A., 2009. Uncorrelated evolution between vocal and plumage coloration traits in the trogons: A comparative study. J. Evol. Biol. 22, 471–484.

Paiva, L.V., Marini, M.Â., 2013. Timing of migration and breeding of the lesser Elaenia (*Elaenia chiriquensis*) in a neotropical savanna. Wilson J. Ornithol. 125, 116–120.

Patricelli, G.L., Krakauer, A.H., 2010. Tactical allocation of effort among multiple signals in sage grouse: An experiment with a robotic female. Behav. Ecol. 21, 97–106.

Petrie, M., Kempenaers, B., 1998. Extra-pair paternity in birds: Explaining variation between species and populations. Trends Ecol. Evol. 13, 52–58.

Podos, J., 1996. Motor constraints on vocal development in a songbird. Anim. Behav. 51, 1061–1070.

Podos, J., 1997. A performance constraint on the evolution of trilled vocalizations in a songbird family (Passeriformes: Emberizidae). Evolution 51, 537–551.

Podos, J., Lahti, D.C., Moseley, D.L., 2009. Vocal performance and sensorimotor learning in songbirds. In: Naguib, M., Janik, V.M. (Eds.), Advances in the Study of Behavior, Vol. 40. Elsevier Inc., pp. 159–195.

Price, T., Schluter, D., Heckman, N.E., 1993. Sexual selection when the female directly benefits. Biol J. Linn. Soc. 48, 187–211.

Proctor, H.C., 1991. Courtship in the water mite *Neumania papillator*: Males capitalize on females adaptations for predation. Anim. Behav. 42, 589–598.

Ryan, M.J., 1998. Sexual selection, receiver biases, and the evolution of sex differences. Science 281, 1999–2003.

Santos, E.S.A., Maia, R., Macedo, R.H., 2009. Condition-dependent resource value affects male–male competition in the blue-black grassquit. Behav. Ecol. 20, 553–559.

Sheldon, B.C., 1993. Sexually transmitted disease in birds: Occurence and evolutionary significance. Philos. Trans. R. Soc. B 339, 491–497.

Sick, H., 2001. Ornitologia Brasileira. Editora Nova Fronteira, Rio de Janeiro.

Sicsú, P., Manica, L.T., Maia, R., Macedo, R.H., 2013. Here comes the sun: Multimodal displays are associated with sunlight incidence. Behav. Ecol. Sociobiol. http://dx.doi.org/10.1007/s00265-013-1574-x.

Stutchbury, B.J.M., Morton, E.S., 1995. The effect of breeding synchrony on extra-pair mating systems in songbirds. Behaviour 132, 675–690.

Stutchbury, B.J.M., Morton, E.S., 2001. Behavioral Ecology of Tropical Birds. Academic Press, San Diego.

Székely, T., Cuthill, I.C., 2000. Trade-off between mating opportunities and parental care: Brood desertion by female Kentish plovers. Proc. R. Soc. B 267, 2087–2092.

Tarof, S.A., Ratcliffe, L.M., Kasumovic, M.M., Boag, P.T., 2004. Are least flycatcher (*Empidonax minimus*) clusters hidden leks? Behav. Ecol. 16, 207–217.

Taylor, P., Roberts, J., Uetz, G., 2006. Compensation for injury? Modified multi-modal courtship of wolf spiders following autotomy of signalling appendages. Ethology. Ecol. Evol. 18, 79–89.

Tori, W.P., Durães, R., Ryder, T.B., Anciães, M., Karubian, J., Macedo, R.H., Uy, J.A.C., Parker, P.G., Smith, T.B., Stein, A.C., Webster, M.S., Blake, J.G., Loiselle, B.A., 2008. Advances in sexual selection theory: Insights from tropical avifauna. Ornitol. Neotrop. 19, 151–163.

Trivers, R., 1972. Parental investment and sexual selection. In: Campbell, B.G. (Ed.), Sexual Selection and the Descent of Man, 1871–1971, Aldine Press, Chicago, Illinois, pp. 136–179.

Veiga, J.P., Puerta, M., 1996. Nutritional constraints determine the expression of a sexual trait in the house sparrow, *Passer domesticus*. Proc. R. Soc. B 263, 229–234.

Wagner, R.H., 1998. Hidden leks: Sexual selection and the clustering of avian territories. In: Parker, P., Burley, N. (Eds.), Avian Reproductive Tactics: Female and Male Perspectives. Ornithological Monographs No 49, Allen Press, Lawrence, pp. 123–145.

Webber, T., 1985. Songs, displays, and other behavior at a courtship gathering of blue-black grassquits. Condor 87, 543–546.

Westneat, D.F., Sherman, P.W., 1997. Density and extra-pair fertilizations in birds: A comparative analysis. Behav. Ecol. Sociobiol. 41, 205–215.

Westneat, D.F., Stewart, I.R.K., 2003. Extra-pair paternity in birds: Causes, correlates, and conflict. Annu. Rev. Ecol. Evol. Syst 34, 365–396.

Zahavi, A., 1975. Mate selection – A selection for a handicap. J. Theor. Biol. 53, 205–214.

Zeh, J.A., Zeh, D.W., 1996. The evolution of polyandry I: Intragenomic conflict and genetic incompatibility. Proc. R. Soc. B. Biol. Sci. 263, 1711–1717.

Chapter 16

Sexual Selection in Neotropical Bats

Christian C. Voigt

Leibniz Institute for Zoo and Wildlife Research, Berlin, Germany

INTRODUCTION

Bats (order Chiroptera) form one of the largest groups of mammals, only out-numbered, with respect to species numbers, by rodents. In many tropical areas, particularly in Latin America, bats are the most abundant mammals in local assemblages (Bass *et al.*, 2010; Rex *et al.*, 2010). The species-rich Phyllostomi-dae family comprises more than 160 species, and is endemic to the New World (Simmons, 2005). This group is particularly outstanding among bats because of their diverse feeding habits (Dumont *et al.*, 2012). Despite this taxonomic and ecological diversity, neotropical bats are poorly studied, particularly their social behaviors and aspects of sexual selection. The underlying reason for this remains unclear. I hope to convince the reader that neotropical bats are indeed most remarkable animals, and that the study of their ecology, physiology, mor-phology, and social behavior may reveal important new insights into the mecha-nisms of sexual selection in mammals. In the following section, I list some features of Chiroptera that make them particularly interesting study organisms.

BATS: LARGE MAMMALS TRAPPED IN SMALL BODIES

First and foremost, bats have a variety of highly specialized adaptations. They are the only mammals capable of powered flight, and they orient in space and time by active biosonar (Griffin, 1958). These two major evolutionary inno-vations have largely influenced the biology and life history of bats (see, for example, Barclay and Harder, 2003; Voigt and Lewanzik, 2011). Active flap-ping flight enables bats to make use of large areas within their habitat, and to occupy larger than average home ranges compared with similar-sized terres-trial mammals. Yet powered flight involves high metabolic rates to produce lift and thrust (e.g., Voigt and Winter, 1999; von Busse *et al.*, 2013), and therefore any social behavior on the wing – for example, mate-guarding flights – entails high metabolic costs. Also, powered flight imposes a comparatively long period

Sexual Selection. http://dx.doi.org/10.1016/B978-0-12-416028-6.00016-5

of suckling on female bats, because juveniles have to reach almost adult size before being able to take off independently for their first flights (Barclay and Harder, 2003). This requires a large maternal investment, particularly since male bats usually do not contribute to the rearing of young (McCracken and Wilkinson, 2002). This large maternal investment reduces the female potential to produce many offspring. Indeed, female bats usually give birth to only one or two offspring annually (Barclay and Harder, 2003), and females of some tropical species may consequently produce only four to six young during their whole lifetime (Voigt *et al.*, 2008). Contrary to this, male bats may sire a far larger number of offspring. For example, males of some species have long-lived sperm that can be stored for several months in the male's epididymis or in the female reproductive tract (Crichton, 2002; Wilkinson and McCracken, 2003). Since sperm longevity may extend the reproductive period, males with long-lived sperm may have a large potential reproductive rate. Also, some species live in highly gregarious colonies where promiscuous males potentially have access to thousands of females. Accordingly, there is a strong contrast in potential reproductive rate between male and female bats. This difference in sex-specific potential reproductive rates is a key factor promoting sexual selection in general, and female choice in particular (Clutton-Brock and Parker, 1992). Also, bats live on average 3.5 times longer than similar-sized mammals (Wilkinson and South, 2002; Munshi-South and Wilkinson, 2010). Consequently, selection facilitated cognitive capacities that enable bats to recognize social partners over a long period, and to establish long-term social relationships (Kerth *et al.*, 2011). Overall, bats share many life-history traits, such as large home range, low reproductive rate, and longevity, which make them special within the group of small mammals (Barclay and Harder, 2003). Therefore, one may refer to bats as large mammals trapped in small bodies.

With respect to sexual selection, sex-specific differences in potential reproductive rates, and some physiological traits, such as their ability to store sperm, facilitate sexual selection in Chiroptera. Yet to date only a few studies have been conducted on this topic. Most of these studies focused on sperm competition in temperate zone bats (Wilkinson and McCracken, 2003). Some striking studies looked at behaviors that are conspicuous to human observers. For example, lek-mating behavior has been investigated in detail in *Hypsignathus monstrosus*, a tropical flying fox from Africa. Males of this species use specialized vocal tracts for producing loud calls of low frequency to attract females (Bradbury, 1977a). However, in comparison with the vast literature on sexual selection in birds, for example, virtually nothing is known about sexual selection in bats.

SENSORY CONSTRAINTS FOR COMMUNICATING IN THE DARKNESS OF THE NIGHT

The nocturnal niche imposes many constraints on bats with respect to, for example, their activity patterns and sensory modalities used for orientating and communicating. Because of the nightly darkness and low light levels in daytime roosts, bats use mostly acoustic and olfactory cues/signals, unless they roost in well-lit daytime roosts or are active at dawn and dusk (Bradbury, 1977b). The fact that humans have a poorly developed sense of smell and lack the ability to hear in the ultrasonic range may explain, at least partly, why investigators have largely ignored sexually selected traits and signals in bats.

Bats are known to use olfaction for social communication – for example, for recognizing colony members (Bloss *et al.*, 2002; Safi and Kerth, 2003), for discriminating between the sexes (Bouchard, 2001), and for maternal–offspring interaction (Gustin and McCracken, 1987). Nonetheless, the over-all importance of olfaction is under-appreciated in the study of bat behavior (Bloss, 1999). Studies regarding the use of olfaction in neotropical bats are outstanding in a quite positive way, because these studies – some of which will be discussed later in this chapter – have provided important insights into the mechanisms of olfactory communication in bats. Volatiles are ideal candidates for mate choice recognition because they are truly honest signals that are almost impossible to fake (Penn and Potts, 1998; Wyatt, 2003). Small, non-polar volatiles are highly elusive and temporary because they vanish quickly due to diffusion or chemical degradation. Thus, small organic compounds may provide short-term information for a receiver (Wyatt, 2003). Larger and/or oily compounds may persist for longer periods, and are therefore suitable for scent-marking of substrates or individuals (Wyatt, 2003). Given the complexity and seemingly infinite number of organic molecules avail-able for sender individuals, olfaction represents a perfect sensory modality for any interaction influenced by sexual selection (Blaustein, 1981; Brennan and Kendrick, 2006), and would be particularly appropriate for nocturnal animals, such as bats, that may not be able to use vision.

The acoustic domain is particularly relevant for bats, since they use active biosonar for orientating in space and time (with the exception of flying foxes; Griffin, 1958). Bats emit high-frequency calls, mostly not audible to humans, and then listen to the faint reflected echoes to detect obstacles in their flight paths and to find food. Consequently, bats are rich in their vocal repertoire and in their auditory capacities (Fenton, 1985). Echolocation calls of bats are typ-ically above 20 kHz in frequency, and they usually have high amplitudes of more than 100 dB SPL (Griffin, 1958; Surlykke *et al.*, 1993; Holderied and von Helverson, 2003). Yet once a bat has uttered a call, the signal is prone to dis-tortion and strong attenuation (Lawrence and Simmons, 1982), and the higher the frequency of echolocation calls, the stronger their attenuation. Therefore, high-frequency echolocation calls do not carry far, particularly when emitted in or around vegetation. Bats may only be able to communicate via high-frequency calls within hearing distance of a listening bat, and this distance is defined by the echolocation frequency and amplitude, environmental conditions (i.e., background noise, ambient temperature, relative humidity, vegetation cover), and the bat's auditory capacities. Thus, bat social vocalizations that have to be broadcast by a sender over longer distances to a receiver are usually of lower frequencies (Bradbury, 1977b).

EXAMPLES OF SEXUAL SELECTION IN NEOTROPICAL BATS

In the following five sections, I will review case studies of neotropical bats to highlight traits and behaviors that have been shaped by sexual selection. Also, I will discuss what the underlying mechanisms of sexual selection might be in these specific cases. I have restricted my survey to species with a sufficiently large body of literature, and therefore my review is incomplete with respect to species covered. The discussed species are *Noctilio leporinus* and *N. albiventris* (Noctilionidae), *Phyllostomus hastatus* (Phyllostomidae), *Leptonycteris curasoae*, and *Lophostoma silvicolum* (Phyllostomidae), and *Saccopteryx bilineata* (Emballonuridae). Each case study sheds light on a specific aspect of sexual selection in bats, such as female mate choice, including the specific sensory modalities used by a species for mate-choice decisions, or mechanisms of male–male competition.

Garlic Smell in Fish-Eating Bats (Genus *Noctilio*)

Two fish-eating bat species, also called bulldog bats because of their face morphology, are currently known for the lowland regions of the Neotropics. Although *Noctilio leporinus* is larger than *N. albiventris*, they share many morphological and behavioral features. Radio-tracking studies have confirmed that females of both species forage in groups for swarming insects or schools of small fishes (Brooke, 1997; Dechmann *et al.*, 2009a). Group foraging is achieved by eavesdropping on the echolocation calls of successfully hunting conspecifics (Dechmann *et al.*, 2009a). A prerequisite for this behavior is that bats are able to distinguish not only between conspecific and heterospecific calls, but also between calls of group and non-group members; such an ability has recently been confirmed experimentally for *N. albiventris* (Voigt-Heucke *et al.*, 2010). The structure of female groups is also maintained in daytime roosts, where three to eight individuals roost in dense clusters at distances of about 20 cm from other clusters (Brooke, 1997). Females may benefit from clustering in daytime roosts by saving thermoregulatory costs (Roverud and Chappell, 1991). Brooke (1997) argued that females may recognize each other by smelling the so-called subaxial glands (Fig. 16.1A). These olfactory organs are located ventrally at the base of the wings and, according to histological studies, they do not include epithelia with secreting cells (Dunn, 1934). Seemingly, females transfer olfactory compounds from the head region to the axial region, where they are decomposed by bacteria (Studier and Lavoie, 1984). This bacterial decomposition presumably alters the odor of the compounds deposited on the subaxial gland, similar to the bacterial degradation of armpit smell in humans (Rennie *et al.*, 1990, 1991; James *et al.*, 2004). The sweet-smelling compounds of subaxial glands are sticky so that they remain detectable on a contaminated surface for several days, even for microsmatic humans. Group members establish a distinct group-specific odor by rubbing their heads against the axial region of others. Brooke and Decker (1996) found more than 372 lipid compounds in the subaxial gland of females; 52 of these were glycolipids, 186 non-polar lipids, and 124 phospholipids. Individuals differed in their odor profile according to sex and colony membership (Brooke and Decker, 1996).

The social system of *N. leporinus* is best described as a female defense polygyny, with males defending groups of females. Accordingly, males are larger than females and superior during direct physical encounters. Brooke (1997) observed that guarding males changed frequently from year to year, despite female groups remaining relatively stable in composition. The social systems of *N. albiventris* and *N. leporinus* are probably similar, yet we lack observations from daytime roosts of *N. albiventris*. Males of both species also possess subaxial glands which emanate a sweetish odor similar to that of females, yet their specific function in the male sex is unknown. Brooke (1997) observed male *N. leporinus* using their wings to fan odor towards other males, suggesting that the odor of male subaxial glands carries information about individuality and/or dominance. More importantly, in the context of this chapter, males of both species possess the so-called inguinal glands (Fig. 16.1B). This remarkable organ is located around the whitish scrotum, and consists of several tubes that are inverted when the ascended testicles are hidden in the body cavity. The whitish

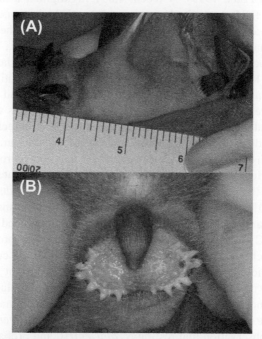

FIGURE 16.1 (A) Subaxial gland of female *Noctilio albiventris* from Panama, located in the reddish armpit region of females. (B) Whitish inguinal gland of male *N. albiventris* from Costa Rica, which contrasts strongly with the blackish penis. See color plate at the back of the book *Photographs courtesy of Christian C. Voigt; the scale indicates inches.*

fingers of the inguinal gland evert when the testicles descend into the scrotum. This process resembles what happens when the fingers of a laboratory glove turn inside-out after a person blows into the glove. The everted inguinal glands emanate a strong garlic odor, probably as a result of bacterial fermentation. In addition to their function as an odor source, inguinal glands may also serve as a visual signal because the whitish scrotum and inguinal gland contrast strongly with the dark penis (Fig. 16.1B). Male *N. leporinus* have been observed rubbing the inguinal gland against the wall of their daytime roosts, possibly to scent-mark their territory and signal the ownership of female groups (Brooke, 1997). It is also likely that male *N. albiventris* rub the oily scents of the inguinal glands on females to scent-mark them as well. This implies that the scents of inguinal glands encode for the individuality and dominance of males. Females could use the scents of male inguinal glands to assess the quality of potential mates, yet whether or not inguinal glands are important for female choice is unknown.

It is widely assumed that immune genes of the highly variable major histocompatibility complex (MHC) alter individual body odor (Penn and Potts, 1998; Kwak *et al.*, 2009), and that females prefer potential mating partners with a dissimilar MHC constitution to generate immune-competent offspring (e.g., Penn and Potts, 1999). A recent study highlighted a highly variable MHC Class II DRB locus in *N. albiventris*, with high sequence variation between alleles and clear signs of positive selection shaping the diversity pattern in functionally important parts of the gene (Schad *et al.*, 2011, 2012a). Interestingly, roosting colonies were not genetically differentiated, yet males showed higher levels of heterozygosity than females, indicating that selection pressure on MHC may differ between the sexes (Schad *et al.*, 2011). Furthermore, we confirmed that ectoparasite infestation rates were linked to specific MHC alleles (Fig. 16.2). This observation is in agreement with the idea that individual immune gene constitution affects ectoparasite susceptibility in *N. albiventris*. Lastly, we demonstrated that non-reproductive males more often carried an allele (Noal-DRB*02) that was associated with higher tick infestation than did reproductively active males or subadults (Fig. 16.2; Schad *et al.*, 2012b). Thus, the MHC allele composition seems to be related to fitness-relevant parameters. However, further studies need to test if the individual MHC allele composition is reflected in the odor profile of males, and if females choose to roost and mate with males that are complementary in the MHC allele constitution. Also, it would be important to understand which olfactory compounds relate to certain MHC alleles to elucidate the link between immunogenetics, olfactory phenotype, and functional mechanisms underlying mate choice decisions in this and other species. A very recent study highlights the observation that resting *N. albiventris*, and probably also other bat species, incur high metabolic costs when echolocating (Dechmann *et al.*, 2013). It would be interesting to see if the rate of echolocation calls emitted by roosting bats is under sexual selection; such a metabolic handicap mechanism

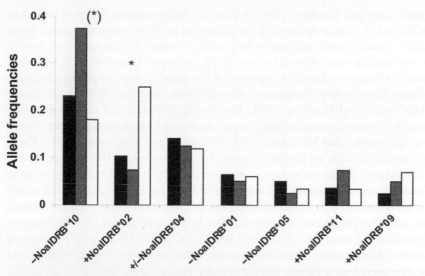

FIGURE 16.2 Relationship between ectoparasite infestation rate and allele frequencies of MHC class II DRB exon 2 (Schad *et al.*, 2012b) in reproductive male (black bars), subadult (gray bars), and non-reproductive male (white bars) *Noctilio albiventris*. Some alleles were associated with an increased (allele indicated with +) and some with a decreased (allele indicated with −) ectoparasite abundance, suggesting that MHC allele constitution defines the degree of individual susceptibility towards ectoparasites, which is an important criterion for mate choice decisions.

has been previously suggested and confirmed for sexually selected bird vocalizations (Zahavi, 1975; Ophir *et al.*, 2010).

Protective Greater Spear-Nosed Bats (*Phyllostomus hastatus*)

The greater spear-nosed bat is the second largest bat species of the Neotropics and a common inhabitant of lowland regions in Central and South America. The diet of *P. hastatus* includes mostly fruits, but they may also feed on pollen, nectar, and insects, and sometimes even on other vertebrates (Santos *et al.*, 2003). *Phyllostomus hastatus* is a sexually dimorphic bat, with males being larger (body mass: 90 g) than females (75 g; Kunz *et al.*, 1998). Individuals typically roost in small and medium-sized colonies, mostly in caves or hollow trees. The social system has been described as a female defense polygyny, with one male tending a cluster of females – a so-called harem (McCracken and Bradbury, 1977, 1981; Kunz *et al.*, 1998). It is important to note that throughout this chapter I do not use the term "harem" in an anthropocentric way, implying that males monopolize copulations; instead, I use it, for reasons of simplicity, to describe stable one-male/multiple-female social units. On the island of Trinidad, harems comprise about 17 females on average. In addition to clusters of females that are defended each by a single male, colonies also harbor groups of bachelor males. In contrast to female clusters, bachelor groups are unstable in composition

(McCracken and Bradbury, 1981). A paternity study revealed that male defenders of harems are able to monopolize the paternities in "their" respective group. Most likely, female clusters originate from yearling females of the same cohort, similar to the establishment of female groups in *N. leporinus*. Analogous to *N. leporinus* and *N. albiventris*, female *P. hastatus* of the same cluster forage together, suggesting some form of cooperation among female members of the same cluster (McCracken and Bradbury, 1981; Boughman, 2006). Social calls seem to be used by females to discriminate group mates and to coordinate group foraging (Boughman and Wilkinson, 1998; Wilkinson and Boughman, 1998). These group-distinct calls are probably learned by group members (Boughman, 1998). In summary, the social system of this species seems to be built largely upon female cooperation among members of the same cluster. This cooperation is presumably mediated by vocalizations. Female cooperation may even encompass the timing of parturition, because females of the same cluster give birth almost exactly at the same time (Porter and Wilkinson, 2001). However, it is unknown whether females of a cluster enter estrous at the same time as well, which could be interpreted as a female counterstrategy against male monopolization. Yet, given that guarding males sire almost all offspring of their clusters, male dominance over females seems to be effective. Behavioral observations in a cave colony indicated that guarding males departed and returned more frequently than other males (Kunz *et al.*, 1998). Thus, males seem to allocate considerable time to defending females against male competitors. Physical interactions among males over access to female groups may therefore select for large-sized males in this and also other bat species.

Nest-Excavating Round-Eared Bats (*Lophostoma silvicolum*)

Male body size and sexually dimorphic glands may not be the only morphological features under sexual selection in bats. In a compelling series of studies on the social behavior of white-throated round-eared bats, it was demonstrated that the teeth and jaw muscles of male *Lophostoma* may represent a sexually selected trait as well. Bats of the genus *Lophostoma* have a stunning roosting behavior in using active arboreal nests of the termite *Nasutitermes corniger* as their preferred roost substrate (Fig. 16.3; Kalko *et al.*, 1999, 2006; Dechmann *et al.*, 2004). Females benefit from roosting in active termite nests because they encounter a more stable and warmer microclimate than, for example, in hollow trees (Dechmann *et al.*, 2004). It is noteworthy that a suitable roost microclimate requires the presence of termites, although bats may eventually destroy the nest. Thus, *Lophostoma* are probably parasitizing termites, given the likelihood that the fitness and survival of the termite colony may suffer from the presence of bats.

Video observations confirmed that single males excavate nests so that females may use the created space as a roost (Kalko *et al.*, 2006). Groups within such termite nests include only a single male and a group of females, suggesting a

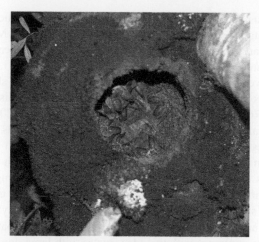

FIGURE 16.3 **Group of *Lophostoma silvicolum* roosting in an excavated termite nest (La Selva Biological Station, Costa Rica)**. See color plate at the back of the book. *Photograph courtesy of Karin Schneeberger.*

resource-defense polygyny (Dechmann *et al.*, 2005). Nest-holding males may use an excavated termite nest for up to 30 months. Male offspring disperse from their natal colony because they are probably forced to do so by the harem holder. Female offspring disperse as well, probably to avoid inbreeding with the only dominant male in the roost, likely to be their father (Dechmann *et al.*, 2007). Even though nest-holding males control the access to roosts, male *L. silvicolum* are not able to monopolize copulations with harem females. About 54% of the offspring in a given female group are not sired by the corresponding dominant harem male. Interestingly, *L. silvicolum* is the only bat species for which the case of infanticide by a dominant male was documented (Knörnschild *et al.*, 2012) – a behavior that may have evolved as a male strategy to reduce the effect of extra-harem paternities and to increase individual reproductive success of nest-holding males by triggering an earlier estrous in a female that loses her offspring.

Termite nests excavated by male *Lophostoma* have been suggested to represent an extended male phenotype formed under the influence of sexual selection (Dechmann and Kerth, 2008). Excavated nests could encode for a good male body condition, since excavating a hollow into the hard substrate of a termite nest requires a high bite force (Dechmann *et al.*, 2009b). Indeed, the teeth, jaw muscles, and specific parts of the skull of male *L. silvicolum* are stronger than might be expected if evolved under the sole influence of natural selection (Santana and Dumont, 2011).

Self-Flagellating Long-Nosed Bats (*Leptonycteris curasoae*)

Leptonycteris curasoae is a nectar-feeding bat that roosts in caves of semi-desert areas in tropical Latin America. Usually, more females than males inhabit these highly gregarious colonies (Martino *et al.*, 1998). The reproductive pattern of females is strongly seasonal, with pregnancies occurring in May and lactation in June (Martino *et al.*, 1998). During the short estrous period of females in November and December, reproductively active males (i.e., those with large testes) develop a patch of about 1–2 cm^2 between their shoulder blades (Fig. 16.4). This so-called dorsal patch or gland produces a distinct, sweetish smell (Muñoz-Romo and Kunz, 2009; Muñoz-Romo *et al.*, 2011a). The epithelium of this area contains odor-producing sebaceous glands (Nassar *et al.*, 2008), but males also transfer body fluids to their dorsal patch (Muñoz-Romo *et al.*, 2011a). In addition, males may scratch their dorsal patch with their hind legs, sometimes causing ruptures in the skin and even bleeding. Dual-choice experiments confirmed that female *L. curasoae* are attracted to males that carry the dorsal patch smell (Muñoz-Romo *et al.*, 2011a). Accordingly, Muñoz-Romo and colleagues argued that the size and extent of the dorsal patch may signal male health. In compliance with this idea, males with large dorsal patches carried fewer ectoparasites than males with small dorsal patches (Muñoz-Romo *et al.*, 2011b). Using gas chromatography and mass spectrometry, it was shown that males carry a large number of volatiles on the dorsal patches; however, it is unknown which of these substances may help female *Leptonycteris* to evaluate male quality (Muñoz-Romo *et al.*, 2012). It is also unclear whether the physical

FIGURE 16.4   **Male *Leptonycteris curasoae* with dorsal patch.** The dorsal patch is the wet area located between the shoulder blades. Males often scratch the dorsal patch using their hind legs so that the skin may get irritated. Sometimes, dorsal patches may even start to bleed. See color plate at the back of the book. *Photograph courtesy of Mariana Muñoz-Romo.*

rupture of the dorsal skin which males inflict upon themselves causes infection, and whether or not the extent of such infections, and associated volatiles, could help females find mates with good immunocompetence.

Luring Songs and Scents in Greater Sac-Winged Bats (*Saccopteryx bilineata*)

In the final case study on sexual selection in neotropical bats, I will review our studies on the mating system and social behavior of greater sac-winged bats, *Saccopteryx bilineata*, a 7- to 9-g bat from lowland areas of Latin America. This bat is, so far, the best-studied species with respect to its social and mating system, including sexually selected signals in bats.

Greater sac-winged bats are sexually dimorphic, with males weighing about 15% less than females (Bradbury and Emmons, 1974; Voigt *et al.*, 2005a) and having a wing sac in each of their antebrachial membranes – the anterior wing membrane that extends from the thumb along the upper arm and the forearm towards the thorax (Fig. 16.5; Starck, 1958; Scully *et al.*, 2000). Females have only a non-functional rudiment of wing sacs in their antebrachium. Colonies of greater sac-winged bats are usually found between buttress roots of large trees, and occasionally in the well-lit parts of hollowed trees or on the outside of buildings (Bradbury and Emmons, 1974; Bradbury and Vehrencamp, 1976). Colony members roost on vertical surfaces and maintain a minimum distance of 5–8 cm to their neighbors (Fig. 16.6; Bradbury and Emmons, 1974). The persistent use of roosts over many years seems to be related to the scarcity of large trees with buttresses (Bradbury and Emmons, 1974). Individuals of a colony are very loyal to their daytime roost; for example, adult males have been observed in 98% and females in 91% of all censuses during several months of field work

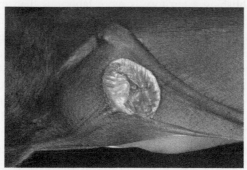

FIGURE 16.5   Dorsal view of a slightly opened left wing sac in the antebrachium of a male greater sac-winged bat. The wing sac is located parallel to the forearm and its interior is pink to whitish. When not in use, wing sacs are closed. However, males may open the sac via a muscular ligament that reaches from the edge of the wing sac towards the thorax, and they may close it via a second ligament that reaches from the tip of the wing sac towards the thumb. See color plate at the back of the book. *Photograph courtesy of Christian C. Voigt.*

FIGURE 16.6   **Two harems of greater sac-winged bats in one of the study colonies that uses an abandoned building as their daytime roost.** The two harem males were marked with colored bands on the right forearms, whereas females were marked on the left forearms. The central individual is an unmarked male juvenile that roosts next to its mother. Metal gauze troughs were installed to collect droppings for monitoring fecal sexual hormone metabolites. See color plate at the back of the book. *Photograph courtesy of Christian C. Voigt.*

(Voigt *et al.*, 2007; Voigt and Schwarzenberger, 2008). Neighboring colonies almost never exchange individuals (Bradbury and Emmons, 1974).

The social system of the greater sac-winged bat has been described as a female-defense polygyny – i.e., males each defend a group of females, a so-called harem (Bradbury and Emmons, 1974; Bradbury and Vehrencamp, 1976, 1977). Harem territories are about 1–2 m² in size, and include two or three females (Fig. 16.6). So-called non-harem males may roost next to harems. Consequently, colonies may consist of several harems and non-harem males. Although breeding in *S. bilineata* is strongly seasonal, colonies are stable in size and composition over the whole year; i.e., daytime roosts are not abandoned during the non-breeding period, probably because alternative roosts are scarce and because establishing a territory is more laborious than maintaining one. Tenure of colony members is long, given the small size of individuals in this species. During a 6-year study, females remained for an average 2.7 years and harem males for 3.2 years in a Costa Rican colony (Nagy *et al.*, 2007).

In Central America, mating of greater sac-winged bats is restricted to a short period between late November and early December (Voigt and Schwarzenberger, 2008; Greiner *et al.*, 2011a). During this short mating period, females come into estrus for only 1 or 2 days (Voigt and Schwarzenberger, 2008; Greiner *et al.*, 2011a, 2011b). The synchronization of the physiological estrous is possibly a female tactic to reduce sexual harassment and male surveillance. The offspring are born about 170 days after ovulation (Voigt and Schwarzenberger, 2008). Testes of male *S. bilineata* are small and only detectable shortly prior to and during the mating season, suggesting that sperm storage in the female reproductive tract or sperm competition is unlikely (Voigt *et al.*, 2007). After juveniles are fully grown, in September, female offspring disperse from their natal colonies. Consequently, inter-colony genetic heterogeneity is usually higher among males than among females, implying that males of a colony are more related to each other than are females (McCracken, 1984). In agreement with this finding, most males (40 out of 52 males; 78% of all recorded males) that roosted in a colony over a 6-year period belonged to only 4 patrilines (Nagy *et al.*, 2007). Apparently, colonies of *S. bilineata* consist of patrilocal kin groups. As a result, closely related males, such as fathers, sons, nephews, and grandsons, compete over access to harem territories and females, which may result in intense local mate competition (Nagy *et al.*, 2007). Despite this strong competition, males seem to benefit from staying in large colonies, because they are likely to obtain more indirect fitness benefits when male relatives reproduce (Nagy *et al.*, 2012). Female *S. bilineata* breed first at the age of 6 months, and male tenure lasts for about 3 years (Nagy *et al.*, 2007). Inbreeding avoidance may therefore promote female dispersal in this species (Clutton-Brock, 1989), when almost all male offspring remain in their natal colonies. Immigration of foreign males into colonies is, overall, rare (about 2 out of 29 males over a 6-year period; Nagy *et al.*, 2007) – a fact that may reflect the low likelihood of males of finding suitable alternative roosts. Harem succession follows a queuing pattern; i.e., the longest tenured male will occupy the closest vacant harem territory (Voigt and Streich, 2003). Most likely, males need to enter a queue and consequently establish site dominance very early in their lives.

In larger colonies, females may switch harem territories during the daytime. Harem males often follow these females, but mostly fail to gain complete control over female movements in the colony. At night, male home-range areas overlap by about 60% with those of females of their own harem. Thus, it is likely that harem males do not necessarily know the whereabouts of harem females when foraging at night, a fact that reduces the potential for male monopolization and increases the likelihood of female choice (Hoffmann *et al.*, 2007).

So-called extra-dominant paternities (in the sense of Clutton-Brock and Isvaran, 2006), or extra-harem paternities in the case of *S. bilineata*, are common in the mating system of the greater sac-winged bat (Heckel *et al.*, 1999; Heckel and von Helversen, 2003). Surprisingly, most harem offspring are sired by males other than the defender of the harem territory (Heckel *et al.*, 1999; Heckel and von Helversen, 2002). In a survey of 6 juvenile cohorts (in total 159 juveniles), 56 juveniles were sired by the harem holder, 45 by other harem males in the colony, 34 by non-harem males, 12 by male residents of a nearby colony, and 12 by unknown males (Voigt *et al.*, 2005a). Despite the presence of extra-harem paternities, harem males sire more offspring than non-harem males when all colony offspring are considered (Heckel and von Helversen, 2002). Females with extra-harem offspring were more genetically dissimilar to their extra-harem mates than to their harem males (Nagy *et al.*, 2007), suggesting that female greater sac-winged bats may pursue an active role in selecting a mate and in mating with genetically dissimilar males.

In the remainder of this section I will describe traits of male greater sac-winged bats that are likely under sexual selection. Two types of visual signals are conspicuous to humans observing male *S. bilineata* in the daytime roost: first, the whitish interior of the wing sacs when males fan volatiles towards females (Starck, 1958; Bradbury and Emmons, 1974; Voigt and von Helversen, 1999), and second, the frequent hovering displays of males in front of females (Fig. 16.7; Bradbury and Emmons, 1974; Voigt and von Helversen, 1999). The light interior of male wing sacs causes short flashes of white when wing sacs snap open repeatedly during hovering displays, or when roosting males fan wing-sac odor towards females (Bradbury and Vehrencamp, 1976; Voigt and von Helversen, 1999). Hovering flights are energetically costly (Voigt and Winter, 1999), and they may be particularly exhausting when some wing beats are used to propel air in a vertical direction towards females instead of providing buoyancy. Accordingly, mass-specific field metabolic rates of males increased with increasing harem size – i.e., with increasing frequency of hovering displays (Voigt *et al.*, 2001). The energetic costs of hovering flight may also explain why small males have a higher fitness than large males (Voigt *et al.*, 2005a). Light males may be able to perform more hovering displays than heavy males because light males expend less energy during aerial locomotion (Voigt, 2000) and most likely also during courtship displays. Possibly, females assess the stamina of males by observing the flashes of white when wing sacs open during hovering displays, or by assessing the duration of hovering displays.

Saccopteryx bilineata has a rich repertoire of vocalizations (Fig. 16.8; Bradbury and Emmons, 1974; Bradbury and Vehrencamp, 1976). Males emit long songs with redundant and complex syllables (Bradbury and Emmons, 1974). These songs were thought to attract females to, or retain females in, harem territories (Bradbury and Emmons, 1974). Early studies in Trinidad showed that male *S. bilineata* displayed 21 simple and 62 composite syllables (Davidson and Wilkinson, 2002). In agreement with the postulated function

FIGURE 16.7 **Hovering display of a male sac-winged bat in front of a roosting female.** During hovering flights, males fan the odor of wing sacs towards the female and the female responds to these fanning movements with social calls. See color plate at the back of the book. *Photographs courtesy of Christian C. Voigt.*

of male songs, harem size was positively correlated with the duration of the songs harem males performed. However, harem size was negatively correlated with the number and spectral frequency of the so-called inverted-V calls (also described as the "screech call"; Davidson and Wilkinson, 2004).

Another explanation for the various vocalizations of male *S. bilineata*, although not mutually exclusive with the previous one, is that vocalizations, and also hovering displays, serve as a greeting behavior. In compliance with this idea, inverted-V calls were found to differ among males in peak frequency, duration, and bandwidth (Davidson and Wilkinson, 2002). A later study described the inverted-V calls as "whistles", and assigned these calls to the hovering displays (Behr and von Helversen, 2004). Playback experiments by Behr and von Helversen (2004) suggested that female *S. bilineata* respond with screeches to male whistles, a behavior that resembles antiphonal calling. The same authors also described in detail the long (up to several minutes) and complex song repertoire of male *S. bilineata*. Male songs often contain multiple and often multisyllabic elements, and are used during courtship displays and territorial interactions (Behr and von Helversen, 2004). Multivariate analyses of the most common syllable type ("trill") showed that this song element differed among males. Thus, song variation may encode for male

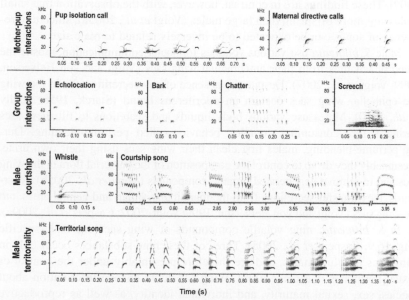

FIGURE 16.8   Examples of vocalizations emitted by adult and juvenile *Saccopteryx bilineata*. in Costa Rican colonies. In each sonogram, frequency (kHz) was plotted as a function of time in seconds (1024 point FFT, Hamming window with 75 % overlap). From top to bottom: mother–pup interactions, group interactions, male courtship, male territoriality (based on categorizations made by Behr and von Helversen, 2004; Behr *et al.*, 2006, 2009; Knörnschild *et al.*, 2012). *Figure reproduced from Voigt* et al. *(2008), by courtesy of Allen Press Publishing Services.*

individuality, a prerequisite for the use of songs during female choice (Behr and von Helversen, 2004).

Recently it has also been suggested that song complexity is learned by juvenile *S. bilineata* (Knörnschild *et al.*, 2012), yet conclusive evidence is missing as to whether or not bats perform true vocal learning similar to birds. So-called babbling behavior of male and female juveniles suggests a sensitive period during ontogeny that may influence the vocal signature of individual *S. bilineata* (Knörnschild *et al.*, 2006). In contrast to courtship songs, territorial songs are fairly short and usually emitted towards male competitors during territorial encounters (Behr *et al.*, 2006). Sometimes two males alternate in emitting territorial calls, most often at dusk before colony members leave, or at dawn when bats return to their daytime roost. The rate at which males utter territorial songs and also the lower end frequency of the fundamental harmonic of long buzz syllables were positively related to the reproductive success of harem males (Behr *et al.*, 2006). Territorial songs, and especially long buzz syllables, are likely to encode for the quality and competitive ability of harem holders, because only high-quality males may bear the supposedly high costs of vocalizations (Behr *et al.*, 2006). Playback experiments using territorial songs as stimuli also confirmed that low frequencies of songs elicited more and longer vocal responses by other harem males (Behr *et al.*, 2009). These findings are in contrast, however, with the observation that small males sire more offspring than large males (Voigt *et al.*, 2005a), since the frequency of songs can be expected to be inversely related to male size.

Male *S. bilineata* use various volatiles for social communication, mostly the blend of volatiles from their antebrachial wing sacs (Voigt and von Helversen, 1999; Voigt *et al.*, 2007). Despite the absence of any secreting cells in the wing sac epithelia, wing sacs contain an odoriferous liquid (Starck, 1958; Scully *et al.*, 2000). Males use various body liquids and secretions to fill these sacs during a time-consuming complex behavior called perfume-blending. During perfume-blending, males first clean their wing sacs using their own urine. Presumably they do so to control the composition of the bacterial flora in the wing sacs and to remove degenerated organic compounds (Voigt *et al.*, 2005b). Afterwards, they transfer secretions from the mandibular, gular, and genital areas into the wing sacs (Voigt and von Helversen, 1999; Voigt, 2002; Caspers *et al.*, 2009a).

In *S. bilineata*, nine volatile compounds of wing sacs are male-specific (Fig. 16.9; Caspers *et al.*, 2008). Three of these nine substances occur only in wing sacs of adult males, and are therefore presumably under the control of sexual hormones (Caspers, 2008). Wing sac odors encode information about species, sex, sexual maturity, and individual identity, as well as reproductive status (Caspers *et al.*, 2008, 2011). Binary odor-choice experiments confirmed that female *S. bilineata* prefer the odor of male conspecifics over that of males from the sister taxon *S. leptura* (Caspers *et al.*, 2009b). In addition to interspecific communication, male *S. bilineata* use chemicals in intraspecific communication (Voigt and von Helversen, 1999; Caspers and Voigt, 2009). For example,

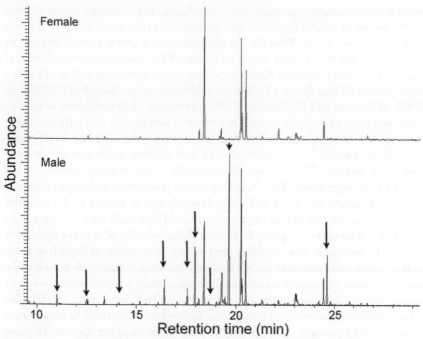

FIGURE 16.9 Odor profiles of male and female *Saccopteryx bilineata*. Each peak of the presented gas chromatographic run presents a single volatile compound. Peaks marked by arrows indicate male-specific substances. *Figure reproduced from Voigt* et al. *(2008), by courtesy of Allen Press Publishing Services.*

male *S. bilineata* scent-mark the borders of their harem territory almost daily by pressing their submandibular region onto the substrate they are hanging on and turning their head alternately left and right, each time releasing a droplet of secretion (Voigt and von Helversen, 1999; Caspers and Voigt, 2009).

Saccopteryx bilineata is an impressive example of just how complex sexually selected behaviors and morphological traits may be in bats. Despite the progress in the understanding of this particular species, many questions remain unanswered. For example, it is as yet unknown whether certain volatiles produced by males serve as pheromones and, most importantly, what criteria female *S. bilineata* use for mate choice, and for what reason.

ARE NEOTROPICAL BATS SPECIAL WITH RESPECT TO SEXUAL SELECTION?

The above question is difficult to separate from the more general issue of whether bats in general are special with respect to sexual selection. As outlined in the introduction, bats are to a certain extent special within the class Mammalia, mostly because of their ability to fly and to echolocate. However, several other features also make bats remarkable. For example, sperm storage over a period of several months is rare in mammals (Crichton, 2002; Wilkinson and McCracken, 2003). Thus far, the phenomenon of sperm storage and sperm competition has been mostly observed in bats of the temperate zone, where it is associated with hibernation. Neotropical bats do not hibernate, and most studies have confirmed that sperm of neotropical species are not long-lived (Crichton, 2002; Wilkinson and McCracken, 2003). However, documentation is incomplete, since, for example, it is unknown whether neotropical vespertilionids also have the sperm-storage capacity possessed by temperate-zone bats in this family. Indeed, it would be very interesting to shed light on sperm storage capabilities in the absence of hibernation. Interestingly, males of many neotropical bat species have large testes, indicating that sperm production and large ejaculates may be important for siring offspring. Possibly this is related to the fact that most of these species live in large colonies, and that male bats are often constrained in their ability to guard females in these colonies or at night when they forage. In particular, low visibility and strong attenuation of high-frequency echolocation calls preclude male bats from tracking, visually or acoustically, the movements of females, unless females reveal their whereabouts to males or unless males follow single females very closely (Hoffmann *et al.*, 2007). The aforementioned case studies highlight the fact that male potential to monopolize female mating partners varies largely among neotropical bat species. In some species, such as *P. hastatus*, males are quite efficient in mate guarding, whereas in others, such as *S. bilineata*, dominant males are clearly incapable of siring all offspring in their harem. These two species seem to lie at the opposite ends of a continuum with respect to monopolization of mating partners. Size dimorphism seems to be related to the potential for males to monopolize females, but not necessarily to male–male competition as exemplified by *S. bilineata*. In greater sac-winged bats, small-sized males seem to be at an advantage even though territorial encounters may involve physical combats between males. Possibly, small males are better able to maneuver and to defend their territory on the wing.

Darwin's vision that sexual selection may lead to the evolution of exaggerated ornaments or weapons in the animal kingdom (Darwin, 1871) has to be extended in bats to include the olfactory phenotype, which is more cryptic and enigmatic to human observers. Some sexually selected morphological features related to olfaction, such as glandular organs and olfactory displays, are obvious in neotropical bats. I assume that many more subtle olfactory ways of choosing the right mating partner or of manipulating sexual or social partners may exist in bats as well. Furthermore, vocalizations, such as social calls and songs, are poorly studied in neotropical bats, with a few exceptions as outlined above. In light of the high diversity of bats in the Neotropics, I speculate that many other remarkable examples of sexually selected traits and behaviors are yet to be discovered. Bats in general, and neotropical bats in particular, may offer interesting model organisms to study sexual selection.

ACKNOWLEDGMENTS

This chapter is dedicated to Otto von Helversen and Jack Bradbury, two inspiring academic teachers who set the ground for the study of sexual selection in neotropical bats. I thank Silke Voigt-Heucke and Julia Schad for commenting on an earlier draft of this manuscript. I also acknowledge the financial support of the German Research Council, which made the long-term study about the social system of *Saccopteryx bilineata* and *Noctilio albiventris* possible (Grants Vo 890/3, 890/11, 890/12).

REFERENCES

Barclay, R.M.R., Harder, L.D., 2003. Life histories of bats: Life in the slow lane. In: Kunz, T.H., Fenton, B.M. (Eds.), Bat Ecology, University of Chicago Press, Chicago, pp. 209–256.

Bass, M., Finer, M., Jenkins, C.N., Kreft, H., Cisneros-Heredia, D.F., McCracken, S.F., Pitman, N.C.A., English, P.A., Swing, K., Villa, G., DiFiore, A., Voigt, C.C., Kunz, T.H., 2010. Global conservation significance of Ecuador's Yasuní National Park. PLoS One 5, e8767.

Behr, O., von Helversen, O., 2004. Bat serenades – complex courtship songs of the sac-winged bat (*Saccopteryx bilineata*). Behav. Ecol. Sociobiol. 56, 106–115.

Behr, O., von Helversen, O., Heckel, G., Nagy, M., Voigt, C.C., Mayer, F., 2006. Territorial songs indicate male quality in the sac-winged bat *Saccopteryx bilineata* (Chiroptera, Emballonuridae). Behav. Ecol. 17, 810–817.

Behr, O., Knörnschild, M., von Helversen, O., 2009. Territorial counter-singing in male sac-winged bats (*Saccopteryx bilineata*): Low-frequency songs trigger a stronger response. Behav. Ecol. Sociobiol. 63, 433–442.

Blaustein, A.R., 1981. Sexual selection and mammalian olfaction. Am. Nat. 117, 1006–1010.

Bloss, J., 1999. Olfaction and the use of chemical signals in bats. Acta Chiropt. 1, 31–45.

Bloss, J., Acree, T.E., Bloss, J.M., Hood, W.R., Kunz, T.H., 2002. Potential use of chemical cues for colony–mate recognition in the big brown bat, *Eptesicus fuscus*. J. Chem. Ecol. 28, 819–834.

Bouchard, S., 2001. Sex discrimination and roost mate recognition by olfactory cues in the African bats, *Mops conylurus* and *Chaerophon pumilus* (Chiroptera: Molossidae). J. Zool. 254, 109–117.

Boughman, J.W., 1998. Vocal learning by greater spear-nosed bats. Proc. R. Soc. B 265, 227–233.

Boughman, J.W., 2006. Selection on social traits in greater spear-nosed bats, *Phyllostomus hastatus*. Behav. Ecol. Sociobiol. 60, 766–777.

Boughman, J.W., Wilkinson, G.S., 1998. Greater spear-nosed bats discriminate group mates by vocalizations. Anim. Behav. 55, 1717–1732.

Bradbury, J.W., 1977a. Lek mating behavior in the hammer-headed bat. Z. Tierpsychol. 45, 225–255.

Bradbury, J.W., 1977b. Social organization and communication. In: Wimsatt, W.A. (Ed.), Biology of Bats, Academic Press, London.

Bradbury, J.W., Emmons, L., 1974. Social organization of some Trinidad bats. I: Emballonuridae. Z. Tierpsychol. 36, 137–183.

Bradbury, J.W., Vehrencamp, S.L., 1976. Social organization and foraging in emballonurid bats. I: Field studies. Behav. Ecol. Sociobiol. 1, 337–381.

Bradbury, J.W., Vehrencamp, S.L., 1977. Social organization and foraging in emballonurid bats. III: Mating systems. Behav. Ecol. Sociobiol. 2, 1–17.

Brennan, P.A., Kendrick, K.M., 2006. Mammalian social odors: Attraction and individual recognition. Phil. Trans. R. Soc. B 361, 2061–2078.

Brooke, A., 1994. Diet of the fishing bat, *Noctilio leporinus* (Chiroptera: Noctilionidae). J. Mammal. 75, 212–218.

Brooke, A., 1997. Social organization and foraging behaviour of the fishing bat *Noctilio leporinus* (Chiroptera; Noctilionidae). Ethology 103, 421–436.

Brooke, A.P., Decker, D.M., 1996. Lipid compounds in secretions of fishing bat, *Noctilio leporinus* (Chiroptera: Noctilionidae). J. Chem. Ecol. 22, 1411–1428.

Caspers, B., Voigt, C.C., 2009. Temporal and spatial distribution of male scent marks in the polygynous greater sac-winged bat. Ethology 115, 713–720.

Caspers, B., Franke, S., Voigt, C.C., 2008. The wing sac odor of male greater sac-winged bats *Saccopteryx bilineata* (Emballonuridae) as a composite trait: Seasonal and individual differences. In: Hurst, J., Beynon, R.J., Roberts, S.C., Wyatt, T.D. (Eds.), Chemical Signals in Vertebrates, Vol. 11. Springer Verlag, Berlin, pp. 151–160.

Caspers, B., Wibbelt, G., Voigt, C.C., 2009a. Histological examinations of facial glands in *Saccopteryx bilineata* (Chiroptera, Emballonuridae), and their potential use in territorial marking. Zoomorphol. 128, 37–43.

Caspers, B., Schröder, F.C., Meinwald, J., Franke, S., Streich, W.J., Voigt, C.C., 2009b. Odor-based species recognition in two sympatric species of sac-winged bats (*Saccopteryx bilineata, S. leptura*): Combining chemical analysis, behavioral observation and odor preference tests. Behav. Ecol. Sociobiol. 63, 741–749.

Caspers, B.A., Schröder, F.C., Franke, S., Voigt, C.C., 2011. Scents of adolescence: The maturation of wing sac odor in greater sac-winged bats, *Saccopteryx bilineata*. PLoS One 6, e21162.

Clutton-Brock, T.H., 1989. Review lecture: mammalian mating systems. Proc. R. Soc. B 236, 339–372.

Clutton-Brock, T.H., Isvaran, K., 2006. Paternity loss in contrasting mammalian societies. Biol. Lett. 2, 513–516.

Clutton-Brock, T.H., Parker, G.A., 1992. Potential reproductive rates and the operation of sexual selection. Q. Rev. Biol. 67, 437–456.

Crichton, E.G., 2002. Sperm storage and fertilization. In: Crichton, E.G., Krutzsch, P.H. (Eds.), Reproductive Biology of Bats, Academic Press, San Diego, pp. 295–320.

Darwin, C., 1871. The Descent of Man, and Selection in Relation to Sex. Murray, London.

Davidson, S., Wilkinson, G.S., 2002. Geographic and individual variation in vocalization by male *Saccopteryx bilineata* (Chiroptera: Emballonuridae). J. Mammal. 83, 526–535.

Davidson, S., Wilkinson, G.S., 2004. Function of male song in the greater white-lined bat, *Saccopteryx bilineata*. Anim. Behav. 67, 883–891.

Dechmann, D.K.N., Kerth, G., 2008. My home is your castle: Roost making is sexually selected in the bat *Lophostoma silvicolum*. J. Mammal. 89, 1379–1390.

Dechmann, D.K.N., Kalko, E.K.V., Kerth, G., 2004. Ecology of an exceptional roost: Energetic benefits could explain why the bat *Lophostoma silvicolum* roosts in active termite nests. Evol. Ecol. Res. 6, 1037–1060.

Dechmann, D.K.N., Kalko, E.K.V., Konig, B., Kerth, G., 2005. Mating system of a neotropical roost-making bats: The white-throated round-eared bat, *Lophostoma silvicolum* (Chiroptera: Phyllostomidae). Behav. Ecol. Sociobiol. 58, 316–325.

Dechmann, D.K.N., Kalko, E.K.V., Kerth, G., 2007. All-offspring dispersal in a tropical mammal with resource defense polygyny. Behav. Ecol. Sociobiol. 61, 1219–1228.

Dechmann, D.K.N., Heucke, S.L., Guggioli, L., Safi, K., Voigt, C.C., Wikelski, M., 2009a. Experimental evidence for group hunting via eavesdropping in echolocating bats. Proc. R. Soc. B 276, 2721–2728.

Dechmann, D.K.N., Santana, S.E., Dumont, E.R., 2009b. Roost making in bats – adaptations for excavating active termite nests. J. Mammal. 90, 1461–1468.

Dechmann, D.K.N., Wikelski, M., van Noordwijk, H.J., Voigt, C.C., Voigt-Heucke, S.L., 2013. Metabolic costs of bat echolocation in a non-foraging context support a role in communication. Front. Physiol. 4, Article 66.

Dumont, E.R., Dávalos, L.M., Goldberg, A., Santana, S.E., Rex, K., Voigt, C.C., 2012. Morphological innovation, diversification and invasion of a new adaptive zone. Proc. R. Soc. B 279, 1797–1805.

Dunn, L.H., 1934. Notes on the little bulldog bat, *Dirias albiventer minor* (Osgood), in Panama. J. Mammal. 15, 89–99.

Fenton, M.B., 1985. Communication in the Chiroptera. Indiana University Press, Bloomington, pp. 161.

Greiner, S., Schwarzenberger, F., Voigt, C.C., 2011a. Predictable estrus timing of a small tropical mammal in an almost seasonal environment. J. Trop. Ecol. 27, 121–131.

Greiner, S., Dehnhard, M., Voigt, C.C., 2011b. Differences in plasma testosterone levels related to social status are more pronounced during mating than non-mating season in the tropical bat *Saccopteryx bilineata*. Can. J. Zool. 89, 1157–1163.

Griffin, D.R., 1958. Listening in the Dark. Yale University Press, New Haven.

Gustin, M.K., McCracken, G.F., 1987. Scent recognition between females and pups in the bat *Tadarida brasiliensis mexicana*. Anim. Behav. 35, 13–19.

Heckel, G., von Helversen, O., 2002. Male tactics and reproductive success in the harem polygynous bat *Saccopteryx bilineata*. Behav. Ecol. 12, 219–227.

Heckel, G., von Helversen, O., 2003. Genetic mating system and the significance of harem associations in the bat *Saccopteryx bilineata*. Mol. Ecol. 12, 219–227.

Heckel, G., Voigt, C.C., Mayer, F., von Helversen, O., 1999. Extra-harem paternity in the white-lined bat *Saccopteryx bilineata*. Behaviour 136, 1173–1185.

Hoffmann, F.F., Hejduk, J., Caspers, B., Siemers, B.M., Voigt, C.C., 2007. In the mating system of *Saccopteryx bilineata* auditory constraints prevent efficient female surveillance by eavesdropping male bats. Can. J. Zool. 85, 863–872.

Holderied, M.W., von Helversen, O., 2003. Echolocation range and wingbeat period match in aerial-hawking bats. Proc. R. Soc. B 270, 2293–2299.

James, A.G., Casey, D., Hyliands, D., Mycock, G., 2004. Fatty acid metabolism by cutaneous bacteria and its role in axillary malodor. World J. Microbiol. Biotechn. 20, 787–793.

Kalko, E.K.V., Friemel, D.D., Handley, C.O., Schnitzler, H.U., 1999. Roosting and foraging behavior of two neotropical gleaning bats, *Tonatia silvicola* and *Trachops cirrhosus* (Phyllostomidae). Biotropica 31, 344–353.

Kalko, E.K.V., Ueberschaer, K., Dechmann, D.K.N., 2006. Roost structure, modification, and availability in the white-throated round-eared bat *Lophostoma silviculum* (Phyllostomidae) living in active termite nests. Biotropica 38, 398–404.

Kerth, G., Perony, N., Schweitzer, F., 2011. Bats are able to maintain long-term social relationships despite the high fission-fusion dynamics of their groups. Proc. R. Soc. B 278, 2761–2767.

Knörnschild, M., Behr, O., von Helversen, O., 2006. Babbling behavior in the sac-winged bat (*Saccopteryx bilineata*). Naturwissenschaften 93, 451–454.

Knörnschild, M., Nagy, M., Metz, M., Mayer, F., von Helversen, O., 2012. Learned vocal group signatures in the polygynous bat *Saccopteryx bilineata*. Anim. Behav. 84, 761–769.

Kunz, T.H., Robson, S.K., Nagy, K.A., 1998. Economy of harem maintenance in the greater spear-nosed bat, *Phyllostomus hastatus*. J. Mammal. 79, 631–642.

Kwak, J., Opiekun, M.C., Matsumura, K., Preti, G., Yamazaki, K., Beauchamp, G.K., 2009. Major histocompatibility complex-regulated odortypes: Peptide-free urinary volatile signals. Physiol. Behav. 96, 184–188.

Lawrence, B.D., Simmons, J.A., 1982. Measurements of atmospheric attenuation at ultrasonic frequencies and the significance for echolocation by bats. J. Acoust. Soc. Am. 71, 585–590.

Martino, A., Arends, A., Aranguren, J., 1998. Reproductive pattern of *Leptonycteris curasoae* Miller (Chiroptera: Phyllostomidae) in northern Venezuela. Mammalia 62, 69–72.

McCracken, G.F., 1984. Social dispersion and genetic variation in two species of Emballonurid bats. Z. Tierpsychol. 66, 55–69.

McCracken, G.F., Bradbury, J.W., 1977. Paternity and genetic heterogeneity in polygynous bat, *Phyllostomus hastatus*. Science 198, 303–306.

McCracken, G.F., Bradbury, J.W., 1981. Social organization and kinship in the polygynous bat *Phyllostomus hastatus*. Behav. Ecol. Sociobiol. 8, 11–34.

McCracken, G.F., Wilkinson, G.S., 2002. Bat mating systems. In: Crichton, E.G., Krutzsch, P.H. (Eds,), Reproductive Biology of Bats., Academic Press, San Diego, pp. 321–362.

Muñoz-Romo, M., Kunz, T.H., 2009. Dorsal patch and chemical signalling in males of the long-nosed bat, *Leptonycteris curasae* (Chiroptera: Phyllostomidae. J. Mammal. 90, 1139–1147.

Muñoz-Romo, M., Burgos, J.F., Kunz, T.H., 2011a. Smearing behaviour of male *Leptonycteris curasoae* (Chiroptera) and female responses to the odor of dorsal patches. Behaviour 148, 461–483.

Muñoz-Romo, M., Burgos, J.F., Kunz, T.H., 2011b. The dorsal patch of males of the Curacaoan long-nosed bat, *Leptonycteris curasoae* (Phyllostomidae: Glossophaginae) as a visual signal. Acta Chiropt. 13, 207–215.

Muñoz-Romo, M., Nielsen, L.T., Nassar, J.M., Kunz, T.H., 2012. Chemical composition of the substances from dorsal patches of males of the Curaçaoan long-nosed bat, *Leptonycteris curasoae* (Phyllostomidae: Glossophaginae). Acta Chiropt. 14, 213–224.

Munshi-South, J., Wilkinson, G.S., 2010. Bats and birds: Exceptional longevity despite high metabolic rates. Ageing Res. Rev. 9, 12–19.

Nagy, M., Heckel, G., Voigt, C.C., Mayer, F., 2007. Patrilineal social system drives female dispersal in a polygynous bat. Proc. R. Soc. B 274, 3019–3025.

Nagy, M., Knörnschild, M., Voigt, C.C., Mayer, F., 2012. Male greater sac-winged bats gain direct fitness benefits when roosting in multi-male colonies. Behav. Ecol. 23, 597–606.

Nassar, J.M., Salazarm, M.V., Quintero, A., Stoner, K.E., Gomez, M., Cabrera, A., Jaffe, K., 2008. Seasonal sebaceous patch in the nectar-feeding bats *Leptonycteris curasoae* and *L. yerbabuenae* (Phyllostomidae: Glossophaginae): Phonological histological and preliminary chemical characterization. Zoology 111, 363–376.

Ophir, A.G., Schrader, S.B., Gillooly, J.F., 2010. Energetic cost of calling: General constraints and species-specific differences. J. Evol. Biol. 23, 1564–1569.

Penn, D., Potts, W.J., 1998. Chemical signals and parasite mediated sexual selection. Trends Ecol. Evol. 13, 391–396.

Penn, D.J., Potts, W.K., 1999. The evolution of mating preferences and major histocompatibility complex genes. Am. Nat. 153, 145–164.

Porter, T.A., Wilkinson, G.S., 2001. Birth synchrony in greater spear-nosed bats (*Phyllostomus hastatus*). J. Zool. 253, 383–390.

Rennie, P.J., Gower, D.B., Holland, K.T., Mallet, A.I., Watkins, W.J., 1990. The skin microflora and the formation of human axillary odor. Int. J. Cosm. Sci. 12, 197–207.

Rennie, P.J., Gower, D.B., Holland, K.T., 1991. In vitro and in vivo studies of human axillary odor and the cutaneous microflora. Brit. J. Dermatol. 124, 596–602.

Rex, K., Czaczkes, B.I., Michener, R., Kunz, T.H., Voigt, C.C., 2010. Specialization and omnivory in diverse mammalian assemblages. Ecoscience 17, 37–46.

Roverud, R.C., Chappell, M.A., 1991. Energetic and thermoregulatory aspects of clustering behavior in the neotropical bat *Noctilio albiventris*. Physiol. Zool. 64, 1527–1541.

Safi, K., Kerth, G., 2003. Secretions of the interaural gland contain information about individuality and colony membership in the Bechstein's bat. Anim. Behav. 65, 363–369.

Santana, S.E., Dumont, E.R., 2011. Do roost-excavating bats have stronger skulls? Biol. J. Linn. Soc. 102, 1–10.

Santos, M., Aguirre, L.F., Vazquez, L.B., Ortega, J., 2003. *Phyllostomus hastatus*. Mamm. Spec. 722, 1–6.

Schad, J., Dechmann, D.K.N., Voigt, C.C., Sommer, S., 2011. MHC class II *DRB* diversity, selection pattern and population structure in a neotropical bat species, *Noctilio albiventris*. Heredity 107, 115–126.

Schad, J., Voigt, C.C., Greiner, S., Dechmann, D.K.N., Sommer, S., 2012a. Independent evolution of functional MHC class II *DRB* genes in neotropical bat species. Immunogenetics 64, 535–547.

Schad, J., Dechmann, D.K.N., Voigt, C.C., Sommer, S., 2012b. Evidence for the 'good genes' model: Association of MHC Class II DRB alleles with ectoparasitism and reproductive state in the neotropical lesser bulldog bat, *Noctilio albiventris*. PLoS One 7, e37101.

Scully, W.M., Fenton, M.B., Saleuddin, A.S.M., 2000. A histological examination of holding sacs and scent glandular organs of some bats (Emballonuridae, Hipposideridae, Phyllostomidae, Vespertilionidae and Molossidae). Can. J. Zool. 78, 613–623.

Simmons, N.B., 2005. Chiroptera. In: Wilson, D.E., Reeder, D.M. (Eds.), Mammal Species of the World. Johns Hopkins Unviersity Press, Baltimore, pp. 312–529.

Starck, D., 1958. Beitrag zur Kenntnis der Armtaschen und anderer Hautdrüsenorgane von *Saccopteryx bilineata* Temminck 1838 (Chiroptera Emballonuridae). Gegenbaur Morphol. Jahrb. 99, 3–25.

Studier, E.H., Lavoie, K.H., 1984. Microbial involvement in scent production in noctilionid bats. J. Mamm. 65, 711–714.

Surlykke, A., Miller, L.A., Mohl, B., Andersen, B.B., Christensen-Dalsgaard, J., Jorgensen, M.B., 1993. Echolocation in two very small bats from Thailand: *Craseonycteris thonglongyai* and *Myotis siligorensis*. Behav. Ecol. Sociobiol. 33, 1–12.

Voigt, C.C., 2000. Intraspecific scaling of flight costs in the bat *Glossophaga soricina* (Phyllostomidae). J. Comp. Physiol. B. 170, 403–410.

Voigt, C.C., 2002. Individual variation of perfume-blending in male sac-winged bats. Anim. Behav. 63, 907–913.

Voigt, C.C., Lewanzik, D., 2011. Trapped in the darkness of the night: Thermal and energetic constraints of daylight flight in bats. Proc. R. Soc. B 278, 2311–2317.

Voigt, C.C., Schwarzenberger, F., 2008. Reproductive endocrinology of a small tropical bat (female *Saccopteryx bilineata*; Emballonuridae) monitored by fecal hormone metabolites. J. Mammal. 89, 50–57.

Voigt, C.C., Streich, W.J., 2003. Queuing for harem access in colonies of the sac-winged bat. Anim. Behav. 65, 149–156.

Voigt, C.C., von Helversen, O., 1999. Storage and display of odor by male *Saccopteryx bilineata* (Chiroptera; Emballonuridae). Behav. Ecol. Sociobiol. 47, 29–40.

Voigt, C.C., Winter, Y., 1999. The energetic costs of hovering flight in nectar-feeding bats (Phyllostomidae, Glossophaginae) and its scaling in sphingid moths, hummingbirds and bats. J. Comp. Physiol. B 169, 38–48.

Voigt, C.C., von Helversen, O., Michener, R., Kunz, T.H., 2001. The economics of harem maintenance in the sac-winged bat, *Saccopteryx bilineata*. Behav. Ecol. Sociobiol. 50, 31–36.

Voigt, C.C., Heckel, G., Mayer, F., 2005a. Sexual selection favours small and symmetric males in
the polygynous greater sac-winged bat *Saccopteryx bilineata* (Emballonuridae, Chiroptera).
Behav. Ecol. Sociobiol. 57, 457–464.

Voigt, C.C., Caspers, B., Speck, S., 2005b. Bats, bacteria and bat smell: Sex-specific diversity of
microbes in a sexually selected scent organ. J. Mammal. 86, 745–749.

Voigt, C.C., Heckel, G., von Helversen, O., 2006. Conflicts and strategies in the harem-polygynous
mating system of the sac-winged bat *Saccopteryx bilineata*. In: McCracken, G., Zubaid, A.,
Kunz, T.H. (Eds.), Functional and Evolutionary Ecology of Bats. Oxford University Press,
Oxford, United Kingdom, pp. 269–278.

Voigt, C.C., Streich, W.J., Dehnhard, M., 2007. Assessment of fecal testosterone metabolite analysis
in free-ranging *Saccopteryx bilineata* (Chiroptera; Emballonuridae). Acta Chiropt. 9, 463–476.

Voigt, C.C., Behr, O., Caspers, B., von Helversen, O., Knörnschild, M., Mayer, F., Nagy, M., 2008.
Songs, scents, and senses: Sexual selection in the mating system of the greater sac-winged bat,
Saccopteryx bilineata. J. Mammal. 89, 1401–1410.

Voigt-Heucke, S.K., Taborsky, M., Dechmann, D.K.N., 2010. A dual function of echolocation: Bats
use echolocation calls to identify familiar and unfamiliar individuals. Anim. Behav. 80, 59–67.

Von Busse, R., Swartz, S., Voigt, C.C., 2013. Flight metabolism in relation to speed in Chiroptera:
Testing the U-shape paradigm in the short-tailed fruit bat *Carollia perspicillata*. J. Exp. Biol.
http://dx.doi.org/10.1242/jeb.081760.

Wilkinson, G.S., Boughman, J.W., 1998. Social calls coordinate foraging in greater spear-nosed
bats. Anim. Behav. 55, 337–350.

Wilkinson, G.S., McCracken, G.F., 2003. Bats and balls: Sexual selection and sperm competition in
the Chiroptera. In: Kunz, T.H., Fenton, M.B. (Eds.), Bat Ecology. University of Chicago Press,
Chicago, Illinois, pp. 128–155.

Wilkinson, G.S., South, J.M., 2002. Life history, ecology and longevity in bats. Aging Cell 1,
124–131.

Wyatt, T., 2003. Pheromones and Animal Behaviour: Communication by Smell and Taste.
Cambridge University Press, Cambridge.

Zahavi, A., 1975. Mate selection: a selection for a handicap. J. Theor. Biol. 53, 205–214.

Index

Note: Page numbers with "f" denote figures; "t" tables; "b" boxes.

Color Plates

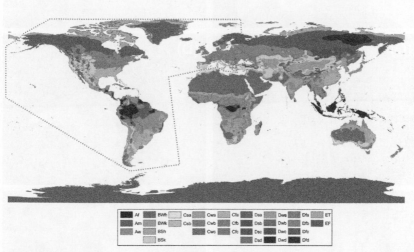

Climate	Tropical	Arid	Temperate	Cold	Polar
Area	11.9	6.0	8.9	14.4	6.0
% Area	25	13	19	31	13
Sites	35	26	92	109	10
% sites	13	10	34	40	4
% bias	-95	-34	45	24	-246

PLATE 1 The Köppen–Geiger climate classification map (modified from Peel *et al.*, 2007) and frequency of field studies on sexual selection in America and Europe (red dotted line). The area where each climate type occurs is expressed in millions of km^2 (see Table 1.1, Chapter 1, for detailed descriptions).

PLATE 2 Reproductive biology of *Serracutisoma proximum*. (A) At the beginning of the reproductive season, males patrol territories on the vegetation and (B) repel other males in seemingly ritualized fights in which territorial males hit each other with their elongated second pair of legs. (C) Females visit the territories, copulate with the territorial males (penis indicated by white arrow) and (D) lay their eggs on the undersurface of the leaves (egg at the tip of the female's ovipositor indicated by white arrow). (E) While female oviposits, the territorial male guards his mate with the second pair of legs extended towards her. (F) Minor males (sneakers) have a short second pair of legs and usually do not defend territories, but instead invade harems and sneak copulations with egg-guarding females, sometimes even when the unaware territorial male (top right) is in the vicinity of the female.

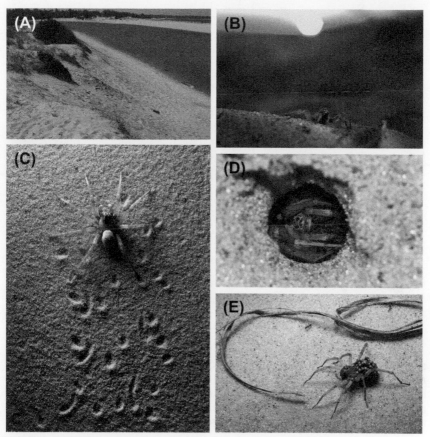

PLATE 3 (A) Typical sand dune habitat of *Allocosa brasiliensis*. (B) Male *A. brasiliensis* walking on the sand dunes during sunset. (C) Traces left on the sand by a male of the white sand dune spider. (D) Male inside his burrow. (E) Female with spiderlings riding on her abdomen. *Photos: M. Casacuberta.*

PLATE 4 (A) Male gift-offering position with male on the right and female on the left in *Paratrechalea ornata*. (B) Male gift-offering position with male on the left and female on the right in *Pisaura mirabilis*. (C) Female on left and male on right in the face-to-face position grasping the gift, during mating in *P. ornata*. (D) Female on left and male on right in the face-to-face position grasping the gift, during mating in *P. mirabilis*. *Photos: (A) M. C. Trillo; (B–D) M. J. Albo.*

PLATE 5 Arthropod species exhibiting exclusive paternal care. (A) Male *Abedus breviceps* carrying eggs attached to his back. (B) Male *Leytpodoctis oviger* carrying eggs attached to the left fourth leg. (C) Male *Lethocerus* sp. guarding a clutch on the emergent vegetation. (D) Male *Zygopachylus albomarginis* inside his mud nest built on a fallen trunk. (E) Inside view of an artificial nest of *Magnispina neptunus*. (F) Mating pair of *Edessa nigropunctata* on the host plant. (G) Male *Rhinocoris tristis* guarding a multiple clutch on the host plant. (H) Male *Bachycybe nodulosa* curled around a mass of eggs in the laboratory. (I) Male *Iporangaia pustulosa* guarding a multiple clutch on the undersurface of a leaf. (J) Male *Cryptopoecilaema almipater* guarding eggs. (K) Male *Stenostygnellus* aff. *flavolimbatus* guarding a multiple clutch. *Photos: J. Martens (B); R. Macías-Ordóñez (C); T. M. Nazareth (E); L. K. Thomas (G); S. Kudo (H); C. Víquez (J); O. Villareal Manzanilla (K). Scale bars ~ 1 cm.*

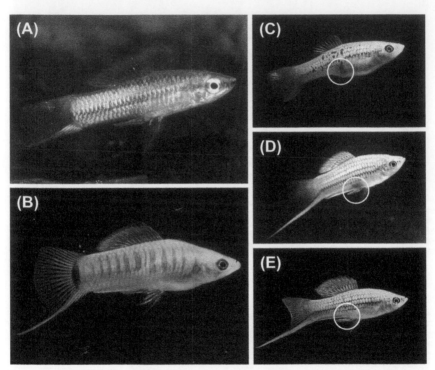

PLATE 6 Alternative reproductive tactics in the swordtail *Xiphophorus multilineatus*. There are four genetically influenced size classes of males in this species, where (A) the smallest size class of males are reversible in their use of mating behaviors (parasitic tactic), using both sneak-chase behavior and courtship depending on social context. (B) The three larger size classes (only Y-L shown) are irreversible in their use of courtship behavior (bourgeois tactic). (C–E) Female mimicry in *X. nezahualcoyotl* where circles indicate (C) true brood spot on female; (D) false brood spot (FBS) on a male; and (E) location of where FBS would be, on male without FBS. *From Rios-Cardenas* et al. *(2010). Photos: K. de Queiroz.*

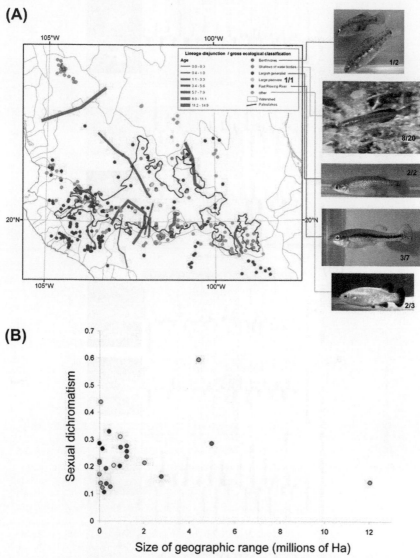

PLATE 7 (A) Distribution of Goodeinae records by ecological type, and "lines of disjunction" between sister clade ranges indicating possible allopatric diversification. Representative species of different ecological types illustrated at the right with respective number of genera/species. (B) Large geographic ranges tend to contain a great diversity of habitats, imposing divergent selective pressures on male and female color patterns, and thus it was hypothesized that larger geographic ranges would provide greater opportunities for the evolution of sexual dimorphism. A quantification of sexual dichromatism shows that it evolves independently of (current) size of geographic range.

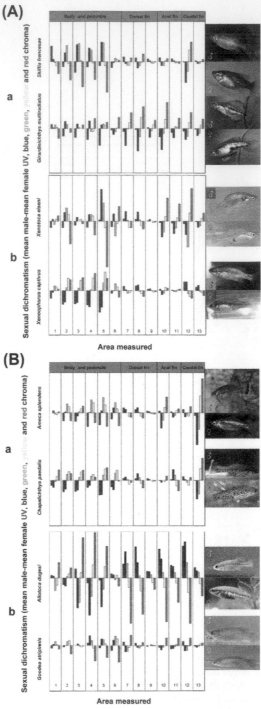

PLATE 8 (A) For 25 species, the mean reflectance at 13 points on flanks and fins of males and females was used to calculate mean UV, blue, green, yellow, and red chroma at each point for each species and sex. Female chromas were subtracted from the males'; positive values indicate points where males have a larger (positive) or smaller (negative) value of chroma than females. (B) Sometimes closely related species share both the color pattern of both sexes and the magnitude of the dimorphism (relative size of bars), whereas in one tribe it is possible to find the largest (*Allotoca dugesi*) and the smallest (*Goodea atripinnis*) degree of sexual dichromatism.

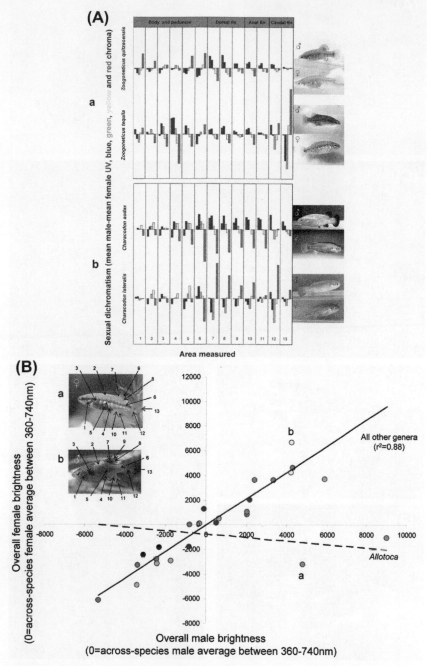

(colours indicate closely related species; see text)

PLATE 9 (A) Some species pairs appear to have diverged primarily in color patterns of their males, which may implicate the action of intersexual selection, with the females selecting from attributes that are increasingly different between species. *Characodon* species (a) are genetically compatible, but *Zoogoneticus* species (b) are not. (B) The observed correlation indicates that males and females respond to similar selective pressures, but does not tell us anything about the degree of dimorphism in brightness. Species share color code if they belong to the same or a closely related genus according to Webb *et al.* (2004), thus *Hubbsina turneri* shares color with the two species of *Girardinichthys*; *G. atripinnis* has the same color as *Ataeniobius toweri*, etc.

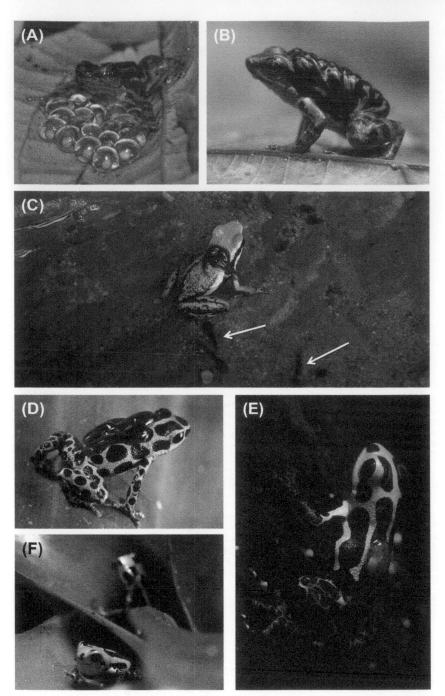

PLATE 10 Parental care in Neotropical poison frogs. (A, B) *Hyloxalus nexipus* is a stream-breeding species with male parental care. (A) A male *H. nexipus* attends an egg clutch of ~16 eggs laid on a leaf on the forest floor, and (B) transports all the tadpoles from one clutch on his back. (C) *Ameerega bassleri* is a terrestrial breeder. Here, a male deposits tadpoles in a terrestrial pool; tadpoles already in the water are indicated with arrows. (D–F) Frogs in the genus *Ranitomeya* breed in small phytotelmata. (D) A male *Ranitomeya variabilis* transports three tadpoles that will be deposited in phytotelmata such as bromeliad axils. (E) A male *R. imitator* transports each tadpole from a clutch individually to very small, nutrient-poor phytotelmata. This species is unusual among poison frogs in that it has biparental care of tadpoles, and in (F) the male calls to the female after leading her to phytotelmata where he has deposited tadpoles, she then provisions tadpoles with unfertilized trophic eggs. *Photos: Jason Brown (B, D), Adam Stuckert (A, F), Evan Twomey (C), and James Tumulty (E).*

PLATE 11 (A) Female blue-black grassquit with cryptic plumage. (B) Male blue-black grassquit with iridescent nuptial plumage. (C) Complete display of male blue-black grassquit where: C1, male is perched; C2–C3, male initiates motor display with exhibition of white underwing patches during wing beats and initializes acoustic output; C4, at peak of leap with vertical body rotation; and C5–C8, descending part of the leap with end of vocalization.

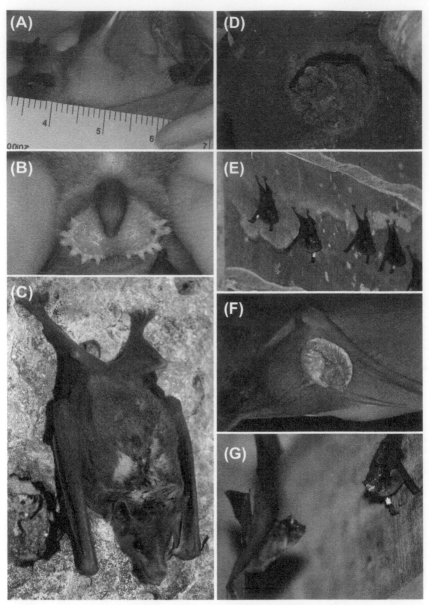

PLATE 12 (A) Subaxial gland of female *Noctilio albiventris*; (B) Inguinal gland of male *N. albiventris*; (C) Male *Leptonycteris curasoae* with dorsal patch; (D) Group of *Lophostoma silvicolum* roosting in an excavated termite nest; (E) Two harems of greater sac-winged bats (*Saccopteryx bilineata*); (F) Dorsal view of a slightly opened left wing sac in the antebrachium of a male greater sac-winged bat; (G) Hovering display of a male sac-winged bat in front of a roosting female. *Photos: (A, B, E–G) Christian C. Voigt; (C) Mariana Muñoz-Romo; (D) Karin Schneeberger.*

Printed and bound by CPI Group (UK) Ltd, Croydon, CR0 4YY

03/10/2024

01040422-0013